CATIA V5 Workbook

Releases 10 & 11

By:
Richard Cozzens
Southern Utah University

ISBN: 1-58503-138-0

Schroff Development Corporation

---

www.schroff.com
www.schroff-europe.com

Preface

This workbook is a basic introduction to the main "Work Bench" functions that are offered by CATIA V5. The work benches covered in this workbook are Sketcher, Part Design, Drafting, Assembly Design, Generative Shape Design, DMU Navigator and Rendering/Real Time Rendering.

This workbook is not meant to be a CATIA V5 reference guide. It is intended to be an organized, planned process to learn the basics of CATIA V5. The CATIA V5 Help Menu is a great reference and The CATIA Companion makes a good supplement to this workbook. The contents of this workbook are a combination of materials gathered and adapted from CATIA V4 basic training conducted by Metalcraft Technologies, Inc. for its engineering personnel. The material was adapted to CATIA V5 Release 1 for Southern Utah University during the spring semester 1999. The material has evolved from CATIA V5 R 1, R11 since that time. The workbook has continued to evolve with each CATIA V5 release. This workbook was the first CATIA V5 classroom curriculum developed.

Workbook Revisions

The first edition of this workbook was developed on CATIA V5 R3 during the summer of 2000. This is the Fifth Edition of the CATIA V5 Workbook developed on CATIA V5 R11.

There are a few subtle differences between the last few releases; for the most part the functionality has not changed. For release changes reference the CATIA V5 Website.

The publisher Schroff Development Corporation and I are dedicated to keeping this workbook current and accurate to the latest CATIA V5 release level. Suggestions are not only welcome but requested. For workbook updates, additional information and sample models log on to **www.schroff1.com/catia**. Schroff Development Corporation home website is **www.schroff.com**.

Acknowledgments

The concept of this book was derived from my attempt to create a classroom curriculum for Southern Utah University's Basic CATIA V5 class. It was Stephen Schroff at Schroff Development Corporation that encouraged me to make it available to the CATIA V5 world.

I would like to thank the following:

Tracy Garrett, SUU, for her excellent work in testing/checking the quality of the Lessons.

Virgil Seaman, Ph.D. at California State University, Los Angeles for his contribution to Lessons 9 and 10.

Don Crenshaw, Everett Community College, consultation and help with updating Lesson 7.

Daniel Ochoa, CSULA, SuperMileage Designer used in Lesson 9 & 10.
Tracy Day, SUU, created the Arbor Press used in Lesson 6.
Jared Mortenson, SUU, model creation and updating the lessons to Release 11.
Brandon Griffiths, SUU, developing a portion of the graphics.

I appreciate all the positive feed back and suggestions I have received from the users of the CATIA V5 Workbook. Many of the updates and changes are a result of the feed back I have received.

I want to thank my wife, for her support and suggestions. I also want to thank my four boys for putting up with me being gone so much while developing/updating this workbook.

Thank You!

Download Site

To access and download workbook CATParts, CATProducts, CATdrawings, updates, additional information and sample models, as well as the model required for Lesson 9 and Lesson 10 log on to

www.schroff1.com/catia

****

Online Supplement

For CATIA V5 Workbook training supplement log on to:

www.catiav5workbook.com

****

Related Sites

www.schroff.com

www.catiav5workbook.com

www.suu.edu/cadcam

www.coe.org

www.CatiaSolutions.com

www.catia.com

****

Table of Contents

Introduction To CATIA V5

Lesson 1 Sketcher Work Bench

Lesson 2 Part Design Work Bench

Lesson 3 Drafting Work Bench

Lesson 4 Drafting Work Bench

Lesson 5 Complex & Multiple Sketch Parts

Lesson 6 Assembly Design Work Bench

Lesson 7 Generative Shape Design Work Bench

Lesson 8 Generative Shape Design Work Bench

Lesson 9 DMU Navigator

Lesson 10 Rendering Work Bench

Lesson 11 Parametric Design

Terms And Definitions

Index

Introduction To CATIA V5

CATIA is one of the worlds leading high-end CAD/CAM/CAE software packages. CATIA V5 takes the power of an industry and technology-leading legacy CAD/CAM/CAE program and updates the programming (a total rewrite) to take advantage of the new Windows technology available to CATIA. CATIA V5 programming also allows you the flexibility of using sketched and parametric based design. CATIA V5 is the power you expect from CATIA with a greatly reduced learning curve. CATIA V5 makes a lot of processes practically automatic. You, the user, define the variables and CATIA V5 creates it for you. If it isn't exactly what you wanted, you adjust your variables and CATIA V5 will update the creation. CATIA V5 functionality is growing with each release; it will be the program that does it all. It is the leading edge technology starting with its product concept, continuing through design, assembly, testing, manufacturing and modeling, to its rendering capabilities. For more information on CATIA V5, refer to the CATIA V5 Home page on the Internet. The address is: http://www.catia.com

Workbook Objectives

The objective of this workbook is to instruct anyone who wants to learn CATIA V5 through organized, graphically rich, step-by-step instructions on the basic processes and tools provided by CATIA V5.

Although most of the steps are detailed for the beginner, the steps and processes are numbered and bolded so the more experienced user can go directly to the subject and/or area of interest.

All of the lessons follow the same format:

a.) A basic **Introduction** to the work bench being covered.

b.) A list of **Lesson Objectives**.

c.) The **Work Bench Tool Bar** (icons) with the tool titles and a brief definition.

d.) A brief introduction to what you are expected to create and learn while completing the lesson.

e.) Step-by-step instruction on how to use the tools and to successfully create a part. There are **Notes**, **Hints** and **Comments** along with links to all of the graphics for visual explanations.

f.) A **Lesson Summary** brings everything you have learned together.

g.) The **Lesson Review** consists of 20 questions taken from the lesson. The purpose of the review is to help solidify the concepts and tools taught in the lesson.

h.) In the last section of each lesson you will find **Practice Exercises**. Being able to answer questions is one thing, being able to create is another! The **Practice Exercises** are problems that require the use of the tools and processes covered in the lesson. Some of the **Practice Exercises** have helpful hints on how to create the part.

The workbook tells you what you are about to learn. The workbook then tells you and shows you (graphically) how and what you are supposed to accomplish. The workbook then follows up by giving you the opportunity to explain what you just learned with the **Lesson Review** questions. The workbook then gives you the opportunity to solidify your newfound knowledge by completing the **Practice Exercises**.

To get the most out of each lesson, it is suggested that you preview the entire lesson so you get an idea of what you are going to be learning. This preview will also give you an idea of what it will take to complete the lesson. Read through the **Lesson Review** questions before you begin, this way you can be searching for the answers as you go through the lesson. Preview the **Practice Exercises**, as you go through the lesson you can be looking for the tools and processes required to complete the **Practice Exercises**.

CATIA V5 Workbook Website

Schroff Development Corporation (SDC) provides a website as a companion to the CATIA V5 Workbook. The website contains the following information.

- Information about the workbook
- Updates on the workbook and CATIA V5
- Selected AVI training files
- Additional Problems and Exercises
- Helpful hints on selected Problems and Exercises
- Selected sample models
- Shared models

The Website is **www.schroff1.com/catia**
An additional online CATIA V5 resource is **www.catiav5workbook.com**

Overview Of Workbook Lessons

Introduction,
CATIA Standard Menus and Tools

This section is critical, because it gives you the foundation to start building your CATIA V5 knowledge of general screen layout, standard tools and mouse functions.

Lesson 1,
Introduction To The Sketcher Work Bench

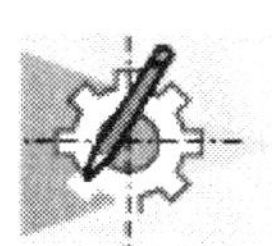

How to start CATIA V5; selecting work benches; brief introduction to the **Sketcher Work Bench** screen layout, tools and tool bars; customizing and setting standards; creating a new part; create sketcher geometry such as lines, arcs etc.; constraining the sketcher geometry; exiting sketcher; extruding a profile; saving the newly created file and exiting CATIA V5; **Lesson 1 Review** and **Lesson 1 Practice Exercises**.

Lesson 2,
Introduction To The Part Design Work Bench

Review of Lesson 1; pulling up an existing file; brief introduction to the **Part Design Work Bench** screen layout, tools and tool bars; customizing and setting standards; review extruding a profile; creating fillets, chamfers, holes and patterns, translating, rotating, symmetry, mirror, scaling; adding constraints; applying material to a solid; modifying constraints and properties; managing the **Specification Tree**; review saving and exiting CATIA V5; **Lesson 2 Review** and **Lesson 2 Practice Exercises**.

Lesson 3,
Introduction To The Drafting Work Bench, Creating Sheets & Views

Brief introduction to the **Drafting Work Bench** screen layout, tools and tool bars; creating sheets and views; creating sheet and view layouts using an existing **CATPart**; customizing default values; creating new sheets; creating detail views, section views and auxiliary views; saving the new drawing file; **Lesson 3 Review** and **Lesson 3 Practice Exercises**.

Lesson 4,
Introduction To The Drafting Work Bench, Creating Text & Dimensions

A continuation of Lesson 3; review screen layout, tools and tool bars; creating text and dimensions; creating and modifying text; creating and modifying leaders; creating borders; importing picture files; printing the drawing; **Lesson 4 Review** and **Lesson 4 Practice Exercises**.

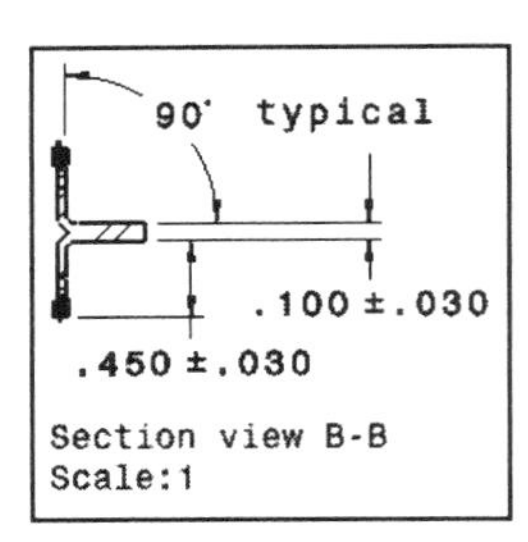

Lesson 5,
Introduction To Creating Complex & Multiple Sketch Parts

Review the **Drafting** and **Part Design Work Bench** screen layout, tools and tool bars; brief explanation of multiple sketch and **Boolean Geometry**; explain the process of creating parts that require multiple sketches; creating parts that require **Boolean Geometry**; **Lesson 5 Review** and **Lesson 5 Practice Exercises**.

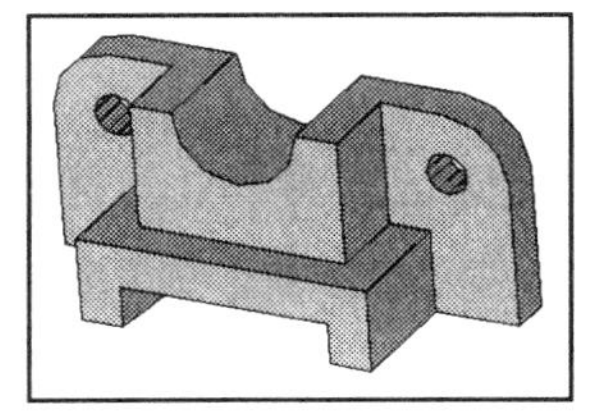

Lesson 6,
Introduction To The Assembly Design Work Bench, Creating An Assembly

Brief introduction to the **Assembly Design Work Bench** screen layout, tools and tool bars; opening the **Assembly Design Work Bench**; setting the standards; inserting components into the assembly; moving parts in the assembly; assembling parts; constraining the assembly; assembly analysis; **Lesson 6 Review** and **Lesson 6 Practice Exercises**.

Lesson 7,
Introduction To The Generative Shape Design Work Bench, Creating Wireframe & Surface Geometry

Brief introduction to the **Generative Shape Design Work Bench** screen layout, tools and tool bars; creating a local axis; creating points; creating lines; creating corners; creating surfaces; filleting surfaces; splitting/trimming surfaces; creating a solid from a surface; **Hide/Show** the wireframe; **Lesson 7 Review** and **Lesson 7 Practice Exercises**.

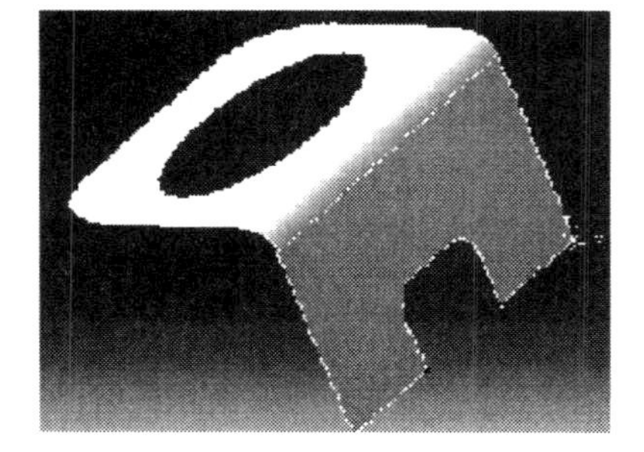

Lesson 8,
Introduction To The Generative Shape Design Work Bench, Creating Surfaces

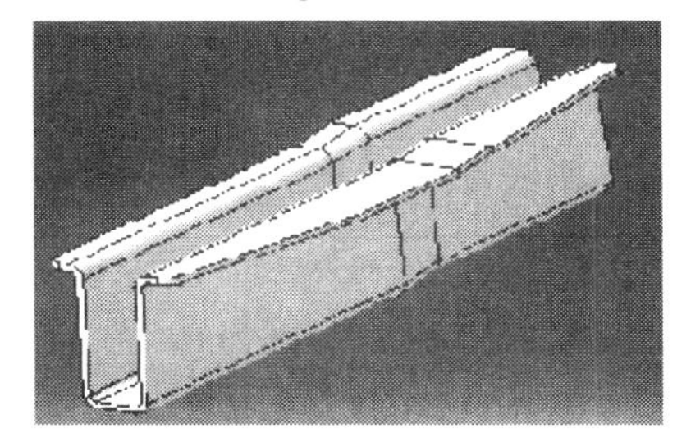

A continuation of Lesson 7 using additional tools; creating surfaces using sweeps; joining surfaces; splitting surfaces, creating a surface offset, extruding a surface, closing a surface; creating a solid from a closed surface; **Hide/Show** a surface; **Lesson 8 Review** and **Lesson 8 Practice Exercises**.

Lesson 9,
Introduction To The DMU Navigator Work Bench Using The Fly Mode

A brief introduction to the **DMU Navigator Work Bench** screen layout, tools and tool bars; downloading the **SuperMileage.CATProduct**; navigating the assembly; inserting objects; creating scenes and annotated views; **Lesson 9 Review** and **Lesson 9 Practice Exercises**.

Lesson 10
Introduction To The Real Time Rendering Work Bench

A brief introduction to the **Rendering Work Bench** screen layout, tools and tool bars; loading models into the **Rendering Work Bench**; creating an **Environment**; defining a light; adding images to the walls; creating a **Camera** and a **Shooting**; creating simulations, turn table and AVI files; **Lesson 10 Review** and **Lesson 10 Practice Exercises**.

Lesson 11
Introduction To Creating And Maintaining Basic Part Intelligence

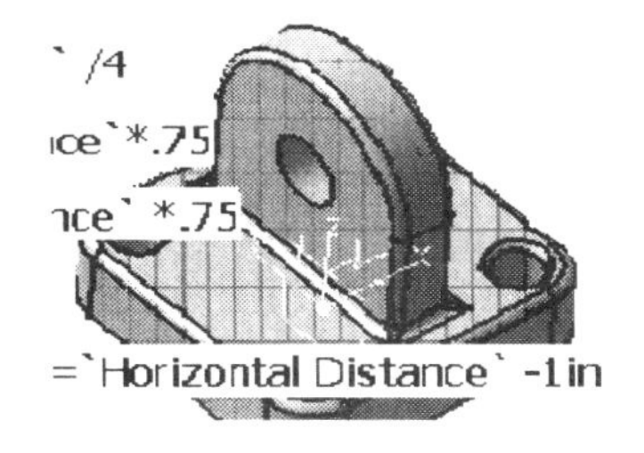

A brief introduction to the **Parametric Design** using Formulas, Dimensional Constraints, Geometrical Constraints and Design Tables; analyzing the part; planning the design; executing the plan; creating the primary feature; creating the secondary feature; creating the secondary feature; putting the plan to the test; additional notes; summary; **Lesson 11 Review** and **Lesson 11 Practice Exercises**.

Workbook/CATIA V5 Terms and Definition

This is another critical section, because it adds to your CATIA V5 foundation of knowledge. This section will also help you navigate and understand the terms used in this workbook.

CATIA V5 Running On Windows

This book was developed with CATIA V5 running on **Microsoft Windows 2000**. CATIA V5 running on UNIX is almost identical. The biggest difference is UNIX will not have all of the **Windows** functionality. The **Windows** functionality gives the user the flexibility of having several options to completing almost any task. For example, you have the **Cut/Copy** and **Paste** tools in the **Windows NT** pull down menu, you have the quick keys that accomplish the same thing (**Ctrl c** and **Ctrl v**) or you could use the CATIA V5 **Cut** and **Paste** tool found in the CATIA V5 **Standard** tool bar. Another example is you have the choice to highlight an entity then select the tool to apply or you can reverse the choices and select the tool, then select the entity. Some of these options are not available on the UNIX operating system. This workbook assumes you have at minimum a basic knowledge of the **Windows** Operating System.

This workbook has been developed and updated on CATIA V5 R11. CATIA V5 R11 was installed using standard defaults. A few differences in windows, prompts and results may arise due to service packs and possible customization. When a change in the default standard is required the lessons in the workbook states the change.

CATIA V5 Standard Menus And Tools

The following standard screen layout shows you where different tools and tool bars are located. The numbers coordinate with the following pages where the tool label is bolded. The tool label is followed by a brief explanation and in some cases, steps on how to use and/or access the tool.

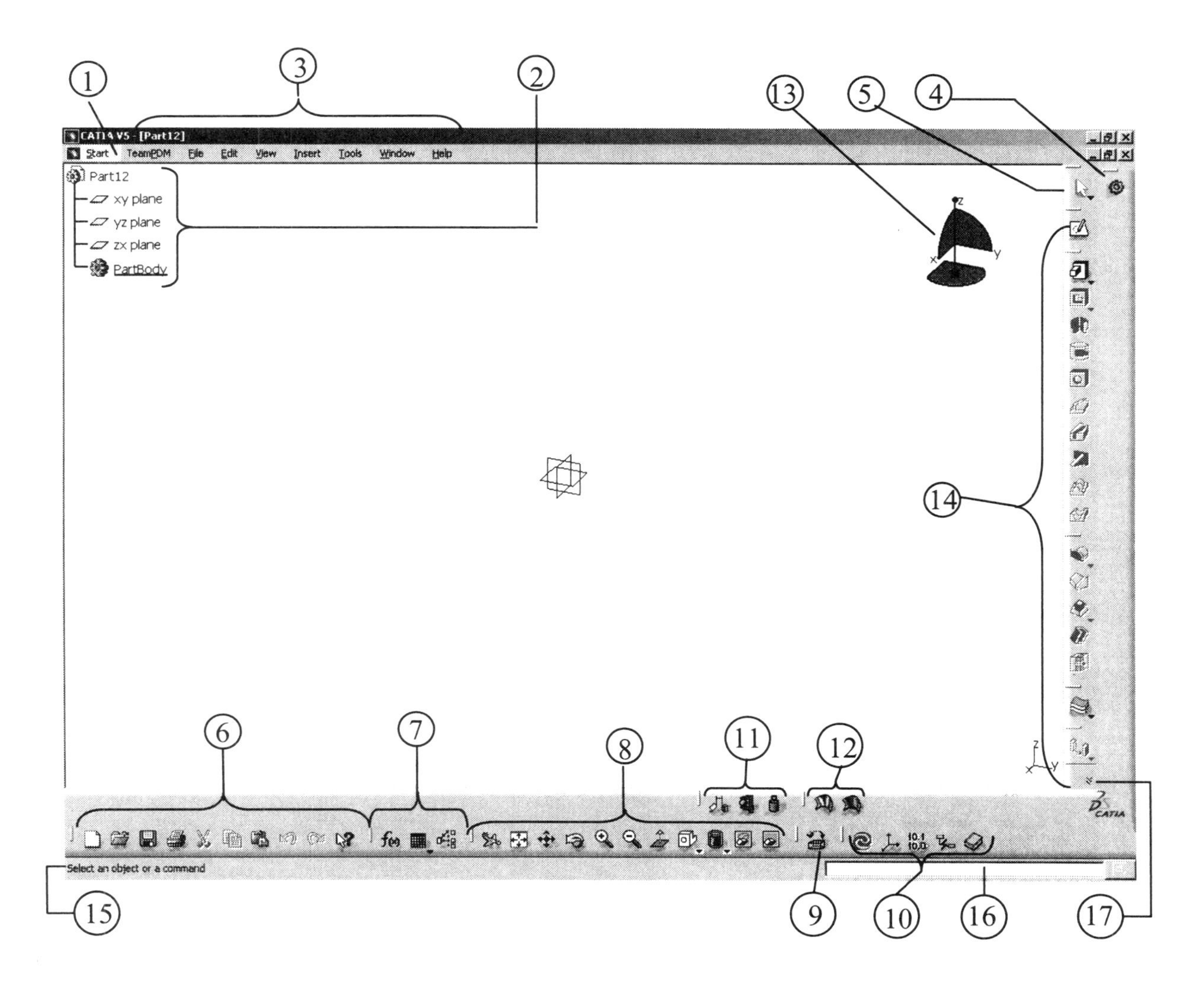

The following list of menus is not meant to be a comprehensive definition of every tool on the standard CATIA V5 screen. The purpose is to provide a quick reference and explanation only. If more detailed information is needed and/or required, refer to the CATIA V5 **Help** menu and/or Internet homepage.

Menu 1
The **Start Menu** Tool Bar

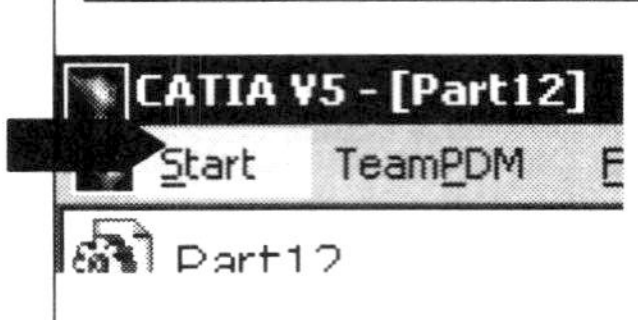

The Start pull down menu gives you access to all of the CATIA V5 work benches. Lesson 1 explains how to select a specific work bench. This workbook only covers the Sketcher, Part Design, Drafting, Assembly Design, Generative Shape Design and DMU Navigator Work Benches.

Menu 2

The **Specification Tree**

The **Specification Tree** contains the history of tools and processes used to create a part. For example, you can look at a completed part and see that there were fillets and holes applied to it. At what point in the part creation were the fillets and holes added? Are there redundant processes and extra elements? Can the process for part creation be improved? Looking at the resultant part will not answer any of these questions. The **Specification Tree** on the other hand has all of this information. The **Specification Tree** contains the entire history of the part creation. For a complex part, the **Specification Tree** could get large. Select the **Tools**, **Options**, **General**, **Display, Tree** option to specify what you want the **Specification Tree** to show and how you want it to appear. The branches of the **Specification Tree** can be expanded and contracted by selecting the – and + symbols located on each branch. You can **Zoom In** and **Pan** the **Specification Tree** the same way you would a part. You must double click on a **Specification Tree** branch to make the workspace go dim. Once the workspace is under intensified, all of the screen manipulation tools will apply to the **Specification Tree**. This means you can move and zoom the **Specification Tree** as you do the part in the workspace. Double clicking a **Specification Tree** branch will bring the part back to normal (the active workspace). The F3 key will hide the **Specification Tree** from view (a toggle key). CATIA V5 allows you to make modifications to the part by using the part itself and/or by using the **Specification Tree**. The **Specification Tree** is used in all nine lessons, but at a basic level. The **Specification Tree** is a very powerful tool, but you must know how to use it to your advantage. The **Specification Tree** shown in this section represents most of the branches and applications used in this workbook. The presentation of the tree will vary depending installation and customization. The tree shown below was created with a standard installation. The tree was customized to show all branches such as **Relationships**, **Formulas** and **Applications** using the **Tools**, **Options** window. This brief explanation is not nearly enough information to fully appreciate its potential uses. For more detailed information, reference the **Help** menu.

The **Specification Tree** (Continued)

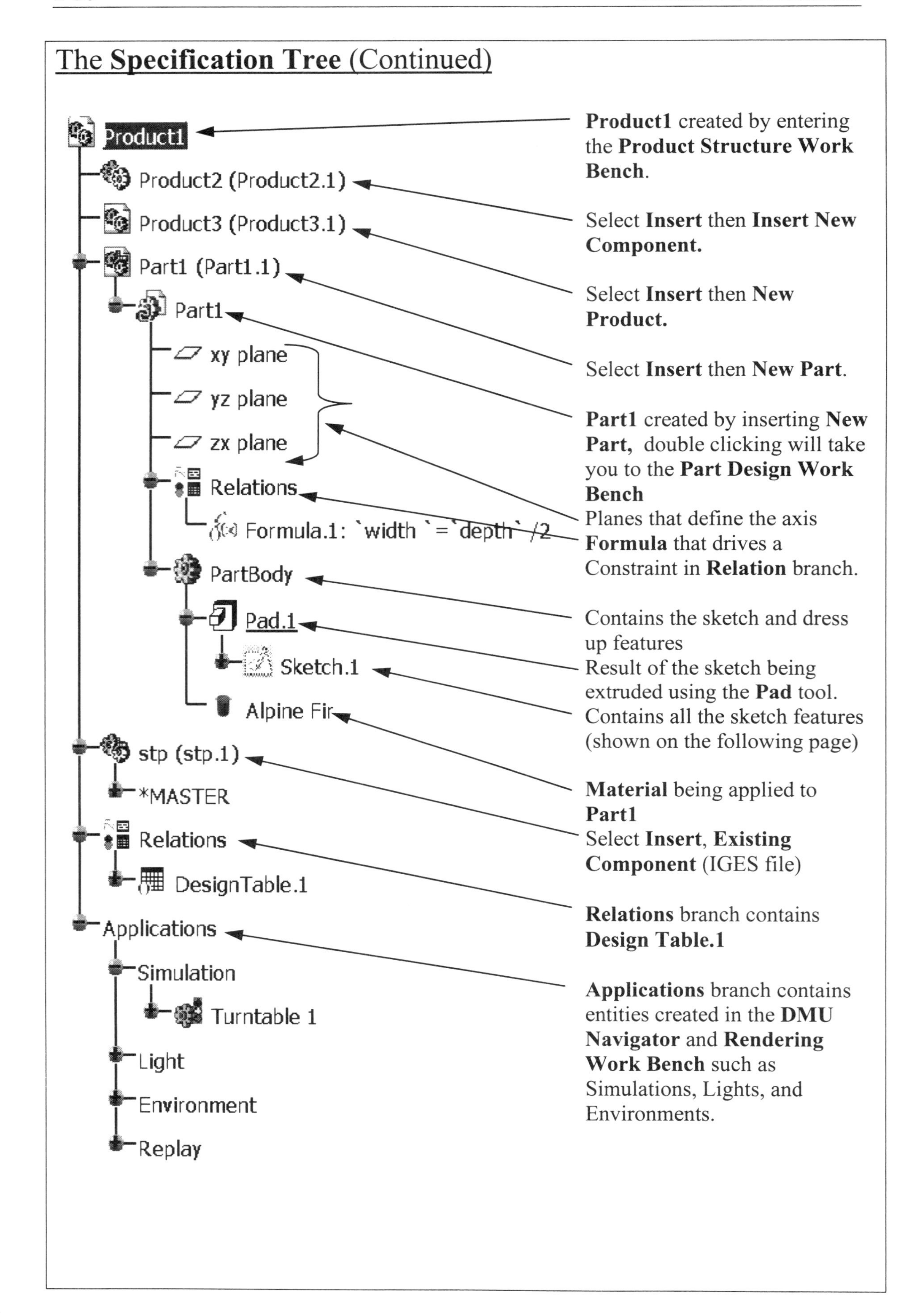

The **Specification Tree** (Continued)

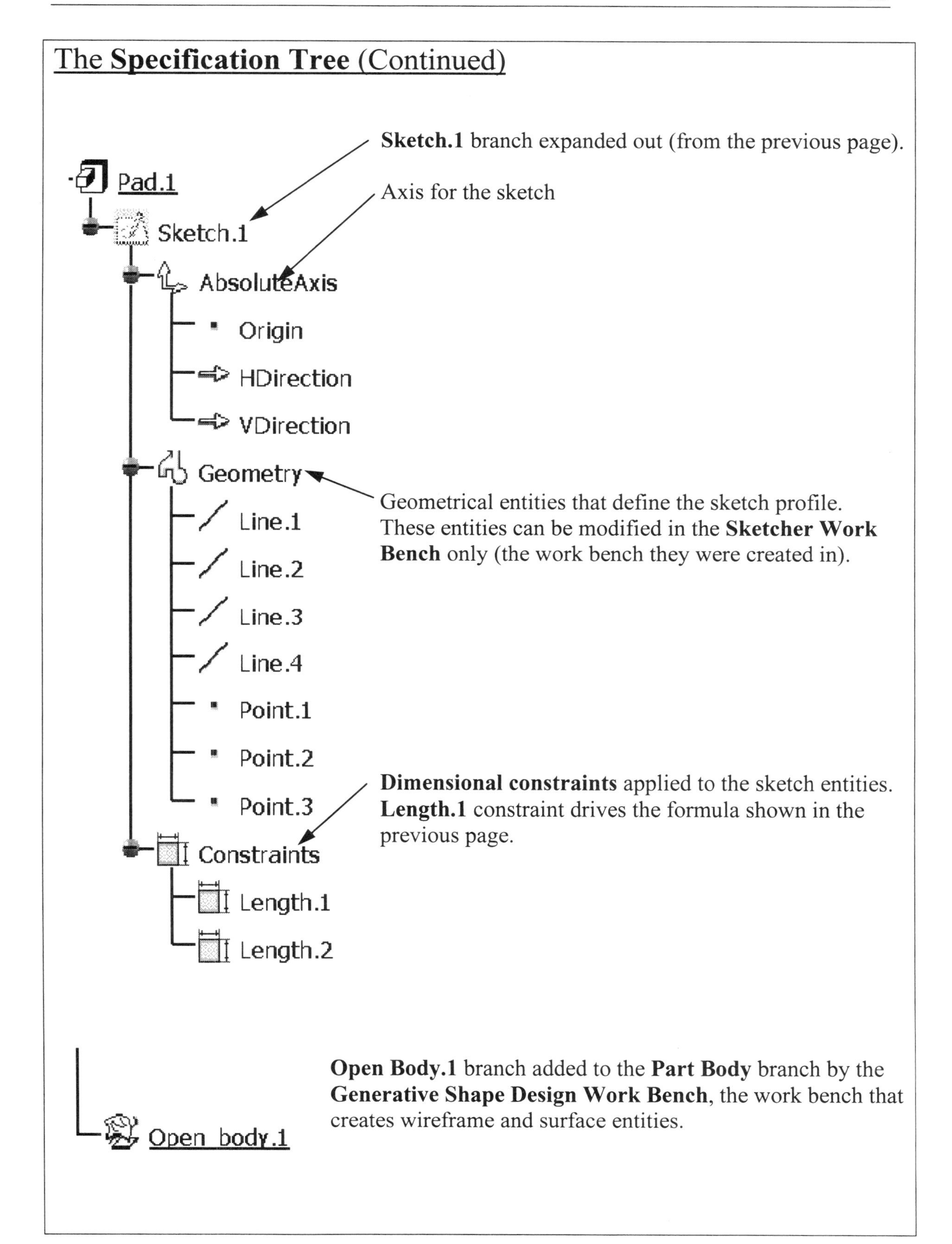

Menu 3
The **Standard Windows** Tool Bar

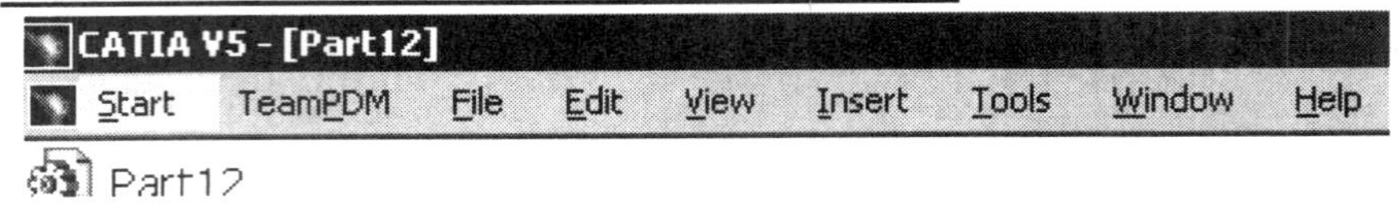

This tool bar contains your standard **Windows** pull down menus.

There are specific CATIA V5 tools found in the different pull down menus. The tools you will be required to use in this workbook will be defined in the lesson that they are used in.

Menu 4
The **Active Work Bench** Tool Bar

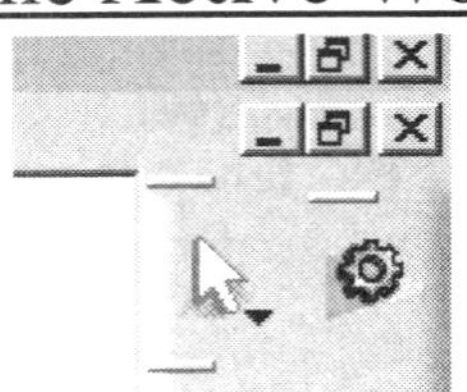

The work bench icon shown in this location signifies which work bench is active. This is especially critical when you have several windows open with different work benches. If you have them displayed on a split screen, the active window will be the one with the blue border. This work bench icon can also be selected to bring up the "**Welcome To CATIA V5**" window. This window allows you to select a new work bench. The selected work bench icon will become the active work bench. Notice, as you select a different work bench, the tools and tool bars on the right side of the screen change with the work bench. Particular branches on the **Specification Tree** will also be added depending on what work bench you enter. For example, the **PartBody** branch is added when entering the **Part Design Work Bench** and the **Open_Body** branch is added when the **Generative Shape Design Work Bench** is entered.

Menu 5
The **Select** Tool Bar

Selecting this tool allows you to select entities in the workspace. The default selector is the **Select Arrow**, which allows single point and left click selections. If you select the small arrow to the right of the icon, it will reveal the other selection tools, they are:

Tool Bar	Tool Name	Tool Definition
	Select	This is the default tool, point and click (left mouse button) to select the desired entity. Multiple entity selection can be done by holding down the **Ctrl** key while selecting.

	Selection Trap	This allows you to draw a box around the entities that you want to select. The box is exclusive to entities that intersect with the selection box. This is a quick and easy multi-select tool.
	Intersecting Trap	This allows you to draw a box around the entities, but will also select the entities that are intersected with the box. The selection box is inclusive.
	Polygon Trap	This selection is similar to the box selection trap, but allows you to sketch a more defined area of inclusion and exclusion of entities. This selection tool is quick, and allows you to be more exclusive in the multi-selection process.
	Paint Stroke Selection	This selection tool allows you to paint a line across the screen and any entity that the paint stroke crosses is selected.

Menu 6
The **CATIA V5 Standard** Tool Bar

This tool bar has nine tools in it, some offer an alternative method of accomplishing a similar task found in the **Standard Windows NT** tool bar. The tools are listed below, with some definition, if it applies to this workbook.

Tool Bar	Tool Name	Tool Definition
	New	Creates a new file (document).
	Open	Opens an existing file (document).
	Save	Saves the active file (document).
	Quick Print	Prints the active file (document).
	Cut	Deletes the selected element and/or elements. This tool has the **Windows NT** functionality of select, drag and drop.
	Copy	Another method of copying a selected element and/or elements. The tool places the copied element and/or elements onto the Windows NT clipboard.
	Paste	Another method of pasting an element and/or elements from the Windows NT clip board.
	Undo	The greatest OOPS tool developed since the invention of the computer! This tool allows you to step backwards one mistake (function) at a time!

	Redo	Make that a double OOPS! This tool allows you to undo your undo! If your last operations weren't so bad and you don't remember all of the parameters you entered, this tool is for you.
	What's This?	Direct link to the help file. Select the item you have a question about then select this tool. CATIAV5 will search the help files for information on the selected item.

Menu 7
The **Knowledge** Tool Bar

This tool bar allows you to use formulas and spread sheets to parameterize your sketches, parts and assemblies.

Tool Bar	Tool Name	Tool Definition
	Formula	This tool allows you to use a formula to drive parameters.
	Check Analysis Toolbox	Allows the user to define design standards and check parts against the standards.
	Design Table	This tool allows you to use data from an existing spread sheet to drive assigned parameters within a design
	Law	Access the law editor.
	Knowledge Inspector	This tool allows you to preview a design change prior to committing to the change. This is an advanced tool.

Menu 8

The View Tool Bar Tool Bar

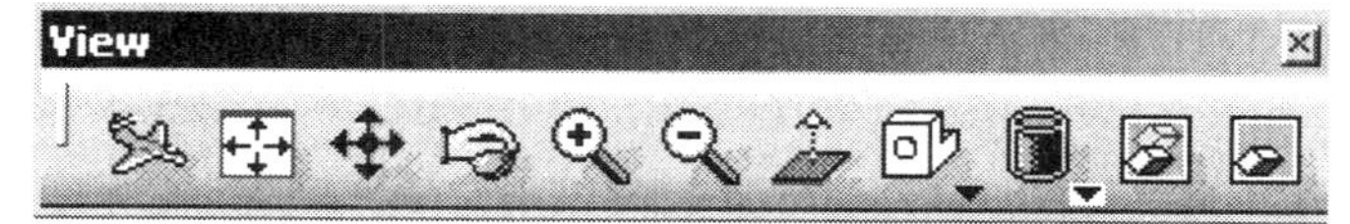

This tool bar contains CATIA V5 specific functions. This workbook will have you use most of them in one lesson or another. Hopefully you will use the tool on your own once you learn how to use it. Most of the tools apply to all of the work benches.

Tool Bar	Tool Name	Tool Definition
	Fly mode	Sets the fly mode. This is a very powerful and fun tool. Reference Lesson 9 on how to use this tool.
	Fit All In	This is similar to other graphics programs. This tool will show the extent of all the graphics currently on the screen. It is a quick way to see what elements are on the screen and where they are in relationship to one another.
	Pan	This is similar to other graphics programs. It allows you to move the part around on the screen. The part does not change its location in the XYZ coordinate system, only in relationship to the screen. Every time you want to **Pan** the part you must select this tool first, unless you have a three-button mouse. **Quick Key**: With a three-button mouse you can press the middle mouse button down and drag the part to the desired location on the screen.
	Rotate:	This tool allows you to rotate the part in three-dimensional space and in real time. It will place a representation of a space ball (sphere) in the center of the screen. There is a three-dimensional X on the space ball, you drag the X to where you want on the space ball and the part will rotate accordingly. This tool is critical to part manipulation. It is important that you get the hang of rotating the part to the orientation you want. This tool must be selected every time you want to rotate the part. This process is explained and shown step-by-step in Lesson 2, Step 7. **Quick Key:** A quicker method is using the mouse. Press the middle mouse button first, while holding the middle button down, press the left mouse button and drag the mouse around on the sphere. This brings up the space ball (sphere). Another method is to press the CTRL key while pressing the middle mouse button and dragging the mouse around the screen.

	Zoom In	This is similar to other graphics programs. This allows you to get a closer look at finer detail. **Quick Key**: Press the middle mouse button, hold it down as you press the left mouse button and release it. Now use the mouse to drag the cursor up the screen and the part will **Zoom In**. Using the mouse to **Zoom In** is a much smoother zooming method; you have more control.
	Zoom Out	This is similar to other graphics programs. This allows you to get the big picture, making the part smaller. **Quick Key**: Press the middle mouse button, hold it down as you press the left mouse button and release it. Now drag the mouse down the screen and the part will **Zoom Out**.
	Normal View	This tool allows you to view a particular plane/surface in a true length view. You specify the plane/surface and CATIA V5 will rotate the plane/surface 90 degrees to your screen view. This will make the geometry on that plane/surface true length. This is a very useful tool. You could try to rotate a plane using the space ball so it is normal to your point of view, but you could only get it "close". This tool gets it "exact". This tool can also be used to flip the direction in which you view a sketch. If in any of the lessons you go into the Sketcher Work Bench and your view is from the wrong direction use this tool to flip your view 180 degrees. It will switch your point of view from looking down on a part, to looking up from the bottom.
	Hide/Show	This tool allows you to select any entity or multiple entities and place them in "no show space". This removes the selected entity/entities from the "working space". Sometimes there are entities that you want to keep for future references but do not want them visually in the way. You can pull the entities back into the "working space" when you are ready for them.
	Swap Visible Space	This tool works hand in hand with the **Hide/Show** tool. Selecting this tool will take you out of the "working space" window and into the "no show space". To pull an element from the "no show space" you would select the **Swap Visible Space** tool icon. This would show the "no show space". You could select the entity you want back in the "working space", and then select the **Hide/Show** tool icon. This would take the entity back to the "working space". You would then need to select the **Swap Visible Space** tool icon to get back to the "working space". This can be confusing; try bringing a part back and forth until you get control of the two tools.

Quick View Mode: This tool icon has the arrow to the bottom right of it, as explained in tool bar 5. The tool options are all of the orthographic view options. **NOTE:** the view projection is dependant on what plane the body was created on.

Quick view

	Isometric View	Select this tool and CATIA V5 will rotate your part to an isometric view.
	Front View	Select this tool and CATIA V5 will rotate your part to an front view.
	Back View	Select this tool and CATIA V5 will rotate your part to an back view.
	Left View	Select this tool and CATIA V5 will rotate your part to an left view.
	Right View	Select this tool and CATIA V5 will rotate your part to an right view.
	Top View	Select this tool and CATIA V5 will rotate your part to an top view.
	Bottom View	Select this tool and CATIA V5 will rotate your part to a bottom view.

View Mode: This tool icon has the arrow to the bottom right of it, as explained in tool bar 5. There are six different options associated with this tool; they are listed below.

View mode

	Wireframe (NHR)	This shows the part as a wireframe, no solid, no shading. The (NHR) means "No Hidden Line Removal". With no hidden line removed, all edges of the part will be visible at all times. This can be confusing at times; you could lose track of what is the front side and what is the back side of a part.
	Dynamic Hidden Line Removal	This is very similar to the **Hidden Line Removal** tool except as you rotate the part, the hidden line removal is real time, where as the **Hidden Line Removal** tool will only update the hidden line removal after the rotating process is complete.
	Shading (SHD)	This tool shows the solid shaded without any edge line representation.
	Shading With Edges	This tool allows you to control how your part is going to be represented, how it looks on the screen. This tool shows the solid shaded and with the edge line representation. The majority of the graphics in this workbook are represented in this format.
	Shading With Edges And Hidden Edges:	This tool shows the solid shaded and the edge line hidden.

	Applies Customized View Parameters:	This tool will bring up a "**Custom View Modes**" window that gives you many different parameters to choose from. If you apply material to your solid you will not see the material represented unless you select the material option in the "**Custom View Modes**" window.

Menu 9
The **Apply Material** Tool Bar

Tool Bar	Tool Name	Tool Definition
	Apply Material	This tool allows you to apply a material to your solid. Applying a material will give it the properties of the material such as the density so CATIA V5 can calculate weight, volume and other part analysis information. Applying material also gives the solid the texture and color of the selected material. CATIA V5 has a library of material. The use of this tool is covered in Lesson 2. Remember, to see the material applied to the solid, you must select **Apply Material** in the **Applies Customized View Parameters** (tool 8.9.6)

Menu 10
The **Tools** Tool Bar

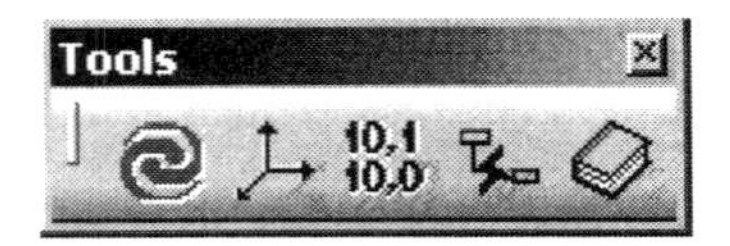

This tool bar changes depending on what work bench you are in. The three tools that are consistently in this tool bar are listed below.

Tool Bar	Tool Name	Tool Definition
	Update All	This tool will be under intensified unless there is an entity in the file that requires updating. If the tool is colored (not under intensified) it is signifying that some entity requires updating. Selecting the tool will update all of the entities. This tool is used most when revisions/changes are made to existing constraints whether it is part design changes and/or assembly changes. If you make a change and the part/assembly does not reflect the change, check this tool, it may require you to select it to force an update.

	Axis System	This tool allows you to create multiple local axis systems. Lesson 7 gives detailed instructions on how to create and orient new axis systems.
	Mean Dimensions	This tool only works if you have previously defined a tolerance to the entity. When tolerances have been applied this tool will compute the actual dimension of the entity being reviewed.
	Create Datum	This tool deactivates the history mode. The entities used to create it will not be linked. The tool, is a toggle tool, if you select it you must unselect it to turn it off.
	Open catalog	This tool allows access to the user-defined catalog. Reference the **Help** menu for detailed instruction and application.

Menu 11
The **Measure** Tool Bar

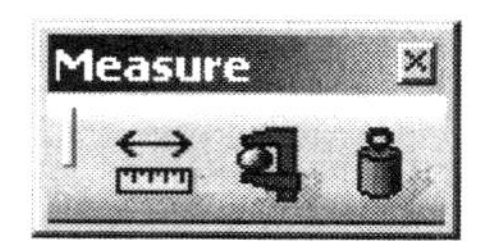

There are three analysis type tools; they are listed below.

Tool Bar	Tool Name	Tool Definition
	Measure Between	This tool allows you to measure the distance between two different entities. You can measure the distance between surfaces, planes, lines, points etc. Select the **Measure Between** tool and then the two entities. This will bring up the **Measure Between** window. This window has more information than most designers would want. In most cases, the dimension created between the two selected entities is enough information.
	Measure Item	This tool is very similar to the **Measure Between** tool except that it measures the length of an individual entity. Select the **Measure Item** tool and then select the item to be measured. This will bring up the **Measure Item** window. As in 11.1, in most cases, the dimension created on the selected entity is all that is needed.
	Measure Inertia	This measures the physical attributes of the selected solid such as volume, mass, centroid, etc. Reference Lesson 6 on how this tool is used.

Menu 12
The **Analysis** Tool Bar

This is another tool bar that is dependant on the active work bench. 12.1 and 12.2 are the most common tools.

Tool Bar	Tool Name	Tool Definition
	Draft Analysis	This tool allows you to analyze draft angles and distances. This tool is particularly helpful when the angles and distances are too small to visually inspect.
	Curvature Analysis	This tool allows you to analyze the curvature of a surface. This is particularly helpful when you have a max and min curvature radius.
	Tap - Thread Analysis	

Menu 13
The **Compass** Tool

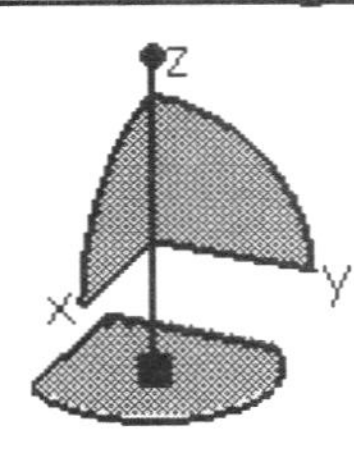

This allows you to modify the location and orientation of a part relative to the **XYZ** coordinates and/or relative to the other parts if they are in an assembly. The application of this tool is described in Lessons 2 and 6.

Menu 14
The **Current Work Bench** Tools

This side bar will be filled with tool bars and tools associated with the current work bench. As you select a different work bench you will notice the tool bars and tools will change. Each lesson covers the specific tools and tool bars found in the work bench being covered by that lesson.

Menu 15

The Prompt Zone

This is a very helpful area. When CATIA V5 is waiting on input from you, the user, there are usually some helpful prompts found in this area. It will let you know what type of entity it is looking for. In this workbook the prompts are bolded.

Menu 16

The Power Input Mode

This input window is for more advanced uses. You will notice as you select a tool, the tool command will appear in this window. In advanced uses, this window can be used similarly to quick keys and scripting. For detailed information, reference the **Help** menu. This tool is not used in this workbook.

Menu 17

The Double Arrows

When the work bench tool bars show these arrows at the bottom or end of the tool bar, it signifies there are additional tools belonging to that particular work bench. Selecting the bottom tool bar and dragging it up on the screen can access the additional tools. The tool bars can be selected and moved anywhere on the screen. Lesson 2 covers this process in detail. If you are looking for a tool that belongs to a particular work bench, it's probably there, just not yet placed in view.

18 **<u>Additional Tools</u>**: The additional tools are not shown in this Introduction. There are other tools that appear depending on the current work bench. The **Drafting Work Bench** and the **Sketcher Work Bench** are the only work benches that have tools that are not shown in this section, but are used in this workbook.

18.1 **Sketcher Grid**: This is for two-dimensional geometry creation. The **Sketcher Grid** tool is a toggle tool, you turn it on and off by selecting it.

18.2 **Snap To Point**: This tool is also a toggle tool. Its function is obvious; it turns the snap on and off.

18.3 **Show Constraint**: This is another toggle tool. It allows you to show or not to show the constraints.

18.4 **Create Detected Constraint**: This is another toggle tool. It allows CATIA V5 to create a constraint automatically if it is detected.

18.5 **Construction/Standard Element**: Construction elements are for construction only, similar to being reference only. Construction elements are dashed and under intensified elements. They are created in the **Sketcher Work Bench** and are not carried over into the **Part Design Work Bench**. They can be selected and switched from standard elements to construction elements and visa versa.

18.6 **Geometrical Constraints**: This is a toggle tool that turns the **Geometrical Constraints** tool on and off. Examples of **Geometrical Constraints** are; parallelism, perpendicular and cylindrical. This tool is more for information and confirmation.

18.7 **Dimensional Constraints**: This is another toggle tool. **Dimensional Constraints** are actual measurable constraints, anything that is controlled by a dimension. This tool is critical in creating a parametric sketch.

Introduction To The Sketcher Work Bench

This lesson will take you through each step in creating a simple sketch and ending with a part that will be referred to as the "**L Shaped Extrusion**." Later in this lesson you will be asked to save this part (file) as the "**L Shaped Extrusion.CATPart**." The completed "**L Shaped Extrusion**" is illustrated in Figure 1.1. In some cases, optional processes will be explained. Referenced illustrations will be used to help explain certain processes and to compare results. It is important that you complete and understand every step in this lesson; otherwise, you will have difficulties in future lessons where much of the basic instruction will not be covered (it will be assumed that you know it). The concepts taught in these steps will give you the tools to navigate through the basics of the **Sketcher Work Bench**. Following the step-by-step instructions, there are twenty questions to help you review the major concepts covered in this lesson. There are practice exercises at the end of this lesson. The practice exercises will help you strengthen and test your newfound CATIA V5 knowledge. This lesson covers the most commonly used tools in the **Sketcher Work Bench**. The less common and/or advanced tools will be covered in later lessons and/or in the Advanced Workbook. It is not the intent of this book to be a comprehensive reference manual, but provide basic instructions for the most common tools and functions in CATIA V5. CATIA V5 in the **Windows NT** environment allows multiple methods of accomplishing the same task. You are encouraged to explore all of the different options.

Figure 1.1

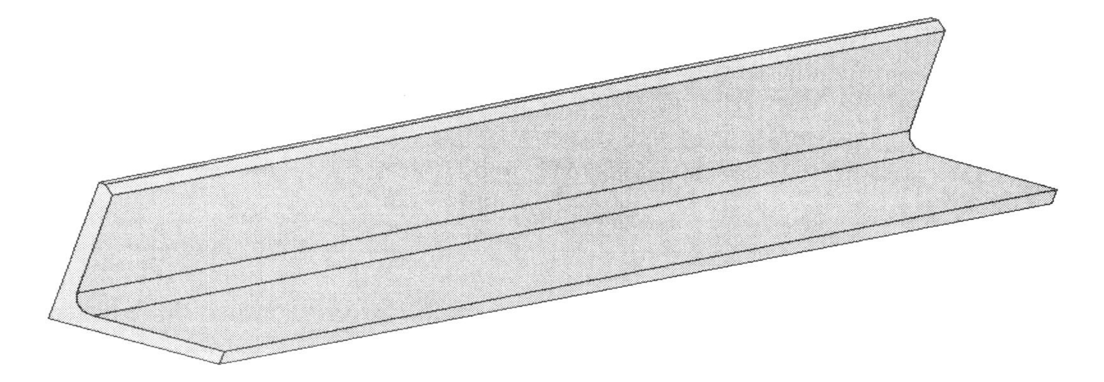

Lesson 1 Objectives

This lesson will show you how to do the following:

- Start CATIA V5
- Select the a specific workbench
- Move around in the CATIA V5 Environment
- Select a working plane
- Enter the **Sketcher Work Bench**
- Basic customization of the **Sketcher Work Bench**
- Create a sketch using the standard sketch tools
- Apply **Constraints** to the sketch
- Modify existing **Constraints**
- Exit the **Sketcher Work Bench**
- Create a **Pad** in the **Part Design Work Bench**
- **Save** the newly create file
- **Exit** CATIA V5

Sketcher Work Bench Tool Bars

There are four standard tool bars found in the **Sketcher Work Bench**. The four tool bars are shown on the following pages. The individual tools found in each of the four tool bars follow with the tool name and a brief definition.

Figure 1.2

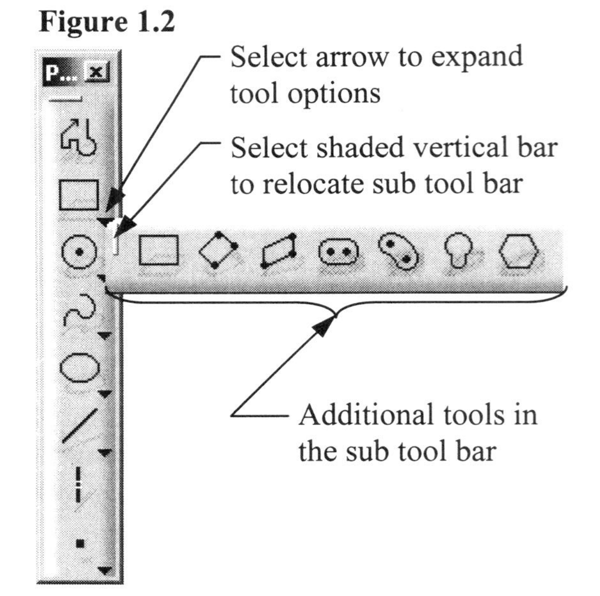

Some tools have an arrow located at the bottom right of the tool icon. The arrow is an indication that there is more than one variation of that particular type of tool. To display the additional tool options (as shown in Figure 1.2) you must move the mouse over the arrow and select it. As the pointer is moved over the arrow the following icon will appear . When the double arrow appears, select it and the additional tool options will appear as shown in Figure 1.2. To select one of the additional tools move your mouse to the desired tool icon and select it. The newly selected tool icon will become the default tool for the additional tool bar. Each additional tool bar has a specific name. The main tool bar names and the submenu tool bars are displayed, labeled and defined in following tables. Selecting the small shaded vertical bar as shown in Figure 1.2 also allows you to tear off the tool bar and relocate it anywhere else on the screen. This is done by holding the left mouse button down, dragging it to the desired location and releasing the mouse button. Relocating the bar will display the name of the tool bar as shown in Figure 1.3. If you use several of the tools on one particular submenu tool bar more than others you might want to expand and relocate the tool bar so you don't have to select through the different tool icons every time.

Figure 1.3

The **Sketcher** Tool Bar (Access The Sketcher Work Bench)

Tool Bar	Tool Name	Tool Definition
	Sketcher	Provides accesses to the **Sketcher Work Bench**. This is where the **Sketch** tools are made available for creating profiles.

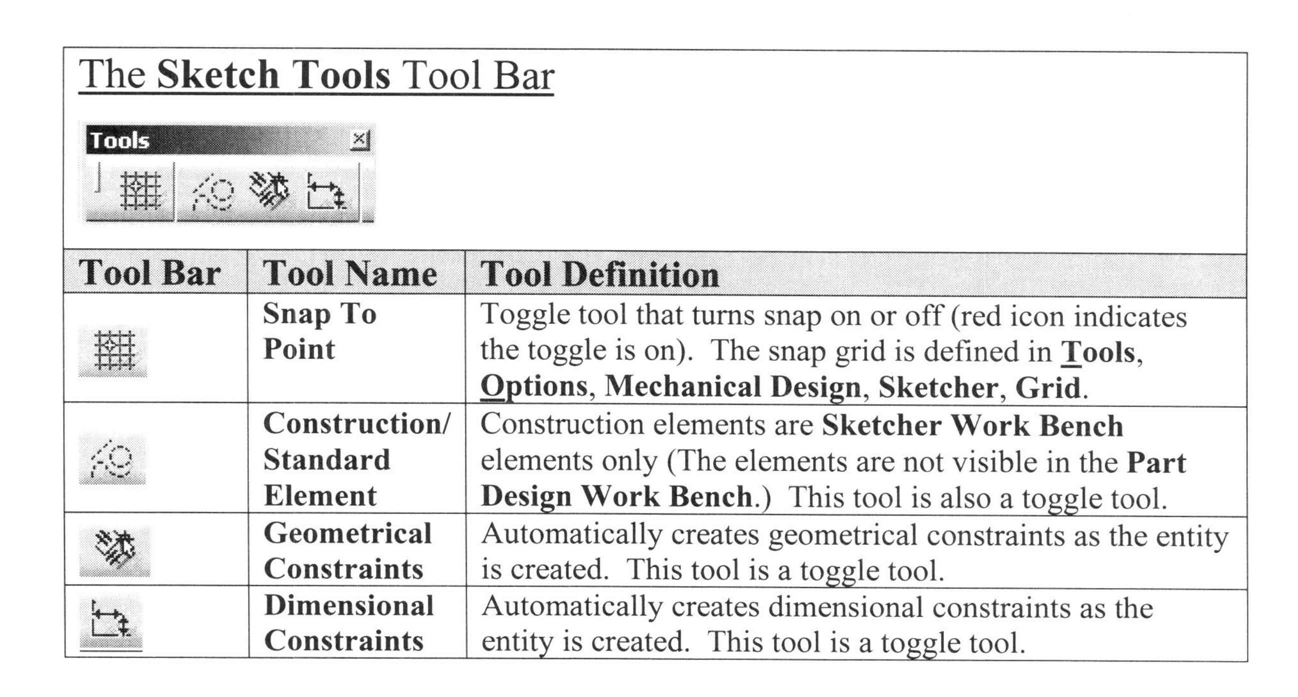

The **Sketch Tools** Tool Bar

Tools

Tool Bar	Tool Name	Tool Definition
	Snap To Point	Toggle tool that turns snap on or off (red icon indicates the toggle is on). The snap grid is defined in **Tools**, **Options**, **Mechanical Design**, **Sketcher**, **Grid**.
	Construction/ Standard Element	Construction elements are **Sketcher Work Bench** elements only (The elements are not visible in the **Part Design Work Bench**.) This tool is also a toggle tool.
	Geometrical Constraints	Automatically creates geometrical constraints as the entity is created. This tool is a toggle tool.
	Dimensional Constraints	Automatically creates dimensional constraints as the the entity is created. This tool is a toggle tool.

The **Sketch Tools** tool bar only appears when you are in the **Sketcher Work Bench**. The four tools found in this tool bar are toggle tools. When the tool is highlighted the tool is on. This particular tool bar changes depending on what other **Sketcher Work Bench** tool is currently selected. For example, if you select the **Rectangle** tool from the **Profile** tool bar you will notice the **Sketch Tools** tool bar will look similar to what is shown in Figure 1.4. The tool bar allows you to type in specific values for the element you are creating. Figure 1.4 is prompting for the **Horizontal** and **Vertical** location of the first corner of the **Rectangle**. Select some of the other **Profile** tools and observe what different options the **Sketch Tools** tool bar prompts you for.

Figure 1.4

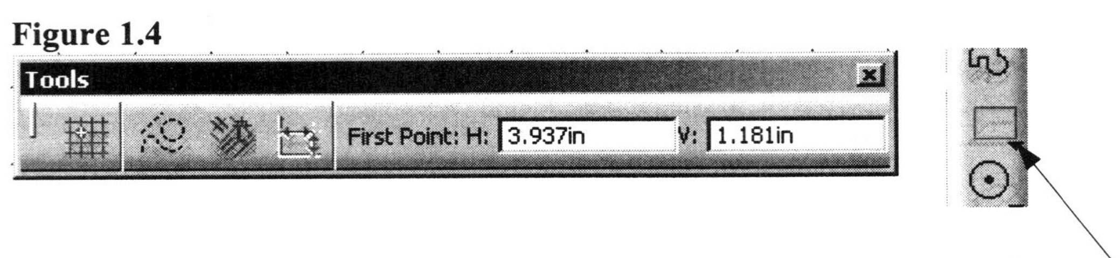

The **Sketch Tools** tool bar with the **Rectangle** tool selected from the **Profile** tool bar.

The **Profile** Tool Bar

Tool Bar	Tool Name	Tool Definition
	Profile *	Creates one continuous profile using lines and/or arcs. Lines: select a starting and ending point. Arcs: hold the left mouse button down and move to the end point of the arc. Terminate profile by double clicking the left mouse button.
	Predefined Profile Tool Bar	
	Rectangle *	Creates a rectangle defined by two points.
	Oriented Rectangle *	Creates a rectangle oriented at a specific angle. This tool requires three points of input as indicated on the tool icon.
	Parallelogram *	Creates a parallelogram oriented at a specific angle. Requires three points of input as indicated on the tool icon.
	Elongated Hole *	Creates an elongated hole. Requires three points of input. 1st point is the center of the first radius. 2nd point is the center of the second radius. 3rd point is the radius of the elongated hole.
	Cylindrical Elongated Hole *	Creates a cylindrical elongated hole. Requires four points of input. 1st point is the center of the cylindrical hole. 2nd point is the start point of the center line. 3rd point is the arc length of the elongated hole. 4th point defines the radius (width) of the elongated hole.
	Keyhole Profile *	Four point selections are required to create the keyhole profile. 1st point defines the center of the large radius. 2nd point defines the center of the small radius. 3rd point defines the size of the small radius. 4th point defines the size of the large radius.
	Hexagon *	Only two points are required to create the hexagon. 1st point defines the center. 2nd point defines the size and angle.
	Circle Tool Bar	
	Circle *	Two points are required to create a circle. 1st point defines the center. 2nd point defines the radius.
	Three Point Circle *	Three points are required to define a circle. CATIA V5 calculates the center and radius using three points defined by the user.

	Circle Using Coordinates *	Creates a circle using polar or cartesian coordinates. The values are typed in by the user.
	Tri-Tangent Circle	This tool requires three existing entities to select from.
	Three Point Arc *	Similar to the **Three Point Circle** tool except that the 1st point defines the start of the arc, the 2nd point defines the center of the arc length and the 3rd point defines the end of the arc length.
	Three Point Arc Starting With Limits *	Similar to the **Three Point Arc** tool except that the 1st point defines the start point, the 2nd point defines the end point of the arc and the 3rd point defines the radius and orientation.
	Arc *	Three points are required to create an arc. 1st point defines the center of the arc radius. 2nd point defines the radius and start point. 3rd point defines the end point.
	Spline Tool Bar	
	Spline *	Creates multiple continuous curves. The radius and location of the curve is determined by the previous point. The location and number of points defining the spline is determined by the user.
	Connect *	Connects two existing splines and/or arcs. The user is required to select the two entities to connect.
	Conic Tool Bar	
	Ellipse *	Creates an ellipse using three points. 1st point defines the center. 2nd point orients the ellipse and defines the major radius. 3rd point defines the minor radius.
	Parabola By Focus *	Creates a parabola using four points. 1st point defines the focus point. 2nd point defines the apex. 3rd point defines the start point. 4th point defines the end point.
	Hyperbola By Focus *	Creates a hyperbola using five points. 1st point defines the focus point. 2nd point defines the center. 3rd point defines the apex. 4th point defines the start point. 5th point defines the end point.
	Creates A Conic *	Creates a conic using five points. 1st point defines the start point. 2nd point defines end point. The next three points define the curve between the first two points.
	Line Tool Bar	
	Line *	Creates a line between two points. The length of the line is determined by the start point and end point.
	Infinite Line *	Creates a line of infinite length. The **Sketch Tools** tool bar gives the user several options, one point horizontal, one point vertical or two points to define the slope.
	Bi-Tangent Line	Requires two existing entities to select from. Note: not all selections will have a possible solution.

	Bisecting Line	Requires two existing lines to select from. Creates a line at the intersection of the two selected lines.
	Axis Tool Bar	
	Axis *	Created the same as a line (start point and end point). The axis can not be used as part of the profile to be extruded (**Pad** tool). It is required to define the line of rotation for the **Shaft** tool and other revolutions.
	Point Tool Bar	
	Point By Clicking *	Creates points by pointing or by typing exact coordinate values. **Note:** Points created as a **Standard Element** can cause problems when creating a solid part (pad). Points are usually created and used as **Construction Elements**.
	Point By Using Coordinates *	Creates a point using polar or Cartesian values.
	Equidistant Points *	Requires an existing element. Allows the user to specify the number of points or the distance between points on an existing element.
	Intersection Point	Requires two existing elements that intersect each other. The intersection can be a projected intersection.
	Projection Point	Requires an existing point and an additional existing element. It projects the existing point onto the selected element.

NOTE: The tools with the * can also be created/defined using the **Sketch Tools** tool bar by typing in the exact values in the appropriate input windows or selecting other options that are offered.

The **Operation** Tool Bar

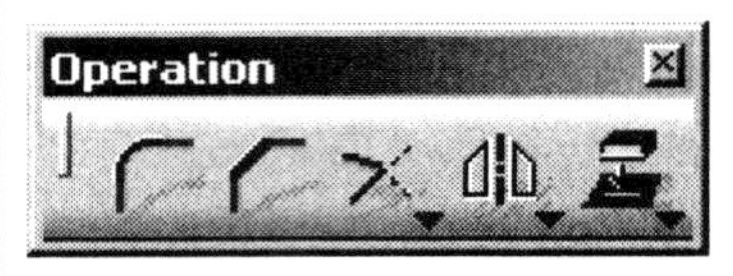

Tool Bar	Tool Name	Tool Definition
	Corner *	Creates corners, the radius is specified by the user.
	Chamfer *	Creates chamfers, the chamfer length is specified by the user.
	Relimitations Tool Bar	
	Trim *	Trims intersecting lines back to the point of intersection. The tool will also stretch a line to the point of intersection. The **Sketch Tools** tool bar allows the user to select **Trim All Elements** or **Trim First Element**.
	Break	Breaks one element into two separate elements.
	Quick Trim *	Can be used as an eraser (delete an element). Selecting this tool gives the user three additional options in the **Sketch Tools** tool bar they are; **Break And Rubber In**, **Break And Rubber Out** and **Break And Keep**.
	Close	Closes all arcs to a complete circle.
	Complement	Creates the missing section of a circle from an existing arc.
	Transformation Tool Bar	
	Symmetry	Creates symmetrical elements using lines or axis.
	Translate *	Translates elements. A **Translation Definition** pop up window allows the user to determine the number of copies and length.
	Rotate *	Rotates elements. A **Rotation Definition** pop up window allows the user to define the number of copies and angle between copies.
	Scale *	Scales elements. A **Scale Definition** pop up window allows the user to define the scale and point to scale from.
	Offset *	Creates elements using offset values specified by the user. The **Sketch Tools** tool bar gives the user four additional options, they are; **No Propagation**, **Tangent Propagation**, **Point Propagation** and **Both Side Offset**.
	3D Geometry Tool Bar	
	Project 3D Elements	Projects 3D elements onto the sketch plane. Allows the user to use existing 3D elements to create a profile. The only difference between the following two tools is the relationship between the sketch plane and the 3D element.

	Intersect 3D Elements	Intersects 3D elements with the sketch plane.
	Project 3D Silhouette Edges	Projects silhouette edges of a 3D element with the sketch plane.

The **Constraint** Tool Bar

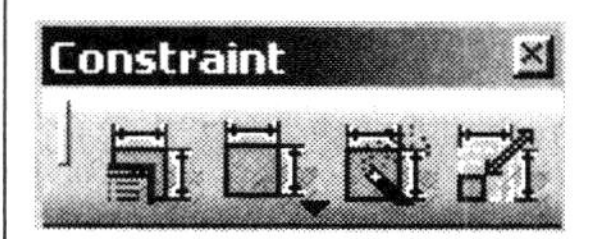

Tool Bar	Tool Name	Tool Definition
	Constraints Defined In Dialog Box	Allows the user to select the type of geometrical constraint to apply to the selected entity. The entity must be selected for the tool to be selected. Only the constraints that apply to the selected entity will be selectable.
	Auto Constraint	Creates geometrical and dimensional constraints automatically.
	Animate Constraint	Animates dimensional constraints. The user specifies a lower and upper limit and the number of steps between the limits.
	Constraint Creation Tool Bar	
	Constraint	Creates dimensional constraints from entities selected by the user.
	Contact Constraint	Creates a contact constraint between two different entities.

When you are in the **Sketcher Work Bench**, the first tool bar (the **Sketch Tools** tool bar) by default will appear floating in the upper left hand corner of the screen. The last three tool bars (**Profile**, **Operation** and **Constraint**) are located by default on the right side of the screen. The four tool bars contain too many tools to show all of them in one lesson. To view and have access to all of the tools, you can select the shaded tab located at the top of each tool bar and drag it anywhere on the screen. This is important, because when you get to Step 12, by the default setup, you will not be able to visually locate the **Operation** tool bar. You will have to select and drag the **Operation** tool bar from the bottom right side of the screen to the location you select.

CATIA V5 allows you an additional method of selecting all of the **Sketcher Work Bench** tools. To select the **Sketcher Work Bench** tools, select **Insert** from the **Standard NT Windows** tool bar. **Insert** contains the three **Sketcher** tool bars. This method displays the names with the tools. Figure 1.5 shows the same selection as shown in Figure 1.2 using this method.

Figure 1.5

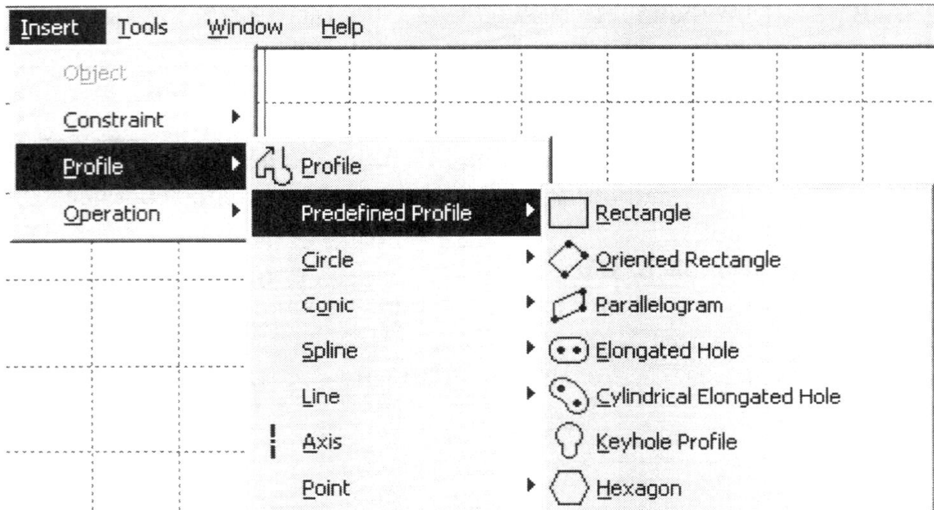

Steps To Creating A Simple Part Using The Sketcher Work Bench

You are now going to use the tools just introduced to you to create an "**L Shaped Extrusion**." The part is referred to as an "**L Shaped Extrusion"** because its profile or shape is similar to an upper case letter L. When you complete all of the steps in this lesson, the result should look similar to the part shown in Figure 1.1.

If you are not able to finish all of the steps in this lesson in one session, you can jump to Steps 23 and 24, which cover saving and exiting CATIA V5. This will allow you to save your work for your next session.

1 Start CATIA V5 

From the **Desktop**, double click on the **CATIA V5** icon. Be patient, it may take a few moments to bring up the CATIA V5 start logo and the actual CATIA V5 working window. CATIA V5 by default will start in **Product Structure Work Bench** as shown in Figure 1.6.

Figure 1.6

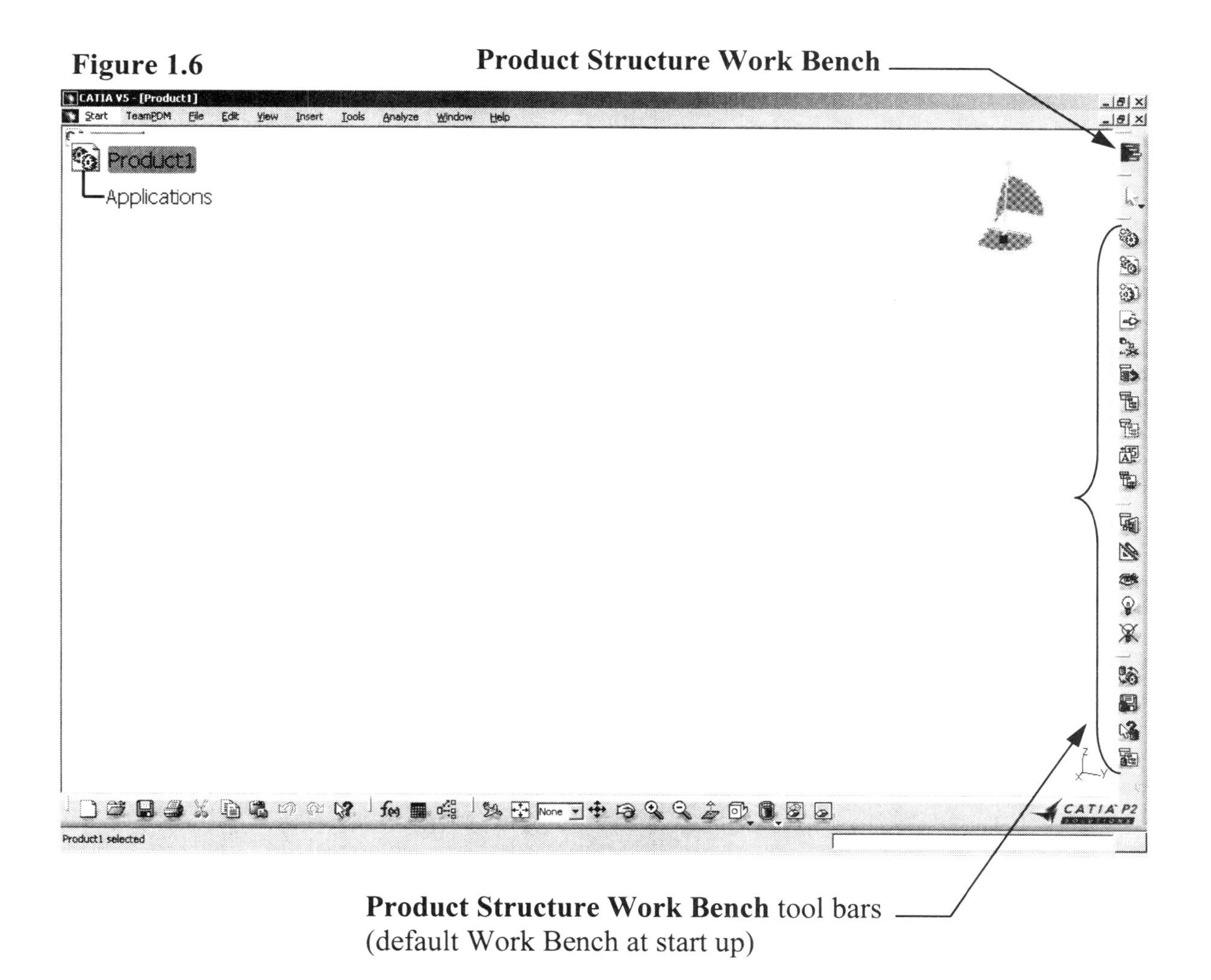

2 Select The Sketcher Work Bench. 

How your CATIA V5 screen looks at start up will depend on how it was setup and/or customized. This step will cover most of the options you could encounter. The CATIA V5 screen will appear as it does in Figure 1.6. The default work bench is the **Product Structure Work Bench**. For this lesson, you will need to select the **Sketcher Work Bench**.

There are two possible methods to selecting the **Sketcher Work Bench**.

2.1 The first method of selecting a work bench is using the **Start** menu located in the **Standard Windows** tool bar. If you have not previously set up your favorite work benches this step will only display the work benches under the CATIA V5 Functions. Step 2.3 explains how to create your favorite work benches.

2.1.1 Select the **Start** tool in the top left side of the screen, reference Figure 1.7. This will bring up a pull down menu that includes all of the CATIA V5 Functions such as **Infrastructure**, **Mechanical Design**, **Shape** etc., and their work benches.

2.1.2 Select the **Mechanical Design** function as shown in Figure 1.7. This will display all the work benches in the **Mechanical Design** function.

Figure 1.7

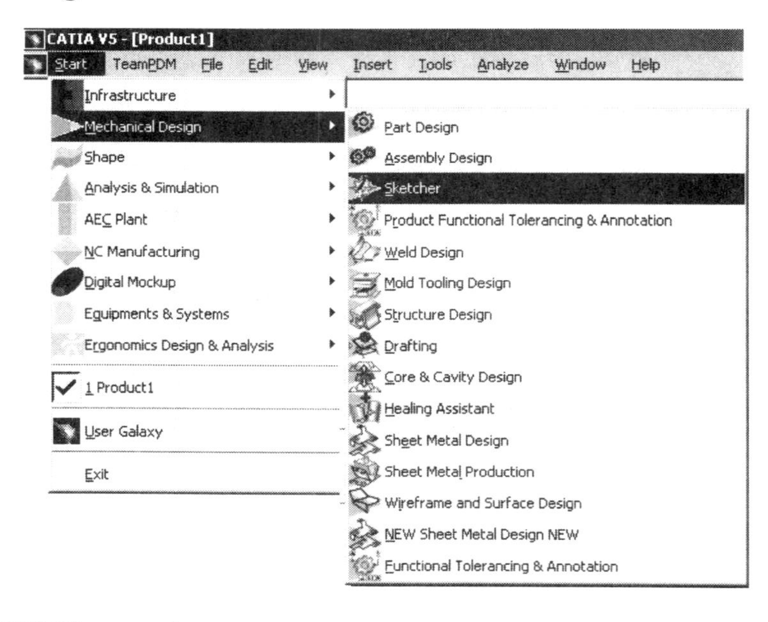

2.1.3 Select the **Sketcher Work Bench** as shown in Figure 1.7. This will bring up the **Part Design Work Bench**. Notice that the Specification Tree changes from **Product1** to **Part1** with the **XY**, **YZ** and **ZX Planes**. Once you're in the **Part Design Work Bench** all you have to do is define the working **Plane**; the plane you want to create the sketch on. Once you select a **Plane**, the work bench will change to the **Sketcher Work Bench**. Selecting the working **Plane** is explained in Step 3.

2.1.4 To start creating the "**L Shaped Extrusion**," continue to Step 3. To learn the other methods of selecting work benches continue to Step 2.2.

2.2 The work benches that appear in the **Welcome To CATIA V5** window depend on how CATIA V5 was installed. To find out which work benches were added to the **Welcome To CATIA V5** window, complete the following steps.

2.2.1 Select the **Product Structure Work Bench** in the top right of your screen, reference Figure 1.9. Selecting the current work bench will activate the **Welcome To CATIA V5** window.

2.2.2 If there are no work benches displayed in the **Welcome To CAITA V5** window then nothing was added at installation. Figure 1.9 shows three work benches, Step 2.3 shows you how to customize this window.

2.3 To add your favorite work benches to the **Welcome To CATIA V5** window complete the following steps.

2.3.1 From the **Standard Windows** tool bar select **Tools**.

2.3.2 Select **Customize**. This brings up the **Customize** window as shown in Figure 1.8.

Figure 1.8

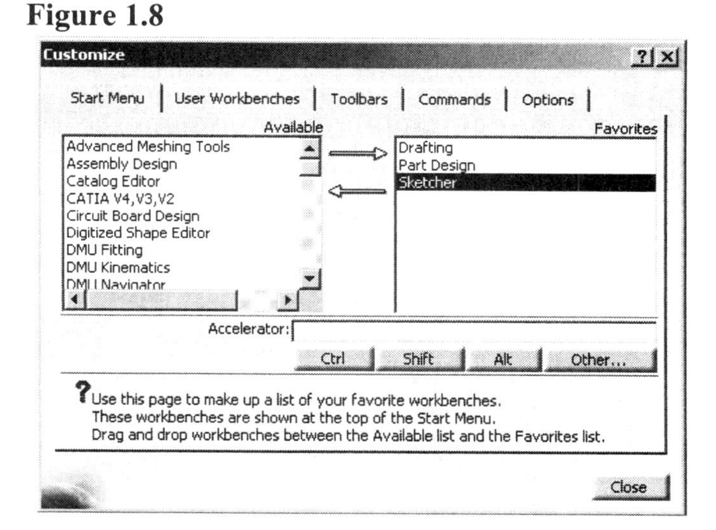

2.3.3 Select the **Start Menu** tab. This brings up two boxes. The first box is labeled **Available**. It contains all of the work benches that are licensed to the computer. The second box is labeled **Favorites.** This is the box where you will place the work benches you want to use most often.

2.3.4 From the **Available** box select the **Sketcher, Part Design** and **Drafting Work Benches.** Use the **Windows** multi-select by holding down the **Ctrl** key.

2.3.5 Select the arrow pointing to the **Favorite** box. This will place the selected work benches in the **Favorite** box.

2.3.6 You can create an **Accelerator** (quick key) by selecting the particular work bench and typing in the chosen **Accelerator** keys in the **Accelerator** box. The keys must be separated by the "+" key. For this step, create an **Accelerator** key for the **Sketcher Work Bench** by completing the following steps.

2.3.6.1 Highlight the work bench in the **Favorites** box that you are going to create the **Accelerator** key for.

2.3.6.2 With the correct work bench highlighted, place the cursor in the **Accelerator** box.

2.3.6.3 Click on the **Ctrl** selection beneath the box.

2.3.6.4 Type in the character "+" with no spaces after the **Ctrl**.

2.3.6.5 Type in the small letter "**k**," again with no spaces between the characters.

2.3.6.6 Hit **Enter**. The **Accelerator** key is now created.

If you attempt to use **Accelerator** keys already assigned, a pop-up window will warn you that it has already been assigned to a particular function. This step has given you one more method of selecting the **Sketcher Work Bench**.

2.3.7 Select the **Close** button. The **Customize** window will disappear.

2.3.8 Select the current work bench as shown in Figure 1.9. This will bring up the **Welcome To CATIA V5** window. This time the **Sketcher**, **Part Design** and **Drafting Work Benches** will be in the **Welcome To CATIA V5** window as shown in Figure 1.9.

2.3.9 You can also select the **Start** tool as shown in Step 2.1.1 and Figure 1.7. This time the three work benches added to the **Welcome To CATIA V5** window will appear in the **CATIA Functions**. The work benches can be selected from this menu also.

2.4 Select the **Sketcher Work Bench** from the **Start** tool bar or the **Welcome To CATIA V5** window. This will bring up the **Part Design Work Bench** and all of the **Part Design** tools. The reason it brings up the **Part Design Work Bench** instead of going directly into the **Sketcher Work Bench** is that you the user must specify the working plane before entering the **Sketcher Work Bench**. The **Sketcher Work Bench** is a tool used to create two-dimensional profiles for use in the **Part Design Work Bench**. You could think of the **Sketcher Work Bench** as a required step to using the **Part Design Work Bench**. Step 3 will show you how to define the working plane and how to actually enter the **Sketcher Work Bench**.

Figure 1.9

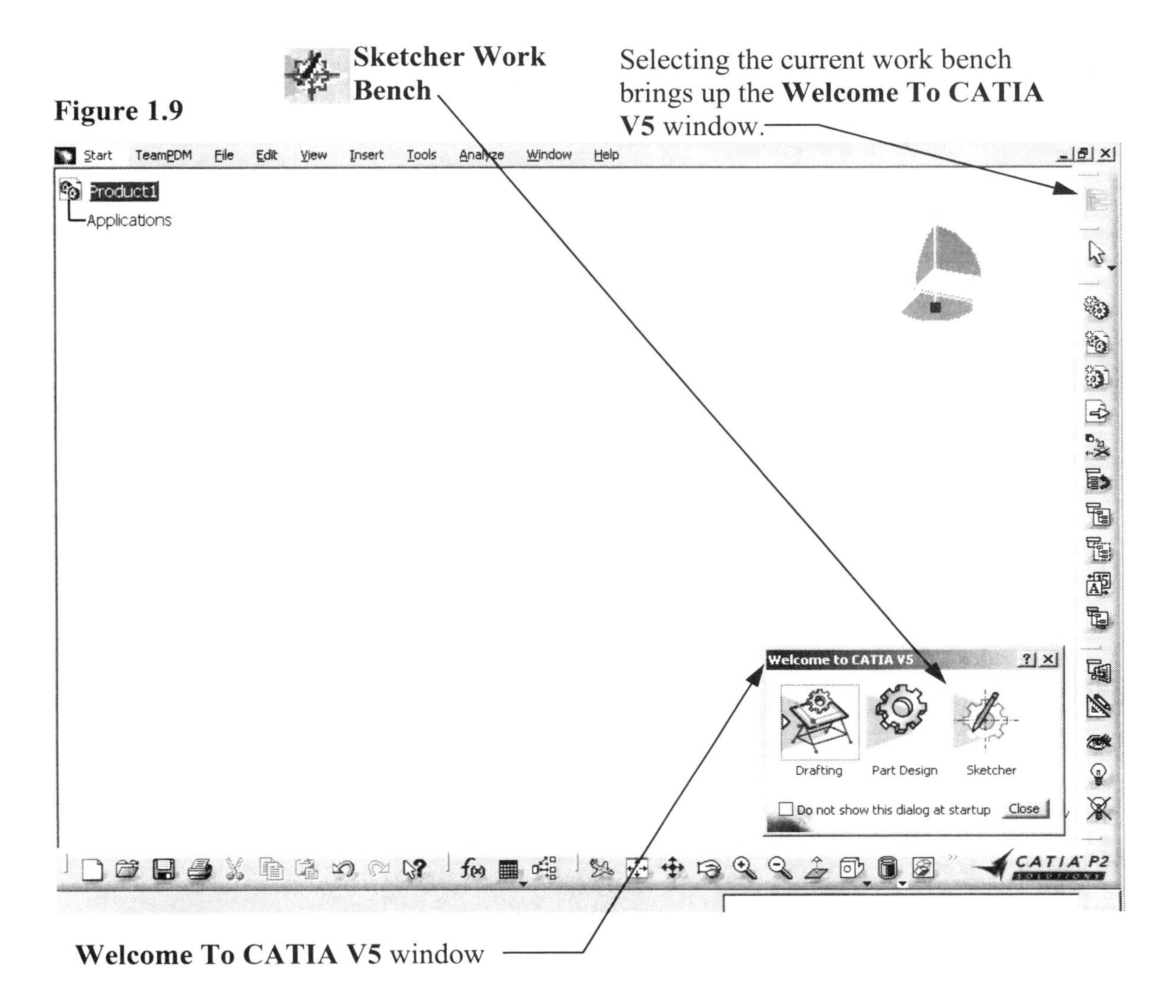

3 Specify A Working Plane

The next step is to create a two-dimensional profile of the part. The **Sketcher Work Bench** is a two-dimensional (planar) work area. To use the **Sketcher Work Bench** you must specify which plane the profile is to be created on. Specifying the plane can be done several different ways.

3.1 Select (highlight) the desired plane from the graphical representation in the center of the screen as shown in Figure 1.10. Notice, as a particular plane is selected, the equivalent plane in the **Specification Tree** is highlighted. If the **Specification Tree** isn't showing the branches with the plane it will need to be expanded. To do this, select the **Plus** symbol ⊕ to the left of the **Specification Tree** or double click on the branch you want expanded.

3.2 The step described above can be reversed. Select the plane in the **Specification Tree** and the coordinating plane in the center of the screen will also be highlighted.

3.3 Other planes, surfaces and/or other planner objects can also be selected to define the **Sketcher Plane**. This option will be covered in more detail later in the book.

3.4 For this lesson select the **ZX Plane** as shown in Figure 1.10.

NOTE: The background color may vary from figure to figure. The lessons do not require you make the changes. The variation is for presentation only.

Figure 1.10

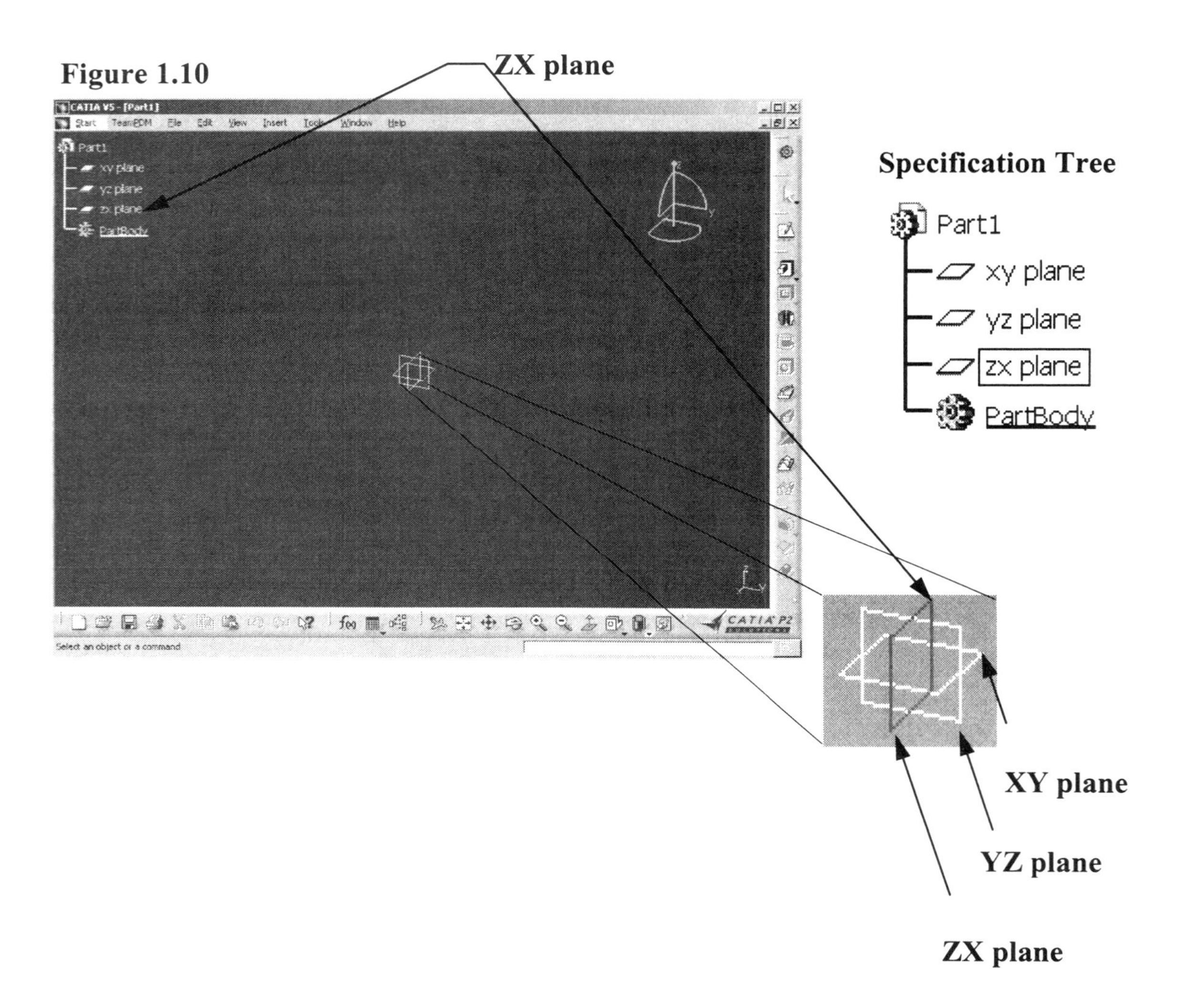

4 Entering the Sketcher Work Bench

Once a plane is selected the screen will animate rotating until the selected plane is parallel to the computer screen (perpendicular to you, true size). The default grid will also appear. You are now officially in the **Sketcher Work Bench**, but before you create the planar profile of the "**L Shaped Extrusion**" you need to customize the grid.

5 Customizing The Grid

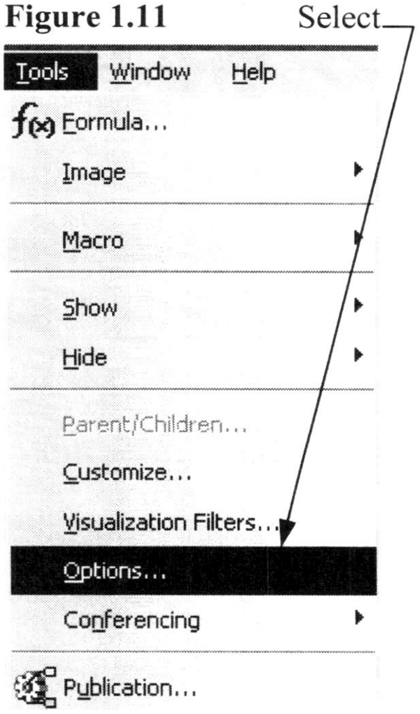

5.1 Go to the top tool bar in the pull down menu and click on **Tools**, **Options** as shown in Figure 1.11. This brings up the **Options** window as shown in Figure 1.12. For Steps 5.2 through 5.6, reference Figure 1.12.

Figure 1.11

5.2 Select the **Mechanical Design** branch to expand it if it is not already expanded.

5.3 Select the **Sketcher** branch under the **Mechanical Design** branch.

5.4 The **Sketcher** tab is the only tab available; therefore, you are not required to select a tab. It will be important for you to remember where these selections are located, as you may want to customize your set up in later lessons.

5.5 The first section under the **Sketcher** tab is the **Grid** section. The first option in the **Grid** section is **Display.** For this particular exercise check the **Display** option, this will display the grid when in the **Sketcher Work Bench**. The **Sketch Tools** tool bar allows you to quickly toggle this tool on/off.

5.6 The second option in the **Grid** section is **Snap To Point** option. For this particular exercise check the **Snap To Point** option. You guessed it; this toggles the **Snap To Point** option to **ON**. The **Sketch Tools** tool bar allows you to quickly toggle this tool on/off.

5.7 The third option that you need to know about is **Primary Spacing**. The user can set the desired spacing. If the default measurement is in metric, the spacing will be in mm. To change this default reference Figure 1.13 and complete the following steps.

5.7.1 In the **Options** window and under the **General** branch select the **Parameters** branch. To get the **Options** window select the **Tools**, **Options** as described in Step 5.1 above.

5.7.2 Select the **Units** tab. The **Options** window on the screen should now look similar to what is shown in Figure 1.13.

Figure 1.12

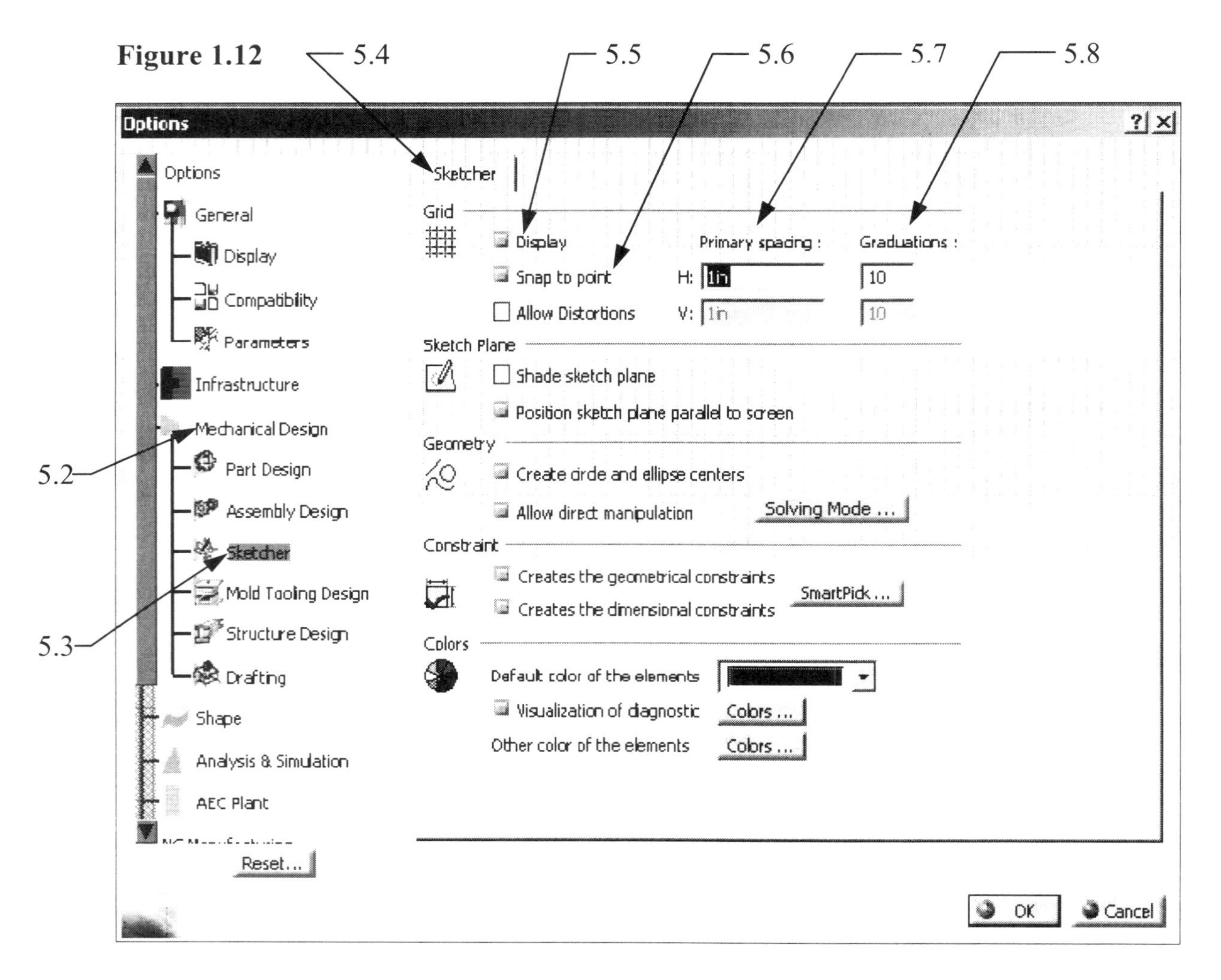

5.7.3 The **Units** box inside the **Options** window allows you to select any type of measurement CATIA V5 recognizes. The scroll bar to the far right of the box allows you to scroll through the many options. For this particular step select the **Length** option.

5.7.4 The **Length** option will appear at the bottom of the window as shown in Figure 1.13. Selecting the arrow at the far right will display all the different types of measurements of length that CATIA V5 allows you to work in. For this particular step select the **Inch (in)**.

5.7.5 Now go back to the **Sketcher** options by selecting the **Sketcher** branch under the **Mechanical Design** branch. Notice the **Primary Spacing** option is now showing in inches, reference Figure 1.12.

NOTE: If you select the **OK** button right after completing Step 5.7.4 you will have to re-select the **Tools**, **Options** tools to get back to the **Sketcher** options.

Figure 1.13

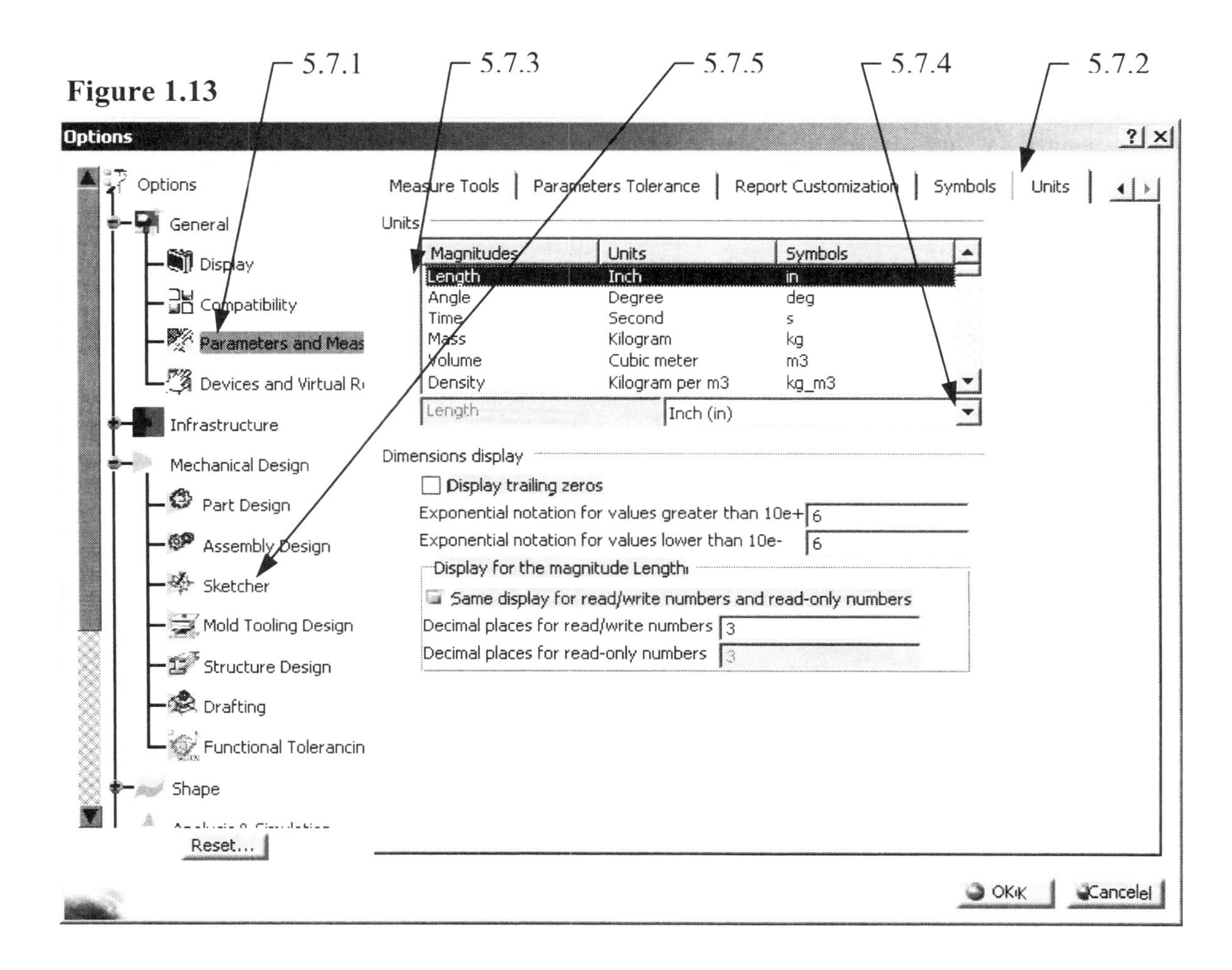

5.8 The fourth option you need to know about under the **Grid** section is **Graduations**. This option divides the **Primary Spacing** into divisions, defined by you. Reference Figure 1.12. As an example, if the **Primary Spacing** is 1″ and the **Graduations** is 1 (division), the grid will remain in 1in grids. If the **Primary Spacing** is 1″ and **Graduations** is set to 2 (divisions), the grid will be a .5in grid. To change the **Primary Spacing** and the **Graduations**, select the value in the window and type in the new value. When entering the values for the **Primary Spacing**, it is not necessary to enter the measurement type. The lowest value allowed for **Graduations** is 1 (zero will not be accepted). For this exercise enter **1** for the **Primary Spacing** and enter **10** for the **Graduations**. Select the **OK** button to apply the **Primary Spacing** and the **Graduations** values. The **Primary Spacing** is represented in the **Sketcher Work Bench** with a solid line while the **Graduations** is a dotted line as shown in Figure 1.14. It is important to remember that the zoomed view on the screen will dictate how the **Primary Spacing** and **Graduations** are represented. If you are zoomed out the **Graduations** and **Primary Spacing** could look very similar to each other, not distinguishable. If you find yourself in this situation use the **Zoom** tool on the tool bar at the bottom of the screen (Figure 1.15). Continue to zoom in until the **Primary Spacing** and **Graduations** are distinguishable.

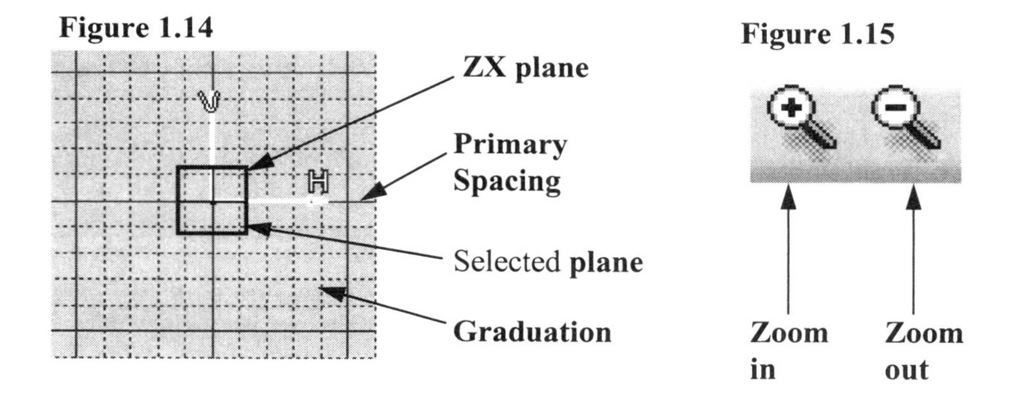

6 Creating Geometry Using The Profile Tools

You are now ready to create the profile (periphery) of the "**L Shaped Extrusion**." The first tool you will use from the **Profile** tool bar is the **Point by Clicking** tool, covered in Step 7. The second tool is the **Line** tool, covered in Steps 8, 9 and 10. The third tool is the **Profile** tool, covered in Step 11.

On the **Sketch Tools** tool bar, make sure the **Snap To Point** is **On**, the **Geometrical Constraints** is **On** and the **Dimensional Constraints** is **On**. If the tools are highlighted in red they are on. The tools in the **Sketch Tools** tool bar can be toggled **On** and **Off** by selecting them. The **Sketch Tools** tool bar is shown in Figure 1.19. Now you are ready to create some sketch geometry!

7 The Starting Point

The (0,0) point in the **Sketcher Work Bench** is the intersection of the **Horizontal (H)** and **Vertical (V)** axis. It can also be described as the intersection of the three planes (**XY**, **ZX** and **YZ**). Reference Figures 1.10, 1.14 and 1.16. The starting point for your profile will be (1,1). You should be able to locate the (1,1) location using the **Primary Spacing** and **Graduations**. To visually verify the location and to **Anchor** your first two lines to the (1,1) location, create a point at the (1,1) coordinate location, using the following steps.

7.1 Select the **Point By Clicking** icon found in the **Profile** tool bar on the right side of the screen. After selecting the **Point By Clicking** icon the mouse will be accompanied by a **Target Selector**. This tool allows you to select and snap to a location on the screen. As Figure 1.16 shows, the target selector also gives you the current location of the **Target Selector.** CATIA V5 will prompt you (in the **Prompt Zone**) to "**Click To Create The Point**." Another way of specifying the location of the point is to type the location in the **Point Coordinates: H:** and **V:** boxes. The **H:** is for horizontal and **V:** is for vertical coordinates. Reference Figure 1.16. **Point By Clicking** allows you to create the point by selecting a location or by typing the **H** and **V** values.

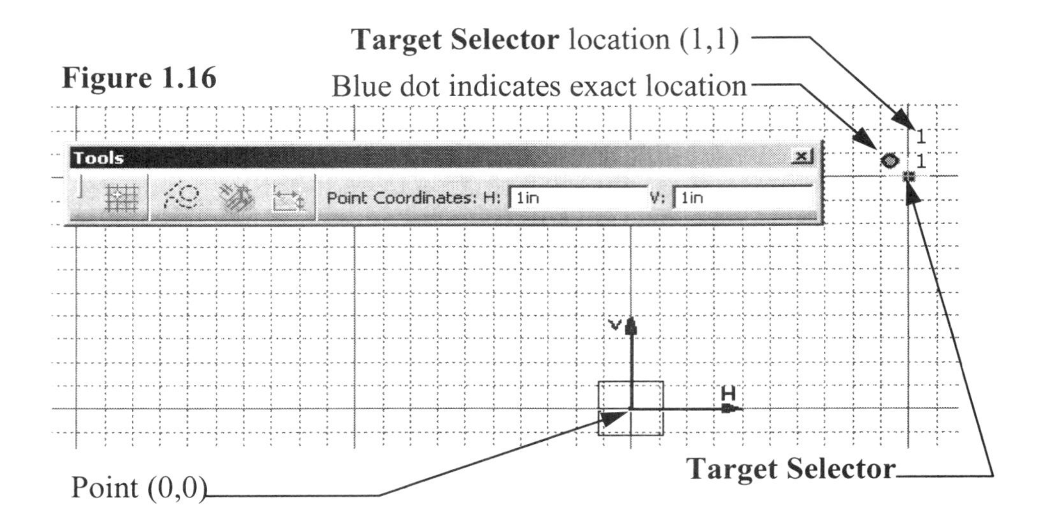

Figure 1.16

7.2 For this step, create the point at the 1,1 location. You could create the point by snapping to the location as shown in Figure 1.16, but for this particular step you need to create the point by typing the value in the **Sketch Tools** tool bar input window as shown in Figure 1.16. To type in the values complete the following.

7.2.1 Type in **1in** for the **Horizontal** coordinate.

7.2.2 Hit the **Tab** key to move the cursor over to the **Vertical** coordinate.

7.2.3 Enter **1in** for the **Vertical** coordinate.

7.2.4 Hit the **Enter** key to have CATIA V5 create the new point.

This method will create constraints that will lock down the location of the point. Figure 1.18 shows what the **Constraint** looks like, similar to green dimensions, locating the newly created point. If you want to create points by coordinates only, select the "**Point By Using Coordinates**" tool, the second **Point** tool option. This will bring up the **Point Definition** widow as shown in Figure 1.17. Notice you are given the opportunity to create the point using a **Cartesian** coordinate or **Polar** coordinate system.

7.3 A **Point** (+) will appear at the (1,1) coordinate. It will remain highlighted until you make another selection. Notice a **Point.1** has been added to the **Specification Tree**, reference Figure 1.18. Remember, you may have to expand the **Specification Tree** to see all of the entities. **Point.1** is under the **Geometry** branch under the **Sketch.1** branch.

Figure 1.17

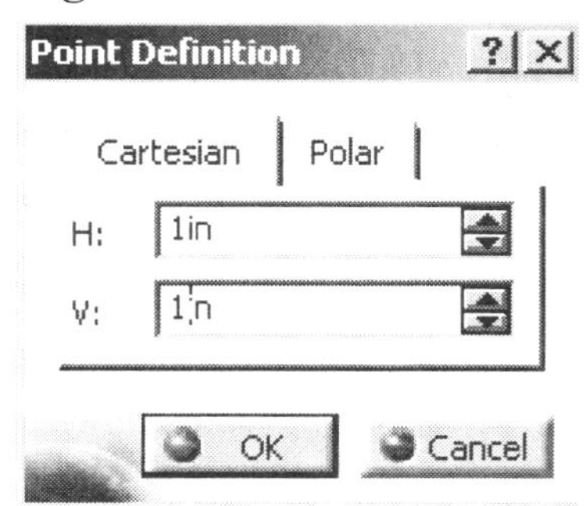

Figure 1.18

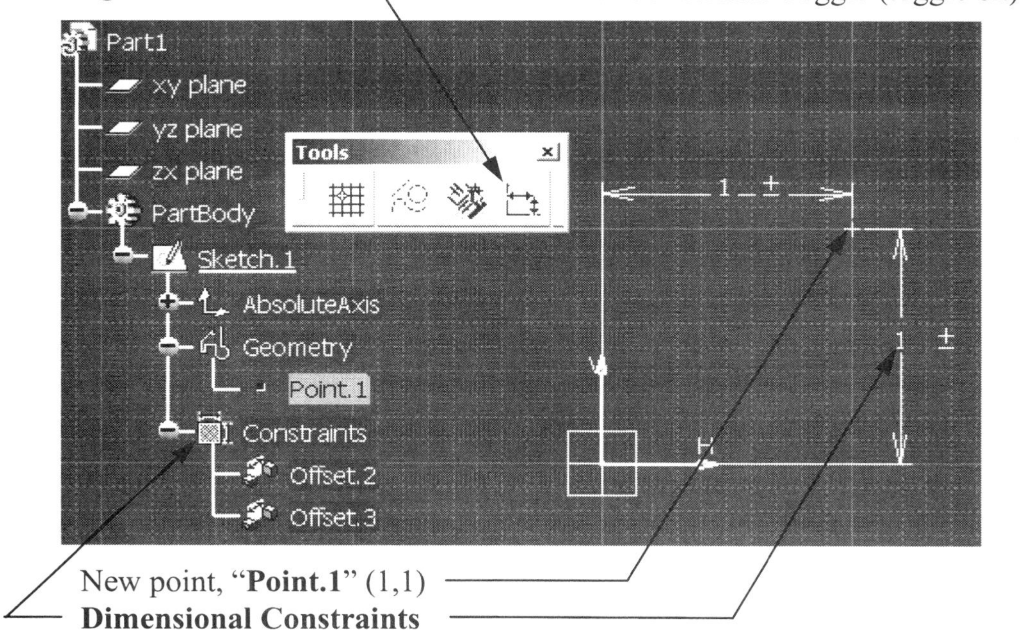

Dimensional Constraint Toggle (toggle on)

New point, "**Point.1**" (1,1)

Dimensional Constraints

8 Creating Line 1 

Remember, the grid you set up is 1in **Primary Spacing** with 10 **Graduations**. This means the dotted lines represent .1 of an inch. Complete the following steps to create **Line.1**.

8.1 Select the **Line** tool from the **Profile** tool bar. The **Sketch Tools** tool bar will be modified to display the line variables; **Start Point**, **End Point**, **Length** and **Angle** as shown in Figure 1.19. You will be prompted to "**Select A Point Or Click To Locate The Start Point**." After selecting a **Start Point** you will be prompted to select an **End Point**. The mouse will be accompanied by a **Target**

Selector and the current location, same as when you created the point. Notice that the **Line** tool bar is similar to the **Point** tool bar.

Figure 1.19

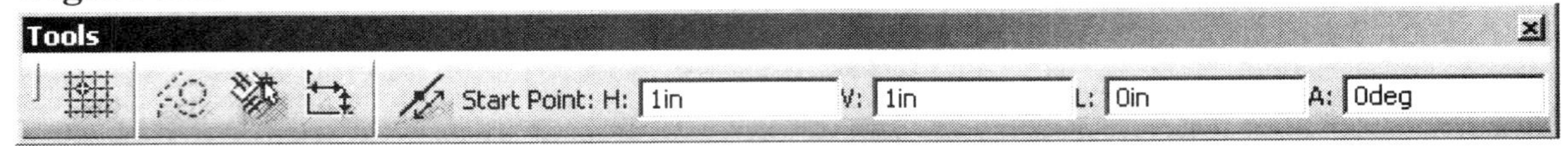

8.2 The starting point for **Line.1** will be **Point.1** that was created in Step 7. Using your mouse, select **Point.1**. You will now be prompted to "**Select A Point Or Click To Locate The End Point**." The **Sketch Tools** tool bar window will also update to prompt for the **End Point**.

8.3 The end point for **Line.1** is (1,2). If you can use the grid to locate the correct location, do so. Move your **Target Selector** up one full grid line, but don't move it to the right or left (0 in the horizontal direction). Click on the grid line intersection (1,2). If you have any doubt where (1,2) is, type in the values using the **Sketch Tools** tool bar. Type in **1** for the **H:** box and **2** for the **V:** box.

8.4 The first line is now created. **Line.1** should look similar to the one shown in Figure 1.21.

Figure 1.20

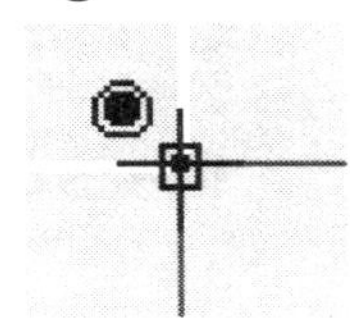

NOTE: Connecting one entity to another is safer and easier when the **Snap To Point** tool is on. When the **Snap To Point** tool is off you must be careful when connecting one entity to another. Both entities must share the same common point. For example, for two connected lines, the end point for the first line must be the same exact starting point for the second line. The lack of a shared point will make the entities unlinked. This broken link will cause problems when moving and/or modifying your profile. The entities will not move together. Another problem with a broken link is that it creates an unclosed profile. Unclosed profiles will be covered later in this lesson. CATIA V5 does supply a visual tool to help you know exactly when the point being selected is shared with another entity. The symbol is shown in Figure 1.20. The blue circle filled with a blue dot signifies the point being selected is the end point of another entity. This will link the two entities together. This is a helpful tool, especially when the **Snap To Grid** tool is off.

NOTE: The **Sketch Tools** tool bar gives you more options than the ones covered in Steps 8.1 thru 8.3. If you are typing in the information to create a line, you have the option of giving **Polar Coordinate** information, reference Figure 1.19. You enter a **Start Point**, **L:** (length of line) and **A:** (for angle). Though this lesson does not require you to use this option, it could be helpful in the future.

9 Creating Line 2

To create the second line, double click on the **Line** tool. Double clicking on the **Line** tool will allow you to create multiple lines without being required to repeatedly select the **Line** tool. With the **Line** tool double clicked, create **Line 2** [**Start Point** (1,1), **End Point** (2,1)]. (Double clicking on the **Line** tool still requires you to select a **Start Point** and an **End Point** every time, but you will not have to select the **Line** tool for every line.) This will create the bottom horizontal line as shown in Figure 1.21.

Figure 1.21

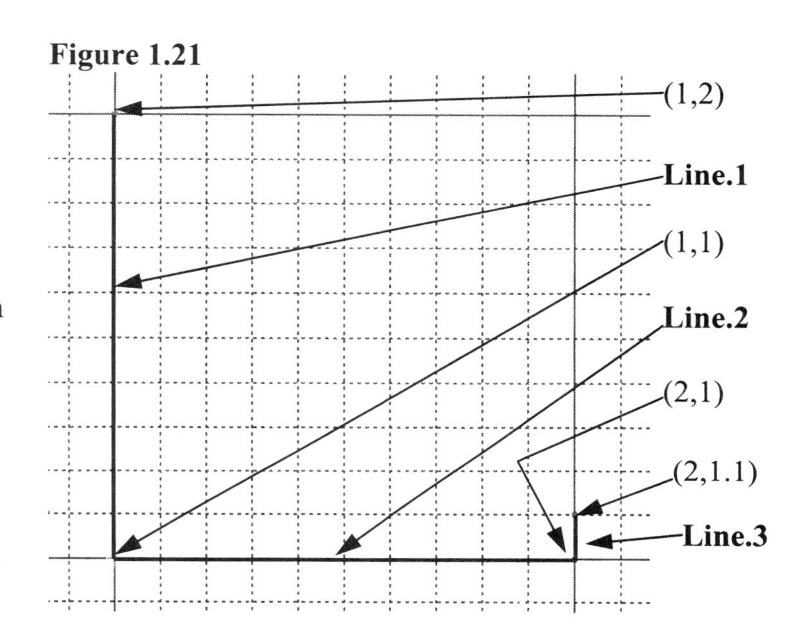

10 Creating Line 3

To create the third line, just select your **Start Point** and **End Point** since the **Line** tool is still active. Create **Line 3** [**Start Point** (2,1), **End Point** (2,1.1)]. Double click on the **End Point** of the last line you want to create, thus disengaging the **Line** tool.

NOTE: If you make a mistake when creating one of the lines you can use the **Undo** tool. The **Undo** tool is located at the bottom of the screen. The **Undo** tool allows you to undo multiple steps. Another option for removing a mistake is deleting it. This can be done using the **Cut** tool also located at the bottom of the screen. Highlight the entity to be deleted then select the **Cut** tool.

11 Creating Lines 4, 5, And 6 Using The Profile Tool 

The 4th, 5th and 6th lines will be created using the **Profile** tool. The **Profile** tool allows true successive line creation. The **End Point** for one line and the **Start Point** for the next line require only one selection. The connected lines will continue to be created with every point selected until you double click. Double clicking the **Ending Point** will end the **Profile** command. The lines created are separate entities, but the command that created them is recognized as one, so if you select the **Undo** command all of the lines created in one **Profile** operation will be undone.

With this tool added to your toolbox of knowledge, finish the "**L Shaped Extrusion**." Create **Lines 4**, **5** and **6** by selecting the following coordinates in succession: select (2,1.1), select (1.1, 1.1), select (1.1,2) and double click on (.6, 2) to end the line creation. The finished profile should similar to the one shown in Figure 1.22.

Figure 1.22

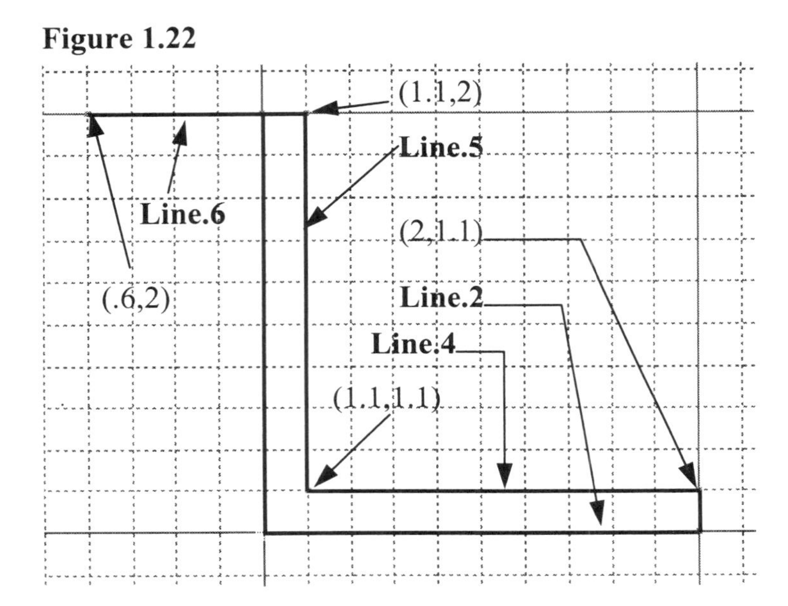

NOTE: This particular exercise does not require any features with radii, but the **Profile** tool has the ability to create them. Instead of selecting an **End Point** and a **Starting Point** for line creation, select the point (where the arc is to begin), hold down the left mouse button and drag it away from the starting point, then release the mouse button. You will notice as you drag the mouse button around, the arc radius and location change. Move the mouse around until you get the radius you want then select that point on the screen.

Steps 12 through 16 give instruction on how to use additional tools to modify the sketch entities you have created.

12 Breaking Line 6 

Step 11 purposely instructed you to create **Line 6** longer than required. In this step you will learn how to **Break** a line. Step 14 will instruct you on how to trim **Line 6** back to **Line 1**. To break **Line 6** simply select the **Break** tool from the **Operation** tool bar. Select **Line 6** as shown in Figure 1.23. The line will highlight then select a location on the line where you want the line broken. For the purpose of this lesson select approximately three **Graduation** lines from the left end point (Figure 1.23). The line is now broken. The easiest way to verify this is to select the broken line; only one of the two line segments will highlight. You could also select the **Measure** tool found at the bottom of the screen (Figure 1.24). Select the **Measure** tool then select (apply to) the line you want to measure. This will tell you how long the selected (broken) line is.

Figure 1.23 (Trimming **Line 6**)

Line.6

Break here

Figure 1.24 (**Measure** tool)

13 Deleting The Broken Line 

This is another easy step, but one that should be remembered. Select the left line fragment of the former line known as **Line 6**; it will highlight. Now select the **Cut** tool (scissors) located at the bottom left of the screen. The highlighted line will be deleted (Figure 1.25). You could also select the **Cut** command from the top pull down menu (under **Edit**) or hit the **Delete** key. This deleting (erase) process is similar in all windows functions and applies to any entity you want to delete (as long as it is selectable).

Figure 1.25

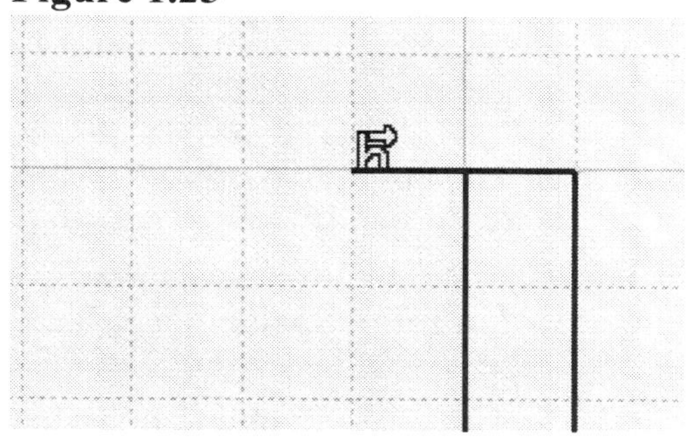

14 Completing The Profile Using The Trim Tool 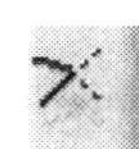

The profile of the "**L Shaped Extrusion**" is now complete, or is it? Extending **Line.6** past **Line.1** does not close the profile properly. If you were to exit the **Sketcher Work Bench** at this point and try to extrude the profile you would get an error, because **Line.6** is over running **Line.1**. To fix this problem select the **Trim** tool and select **Line.6** on the right side of **Line.1**. Now select **Line.1**. **Line.6** is automatically trimmed to the second line selected. Reference Figure 1.26 for line selection and Figure 1.27 for final result after **Trim**.

Figure 1.26 **Figure 1.27** (**Line.6** after trim)

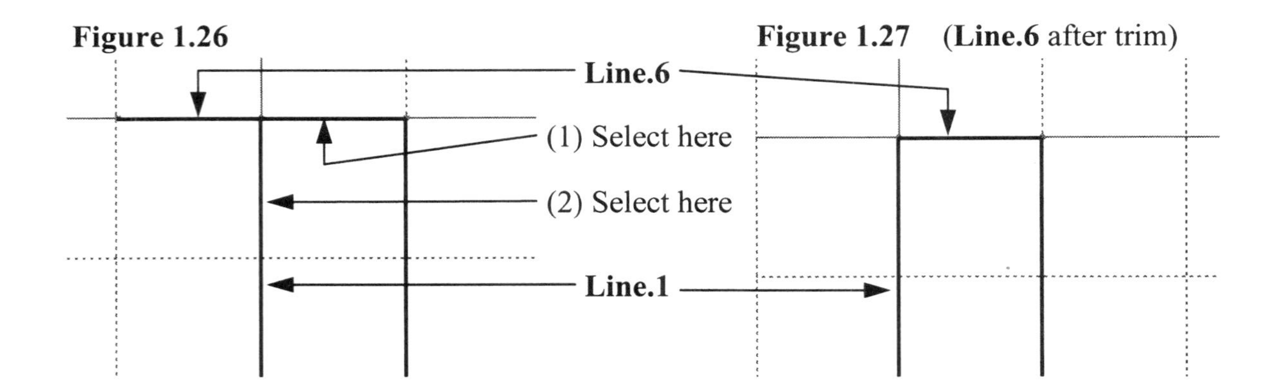

15 Modifying The Profile Using The Corner Tool 

The **Corner** tool is located in the **Operation** tool bar. This tool modifies existing entities; in this case, it will put a specified radius in the place of a square corner. In a later lesson you will learn to use **Dress Up** tools to create this type of geometry in the **Part Design Work Bench**. Lessons 2 will discuss the advantages and disadvantages to creating **Fillets** and **Chamfers** in the **Sketcher Work Bench**. The following instructions take you through the process of creating **Corners** (fillets).

15.1 Select the Corner tool.

15.2 The command prompt at the bottom left hand of the screen will prompt you with the following: "**Select the first curve, or a common point**."

15.3 For this exercise select **Line.4**. Reference Figure 1.28.

15.4 The next command prompt will ask you to "**Select the second curve**."

15.5 For this exercise select **Line.5**. Reference Figure 1.28.

15.6 Now move your mouse around; the radius of the corner you just created will grow and shrink according to the location of your mouse. The command prompt will prompt you to "**Click to locate the corner**;" in other words, move the mouse until the radius of the **Corner** is where you want it and click.

15.7 You now have a radius for that **Corner**. Your part should now look similar to the part shown in Figure 1.28. If your radius dimension does not match the one shown below it is ok, it will be modified later.

Figure 1.28 (Sketch with **Corner** added)

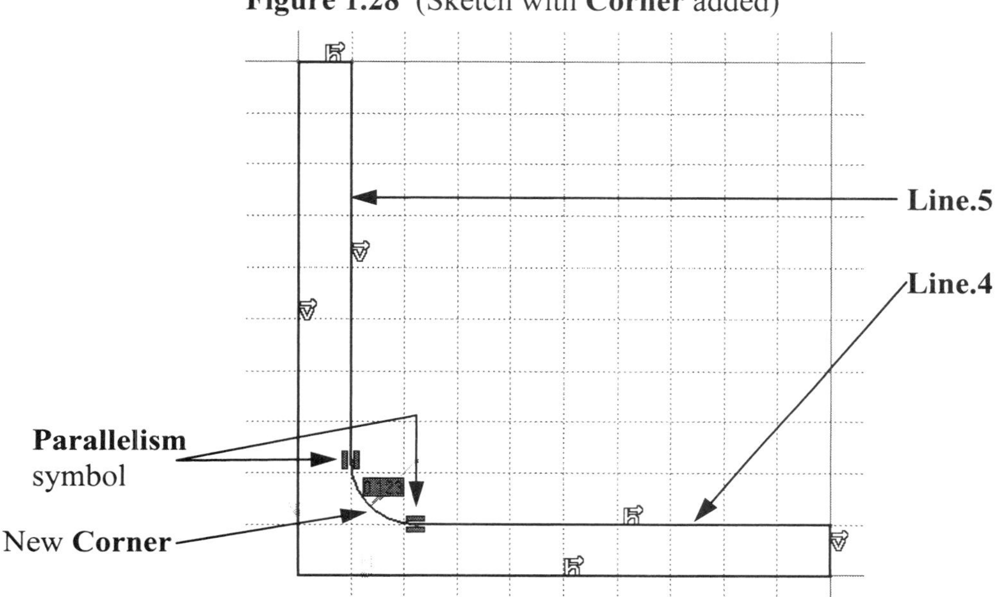

NOTE: The **Corner** will have a green dimension with a value attached to it. The value is the radius of the **Corner** you just created. Step 19 (modifying the constraints) will supply us with the tools to make this radius exact. Lesson 2 will explain a method of creating a **Corner** (radius) on a solid in the **Part Design Work Bench**.

16 Modifying The Profile Using The Chamfer Tool 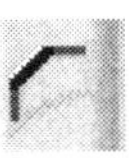

The **Chamfer** tool is also located in the **Operation** tool bar. This procedure assumes you know what a **Chamfer** is. The steps required to create a **Chamfer** are almost identical to creating a **Corner**.

Figure 1.29

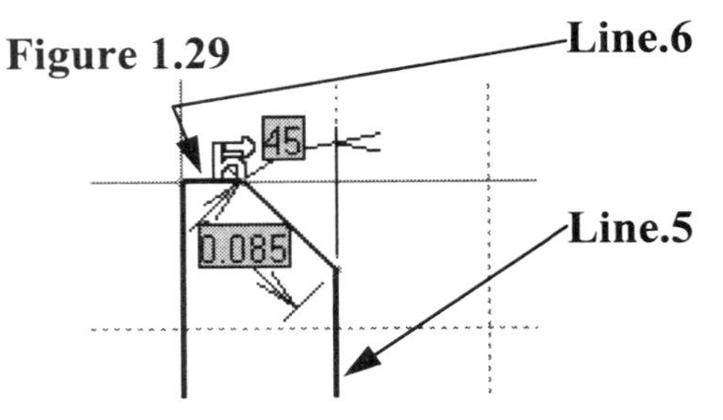

16.1 Select the **Chamfer** tool.

16.2 The command prompt at the bottom left hand of the screen will prompt you with the following: "**Select the first curve, or a common point**."

16.3 For this exercise select **Line.5**. Reference Figure 1.29.

16.4 The next command prompt will ask you to "**Select the second curve**."

16.5 For this exercise select **Line.6**. Reference Figure 1.29.

16.6 Now move your mouse around, the length of the **Chamfer** will grow as you move the mouse away from the intersection of the two selected lines. The length of the **Chamfer** will shrink as you move it back towards the intersection. If you move the mouse to the top left quadrant you will notice the **Chamfer** also moves to that quadrant. CATIA V5 gives you the option of all four quadrants. For this lesson use the bottom left quadrant. The command prompt will prompt you to "**Click to locate the chamfer**."

16.7 You should now have a **Chamfer** that looks like the one shown in Figure 1.29.

NOTE: The **Chamfer** has two green colored dimensions attached to it. Both dimensions have values attached to them. One dimension is the **Chamfer** length and the other is the **Chamfer** angle. Reference Step 19 (modifying the constraints) on how to modify the values to exactly what you require for your **Chamfer**. This **Chamfer** is a two-dimensional entity. Lesson 2 explains a method of creating a **Chamfer** on a solid in the **Part Design Work Bench**.

17 Anchoring The Profile Using The Anchor Tool

Select **Line.6**. As you select the line hold the mouse button down. Now drag the mouse up. Notice that the entire profile expands and contracts as you drag the mouse button around. **Lines 1** and **2** can be modified in length, but the location can't change. All of the other lines can be modified in position, length and angle. You cannot modify the location of **Lines 1** and **2** because they are linked to **Point.1** and **Point.1** is constrained to the location (1,1). The green dimension lines that were created with **Point.1** are constraints. It is the constraint values that tie **Point.1**, **Line.1** and **Line.2** to their current positions. To move the point and/or either line you have to modify the constraint. This will be covered in Step 19.

If there is a particular entity you don't want to move in relationship to another entity, you can constrain it. Constraints are restrictions on one entity to another entity. The **Anchor** tool restricts the entities movement in relationship to the coordinate location only. **Lines 1** and **2** are not truly anchored because the constraint has tied their relationship to **Point.1.** The effect is the same; **Lines 1** and **2** cannot be moved. If you want to constrain the location of an entity without constraining any other entity, the **Anchor** tool is a good option. For example, you may want to modify the "**L Shaped Extrusion**," but you know you don't want **Line.6** to move at all. You can restrict **Line.6** by **Anchoring** it. Elements can be **Anchored** by completing the following steps.

17.1 Select the entity that you want to **Anchor**. For this lesson select **Line 6**.

17.2 Select the **Constraints Defined In Dialog Box** tool. This will bring up the **Constraint Definition** window. Reference Figure 1.30.

17.3 The **Constraint Definition** window gives you a lot of options as far as selecting a constraint. For this lesson select the **Fix** constraint.

17.4 Select the **OK** button to apply the **Fix** constraint. Notice that **Line.6** will turn green, meaning that it is constrained, and the **Anchor** tool also shows up on the line. This signifies what kind of constraint is applied as shown in Figure 1.31.

Figure 1.30

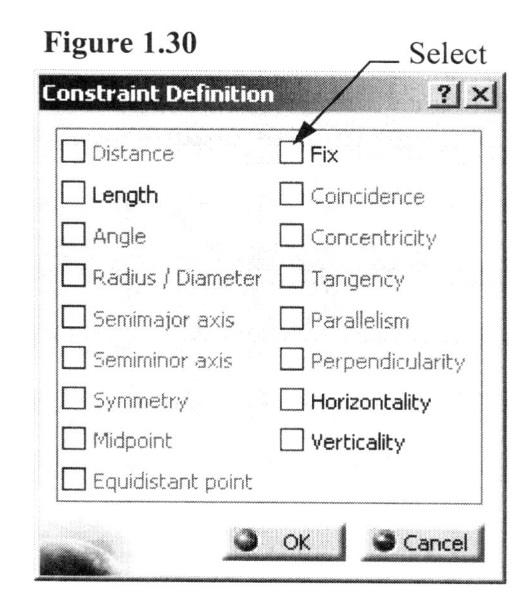

Allowing quick modification to a sketch can be a powerful tool, especially in the beginning stages of a design. As the design nears completion, the ideas are being locked down and there are fewer variables. This is where CATIA V5 constraints come to the aid of the designer. As variables become known constants you can constrain them.

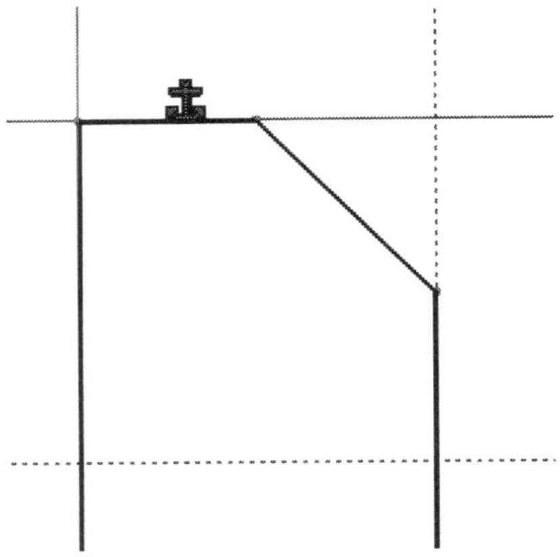

The purpose of this step was to give you a brief introduction to how CATIA V5 allows you to move and modify the sketched entities. It also introduced you to how to constrain the entities. The only way to fully understand all of the tools available to you is to test them yourself. Step 18 covers constraints in more detail.

Figure 1.31

18 Constraining The Profile

There are several reasons why you would want to constrain your profile. One reason is that you or any one else could accidentally select a line and move it out of position as you experienced in Step 17. **Constraints** help to keep the required relationships between the **Sketcher** entities that make up the profile. There are multiple ways of constraining a part in CATIA V5. The nice thing about CATIA V5 is that constraining is optional, not required. Hopefully, this step will convince you that **Constraints** can be a powerful tool.

18.1 Constraint

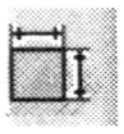

This tool allows you to create individual constraints one at a time. You have already applied a **Constraint** and may not even know it. The **Anchor** tool in Step 17 is a **Constraint**, referred to as a **Geometrical Constraint**. The values attached to the **Chamfer** and **Corner** are **Constraints**, referred to as **Dimensional Constraints**. To apply **Dimensional Constraints** complete the following steps.

18.1.1 Select the **Constraint** tool.

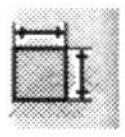

18.1.2 Select the line and/or **Sketcher** element to be constrained.

18.1.3 The **Sketcher** element will turn red indicating it has been selected. Move the cursor away from the element selected and notice the newly created constraint, along with the dimension value box, moves with it. Position the constraint in the desired position and click to drop.

18.1.4 If the location of the **Constraint** is not satisfactory reselect the **Constraint** by double clicking on it and drag and drop it at the desired location.

18.1.5 To edit the value of the **Constraint** double click on the value box. This will bring up the **Constraint Definition** window shown in Figure 1.32. This window shows the existing value for the **Sketcher** element. This value can be edited by typing the new value over the existing value. Then select **OK** or hit the **Enter** key. The entity linked to the **Constraint** will automatically be updated to the new value.

Figure 1.32

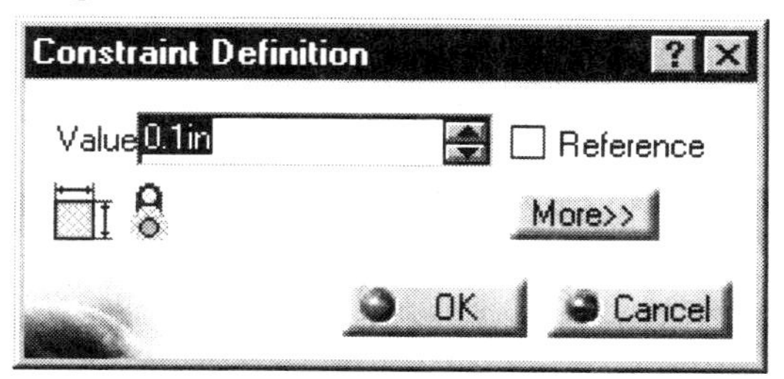

If the **Constraint** is between two different entities, such as lines, select the first line and then the second line. CATIA V5 will constrain the distance between the two entities. The **Constraint** value will appear near the constraint. To move the **Constraint** value, follow Steps 18.1.4. For this lesson constrain your "**L Shaped Extrusion**" similar to the one shown in Figure 1.33.

Figure 1.33

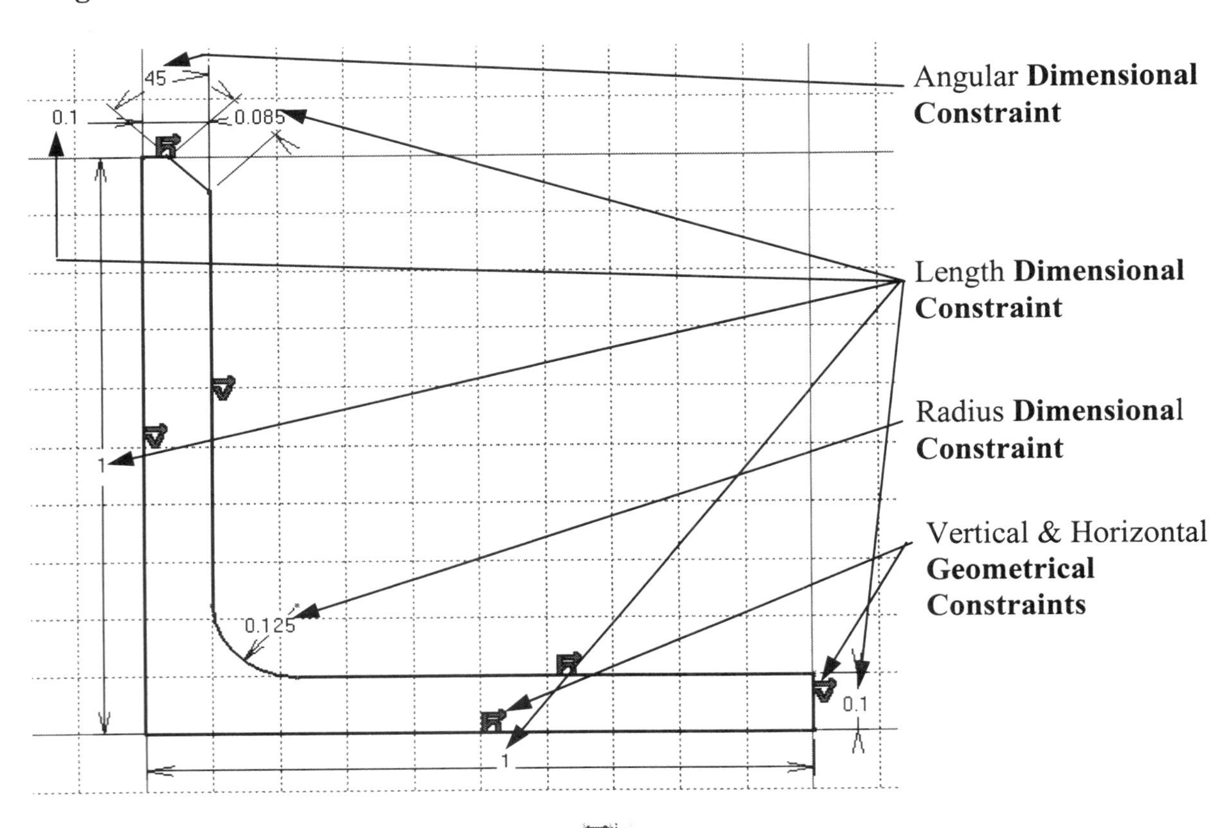

18.2 Auto Constraining The Profile

This method accomplishes the same task as the **Constraint** tool just explained except that **Auto Constrain** can be much quicker (automatic). Once you select the **Auto Constraint** tool CATIA V5 will bring up the **Auto Constraint** window prompting you to select which entities you want to constrain. Figure 1.34 shows what the **Auto Constraint** window looks like. You can select one entity at a time, multi-select or select only a few specific entities that you want constrained. After making your selection select **OK** located at the bottom of the **Auto Constraint** window. The entities selected will show a green **Constraint** attached. Getting complete control of this tool will take some practice and patience. If you feel brave, use this tool to constrain your "**L Shaped Extrusion**" and see if you get the same results shown in Figure 1.33.

Figure 1.34

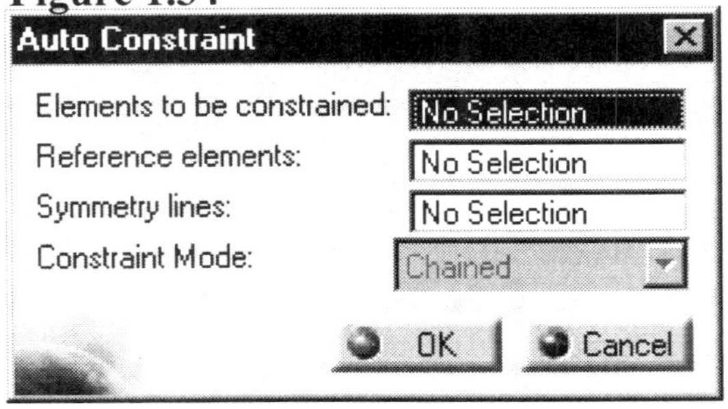

Auto Constraint

Elements to be constrained: No Selection
Reference elements: No Selection
Symmetry lines: No Selection
Constraint Mode: Chained

OK Cancel

18.3 Constraint Defined In Dialog Box

To use this tool you have to select one or more entities and then select the **Constraint Definition In A Dialog Box** tool. This will bring up the **Constraint Definition** window as shown in Figure 1.35. The window will contain all the possible **Constraints**, but not all will be selectable. The only selectable **Constraints** are the ones that apply to the entities selected. For example, if you selected a line you could apply the **Length**, **Fix** and **Horizontality Constraints**; all of the other **Constraints** will be dimmed (meaning they are not selectable). CATIA V5 will not allow you to select the **Radius/Diameter** constraint because it does not apply to lines. Relationships between entities can also be established using this tool. For example, if you wanted **Parallelism** and **Horizontality Constraints** between the top profile line and the bottom profile line on the base leg of the "**L Shaped Extrusion**," you would do the following.

Figure 1.35

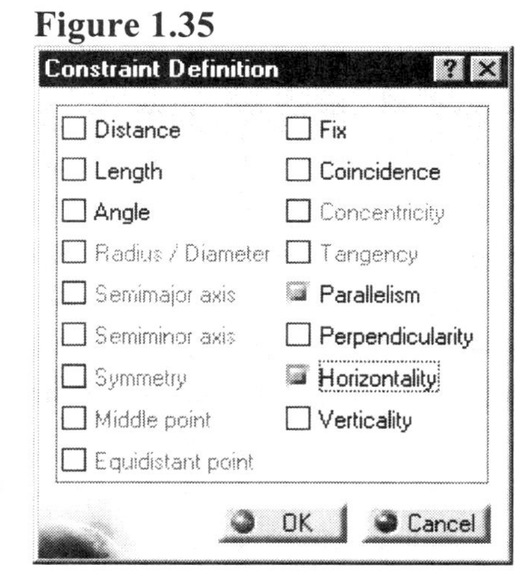

18.3.1 Select both the bottom and top lines of the base leg of the "**L Shaped Extrusion**" (**Lines 2** and **4** shown in Figure 1.36). This is a windows multi-select task, which is accomplished by holding down the **CTRL** key while selecting both lines. Both lines will highlight.

18.3.2 Select the **Constraints Defined In Dialog Box** tool.

18.3.3 The **Constraint Definition** window will pop up as shown in Figure 1.35.

18.3.4 Select the **Parallelism** box and the **Horizontality** box.

18.3.5 Select **OK**.

NOTE: The **Constraints** that appear on the sketch are: the **Parallelism** and **Horizontality** symbols, reference Figure 1.36.

Figure 1.36

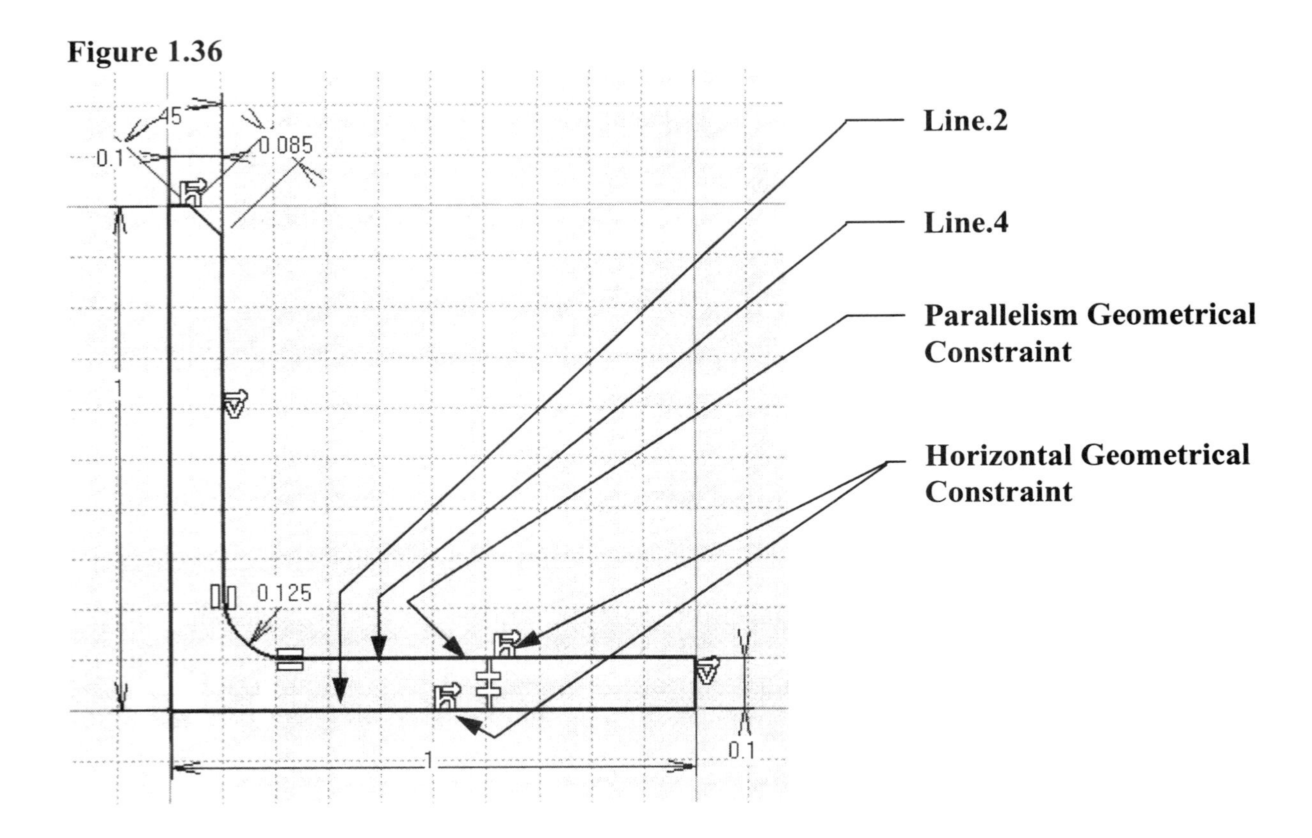

The only way to really get complete control of this tool is to use it, experience it and don't be afraid to make a few mistakes (that's why there is an **Undo** button).

18.4 Animate Constraint 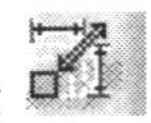

The **Animate Constraint** tool allows you to visualize the effect one **Constraint** has on the entire profile. This is a very helpful tool, but be aware, you may not always end up with what you started with. Remember, entities will not always stay attached as other entity values change. CATIA V5 will remember the relationships the different entities have with each other, if they were created with a relationship. For example, just because the end point of one line is the same as the start point of another line it does not mean there is any relationship between the two lines. To use this tool follow the steps listed below.

18.4.1 Select one existing **Constraint.** Only one **Constraint** can be animated at a time.

18.4.2 Select the **Animate Constraint** tool.

18.4.3 The **Animate Constraint** window will pop up as shown in Figure 1.37.

18.4.4 Modify the parameters as desired/required and/or accept the default values.

18.4.5 Select the **Play** button. This will start the animation from the starting limit to the ending limit.

18.4.6 Watch the profile change as the selected entity animates from the first value to the last value. The **Animate Constraint** window has other options that you can test.

NOTE: If your profile has entities created without relationship to other entities, the **Rewind** button could result in a different profile than what you started with. Be careful.

Figure 1.37

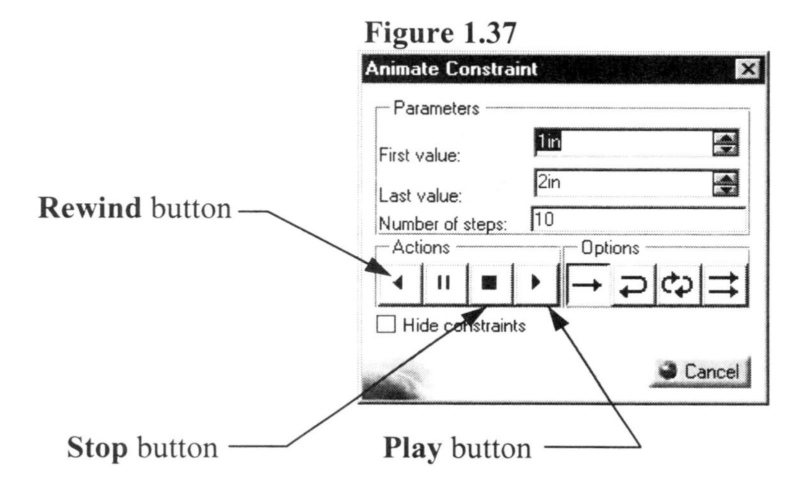

19 Modifying The Constraints

This process was previously described briefly in Step 18. The ability to modify **Constraints** in CATIA V5 is essential, so the following steps are for your review.

19.1 Double click on the value box of the **Dimensional Constraint** you want to modify. The value box is the green dimension line with an attached value.

19.2 The **Constraint Definition** window will pop up as shown in Figure 1.38. This window shows the existing value for the **Sketcher** element.

19.3 Edit the value by typing over the existing value.

19.4 Apply the new value by selecting the **OK** button or pushing the **Enter** key.

19.5 The entity linked to the **Constraint** will automatically be updated to the new value. Your profile also updates automatically.

Figure 1.38

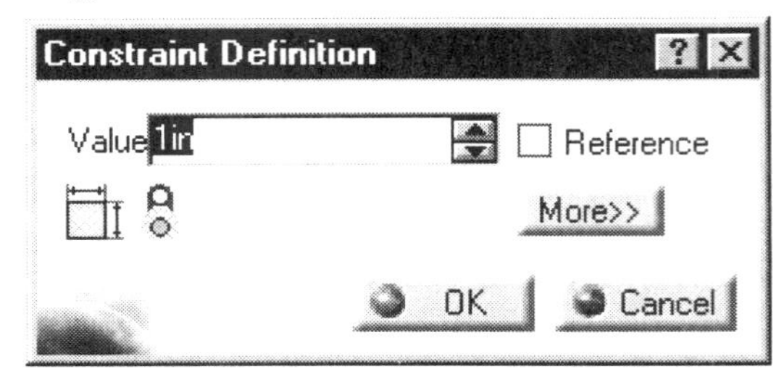

If you want to know more information about a particular **Constraint**, double click on it and the **Constraint Definition** window will pop up. Select the **More** button to get detailed **Constraint** information. Figure 1.39 shows how the **Constraint Definition** window looks when the **More** button is selected. To get back to the default **Constraint Definition** window select the **Less** button.

Figure 1.39 (**Constraint Definition** window with the **More** button selected)

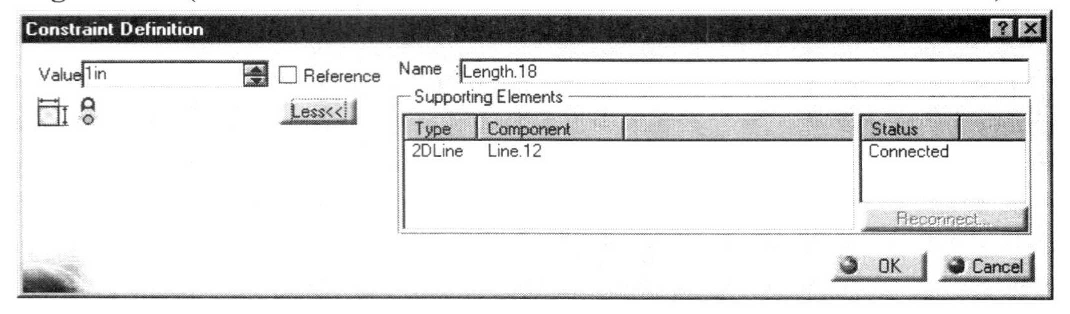

See what you can learn about one of the **Constraints** on the **"L Shaped Extrusion."** Double click on the **Constraint** on the bottom line of the base leg. From the **Constraint Definition** window, select the **More** button. The pop-up window gives you information on other entities the selected **Constraint** is connected (linked) to. It also gives you the opportunity to change the name of the **Constraint** that shows up on the **Specification Tree**.

20 Over Constraining The Profile… Not A Good Thing!

It is possible to over constrain a profile in the **Sketcher Work Bench**. When you over constrain the profile CATIA V5 will inform you that you have a problem. The CATIA V5 definition of over constraining is putting two different **Constraints** on one or more entities. The two **Constraints** can be correct individually, but collectively have conflicting values. When an over constrained condition exists CATIA V5 will turn all of the affected constraining values purple. **Purple is the default color for over constrained sketches!** Remember, an **Over Constraint** condition is not a good thing. CATIA V5 will not allow you to extrude an over constrained profile. The easiest way to get out of the over constrained condition is to **Undo** or **Cut** the last **Constraint** created, the **Constraint** that caused the over constrained condition. You must reconsider which **Constraints** are necessary to accomplish what you want. In the case of the "**L Shaped Extrusion**," you are creating the **Constraints** that are used to maintain the specified dimensions. If your profile is not over constrained, you are ready to move on to the next step. If the instructions were followed an over constrained condition should not exist.

21 Exiting The Sketcher Work Bench

If your "**L Shaped Extrusion**" is similar to the one shown in Figure 1.33, you are ready to move the profile into the 3D world, the **Part Design Work Bench**. As a reminder, the following conditions will not allow you to successfully extrude your profile once out of the **Sketcher Work Bench**.

21.1 An unclosed profile as shown in Figure 1.40a. Notice the profile has a gap in it.

21.2 A profile with floating entities as shown in Figure 1.40b. Notice there is a line not attached to any other entity, it is floating.

21.3 Multiple profiles in one sketch as shown in Figure 1.40c. Notice both profiles are closed profiles, but there are two of them. CATIA V5 allows only one profile per sketch. Lesson 5 covers designs requiring multiple sketches.

21.4 An over constrained profile as shown in Figure 1.40d. Notice this example shows that one line is being dimensioned two different ways.

Figure 1.40 Profiles that cannot be extruded

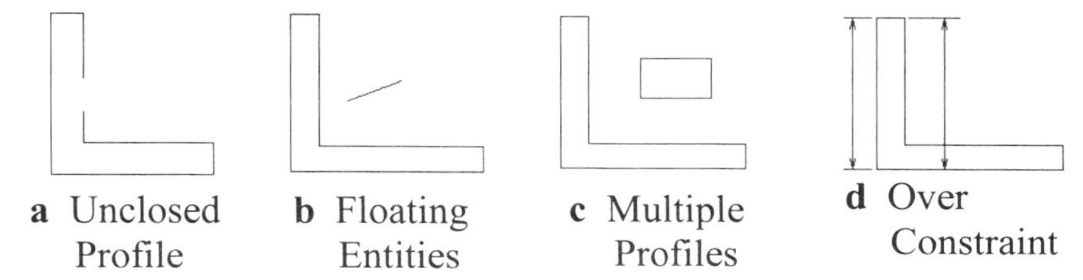

You can exit the **Sketcher Work Bench** with your profile in any of the above conditions, but CATIA V5 will not extrude the profile into a three-dimensional (solid) part.

If you are ready to exit the **Sketcher Work Bench**, select the **Exit** tool . The **Exit** tool is located at the top right of the **Sketcher Work Bench**.

NOTE: The profile rotates back to the original three-dimensional view with your newly created profile of the "**L Shaped Extrusion**." The **Sketcher Work Bench** grid disappears. The tools on the right hand tool bar will change as shown in Figure 1.41. The only tools available for your use at this time are **Pad**, **Shaft**, **Rib** and **Loft**. The **Pad** tool is covered in Step 22 and Lesson 2. The next step will tell you how to use the **PAD** tool.

If your screen looks similar to Figure 1.41, you are now in the **Part Design Work Bench** and ready to go to Step 22.

Figure 1.41

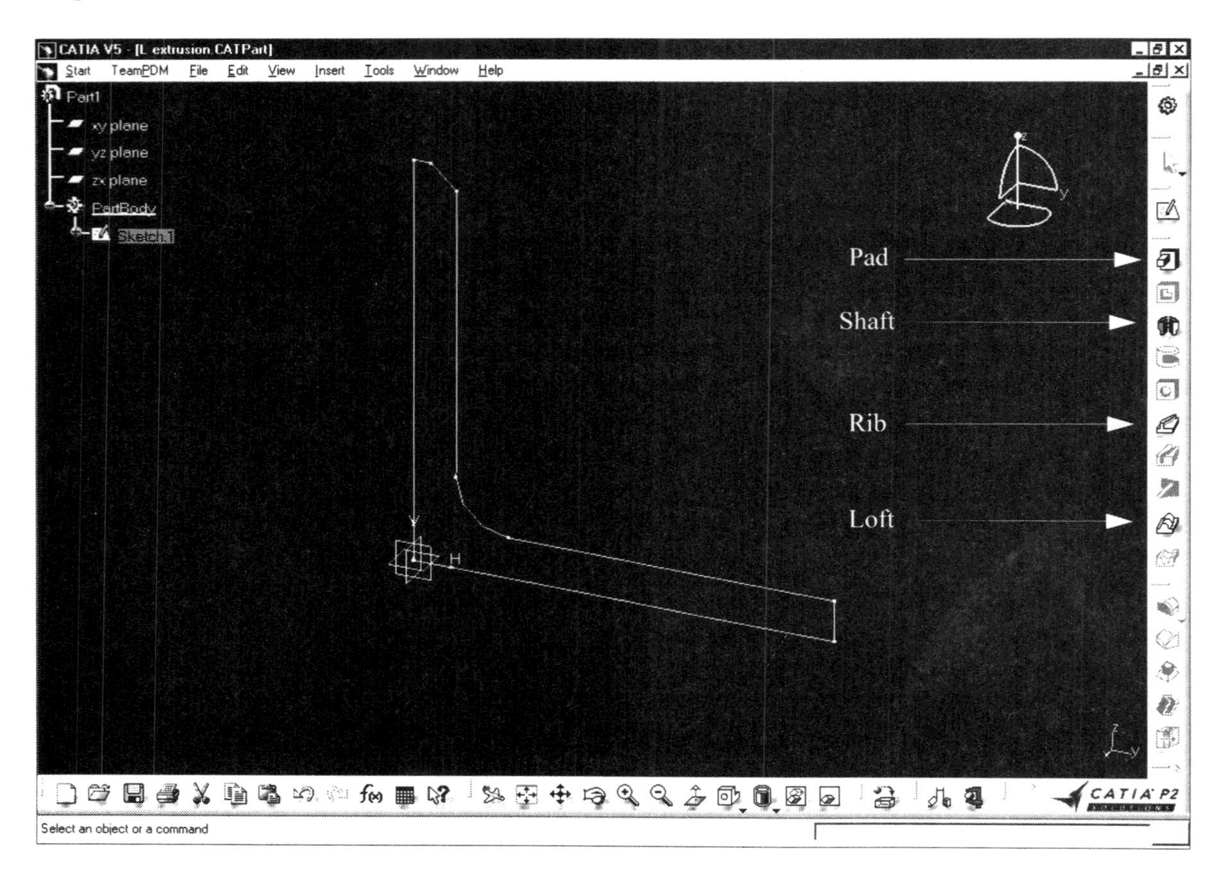

22 Extruding The Newly Created Profile Using The Pad Tool 

This step will put your newly created profile of the "**L Shaped Extrusion**" to the test. This is where you find out if there are any problems with your profile sketch created in the **Sketcher Work Bench**.

If you haven't selected anything in the work area since exiting the **Sketcher Work Bench**, your profile should still be highlighted. If it is not still highlighted, select the profile or select the **Sketch** branch from the **Specification Tree**. When the profile is highlighted you can select the **Pad** tool. This will bring the **Pad Definition** window up as shown in Figure 1.42 and 1.43. As the **Pad Definition** window pops up you should notice your profile becomes three-dimensional. The **Specification Tree** just added another branch, the **Pad.1** branch. At this point you can specify how long to extrude the profile. You can type it in or select the up arrow and watch the part grow. Select the down arrow and watch it shrink. You can reverse the direction and/or mirror the extruded length. If these are not enough options you can select the **More** button in the **Pad Definition** window (Figure 1.43). The **More** button will let you specify the start location, **First Limit,** and the ending plane, **Second Limit,** of the profile being extruded. The **Reverse Direction** button will allow you to select an extruded direction other than the default direction, which would be normal to the sketch plane.

Figure 1.42 Note: View has been rotated. Part remains on ZX plane.

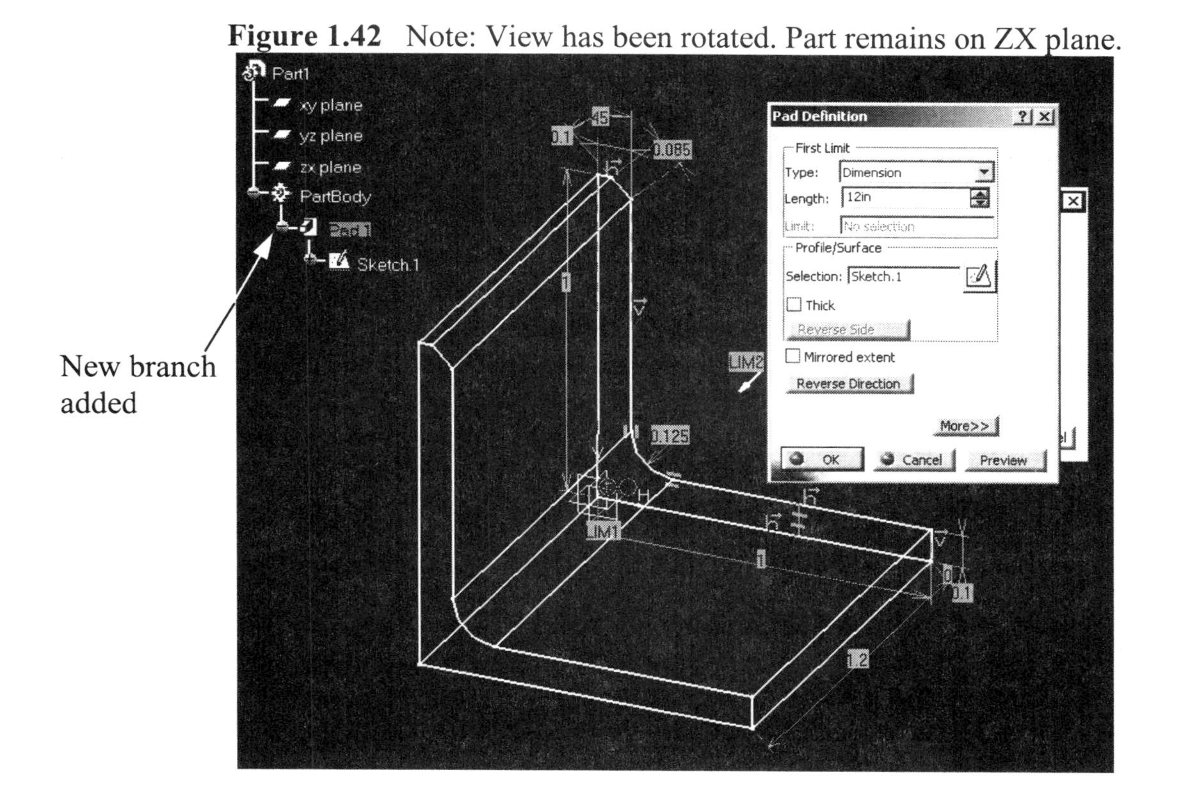

Figure 1.43 (**Pad Definition** window with **More** selected)

Pad Definition
First Limit
Type: Dimension
Length: 12in
Limit: No selection
Profile
Selection: No selection
Reverse Side
Mirrored extent
Reverse Direction
Second Limit
Type: Dimension
Length: 0in
Limit: No selection
Direction
Normal to profile
Reference: No selection
<<Less
OK Cancel Preview

Type options available

Dimension
Dimension
Up to next
Up to last
Up to plane
Up to surface

These options are the same in the first and second limit boxes

Once you have the **Pad Definition** window set up the way you want it select the **Apply** button, this will give you a preview of what you just created. If you are not satisfied with the result select the **Cancel** button. If you are satisfied select the **Ok** button. The **Ok** button will create a three-dimensional part (solid) from your sketch. For the "**L Shaped Extrusion**" extrude the profile 12 inches. Your "**L Shaped Extrusion**" should look similar to what is shown in Figure 1.44.

23 Saving The Newly Created "L Shaped Extrusion"

You can stop what you are doing at any time and save the file you are working on. CATIA V5 also allows the user to set the time period for the automatic save. Before saving and exiting make sure you have finished all operations you have started. If you save and/or exit in the middle of an operation, the operation will not be saved. CATIA V5 allows you to name the file as you wish. The file extension will be named "***.CATPart**." All of the files created in the **Sketcher Work Bench** and finished in the **Part Design Work Benches** will have a "***.CATAPart**" extension. To **Save** a CATIA V5 file complete the following steps.

Figure 1.44

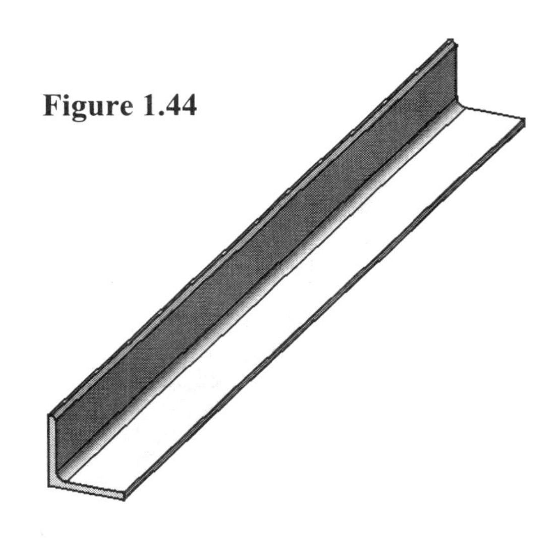

23.1 Verify that all operations are complete and the part (**CATPart**) is the way you want it to be saved.

23.2 Select **File** from the top tool bar (Figure 1.45).

23.3 Select **Save As** (Figure 1.45).

23.4 In the **Save As** window, select the directory you want the **CATPart** to be **Saved in** as shown in Figure 1.46.

23.5 In the same window, type in the **File name**. For this lesson save the file as "**L Shaped Extrusion**."

23.6 Notice CATIA V5 will automatically give the file the extension "***.CATPart**."

23.7 If everything is the way you want it in the **File**, **Save As** window select the **Save** button.

Figure 1.45

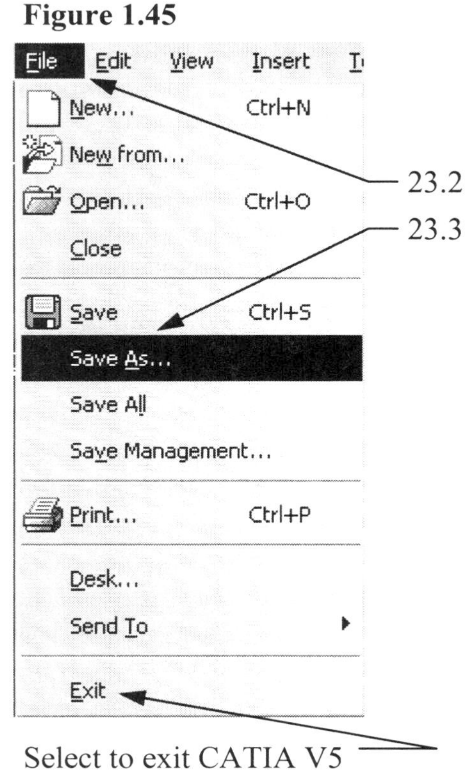

Select to exit CATIA V5 (reference Step 24)

Figure 1.46

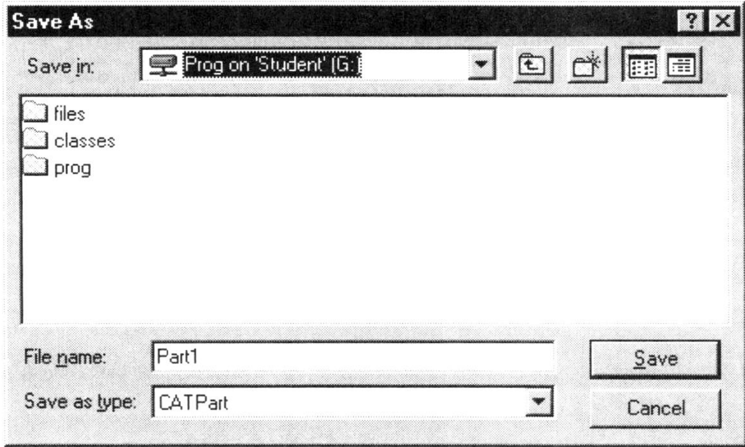

NOTE: Remember the file name and the directory you saved it to, you will need it for Lesson 2.

24 Exiting CATIA V5

To exit CATIA V5, complete the following steps.

24.1 Make sure you saved the **CATPart** (if you wanted it saved). If you have made any changes to the **CATPart** and not saved CATIA V5 will prompt you to save when exiting.

24.2 Select **File** from the top pull down tool bar as shown in Figure 1.45.

24.3 Select **Exit**.

24.4 If the **CATPart** was previously saved CATIA V5 will shut down and your computer will go back to the NT Desktop. As described above, if some changes were made to the **CATPart** without being saved CATIA V5 will prompt you to "**Save**" before allowing you to exit to the NT Desktop.

Lesson 1 Summary

If you are not use to the concepts of a sketch tool and parameterized entities this can be a difficult lesson, a lot of people struggle with it. It is not difficult it is just different! It is critical that you get comfortable with these two concepts before moving on to the next lesson. After you spend enough time using this process you will learn to appreciate the power these concepts possess.

Lesson 1 Review

After completing this lesson you should be able to answer the questions and explain the concepts listed below.

1. What is the definition of a **Constraint**?
2. Does CATIA V5 require **Constraints** to create a profile in the **Sketcher Work Bench**?
3. What is meant by an unclosed profile?
4. **T** or **F** Unclosed profiles can be extruded using the **Pad** tool.
5. **T** or **F** There is more than one way to select the XY plane when defining the XY plane as a sketch plane.
6. **T** or **F** **Over Constraining** a profile is a good thing!
7. Explain your answer to question 6.
8. What does **Anchoring** the profile do in the **Sketcher Work Bench**?
9. Explain how you would change the **Sketcher** units of measurements from mm to inches.
10. The **Sketcher Grid** is made up of two different entities, one is the **Primary Spacing**, name the other.
11. What is the advantage of **Constraining** a profile in the **Sketcher Work Bench**?
12. How do you modify a **Constraint**?
13. What icon do you use to exit the **Sketcher Work Bench** and enter the **Part Design Work Bench**?
14. How can you view all of the default tool bars in the **Sketcher Work Bench**?
15. What tool in the **Part Design Work Bench** is used to extrude a profile created in the **Sketcher Work Bench**?
16. The actual process of extruding a profile adds what branch to the **Specification Tree**?
17. List all four of the **Constraint** tools.

18. Can one **Sketch** have more than one profile?

19. While in the **Sketcher Work Bench** and using the mouse, how would you move (pan) the profile around the screen?

20. When you are connecting one end point of a line to another, how does CATIA V5 let you know you are **Snapping** to the existing end point and not just getting close?

Lesson 1 Practice Exercises

Now that your CATIA V5 tool box has some tools in it, put them to use on the following practice exercises. The shapes are simple and can be completed in one sketch. The dimensions represent the constraints you are to use in the **Sketcher Work Bench**. The first practice exercise has suggested steps to complete the task along with some helpful hints. Each subsequent practice exercise contains less suggested steps and helpful hints. By the last practice exercise you will be on your own!

Each practice exercise has a name to use when saving the exercise. It is critical that you use the suggested name so you can find the correct CATPart if it is used in a later lesson. Good Luck!

1 Using the **Sketcher Work Bench** and the other tools covered in Lesson 1, create the following profile and extrude to the dimensions shown below. When completed save as "**Lesson 1 Exercise 1.CATPart**."

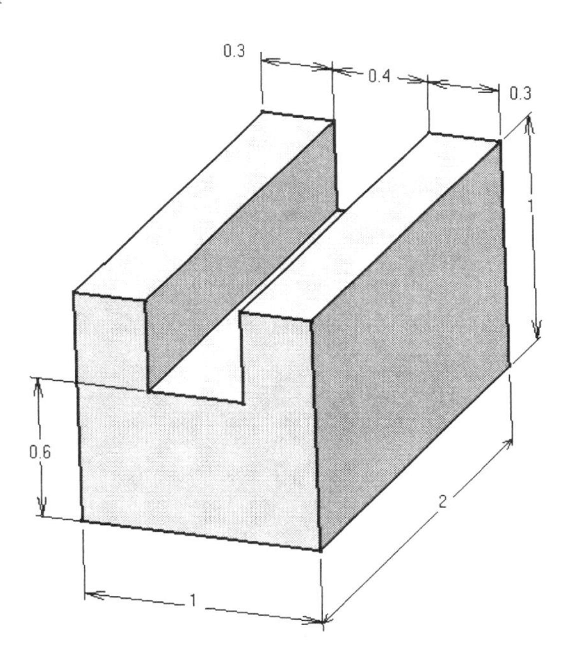

Suggested Steps:

1. Select the XY plane (the plane the profile will be sketched on). Reference Step 3 for information on selecting planes.
2. Enter the **Sketcher Work Bench**. Reference Step 4.
3. Sketch the profile of the part. Hint: use the **Profile** tool.
4. Anchor the lower left hand corner of the sketch. Reference Step 17 for anchoring a profile.
5. Constrain the profile to match the dimensions shown above. Reference Step 18 for constraining a profile.

6. Exit the **Sketcher Work Bench**, return to the **Part Design Work Bench** (the 3D environment). Reference Step 21 for exiting the **Sketcher Work Bench** and entering the **Part Design Work Bench**.
7. Once in the **Part Design Work Bench**, extrude the profile to the dimension shown (2″). Reference Step 22 for extruding a profile.
8. Save the part as "**Lesson 1 Exercise 1.CATPart**." Reference Step 23 for saving a file.

2 This part (profile) should be straightforward. This would be a good exercise to try different methods of constraining and testing the results. Save the shape as "**Lesson 1 Exercise 2.CATPart**."

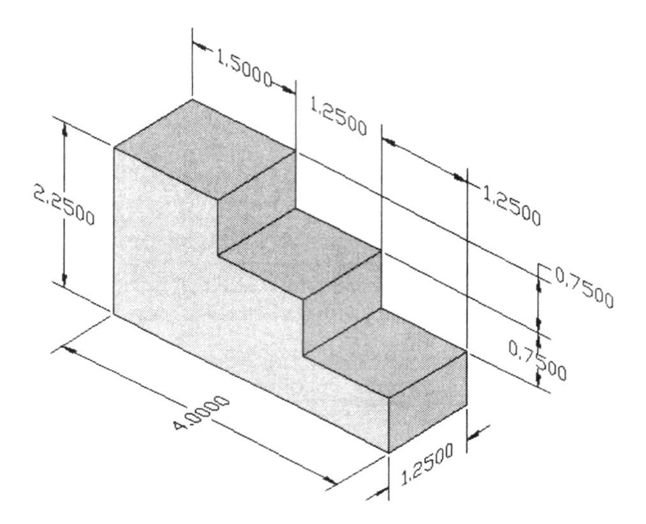

HINT: To help make it easier to sketch this part, set the grid **Primary Spacing** to 1 and the **Graduations** to 4. This will put the grid lines in the **Sketcher** screen to a .25 inch spacing. With that spacing, all you have to do is snap to the intersections of the grid to sketch the part.

3 This practice exercise should challenge you. For this part, use radius values, not angles. Save this CATPart as "**Lesson 1 Exercise 3.CATPart**."

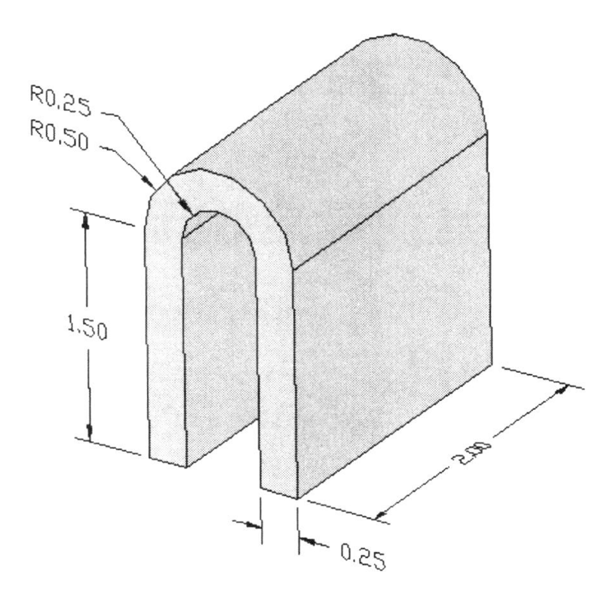

HINT: This part can be done using the radius option in the **Profile** command. Before starting, set the grid **Primary Spacing** to 1 and the **Graduations** to 4.

Sketching with the Profile icon (radius option)

1. Start at the bottom left corner of the part.
2. Select the **Profile** icon from the right menu bar.
3. Sketch the vertical 1.50 inch line that defines the left edge of the part.
4. Now sketch the first arc along the top of the part. To do this, hold down the left mouse button and drag it in the direction you want the arc to go then release the mouse button. The arc will appear and allow you to drag and place it where you want. Place it on the grid intersection 2 inches above the bottom of the part and a half-inch to the right. This will only create half of the arc needed, so the process will have to be repeated to sketch the other half of the arc.
5. Finish sketching the rest of the part. When you reach the inside .25 radius repeat Step 4.
6. When the sketch is done constrain it to double check that all of the dimensions match the part shown above. Make changes if necessary.

4 This will give you more practice with the **Line** and **Profile** icons in the **Sketcher Work Bench**. When you are done save the CATPart as "**Lesson 1 Exercise 4.CATPart**."

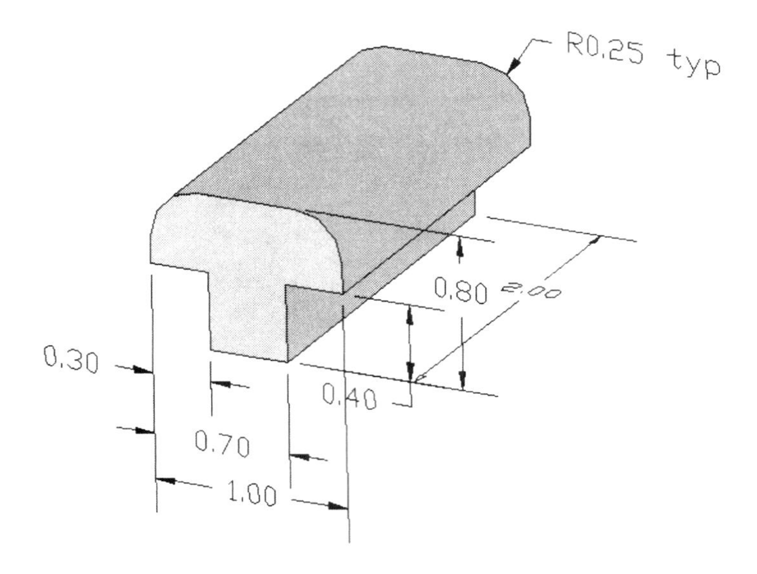

5 This exercise will give you some practice using the **Corner** and **Line** icons. When you are done save the CATPart as "**Lesson 1 Exercise 5.CATPart**."

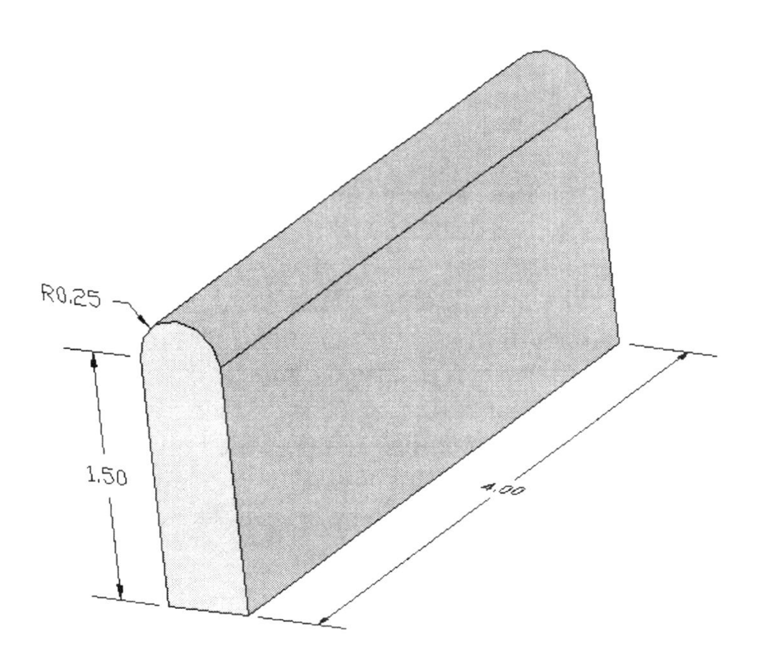

6 This will give you more practice using the **Line** and **Corner** icons. Save this CATPart as "**Lesson 1 Exercise 6.CATPart**."

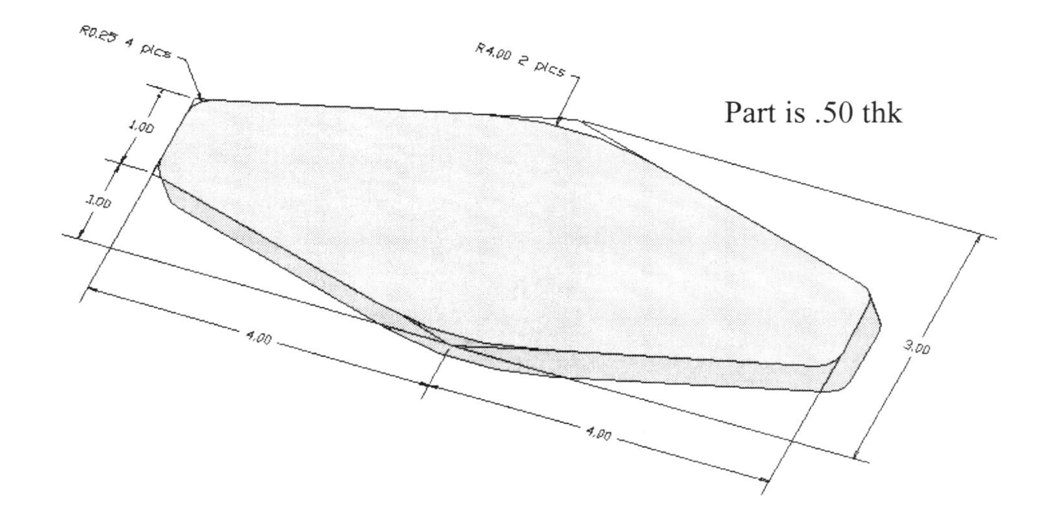

HINT: Use the **Line** or **Profile** icon first to sketch the profile using sharp corners (no radius). Once it is constrained to the dimensions above go back and add in the radii using the **Corner** icon.

7 This practice exercise is a little bit more challenging; see what you can do with it. Save this CATPart as "**Lesson 1 Exercise 7.CATPart**."

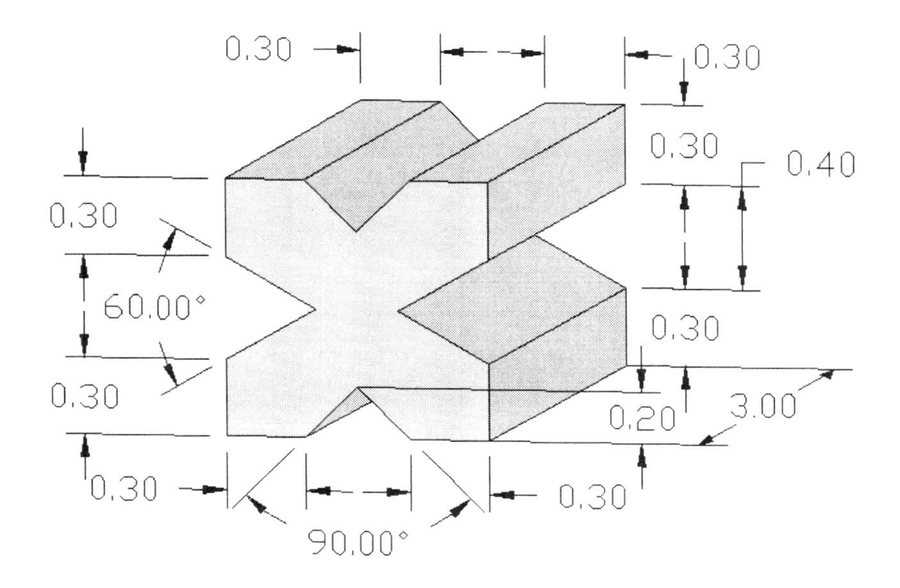

HINT: It is not as complicated as it looks. If your grid **Graduations** are set to 10, just snap to the intersections for the beginning and ending points of your lines. To set the constraint for the angles select the angled lines and the angle constraint will appear. Reference Step 19 for modifying the angle value. If the profile gets over constrained, delete the **Parallel** constraint.

Introduction To The Part Design Work Bench

This lesson uses the "**L Shaped Extrusion.CATPart**" created in Lesson 1. The first part of this lesson introduces you to all of the different default tool bars found in the **Part Design Work Bench** and the specific tools found in them. The object of this lesson is to complete the "**L Shaped Extrusion**" started in Lesson 1. When you complete this lesson the "**L Shaped Extrusion**" will be a "**T Shaped Extrusion**" and look similar to what is shown in Figure 2.1. The steps to complete the part are described in detail. Not all of the tools will be used to complete the part. This lesson is basic; therefore, there will be some overlap from Lesson 1.

Figure 2.1 The completed "**T Shaped Extrusion.CATPart**"

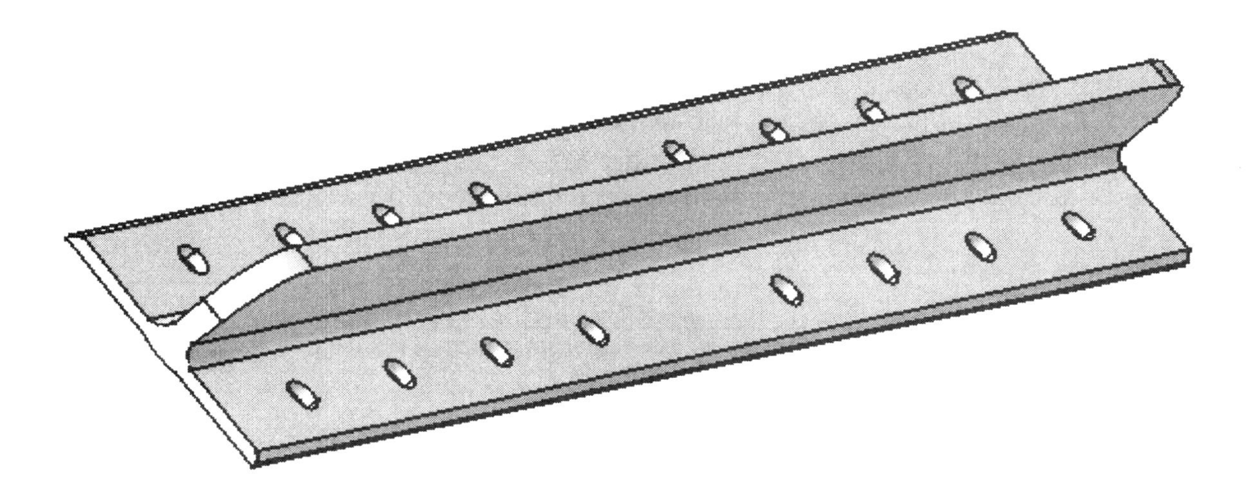

Lesson 2 Objectives

This lesson will show you how to do the following:

- Select the **Part Design Work Bench**
- Open the "**L Shaped Extrusion.CATPart**"
- Review **Pad** creation.
- Create several different **Fillets**
- Create a **Chamfer**
- Create and place **Holes**
- Create **Patterns**
- Modify the existing "**L Shaped Extrusion**"
- **Translate** entities

- **Rotate** entities
- Use the **Symmetry** tool
- **Mirror** entities
- **Scale** entities
- Create **Part Design Constraints**
- **Apply Material** to entities
- Manage the **Specification Tree**
- Customize the **Specification Tree**
- **Review** the design process
- **Save** the file
- **Exit** CATIA V5 (review)

Part Design Work Bench Tool Bars

There are seven standard tool bars found in the **Part Design Work Bench**. The seven tool bars are shown on the following pages. The individual tools found in each of the seven tool bars are also shown and include the tool name and a brief definition.

Tools that have multiple options have an arrow at the bottom right of the tool icon. Reference the Lesson 1 section titled "**Sketcher Work Bench Tool Bars**" for information on how to change the default tool and on tools that have multiple options.

The **Sketcher** Tool Bar (Access The Sketcher Work Bench)

Note: This is just a reminder from Lesson 1.

Tool Bar	Tool Name	Tool Definition
	Sketcher	Provides access to the **Sketcher Work Bench**. This is where the **Sketch** tools are made available for creating profiles.

The **Sketch-Based Features** Tool Bar

Sketch-Based Features

Tool Bar	Tool Name	Tool Definition
	Shaft	Creates a sketch by revolving a sketch around an axis. The **Shaft** tool requires the **Axis** tool (found in the **Sketcher Work Bench**) to define the axis of rotation.
	Groove	Creates a groove in an existing solid by revolving a sketch around an axis. The **Axis** tool is not required; other entities such as edges of other solids can be used.
	Hole	Creates a variety of hole types including; tapered, simple, counterbored, counter sunk and counterdrilled. Although this tool is in the **Part Design Work Bench** it creates a sketch that can be modified in the **Sketcher Work Bench**.
	Rib	Creates a rib by sweeping a profile along a center curve. Create the profile sketch then create another sketch defining the curve (path) the profile will be swept along.
	Slot	Creates a slot by sweeping a profile along a center curve. The profile does not have to be a closed profile. The center curve can be existing entities. The profile is subtracted from the intersecting (existing) solid.
	Stiffener	Creates a stiffener. Requires a sketch, the sketch does not have to be a closed profile.
	Loft	Creates a solid loft. Requires a closed profile and a spine to define the path.
	Removed Loft	Creates a negative loft. Same as the **Loft** tool except that material is subtracted from the existing solid. An existing solid is required to subtract from.
	Pads Tool Bar	
	Pad	Creates a pad (solid) by extruding a profile. The profile can be an open or closed profile. Note: In most cases closed profiles are easier to work with.
	Drafted Filleted Pad	Creates a pad with drafts and fillets. This tool saves several steps by building in two of the **Dress Up** tools.
	Multi-Pad	Creates a multi-pad part with various thicknesses. This requires multiple closed profiles in the same sketch. The tool allows the user to define the thickness of each closed profile.

	Pockets Tool Bar	
	Pocket	Creates a pocket by removing material defined by the selected profile. Opposite of the **Pad** tool, it subtracts material. Requires an existing solid to subtract from.
	Drafted Filleted Pocket	Creates a pocket with drafts and fillets. Another tool with **Dress Up** tools built in.
	Multi-Pocket	Creates multi-pockets with various thicknesses. Opposite of the **Multi-Pad** tool. Requires an existing solid to subtract from.

The **Transformation Features** Tool Bar

Note: Although some of these tools are similar to the **Transformation** tool bar in the **Sketcher Work Bench** these tools are specific to the **Part Design Work Bench** only. Profiles must be modified in the **Sketcher Work Bench**.

Tool Bar	Tool Name	Tool Definition
	Scaling	Scales a body by using a point or a face or a plane.
	Mirror	Mirrors a body using a plane or face.
	Transformations Tool Bar	
	Translation	Translates a specified body. The user must specify the direction, distance and body to translate.
	Rotation	Rotates a specified body. The user must specify the axis, angle of rotation and body to rotate.
	Symmetry	Mirrors a body without keeping the original body.
	Patterns Tool Bar	
	Rectangular Pattern	Creates a rectangular pattern. The user specifies the rectangular pattern values and the feature to repeat.
	Circular Pattern	Creates a circular pattern. The user specifies the circular pattern values and the feature to repeat.
	User Pattern	Creates a pattern defined by the user.

The **Dress-Up Features** Tool Bar

Note: These features could be created in the **Sketcher Work Bench** but it would require a lot more work. **Dress-Up** tools in the **Part Design Work Bench** is a significant time saver in the initial design and modifications.

Tool Bar	Tool Name	Tool Definition
	Chamfer	Creates a chamfer on a part. The type of chamfer and the values are determined by the user in the **Chamfer Definition** window.
	Shell	Shells an existing body. The user specifies the face to remove and the thickness of the shell.
	Thickness	Adds or removes thickness from an existing body. The user specifies the face to thicken and the value of the added thickness.
	Thread/Tap	Creates a thread or tap.
	Fillets Tool Bar	
	Edge Fillet	Creates a fillet using edges or faces.
	Variable Radius Fillet	Creates a variable radius fillet.
	Face-Face Fillet	Creates a face to face fillet using two faces.
	Tritangent Fillet	Creates a fillet by removing a face. The user specifies two faces to create a fillet between and then the face to be removed.
	Drafts Tool Bar	
	Draft Angle	Creates a draft on an existing body. The user specifies the direction to draft, faces to draft and the angle of the draft.
	Draft Reflect Line	Creates a draft reflect line.
	Variable Draft	Creates a variable draft.

The **Annotations** Tool Bar

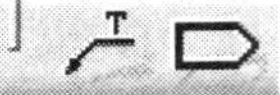

Tool Bar	Tool Name	Tool Definition
	Text With Leader	Creates text with a leader. This allows the user to attach text to a feature in the **Part Design Work Bench**. The text can be placed in **Hide/Show** for future reference.
	Flag Note	Creates a flag note. The flag note can be linked to a file and/or a URL. The flag note can also be placed in **Hide/Show** for future reference.

The **Constraints** Tool Bar

Note: These constraint tools are similar to the constraint tools in the **Sketcher Work Bench** but apply only to the constraints created in the **Part Design work Bench.**

Tool Bar	Tool Name	Tool Definition
	Constraints Defined In Dialog Box	Shows the user the type of constraint that is available for the entities selected. The user must select the entities before the tool is available for selection.
	Constraint	Creates a constraint between to different entities.

The **Surface-Based Features** Tool Bar

	Container4 Tool Bar	
	Thick Surface	Creates a thickness from a face or surface. This tool allows the user to add thickness on both sides.
	Split	Splits a body into two separate entities using a plane, face or surface.
	Close Surface	Closes a surface so a body can be created.
	Sew Surface	Sews a face or a surface into a body.

Figure 2.2

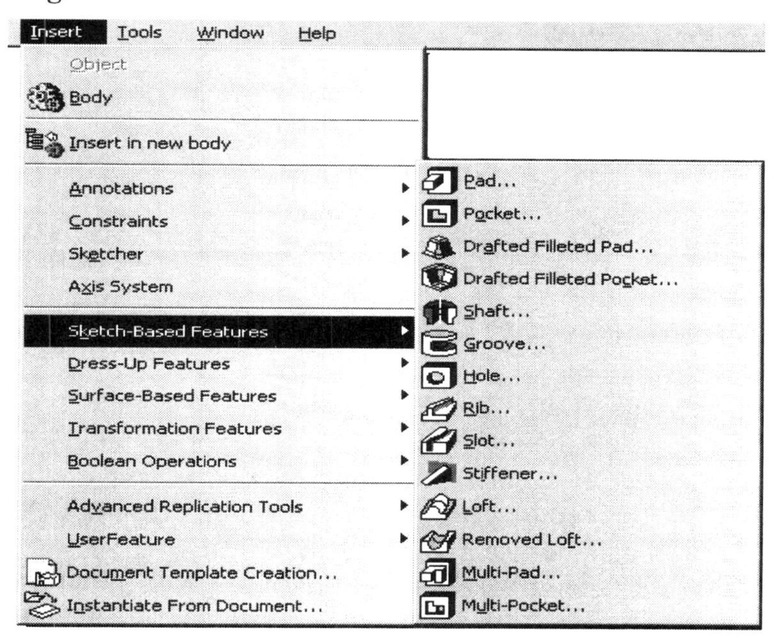

All of the tools listed above in the individual tool bars can also be accessed from the **Insert** pull down menu. This method of tool selection is helpful to the new user because all of the tool names are visible. This method also displays some additional tools as Figure 2.2 shows. The additional tools will not be used in this lesson, but it is important for you to know for future reference.

Steps To Extruding A Profile Using The Part Design Work Bench

The following steps will give detailed step-by-step instructions on taking the profile of the "**L Shaped Extrusion**" created in Lesson 1 and extruding it into a completed three-dimensional part. For every step covered, there are multiple methods that could be used. This lesson, for the most part, will show you one method. You are encouraged to follow the steps and then explore other options on your own.

1 Start CATIA V5

This was covered in Lesson 1. If you need to review this step, refer to Lesson 1 Step 1. Lesson 1 shows the screen you should see when you first enter CATIA V5.

2 Select The Part Design Workbench 

The method of selecting the **Part Design Work Bench** is the same as selecting the **Sketcher Work Bench** in Lesson 1. Make sure you select the **Part Design** tool and not the **Sketcher** tool. For a detailed review on selecting a workbench, refer to Lesson 1 Step 2.

3 Open The "L Shaped Extrusion.CATPart" Document

Opening a CATIA V5 file is the same as any other Windows and/or Windows NT program. From the top menu bar select the **File**, **Open** options. A pop-up window will appear, which gives you a choice of directories to choose from. Select the directory where you saved the **"L Shaped Extrusion"** file created in Lesson 1. All of the CATIA V5 CATPart files should be displayed including the **"L Shaped Extrusion.CATPart."** Double click on the **"L Shaped Extrusion.CATPart"** file. CATIA V5 will now load the **"L Shaped Extrusion.CATPart"** as you last saved it.

4 Review The "L Shaped Extrusion.CATPart"

Before continuing to the next step, review the "**L Shaped Extrusion.CATpart**" CATIA V5 loaded. Verify that everything is the way you expected it to be. If you completed every step in Lesson 1, your "**L Shaped Extrusion.CATPart**" is already extruded into a three-dimensional part and you can skip to Step 6.

5 Extruding The Sketcher Profile Using The Pad Tool 

This step was covered in detail in Lesson 1, even though the **Pad** tool is a **Part Design Work Bench** tool. This is a critical step, so you want to make sure that if this process was not completed in Lesson 1, it is not missed before proceeding to Step 6. If you have not completed this step, refer to Lesson 1 Step 22 for detailed instructions on how to use the **Pad** Tool. Figure 1.44 is a good representation of what your part should look like after completing Step 22.

NOTE: It is critical that your "**L Shaped Extrusion.CATPart**" part look similar to what is shown in Figure 1.44 before proceeding!

6 Creating A Fillet

A **Fillet** is a rounded corner on an outside or inside edge of a part. It can also be defined as a radial transition between two adjacent surfaces. CATIA V5 has several **Filleting** options to choose from. These options are the **Edge Fillet**, **Variable Radius Fillet**, **Face-Face Fillet** and the **Tritangent Fillet**. These options can be seen by holding down the left mouse button on the arrow next to the default **Fillet** tool. The most commonly used **Fillet** command is the **Edge Fillet**. The following steps will take you step-by-step through the process of creating an **Edge Fillet** on your "**L Shaped Extrusion.CATPart**."

Figure 2.3

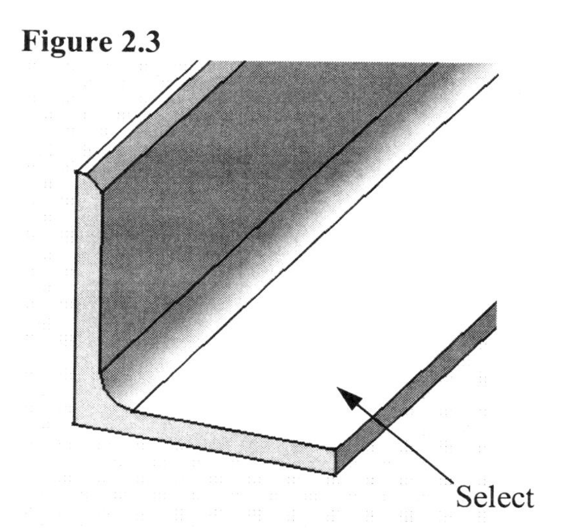

6.1 Make sure the **Units** are set to inches (for a review on how to do this, refer back to Lesson 1 Step 5).

6.2 Select the right edge of the part as shown in Figure 2.3. The line will turn red indicating it has been selected.

6.3 Select the **Edge Fillet** tool from the **Dress-Up Features** tool bar.

6.4 The **Edge Fillet Definition** window will appear on the screen, reference Figure 2.4.

6.5 For the **Radius** type in **.50** then tab. This will set the **Fillet** radius to .50 of an inch. The radius and the dimension now appear on the part. CATIA V5 will preview the fillet before it actually creates it.

6.6 The **Object(s) to fillet** box should indicate that one edge has been selected as shown in Figure 2.4.

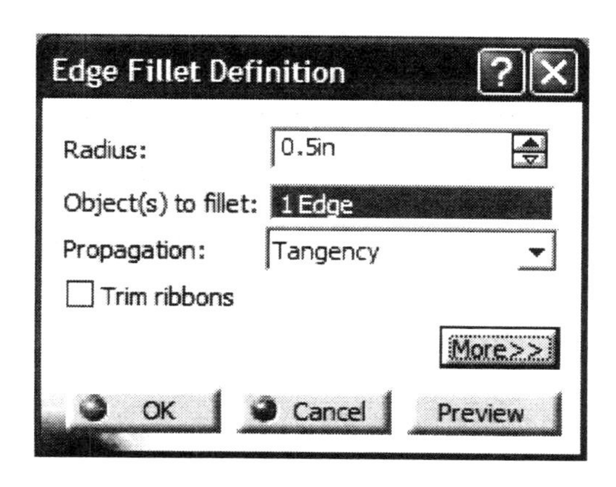

6.7 Leave the **Propagation** box set to **Tangency** as shown in Figure 2.4.

6.8 If you are satisfied with the preview and the options are typed in as shown in Figure 2.4, select the **OK** button and the **Fillet** will be created. The part should now look like the one shown in Figure 2.5.

Figure 2.5

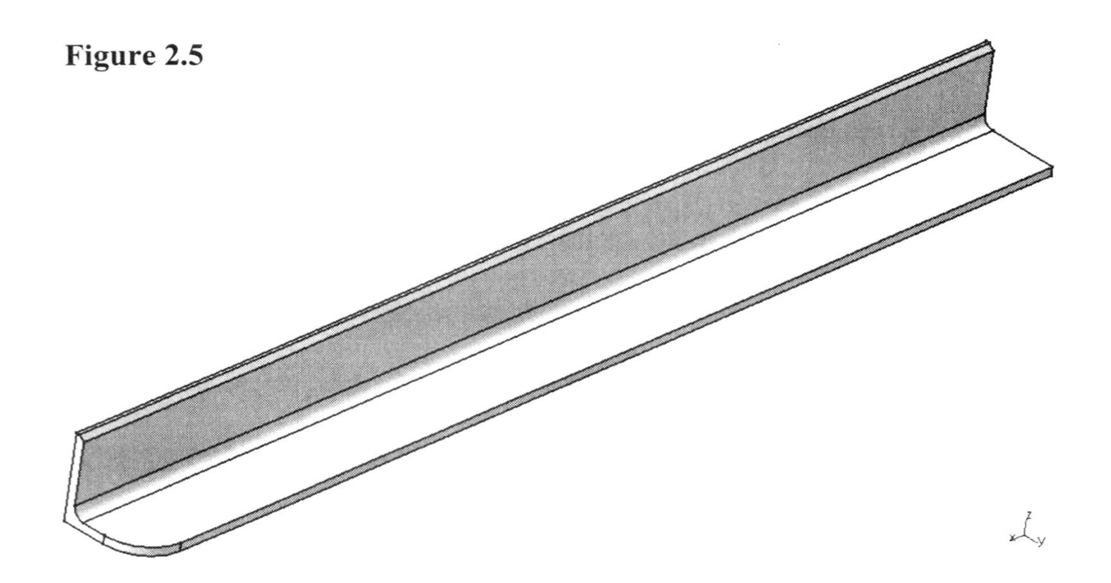

NOTE: The **Fillet** was added to the **Specification Tree** as **EdgeFillet.1** (Figure 2.6).

Figure 2.6

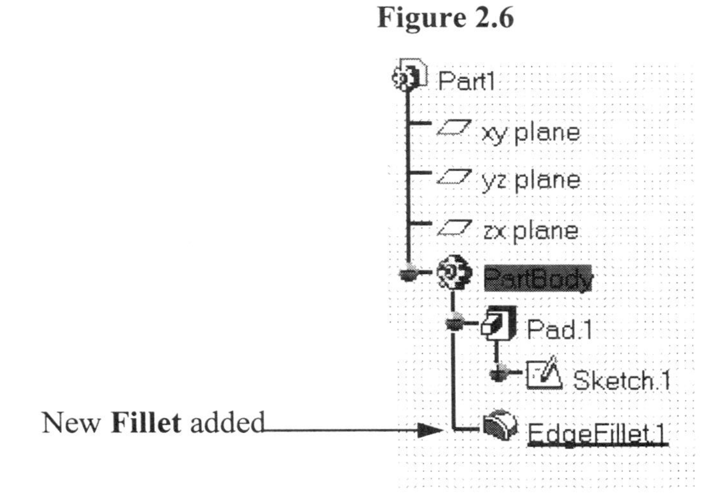

6.9 Follow Steps 6.3 through 6.8 to put a **Fillet** on the other end of the part. The results should look similar to the one shown in Figure 2.7. The new **Fillet** will also show up on the **Specification Tree**.

Figure 2.7

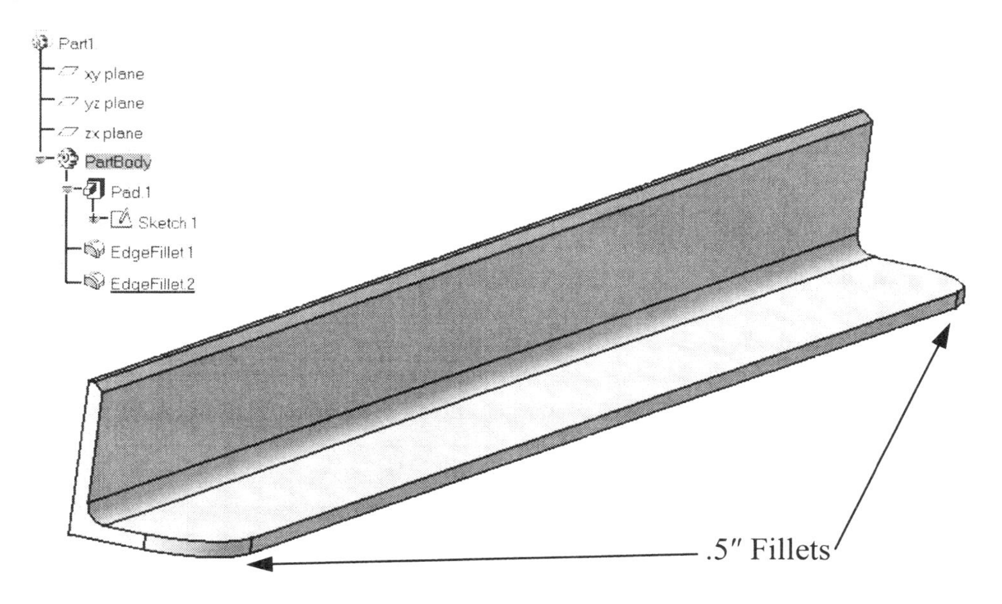

7 Creating A Fillet On The Back Edge Of The "L Shaped Extrusion"

7.1 To select the back edge of the extrusion, you will need to rotate the part until the line that defines the edge is visible. To do this, select the **Rotate** tool from the bottom of the screen. Then select an area in the middle of the screen and drag the mouse in the direction you want the part to be rotated. A dotted sphere will appear on the screen showing the axis of rotation (Figure 2.8).

Figure 2.8

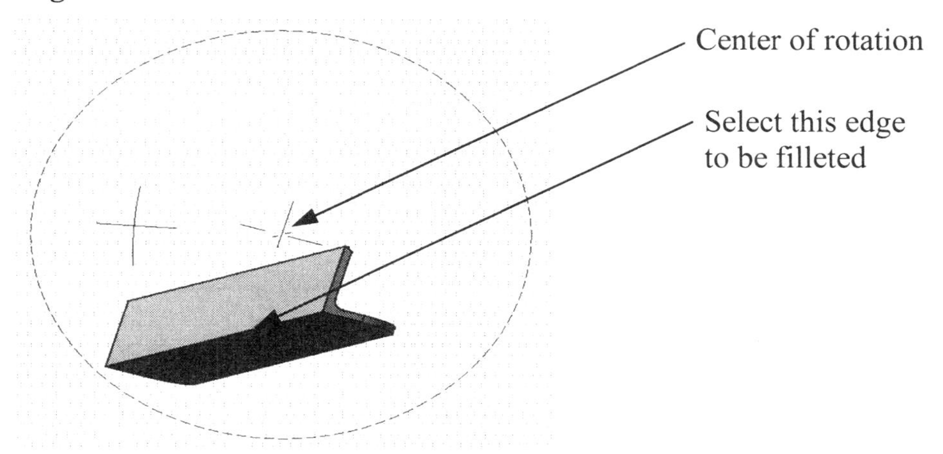

7.2 Once the part is rotated so that the back edge of the extrusion is visible, select the line as shown in Figure 2.8.

Select the **Edge Fillet** tool.

7.3 When the **Edge Fillet Definition** pop-up window appears, change the value in the **Radius** box to **.225″** as shown in Figure 2.9.

7.4 Your "**L Shaped Extrusion**" should now look similar to the one shown below in Figure 2.10.

Figure 2.9

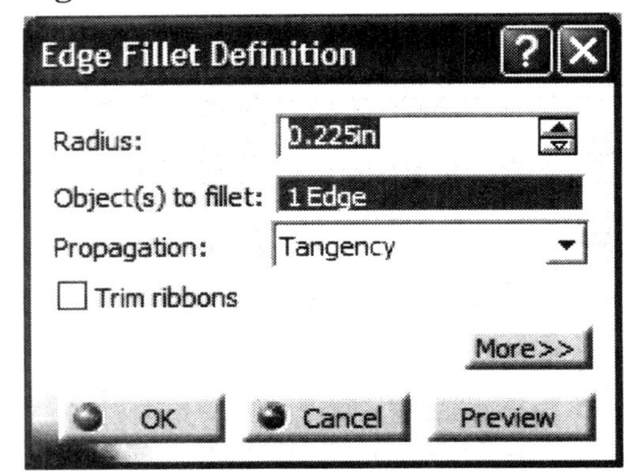

Figure 2.10

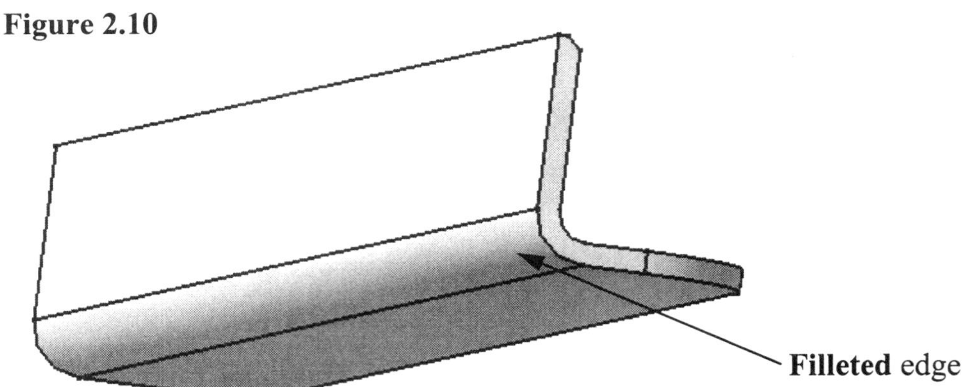

7.5 Save the revised "**L Shaped Extrusion.CATPart**." Select **File**, **Save** and **Yes** replace existing file if previously saved. For a review on saving a **CATPart** see Lesson 1 Step 23.

8 Creating A Chamfer 

Chamfers are used quite often on three-dimensional objects. The main purpose of a chamfer is to break a sharp edge on a part or to add a taper to an edge. You will now add a chamfer to your "**L Shaped Extrusion**."

8.1 This step will show you how to create a chamfer on the back edge of the part. Since Step 7 had you create a fillet on the back edge, you will have to remove it before replacing it with the chamfer. To remove the fillet created in Step 7, complete the following steps.

8.1.1 Highlight the **EdgeFillet.3** from the **Specification Tree** in the upper left hand corner of the screen (Figure 2.11). If you followed the steps exactly, the **Specification Tree** should label the fillet on the back edge of the part as **Edge Fillet.3**. If you have been exploring on your own, your fillet may be labeled another number. You can verify that you selected the correct fillet by checking which fillet is hightlighted on your "**L Shaped Extrusion**." If the highlighted **Fillet** is the same as shown in Figure 2.11, you have selected the correct fillet and may continue.

Figure 2.11

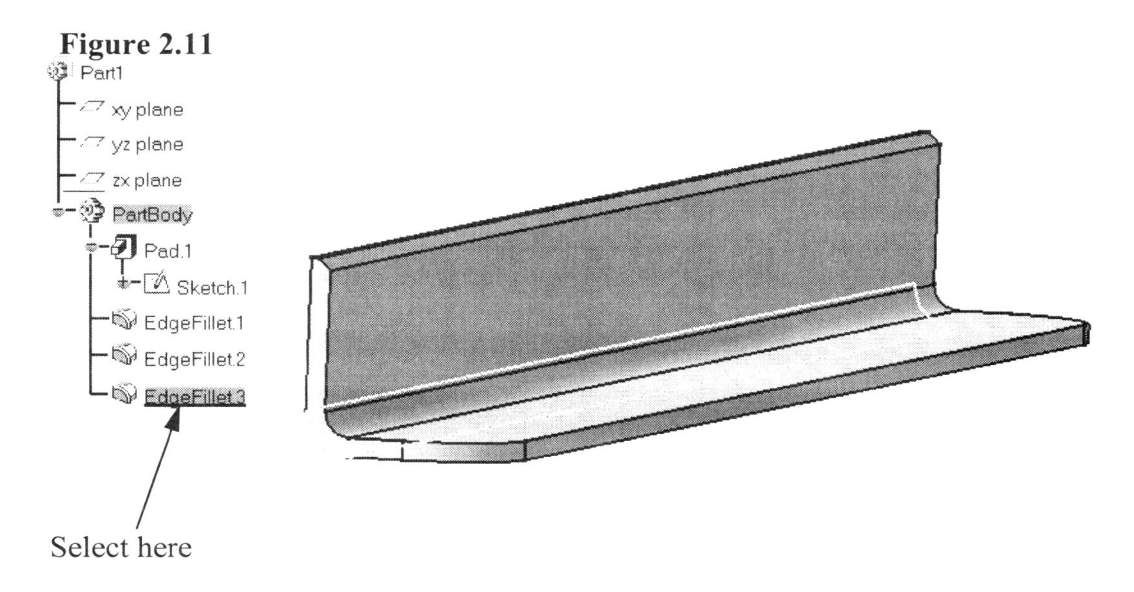

8.1.2 Select the **Cut** tool from the **Standard** tool bar or hit the **Delete** key on the key board. The **EdgeFillet.3** should dissappear from the tree and the corresponding **Fillet** on the "**L Shaped Extrusion**" should also dissappear. The back edge of the part should be a square edge. Your part should now look like the one in Figure 2.7.

8.2 This step is almost identical to what you did in Step 7. Rotate the view to where the back edge is in view. Select the line that defines the back edge of the "**L Shaped Extrusion**." See Figure 2.8 to see which line to select. Once the line is highlighted, select the **Chamfer** tool.

8.3 The **Chamfer Definition** window will appear as shown in Figure 2.12. The first option given is the **Mode** box. You specify the chamfer by determining the length and the angle (**Length1/Angle**) or by determining the length of both sides of the chamfer (**Length1/Length2**). For this lesson, select the **Length1/Angle** option.

Figure 2.12

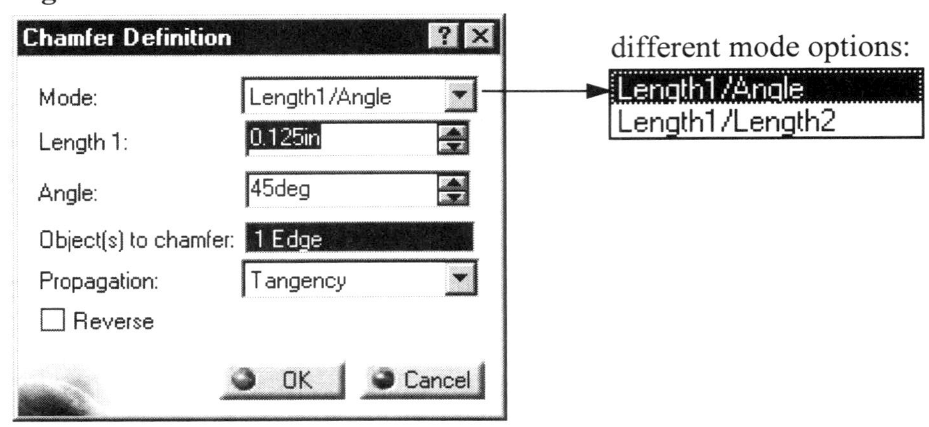

8.4 Enter **.125in** value in for the **Length 1** box and **45deg** in for the **Angle** box. If your **Chamfer Definition** window looks like the one in Figure 2.12, select **OK**.

8.5 Your part should look similar to the one shown in Figure 2.13. Notice that **Chamfer.1** has also been added to the **Specification Tree**.

Figure 2.13

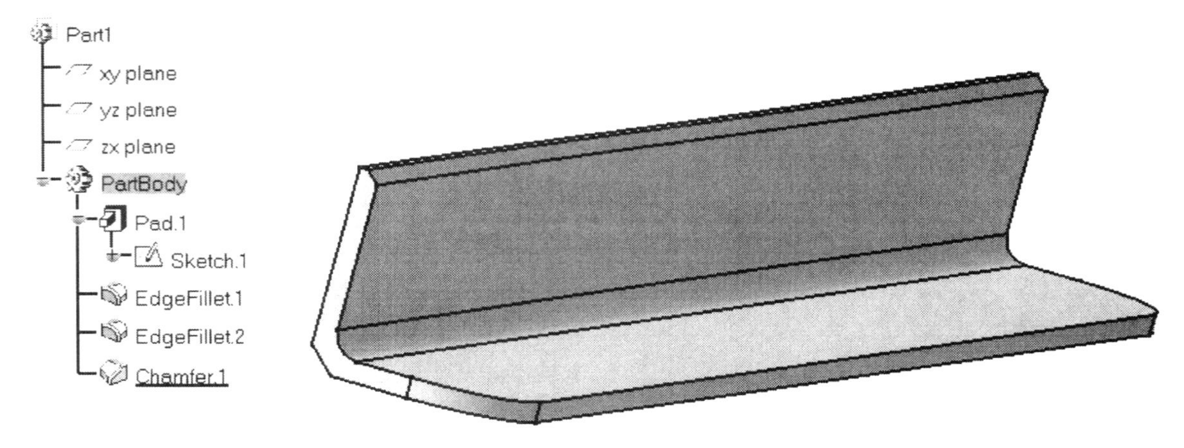

8.6 It would be a good idea to save your current **CATPart** before continuing.

9 Creating Holes

The following steps will take you through the process of creating and locating holes.

9.1 Make sure you have the "**L Shaped Extrusion**" with the radii and chamfer up on your screen as shown in Figure 2.14. As a double check, verify that your default units are set to inches.

9.2 The first hole will be created on the left side of the standing leg. Before the hole can be created, two edges need to be selected. The two edges are required to correctly and accurately locate the holes. The surface the hole is located on will also have to be selected. The two lines (edges) and the surface can be selected at the same time by holding down the **Ctrl** key as you select the lines and surface. Select the lines and surface as shown in Figure 2.14.

NOTE: If you select the surface first, the edges of the surface will highlight. You will still have to select the edges individually when prompted. When an edge has been selected twice, once for the surface and a second time for the edge, the second highlight color is a deeper red color.

Figure 2.14

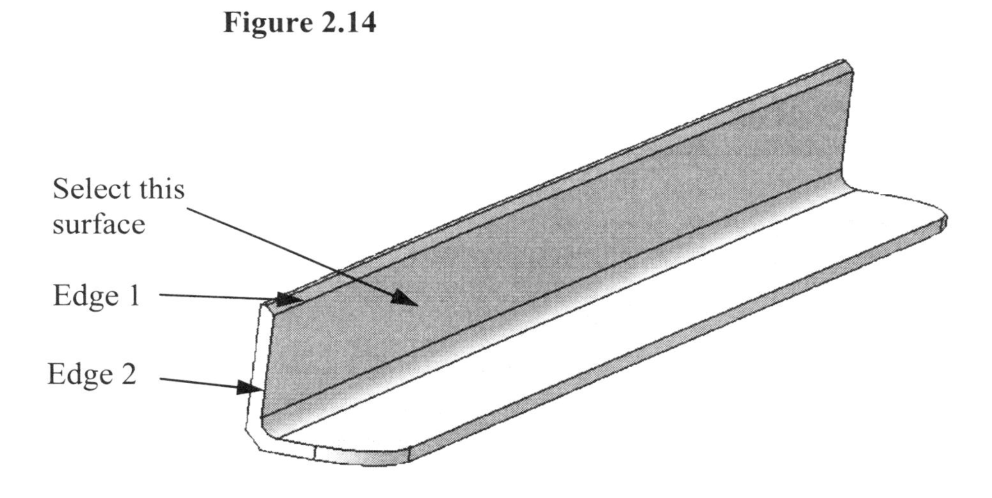

9.3 Select the **Hole** tool from the **Sketch-Based Features** tool bar located on the right side of the screen. The lines and surface selected in Step 9.2 must still be highlighted.

9.4 The **Hole Definition** window will appear on the screen as shown in Figure 2.15. A highlighted outline of the hole will appear on the part with the constraints that show the distance from the two edges selected in Step 9.2.

Figure 2.15

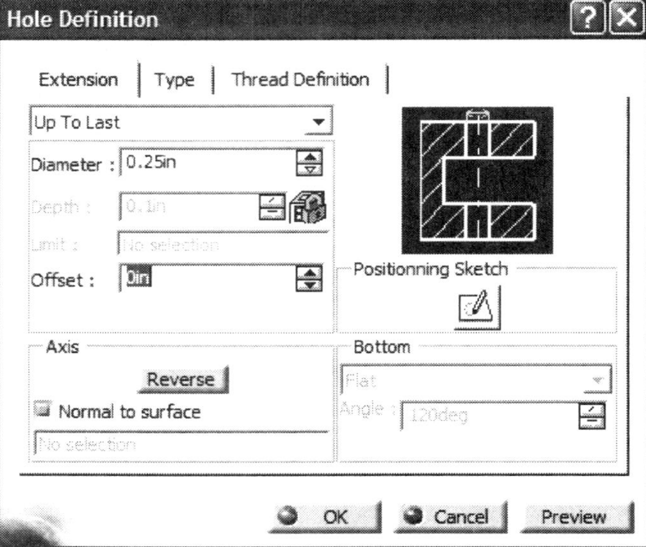

9.5 You now need to accurately position the hole in the correct location. To do this you will use the **Constraints** created in Step 9.4. The **Constraints** you created in Step 9.4 will most likely be crowded together, in fact, you may not even be able to read the **Constraint** values. This can be fixed by zooming up on the **Constraints**, selecting one of the **Constraints** and dragging it away from the collection of **Constraints**. Then select another **Constraint**, dragging and dropping it far enough away from the other constraints that it can be read and selected. The **Constraint** values can be modified by double clicking on the **Constraint**. Double clicking on the **Constraint** brings up the **Constraint Definition** window as shown in Lesson 1. This **Constraint Definition** box allows you to type in the **Constraint Value**. For this step, change the **Value** of the vertical **Constraint** to **.35in** and the horizontal **Constraint** to **1in** as shown in Figure 2.16.

9.6 You now need to specify the type of hole you want to create. CATIA V5 has several different types of holes to choose from. The different hole types can be accessed by selecting the **Type** tab at the top of the **Hole Definition** window as shown in Figure 2.17. The box on the right hand side of the window will display a visual picture of the hole currently highlighted. These different pictures are shown in Figure 2.18. For this step select the "**Simple**" option as shown in Figure 2.17.

Figure 2.16

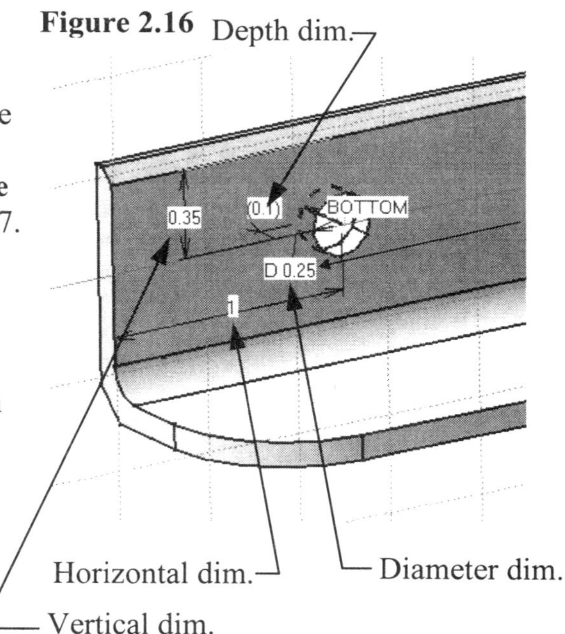

Figure 2.17 (**Types** of holes)

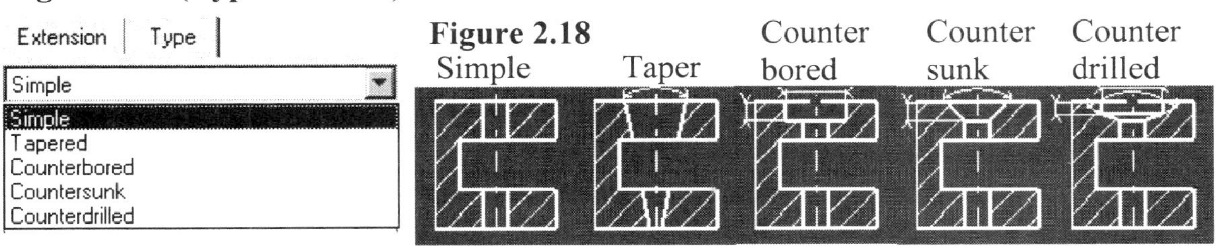

Figure 2.18

Simple Taper Counter bored Counter sunk Counter drilled

9.7 The hole limits will need to be selected next. The limit will specify how deep the hole will go. CATIA V5 has several different options available for hole limits. To specify a limit type, select the **Extension** tab located next to the **Type** tab (Figure 2.15). The first box in the **Extension** tab is an options box. By selecting the down arrow at the right side of the box, you will see the different options you have to choose from. The options are shown in Figure 2.20. For this step, select the **Up To Last** option as shown in Figure 2.19.

Figure 2.19

Figure 2.20

Blind Up To Last Up To Next Up To Plane Up To Surface

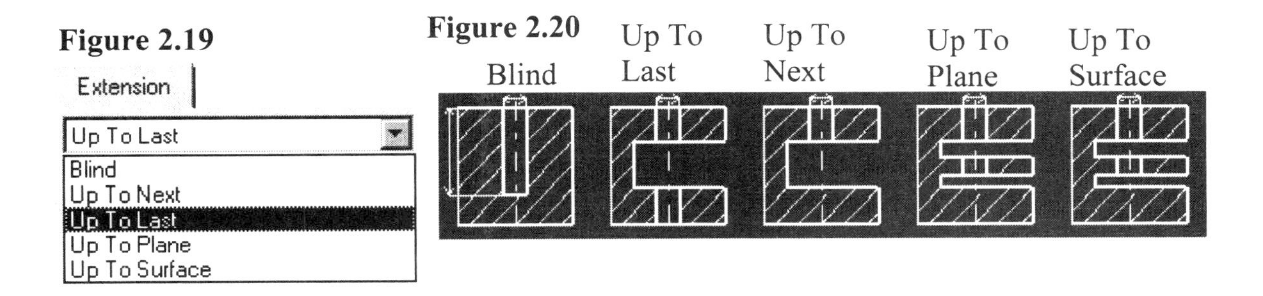

9.8 The last constraint not yet specified is the **Diameter** of the hole. Type in **.25in** for the diameter value as shown in Figure 2.15.

9.9 If your **Hole Definition** window looks like the one in Figure 2.15, select **OK**. You now have a hole in the standing leg of your "**L Shaped Extrusion**." Reference Figure 2.21. Notice that a new branch was added to the **Specification Tree**. The branch is labeled **Hole.1**.

Figure 2.21

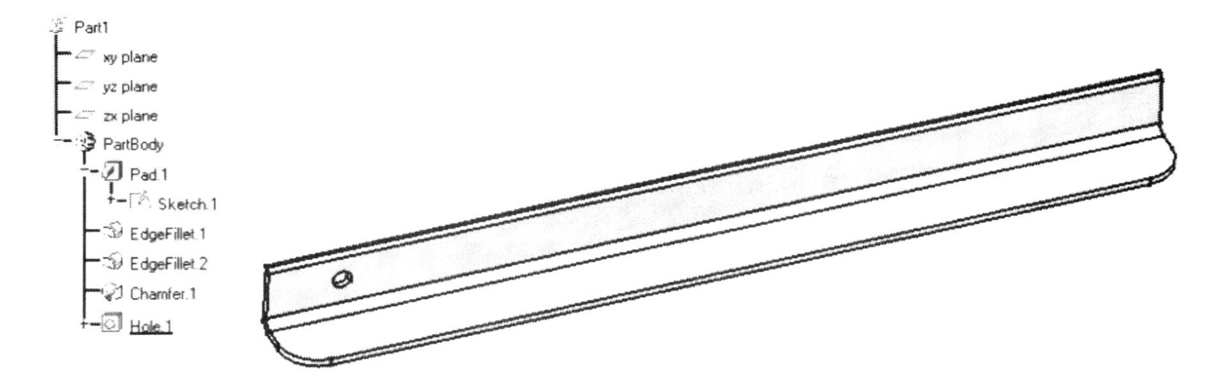

9.10 Now to test the knowledge you have gained. Add a second hole on the other end of the "**L Shaped Extrusion**." Use the same hole to edge dimensions used for the first hole. When you are done, your part should look like the one shown in Figure 2.22.

Figure 2.22

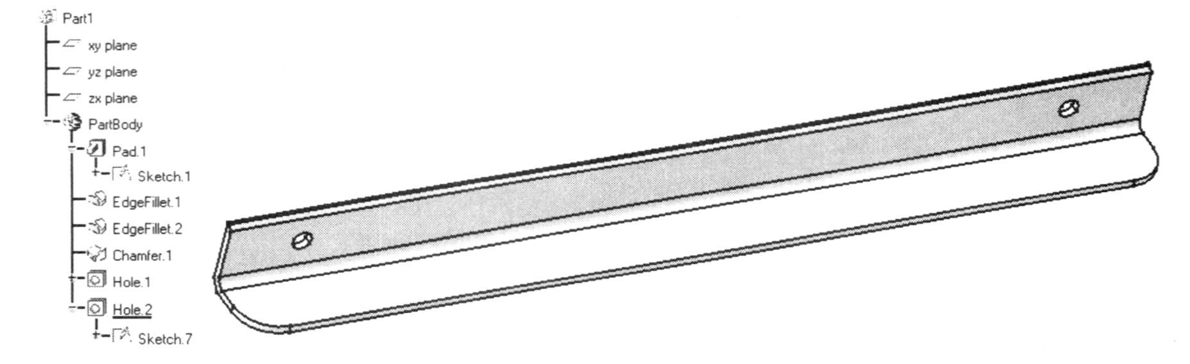

10 Creating A Pattern Of Holes

Creating patterns are easy and fast using the CATIA V5 **Pattern** tool. The following steps will show you how to create a pattern of holes in your "**L Shaped Extrusion**."

10.1 Before you get started, make sure the part on your screen looks like the one in Figure 2.22 and that the **Units** are set to **Inches**.

10.2 The hole pattern will be created on the standing leg of the "**L Shaped Extrusion**." The first hole created in Step 9 will be used as the strating hole.

10.3 Select **Hole.1** from the **Specification Tree** or by selecting the actual hole. The selected hole should be highlighted, signifying it has been selected. The corresponding hole on the part should also be highlighted (Figure 2.23).

Figure 2.23

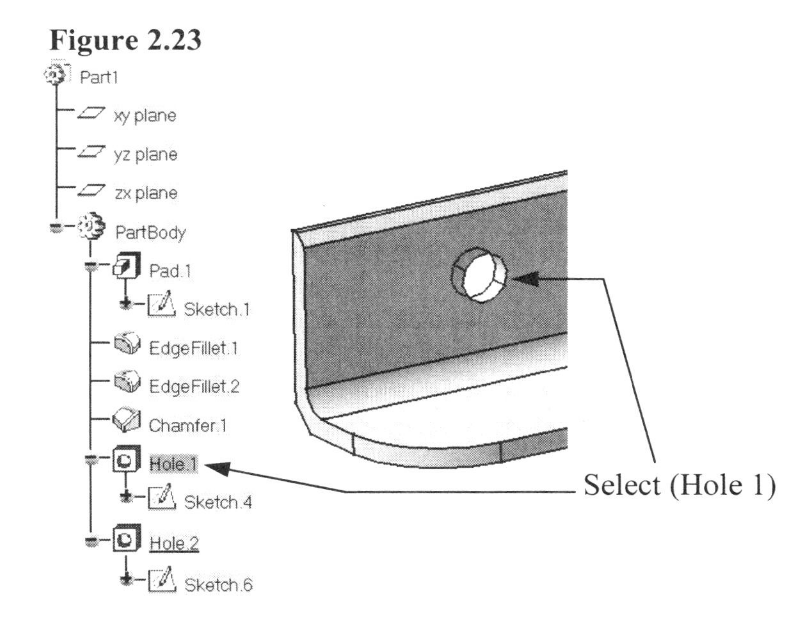

10.4 Select the **Rectangular Pattern** tool from the **Transformation Feature** tool bar located on the right side of the screen. If the tool icon is not visible on the right side of the screen, it may be hidden below the screen. If there is a tool bar hidden on the bottom of the screen, there will be two little arrows on the end of the last tool bar as shown in Figure 2.24. You can view them by clicking on the arrows with the left mouse button and dragging the hidden tool bar up into view. You can use the windows drag and drop method of placement. If you drag one tool bar up and the arrows are still there that means there are still more hidden tool bars. Drag and drop the tool bars until you find the **Transformation Features** tool bar. The **Rectangular Pattern** tool is located on this tool bar.

Figure 2.24 (lower right hand corner of screen)

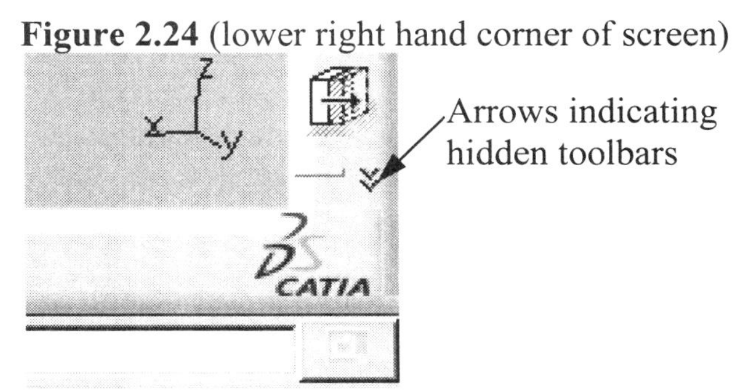

10.5 The **Rectangular Pattern Definition** window will appear on the screen as shown in Figure 2.25. Notice that there is a **First Direction** and **Second Direction** tab at the top of the window. CATIA V5 allows you to specify the hole pattern in two directions. For your "**L Shaped Extrusion**," we will only be using the **First Direction** tab.

Figure 2.25

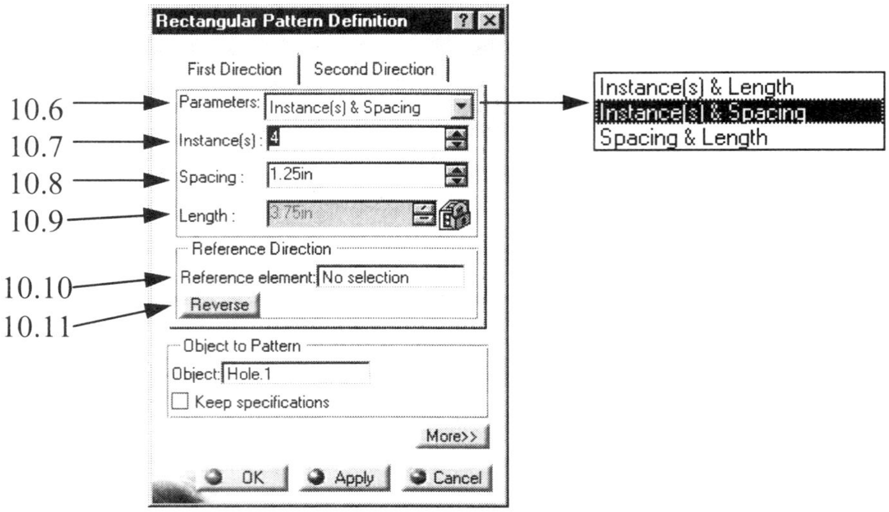

10.6	Parameters: Instance(s) & Spacing
10.7	Instance(s): 4
10.8	Spacing: 1.25in
10.9	Length: 3.75in
10.10	Reference element: No selection
10.11	Reverse

Rectangular Pattern Definition

First Direction | Second Direction

Reference Direction

Object to Pattern

Object: Hole.1

Keep specifications

More>>

OK Apply Cancel

Instance(s) & Length
Instance(s) & Spacing
Spacing & Length

10.6 The first box in the **First Direction** tab is the **Parameters** box. CATIA V5 gives you several options to choose from. Select the **Instance(s) & Spacing** option (Figure 2.25). This will allow you to specify the number of holes you want and the spacing between each hole.

10.7 The **Instances** box specifies how many times the hole is going to be copied in the specified direction. For this step, type "**4**" in the **Instances** box. It is important to know that the value entered in the **Instances** box will count the existing hole as one of the entered value. For example, you just entered the value of **4** in the **Instances** box. A total of 4 holes will be created, one you selected to pattern after and three additional holes.

10.8 For the **Spacing** value type in **1.25in** This will place the holes 1.25 inches apart from each other.

10.9 The **Length** box will display the total length of the hole pattern from the center of the first hole to the center of the last hole. The **Length** box will be dimmed, which means the value can not be changed.

10.10 You now need to specify the direction of the hole pattern. You can specify the direction by selecting the **Reference Element** box. The box will turn blue indicating it is selected.

10.11 Select the top edge of the "**L Shaped Extrusion**" as shown in Figure 2.26. The top edge will highlight signifing it has been selected. At this point CATIA V5 will preview the hole pattern in the part. This highlighted preview allows you to verify that the pattern is what you want. If the hole pattern is previewed in the wrong direction, you can select the **Reverse** button located just below the **Reference Element** box. This reverses the direction of the hole pattern. If your part looks similar the the one in Figure 2.27, select the **OK** button. Notice that **RectPattern.1** branch has been added to the **Specification Tree**.

Figure 2.26

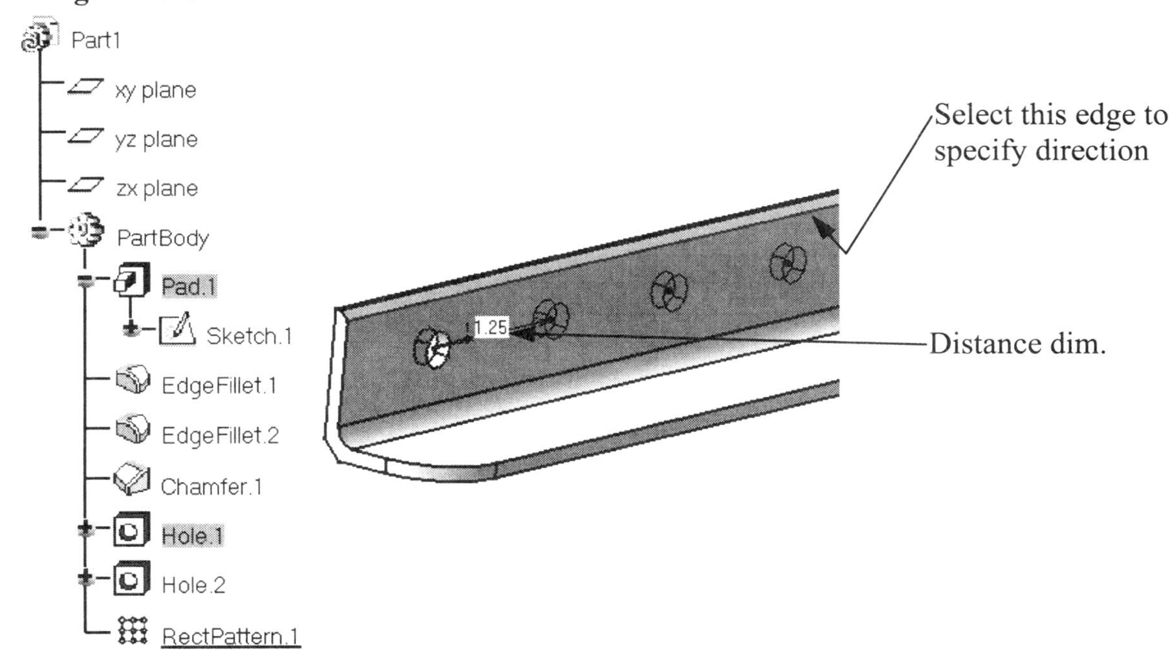

Figure 2.27

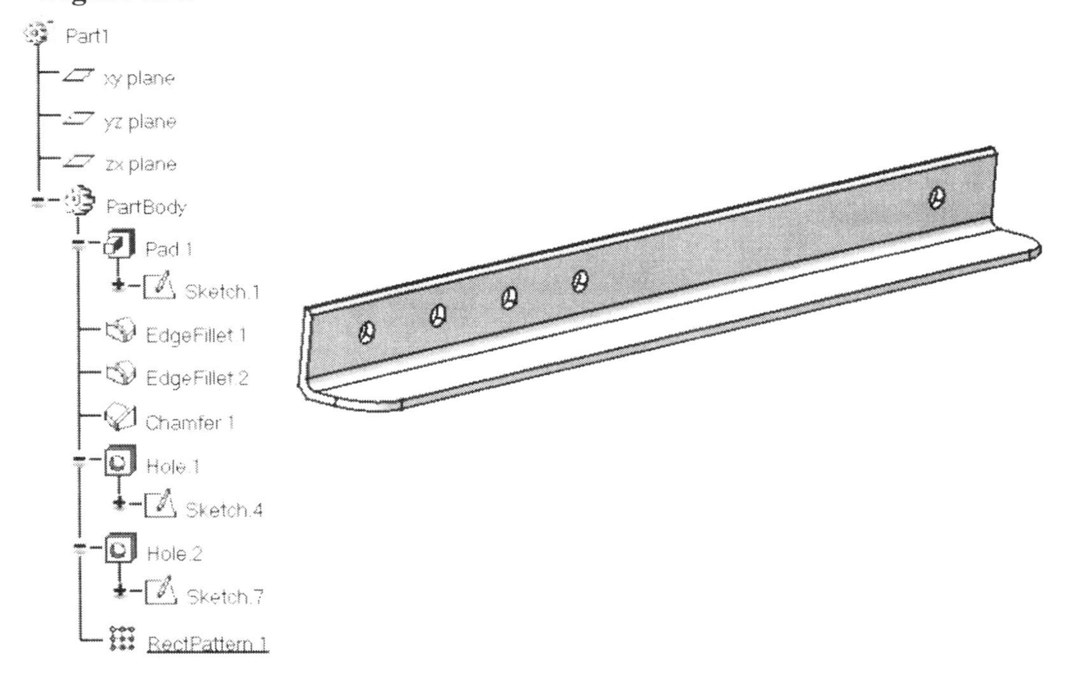

10.12 Now use what you have learned to create the same hole pattern on the other side of the part. Use **Hole.2** to start the hole pattern. Figure 2.28 shows what your part should look like when your second hole pattern is complete. Notice that the **Specification Tree** has added another hole pattern below the **Hole.2** branch.

Figure 2.28

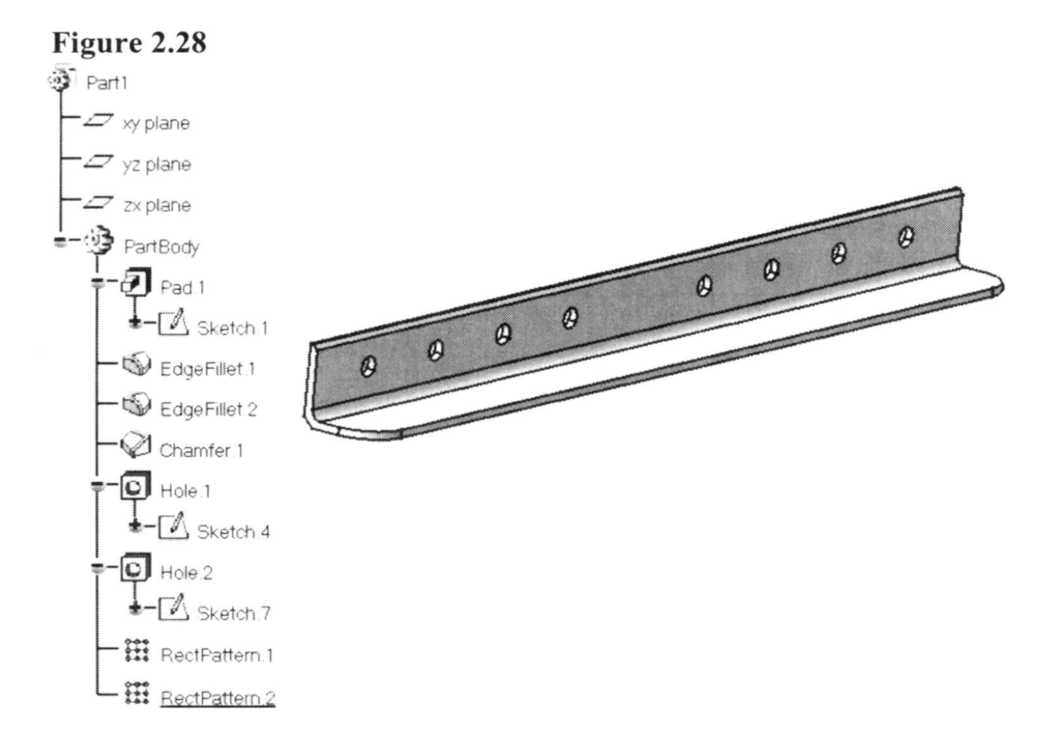

11 Modifying The Width Of The Base Leg

In the past, design modification has been a very time consuming process. Most designs go through mulitiple modifications prior to being manufactured. CATIA V5 has made design modification much quicker and easier. To demonstrate CATIA V5's ability to modify parts, this step will take you through the process of adjusting the width of the base leg of the "**L Shaped Extrusion**."

11.1 Before you get started, make sure the **Units** are set to **Inches**. Make sure the part on your screen looks similar to the one shown in Figure 2.28.

Figure 2.29

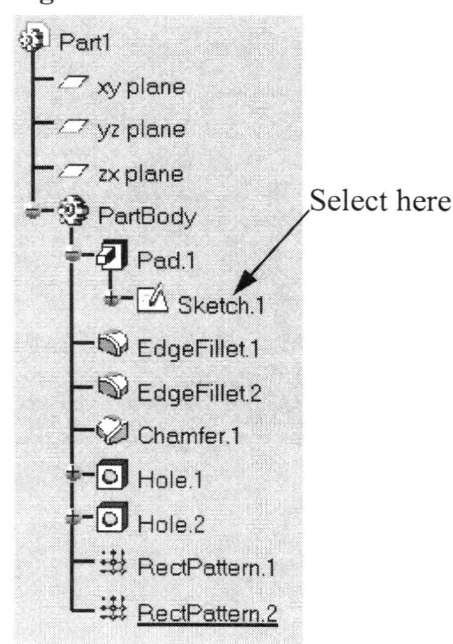

11.2 In order to adjust the base leg of the "**L Shaped Extrusion**," you need to go back to where you specified its current width. This was specified when you created the sketch in the **Sketcher Work Bench** (Lesson 1). You can bring that sketch back and modify the constraint that controls the base leg width. There are several methods to bring up the sketch. For this step use the **Specification Tree**. The **Sketch.1** branch must be visible. If the **Sketch.1** branch is not visible, select the plus to the side of the **Pad.1** branch. The expands the **Specification Tree.** The contracts the **Specification Tree**.

11.3 With the **Sketch.1** branch visible, double click on it. Reference Figure 2.29. The screen will change from the **Part Design Work Bench** (3D environment) to the **Sketcher Work Bench** (2D sketch environment). As the screen changes from the **Part Design Work Bench** to the **Sketcher Work Bench,** the "**L Shaped Extrusion**" will rotate to where the plane selected in Lesson 1 will be normal (perpendicular) to the screen. This is the plane **Sketch.1** was created in. **Sketch.1** is linked to this plane. Your screen should look similar to the one shown in Figure 2.30. You are now ready to modify the width of the base leg.

Figure 2.30

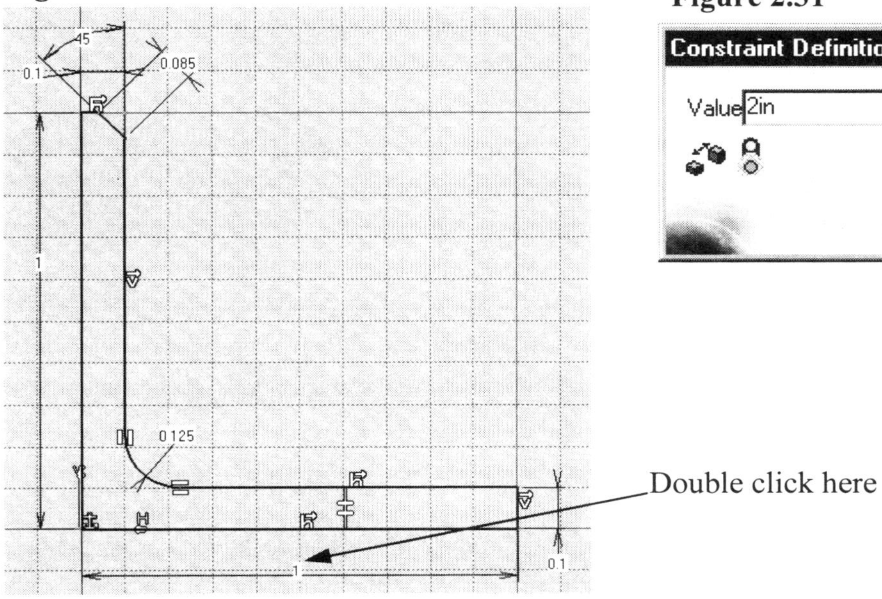

Figure 2.31

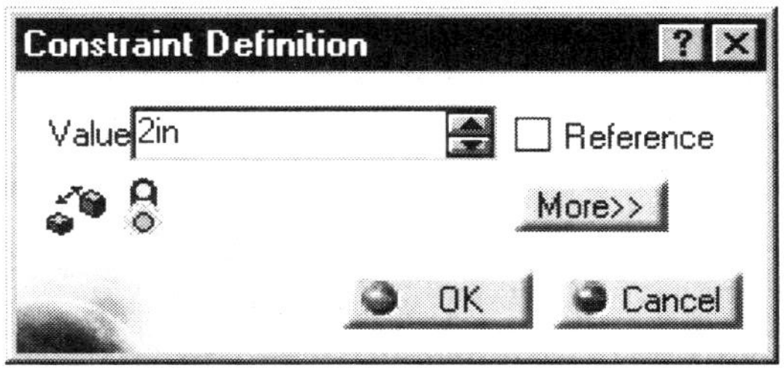

11.4 Double click on the **Constraint** that specifies the width of the base leg. As shown in Figure 2.30, the constraint is currently 1″.

11.5 The **Constraint Definition** window will appear with the 1″ value highlighted in the **Value** box. Change the 1″ value to 2″ as shown in Figure 2.31.

11.6 Select the **OK** button at the bottom of the **Constraint Definition** window. Notice that the sketch automatically modifies the base leg from 1″ to 2″. Figure 2.32 shows the modified sketch.

Figure 2.32

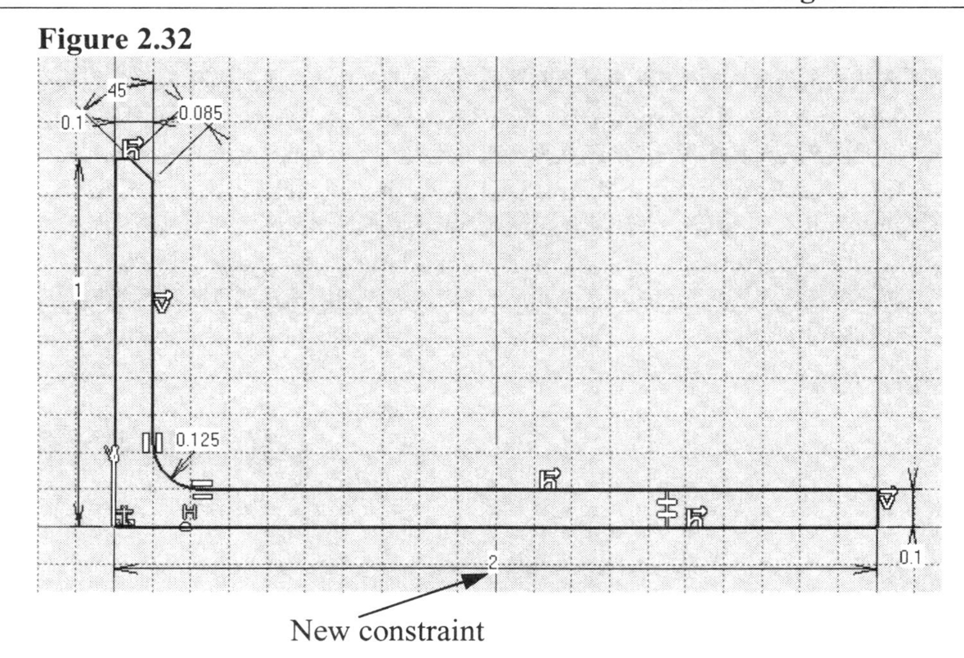

11.7 You have completed the sketch modification in the **Sketcher Work Bench**. It is time to continue this lesson in the **Part Design Work Bench**. To exit the **Sketcher Work Bench,** select the **Exit** tool from the upper right hand corner of the screen. This will take you out of the **Sketcher Work Bench** and back into the **Part Design Work Bench**.

11.8 After returning to the **Part Design Work Bench** the "**L Shaped Extrusion**" part may appear red for a few seconds. This means that CATIA V5 is in the process of updating the part to match the changes made in the **Sketcher Work Bench**. When CATIA V5 is done updating the part, it will return back to its original color and the base leg will be modified from 1″ to 2″ as shown in Figure 2.33.

Figure 2.33

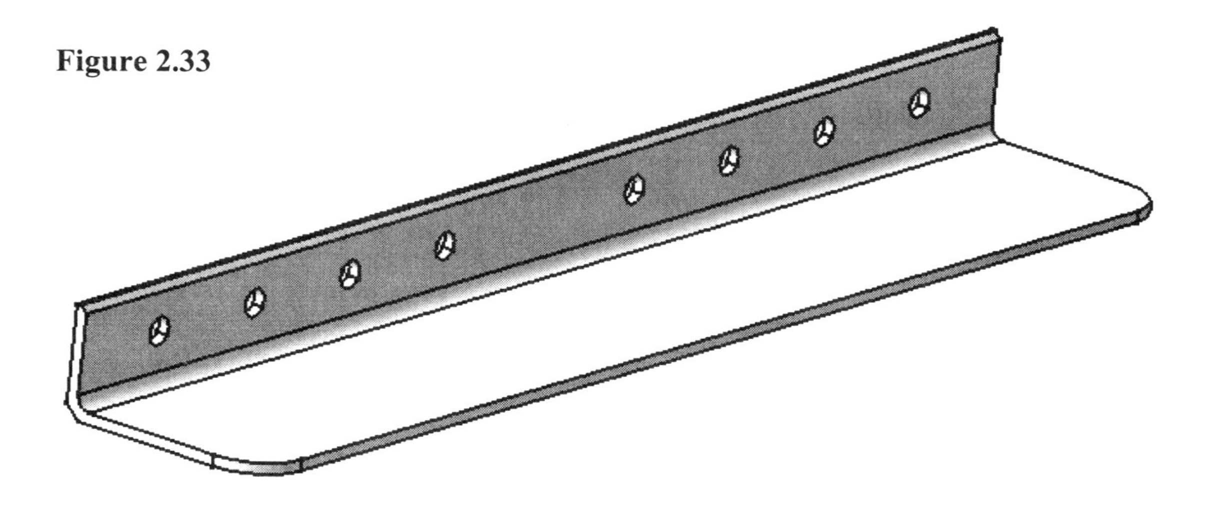

11.9 The purpose of the previous steps was to show you how easy it is to modify a part. Now, after all that work, you are going to find out how easy it is to **Undo** all that work; the work completed in Step 11. There are two ways to accomplish this. The first method is to select the **Undo** button, one step at a time, until your "**L Shaped Extrusion**" base leg is back to 1″. If your part remains red after the **Undo** command, you will have to select the **Update** tool. Selecting the **Update** tool forces CATIA V5 to update the modification. The part should turn to its normal color and the base leg should go back to 1″. The second method is to go back into the **Sketcher Work Bench** and change the constraint back to 1″, then **Exit** back to the **Part Design Work Bench**.

NOTE: The **Update** tool is located at the bottom of the screen. If it is not visible, you will have to pull it out on the screen so it is selectible. For future reference, any time the **Update** tool is highlighted, CATIA V5 is signalling that an update is required. All you have to do is select the **Update** tool and CATIA V5 will update the part.

NOTE: The steps for modifying the width of the base leg of the "**L Shaped Extrusion**" can be applied to all of the other geometrical properties of the part such as radii, chamfers and holes (any constraint value). The values can be modified by double clicking on the **Pad.1** branch in the **Specification Tree**. This makes all **Constraints** visible and pickable. You can double click on a constraint which will bring up the **Constraint Definition** window for that particular constraint. It is in the **Constraint Definition** window that you can modify the particular constraint value. Double clicking on the "**L Shaped Extrusion**" part will also make all of the constraints visible for selecting and modifying.

When you get into parts with multiple sketches, you will need to specify the particular plane the sketch was created in. For example, the values specified for the profile sketch of the "**L Shaped Extrusion**" were created in the **Sketcher Work Bench** in the **ZX Plane**. To modify those values, you would need to specify the same plane, the **ZX Plane**. You could not modify the length of the "**L Shaped Extrusion**" in the **Sketcher Work Bench**. The length was created with the **Pad** tool in the **Part Design Work Bench**.

12 Translating The "L Shaped Extrusion" 

Translating the part is simply moving the object from one place to another. To demonstrate this command, you will move your "**L Shaped Extrusion**" forward five inches. The following steps describe this process.

12.1 Select the **Translation** tool from the **Transformation Features** tool bar located on the right side of the screen. If this tool is not visible, you may need to drag it up from the bottom right corner and into view. If you are not sure how to drag and drop a hidden tool bar, refer back to Step 10.4.

12.2 The **Translate Definition** window allows you to specify what entity to translate and how far to translate it. Reference Figure 2.34.

Figure 2.34

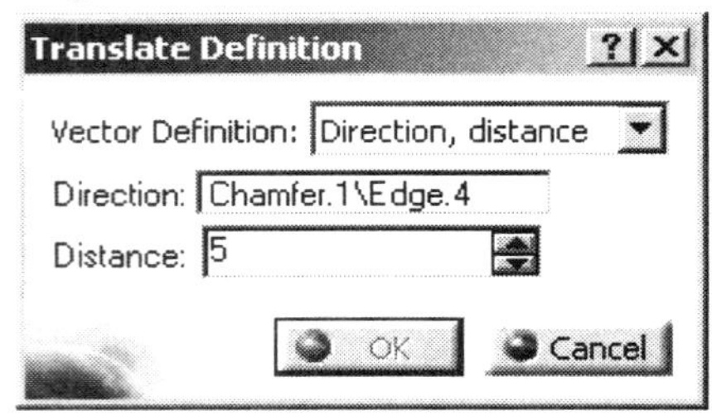

12.3 If no translation **Direction** has been specified, the defalt value will be "**No Selection.**" In order to select a direction for the translation, you must select the the **Direction** box. To do this, select in the middle of the box. It will highlight blue meaning it has been selected.

12.4 Select the bottom edge of the base leg of the "**L Shaped Extrusion**" as shown in Figure 2.35. Notice that the **Direction** box now shows the name of the line selected. The part has also been highlighted meaning that it is the entity that will be translated. Two arrows also appear in the middle of the part pointing in the positive and negative directions of the translation.

Figure 2.35

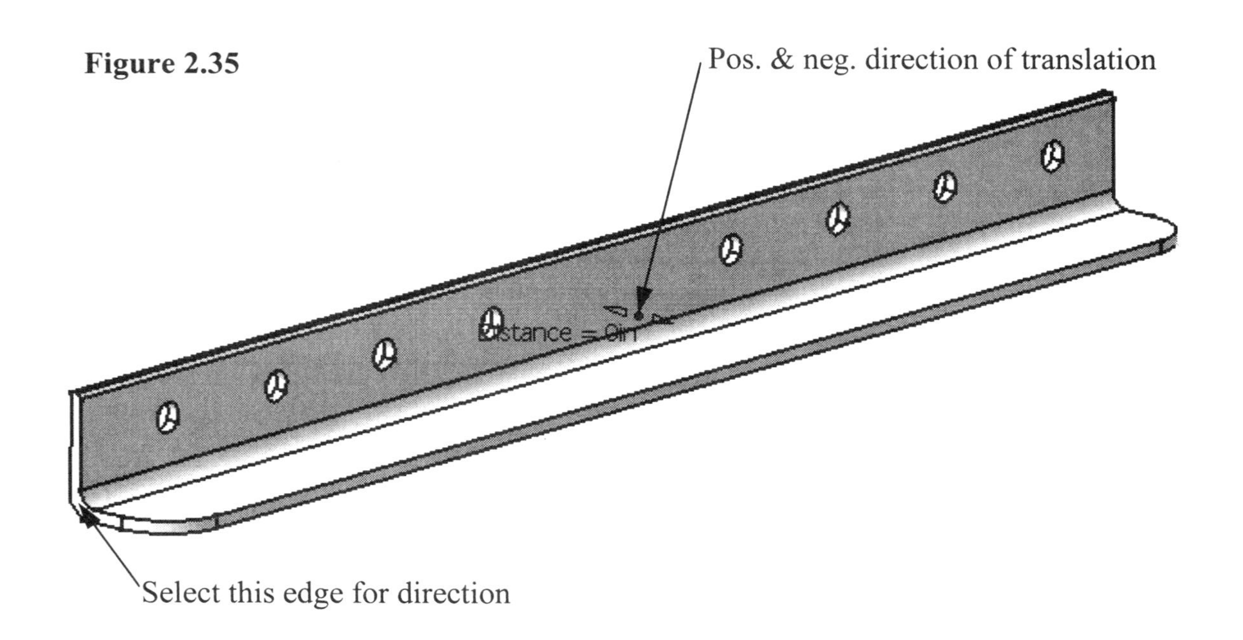

12.5 Now enter **5** in the **Distance** box as shown in Figure 2.34.

12.6 Select the **OK** button. The part will be moved five inches forward. To verify that the part was really translated, double click on the **Translate.1** branch that has just been added to the **Specification Tree**. CATIA V5 will show you where the part was and where it is now, after the translation. The **Translate Definition** window will also appear. This will allow you to modify the selected translation.

12.7 Select the **Cancel** button and the transformation will not be modifed.

13 Rotating The "L Shaped Extrusion"

Figure 2.36

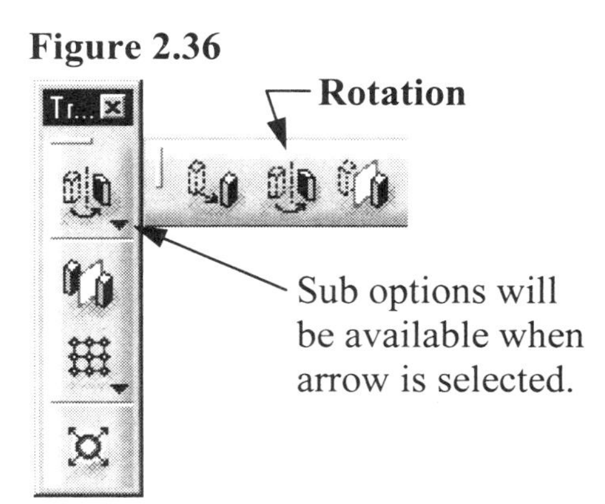

When creating a three-dimensional object, it is very important that you are able to rotate the object to be able to see the side you want to work on. In CATIA V5 there are several ways to rotate the part which will be be discussed below.

13.1 The first way to rotate a part is using the **Rotation** tool from the **Transformation Features** tool bar. As explained previously, this tool bar may be hidden on the bottom right corner of the screen. You may need to drag it up into view. If the **Rotation** tool is still not visible after pulling up the **Transformation Features** tool bar, select the arrow in the bottom corner of the **Translatoin** tool. It will be one of the sub-options available (Figure 2.36). To use this command, you will need to select the object to be rotated and specify an axis of rotation and the degree the object is to be rotated. The steps below will demonstrate this.

Figure 2.37

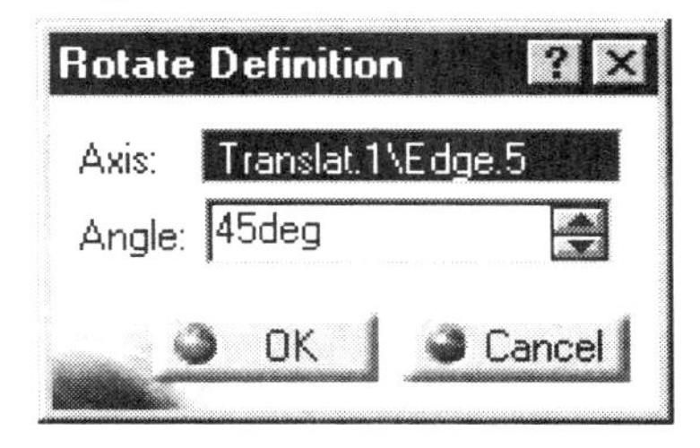

13.1.1 Select the **Rotation** tool from the **Transformation Features** tool bar. The **Rotate Definition** window will appear on the screen as shown in Figure 2.37.

13.1.2 The first option in the window is the **Axis** Specification box. CATIA V5 will only accept a straight line to define the axis of rotation. Select the left edge of the standing leg of the "**L Shaped Extrusion**" as shown in Figure 2.38. The "**L Shaped Extrusion"** will highlight indcating that it is the entity to be rotated.

Figure 2.38

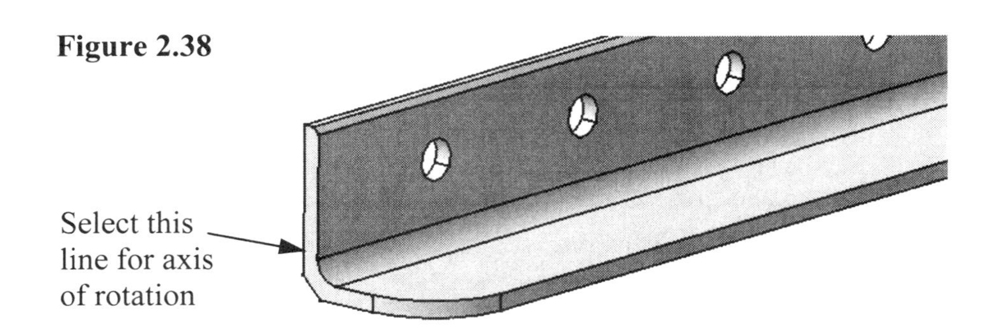

13.1.3 Now, all that is required is an angle entered in the **Angle** box to tell CATIA V5 how many degrees to rotate the part. Enter in **45** degrees and select the **OK** button.

13.1.4 The part has now been rotated 45 degrees around the line that was selected as the axis.

NOTE: The **Rotate.1** will now appear in the **Specification Tree**. If you need to modify the angle, double click **Rotate.1** in the **Specification Tree**. The pop-up window will appear allowing you to modify the values previously specified.

NOTE: This rotation method changes the orientation of the part in relationship to the 0,0,0 axis. The part moves in relationship to the axis not to you the viewer.

13.2 The next way to rotate a part in CATIA V5 is to use the **Rotate** tool found at the bottom of the screen. This tool does not move the part in relationship to the axis, but moves the viewers perspective of the part. The orientation of the geometry, with reference to the axis or with any other entity, is not effected. Since this does not change the geometry at all, there is no change made to the **Specification Tree**. This same rotation can be accomplished with the mouse only. Push the middle mouse button. Continue to hold the middle mouse button down and push down the left mouse button. This sequence of mouse buttons will bring up the same rotation sphere. For detailed instructions on this process, reference the Introduction and Step 7.1 in this lesson.

13.2.1 To use this command, select the **Rotate** tool. The tool will highlight red meaning it has been selected.

13.2.2 Select any point on the screen and hold down the left mouse button.

13.2.3 The rotation sphere will appear and show the axis of rotation (Figure 2.8).

13.2.4 When you have rotated your point of view to where you want it, release the left mouse button and the rotation will be terminated.

13.3 One other way to rotate a part is using the **Compass** located in the upper right corner of the screen as shown in Figure 2.39. The **Compass** can be used to perform many different tasks besides rotation. Figure 2.39 shows the names of the different parts of the **Compass**.

Figure 2.39

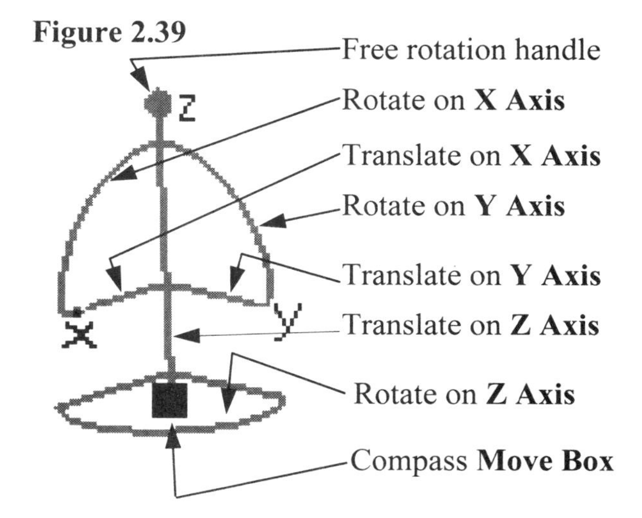

13.3.1 One way the **Compass** can be used to rotate a part, is to hold the left mouse botton down on the free rotation handle and move it around. You will notice that as the **Compass** pivots, the part will do the same thing. This type of rotation is much the same as the rotation process explained in Step 13.2.

13.3.2 Figure 2.39 shows the different components of the **Compass.** Each component allows your perspective of the part to be rotated or translated. You might want to play around with these different rotations and translations to see exactly how the movements change your perspective of the part.

NOTE: The **Rotate** tool and **Space Ball** methods do not change the orientation of the part in relationship to the 0,0,0 axis. As you rotate and/or translate the part, notice that the planes representing the 0,0,0 locaton change with the part.

13.3.3 Another way of using the **Compass** to rotate a part, is by dragging the compass and placing it on the object you want to rotate.

13.3.3.1 Drag the **Compass** over to the "**L Shaped Extrusion**" by holding the left mouse button down on the **Move Box** located at the base of the **Compass,** reference Figure 2.39, and place it on the base leg of the "**L Shaped Extrusion**" as shown in Figure 2.40.

13.3.3.2 Select (highlight) the "**L Shaped Extrusion**" **PartBody** in the **Specification Tree**. This will assign the compass to the part for rotate operations. The **Compass** will turn green (if it wasn't already). Once the **Compass** is green, any rotations and/or translations you make will move the part in relationship to the 0,0,0 axis.

Figure 2.40

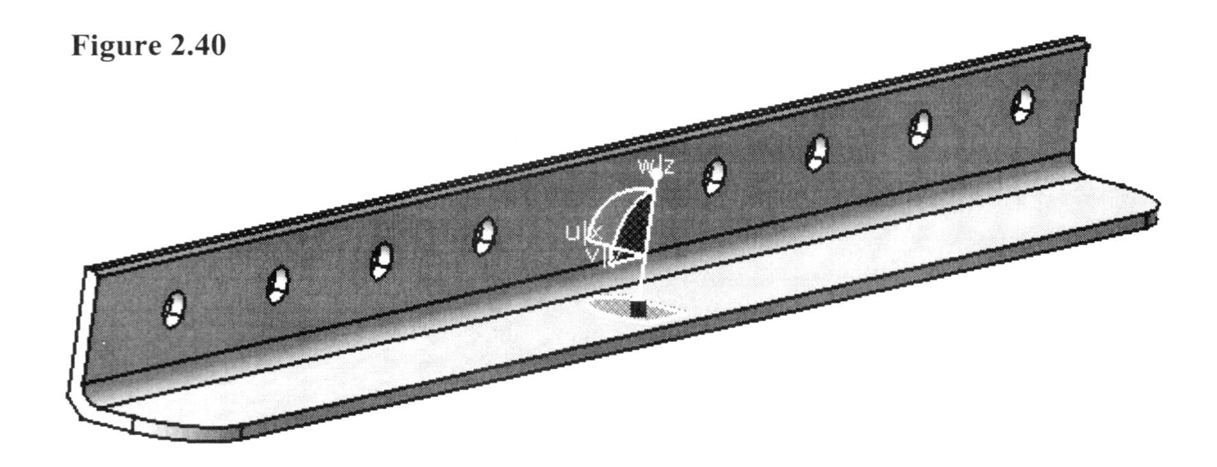

13.3.3.3 At this point there are two ways to manipulate the part. One is to select any of the translation or rotation entities on the **Compass** as shown in Figure 2.39 and explained in Steps 13.3.1. This will translate or rotate the part accordingly. The other option is to click the right mouse button with it on the **Compass.** Select **Edit** at the bottom of the window. This will bring up the **Compass Manipulation** window as shown in Figure 2.41.

Figure 2.41

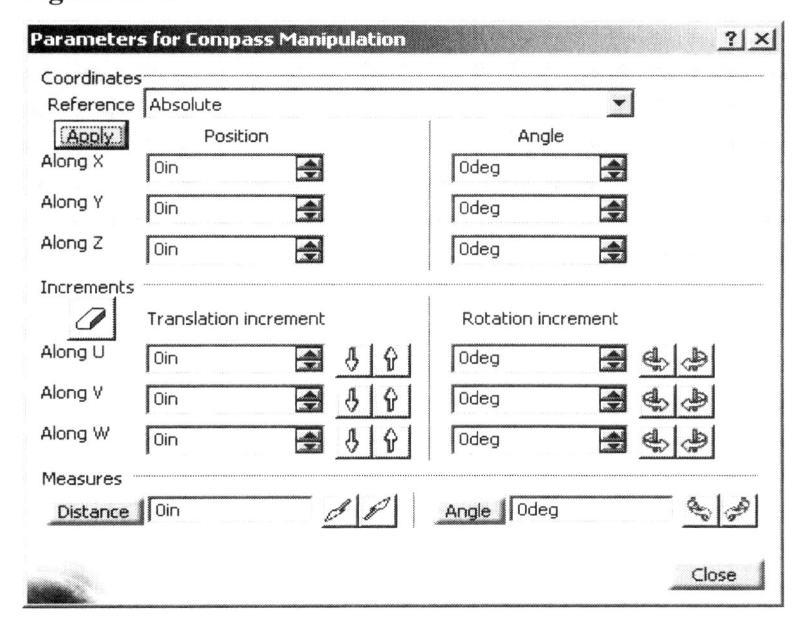

13.3.3.4 The first option is **Coordinates**. This refers to the position of the compass with reference to the (0,0,0) point. Remember, the (0,0,0) is where the **XY**, **YZ**, and **ZX Planes** intersect in the center of the screen. The values that show in these windows reflect the current location of the part. Enter a zero in the **X**, **Y**, and **Z** windows for the **Coordinates** and a zero in the **X**, **Y**, and **Z** windows for the **Angles** then select the **Apply** button. This will move and rotate the **Compass** and the part to the (0,0,0) location as specified.

13.3.3.5 You may receive a **Move Warning** stating that "some sketchs are based on reference planes: Sketch.1 Do you want to make them free?" You will need to choose **OK** for the move to take place. This means that your **Sketch.1** is no longer on the ZX plane where it was originally created. This will not, however, interfere with the **Sketcher Work Bench** representation of **Sketch.1** should you double click on it.

NOTE: This rotation method changes the orientation of the part in relationship to the 0,0,0 axis.

NOTE: Step 13.3.3.4 gives you access to many more **Compass Manipulation** tools. Be aware of the options available to you for future reference.

14 Creating A Symmetrical "L Shaped Extrusion" 

This command is used to transform the part symmetrically from one side of a plane or surface to another. The following steps will take you through this process.

14.1 Select the **Symmetry** tool from the tool bars at the right of the screen. It is one of the sub options found when the arrow is selected on the **Transformation Features** tool bar. Reference Figure 2.36 to see where this tool is located in the tool bar.

14.2 The **Symmetry Definition** window will appear on the screen as shown in Figure 2.42.

Figure 2.42

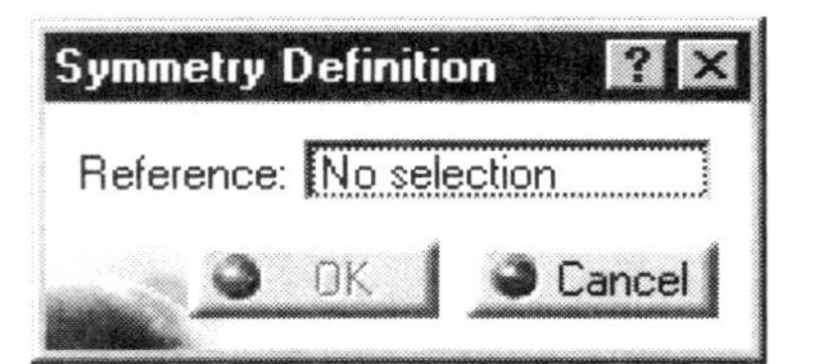

14.3 The **Reference** box is prompting for a plane or surface to be selected. For this lesson, select the left side surface of the "**L Shaped Extrusion**" as shown in Figure 2.43.

Figure 2.43

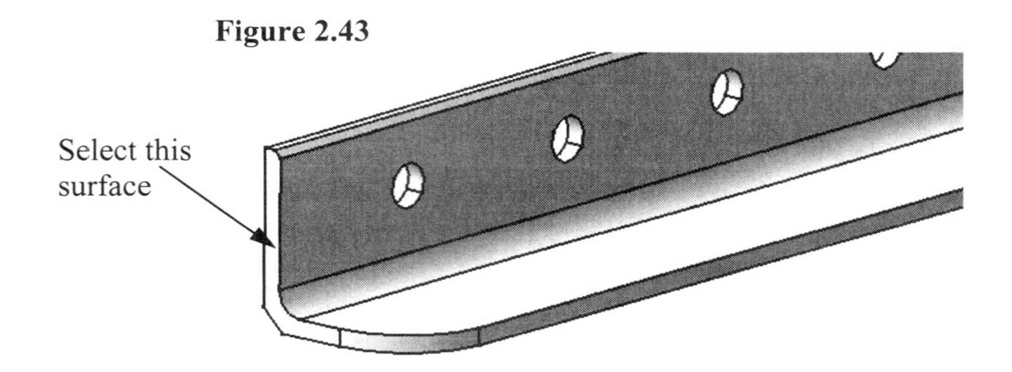

14.4 The outline of the new part will be highlighted in orange to show where the symmetrical part will be created. If your screen looks similar to the one in Figure 2.44, select the **OK** button. The "**L Shaped Extrusion**" should now be replaced with a symmetrical part on the other side of the selected surface. Notice that the **Specification Tree** now has a new branch called **Symmetry.1**.

NOTE: **Symmetry** creates a new part that is symmetrical to the original. Once the **Symmetry** is accepted, the orignal part no longer exists. If you want to keep both the original part and the symmetrical part, you need to use the **Mirror** tool, which is covered in the next step.

Figure 2.44

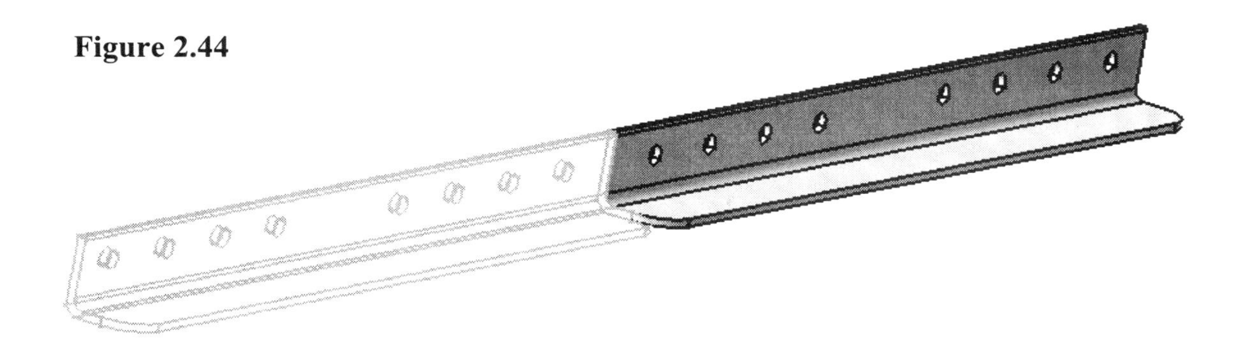

15 Mirroring The "L Shaped Extrusion"

The **Mirror** tool is just like the **Symmetry** tool except when the part is mirrored, the original part will not dissappear. It will be joined to the new mirrored part to make one complete part. The steps below will demonstrate how to mirror the "**L Shaped Extrusion**." The finished product will look more like a "**T Shaped Extrusion**."

15.1 Select the bottom side of the base leg of the "**L Shaped Extrusion**" as shown in Figure 2.45. The entire outline of the surface should highlight in red. If only one line is highlighted, you haven't selected the surface. Reselect the surface until the entire profile is highlighted.

Figure 2.45

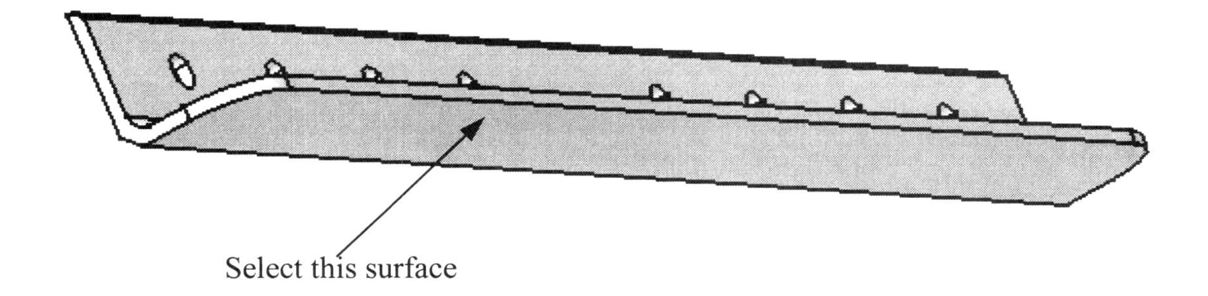

15.2 Select the **Mirror** tool from the right hand side of the screen. The **Mirror Definition** window will appear on the screen. Notice, that in the **Mirroring Element** box, it shows that **Face.3** has been selected. This should be the surface you selected in Step 15.1.

15.3 Select the **OK** button and the part will be mirrored around the bottom surface of the base leg of the "**L Shaped Extrusion**" as shown in Figure 2.46. As with the other tools, a **Mirror.1** branch is added to the **Specification Tree**.

15.4 Mirroring the "**L Shaped Extrusion**" changed the part shape from an **L** to a **T**. Save the **CATPart** as "**T Shaped Extrusion.CATPart**" using the **Save As** command. Now you will have two separate **CATParts**, the "**L Shaped Extrusion**" and the "**T Shaped Extrusion**." **The remainder of this lesson, Lesson 3 and Lesson 4 will reference the "T Shaped Extrusion."**

15.5 The current "**T Shaped Extrusion**" is not the standard mill T extrusion. To make the extrusion represent a standard mill T extrusion remove the chamfer created in Step 8. Select **Chamfer.1** from the **Specification Tree** and press the **Delete** key or select the **Cut** tool. The orginal part along with mirrored part will update with the chamfer removed as shown in Figure 2.47.

Figure 2.46

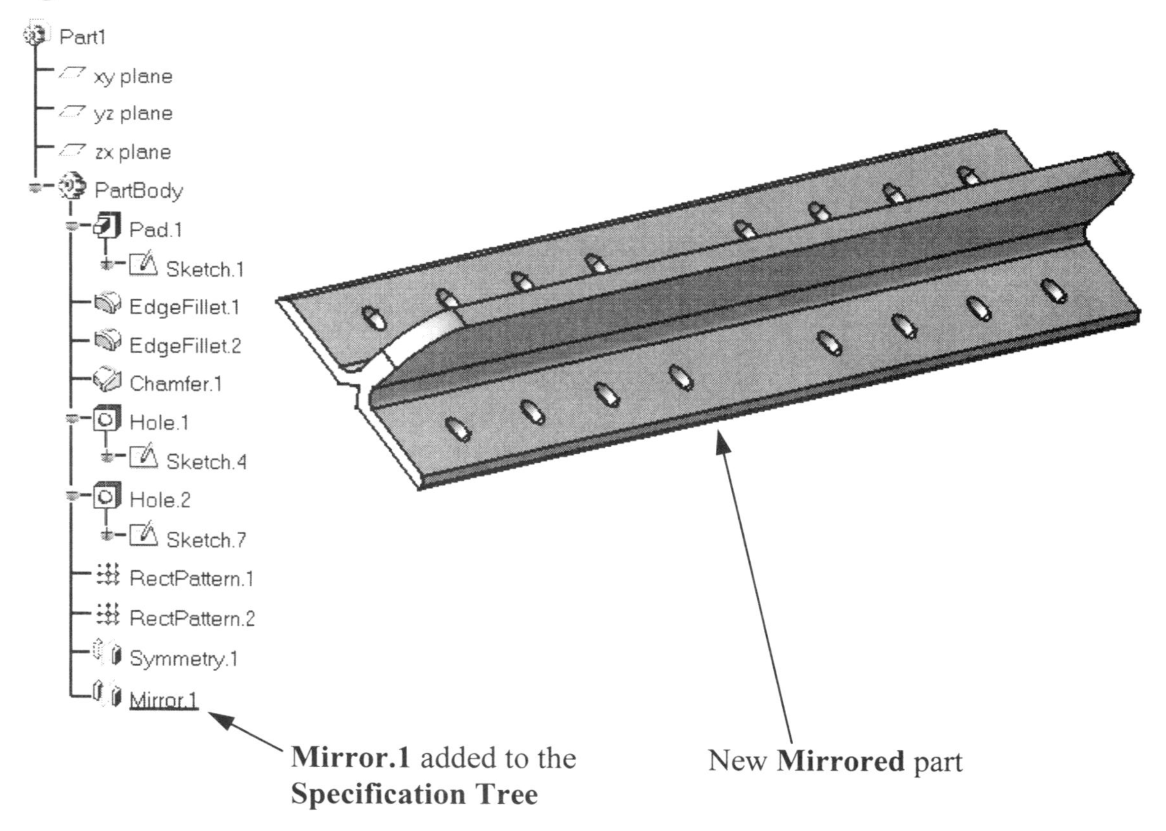

NOTE: The mirrored entity is still part of the same **PartBody**. This means that there is still just one part on the screen (Part1).

16 Scaling The "T Shaped Extrusion"

The **Scaling** tool in CATIA V5 will make an object larger or smaller by a specified scaling factor. A scaling factor greater than 1 will make the part larger and a scaling factor less than 1 will make the part smaller. CATIA V5 does not recognize a negative number as a scaling factor.

CATIA V5 has two methods for scaling a part. The first method is by selecting a point. The second method is by selecting a face or plane. **Scaling** a part by selecting a point will scale the part equally around that point. **Scaling** a part by selecting a face or plane will only scale the parts of the solid that do not lie on the plane or run parallel to the plane. The following steps will show you how to scale the new "**T Shaped Extrusion**" by selecting a point. **Scaling** a part by selecting a plane will not be demonstrated, but it is recommended that you take some time to try it. To scale the "**T Shaped Extrusion**," complete the following steps.

16.1 Select the point at the top of the standing leg where the radius ends as shown in Figure 2.47. Make sure that only the point is highlighted and not the line.

16.2 Select the **Scaling** tool from the **Transformation Features** tool bar.

16.3 When the **Scaling Definition** window appears on the screen as shown in Figure 2.48, enter **.5** into the **Ratio** box. You will not need to enter anything in the **Reference** box, because the point has already been selected as your reference element.

Figure 2.47

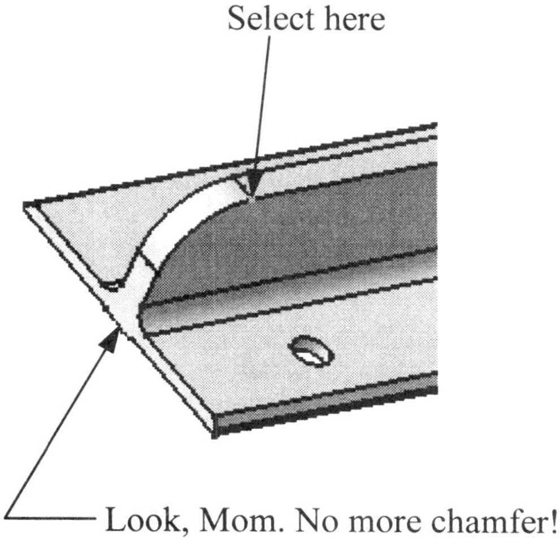

Select here

Look, Mom. No more chamfer!

Figure 2.48

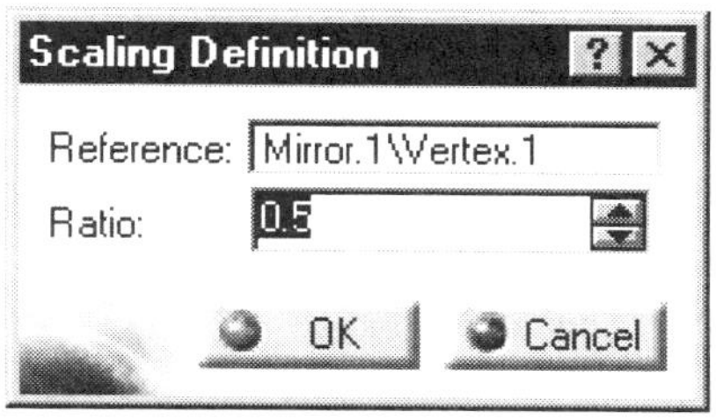

Scaling Definition

Reference: Mirror.1\Vertex.1

Ratio: 0.5

OK Cancel

16.4 Select the **OK** button and the part will be scaled to half it previous size. As in the previous processes, a new branch will be added to the **Specification Tree**. The new branch will be labled **Scaling.1** .

17 Applying Constraints In The Part Design Work Bench

In Lesson 1, **Constraints** were added to the two-dimensional sketch to control the profile of the "**L Shaped Extrusion**." The same type of **Constraints** can be used to control three-dimensional parts in the **Part Design Work Bench**. The following steps will show you how to set and modify **Constraints** in the **Part Design Work Bench**.

In Lesson 1, **Constraints** were placed on the sketch of the "**L**" profile. The **Constraints** were green in color. CATIA V5 allows you to modify the colors and appearance of the **Constraints**. The steps below show where to go to access the **Constraints** modification window.

17.1 Select the **Tools** tab from the top of the screen as shown in Figure 2.49.

17.2 In the **Tools** pull down window select **Options** as shown in Figure 2.49.

17.3 When the **Options** window appears , select the **Parameters** branch as shown in Figure 2.50.

17.4 From the far, right of the window, select the **Symbols** tab.

The **Symbols** window is divided into three boxes. The first box is the **Constraint Style** box, then the **Dimension Style** box, and the last box is the **Display at Creation** box. The following paragraphs will explain the functions of each of these three boxes. This section is intended to be for your information only. There will not be any modifications made to the **Symbols** tab. It will be up to you to experiment with the different settings and decide what works best for you.

Figure 2.49 Select here

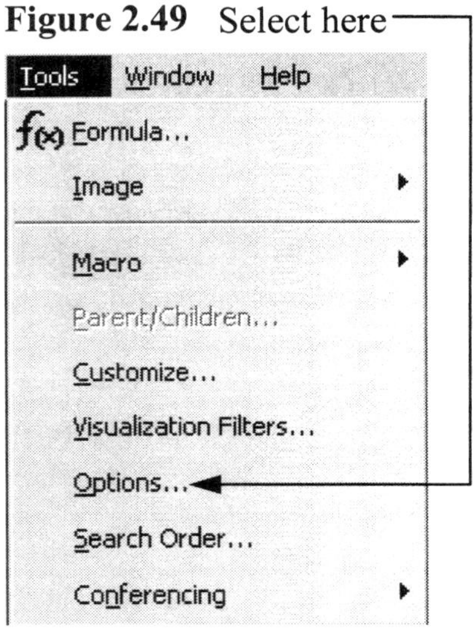

Figure 2.50

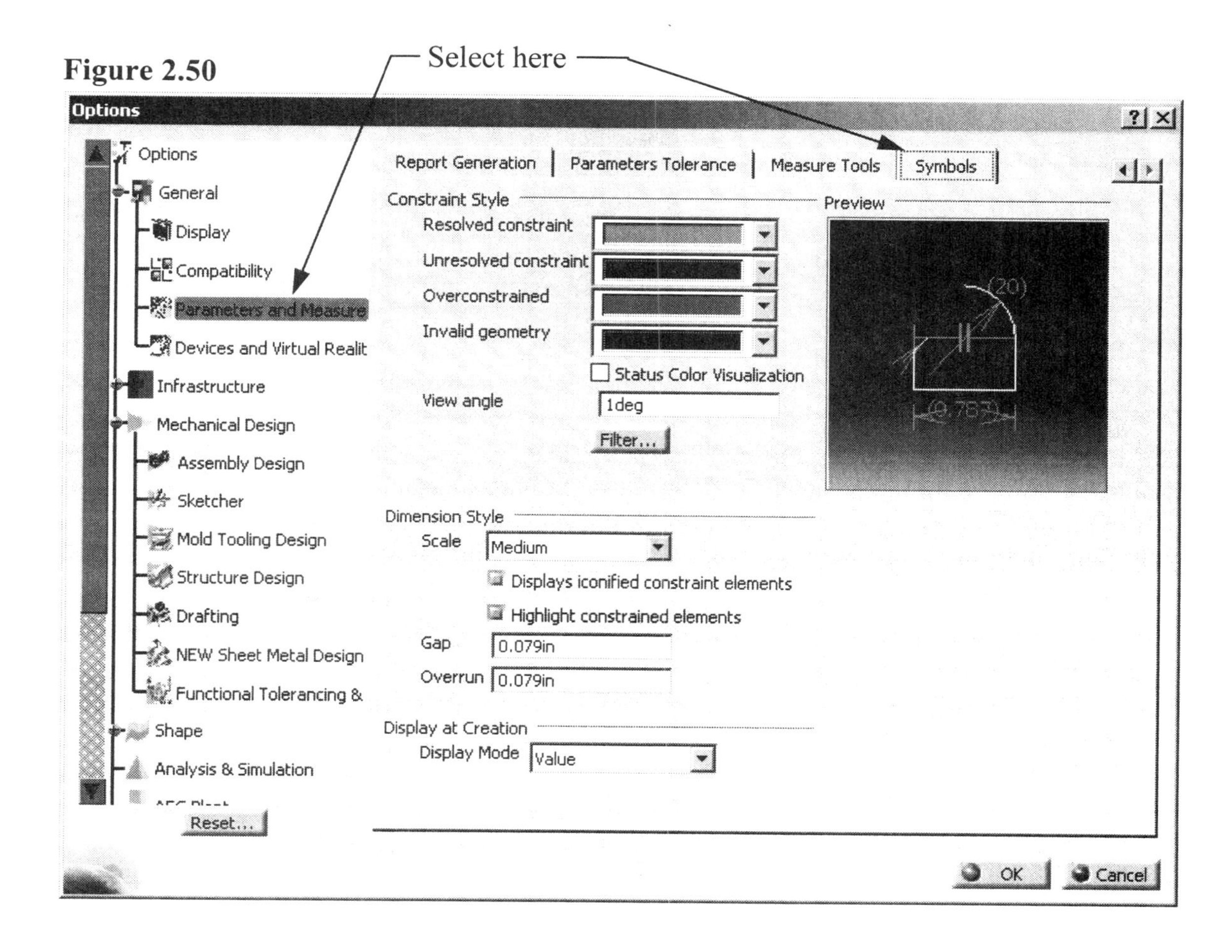

The **Constraint Style** box controls the colors of the different types of **Constraints**. These colors can be changed by selecting the down arrow to the right of the color box and selecting the color of your choice from the list provided. Selecting the **More Colors** option at the bottom of the list will give you additional color choices.

Figure 2.51

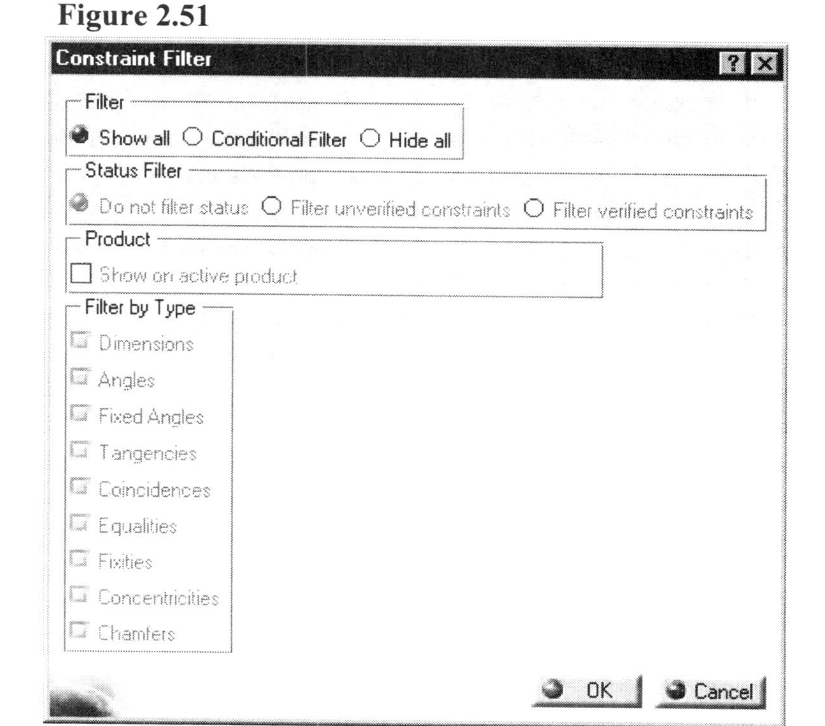

There is also a **Filter** button in the **Constraint Style** box. Selecting the **Filter** button will bring up the **Constraint Filter** box as shown in Figure 2.51. The **Constraint Filter** allows you to specify which **Constraints** you want to appear on the screen and which **Constraints** you don't want to appear.

The **Dimension Style** box controls the actual geometry of the **Constraint**. The **Scale** option allows you to specify the size of the **Constraint.** The **Gap** option allows you to alter the distance between the part and the extension line of the dimension. The **OverRun** option allows you to specify how much of the extension line of the dimension sticks out past the dimension.

The **Display at Creation** provides a visual picture of how the different **Constraints** will look with the current settings.

17.5 Select the **Ok** button to return to the **Part Design Workbench**.

17.6 The first **Constraint** to be placed will be one that defines the height of the standing leg. To place this **Constraint,** select the **Constraint** tool from the right side of the screen. It may be hidden at the bottom of the screen.

17.7 Rotate the part until the bottom surface is visible and select it as shown in Figure 2.52.

Figure 2.52

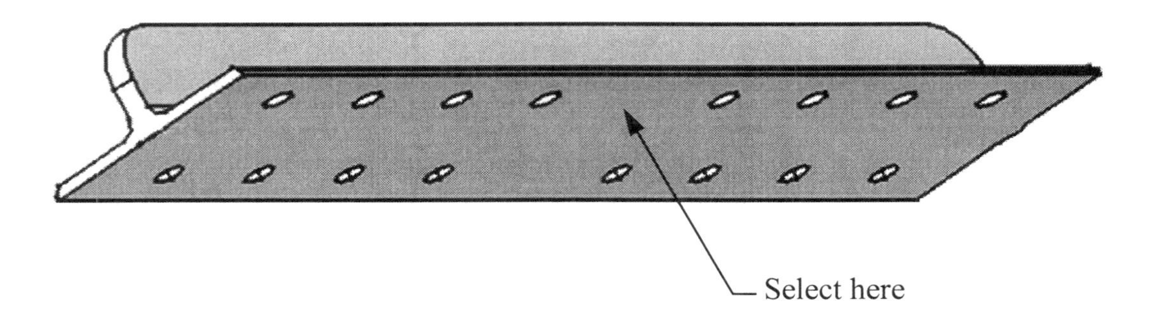

17.8 Rotate the part back until the top of the standing leg is visible and select it as shown in Figure 2.53.

NOTE: If you are having trouble selecting the entire surface (only one line will select), try zooming in closer and then selecting the surface.

Figure 2.53

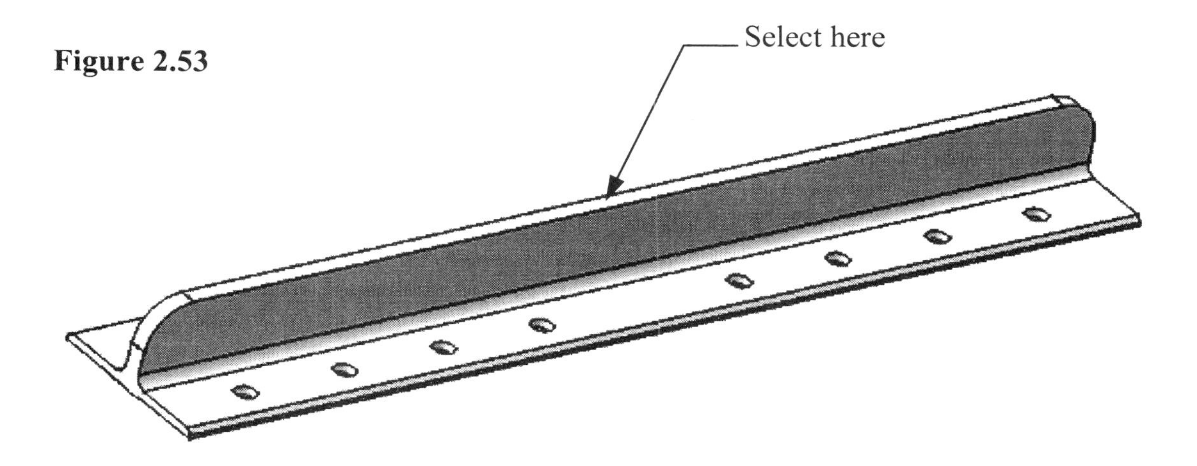

17.9 When the **Constraint** appears on the screen, drag it away from the part and place it as shown in Figure 2.54. The **Constraint** should show **0.5″** in the value box as shown in Figure 2.54.

Figure 2.54

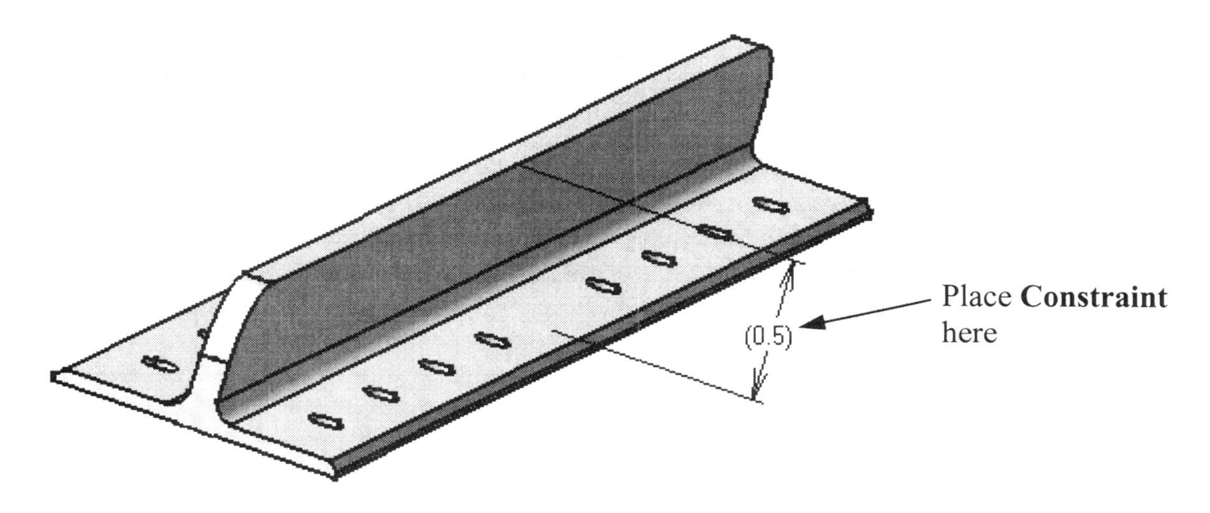

17.10 Now apply what you have learned by creating a constraint that defines the length of the "**T Shaped Extrusion**." When you are done your part should look similar to the one shown in Figure 2.55.

Figure 2.55

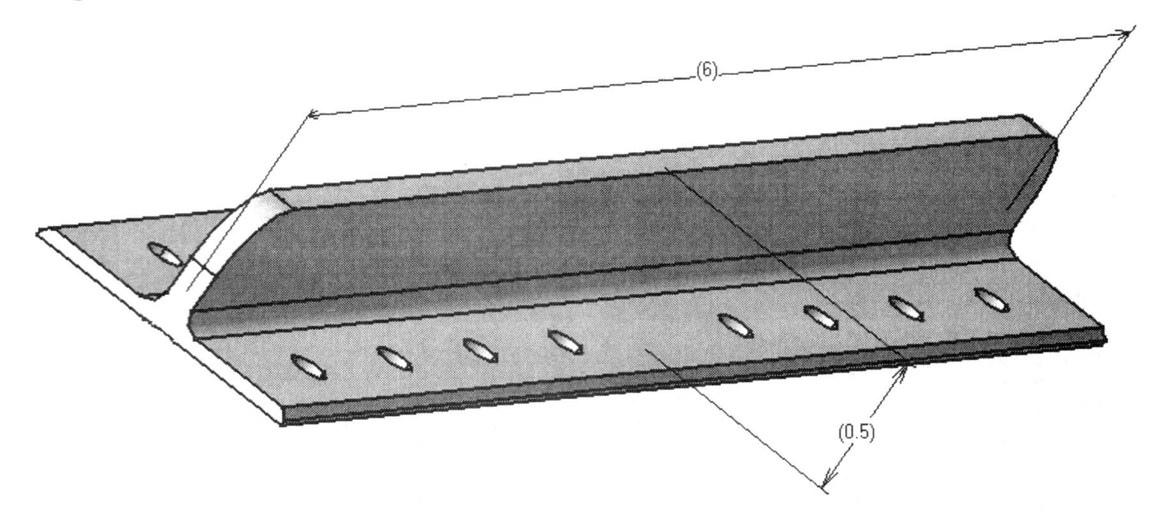

18 Applying Material To The "T Shaped Extrusion"

A special feature of CATIA V5 is the ability to add a material to the solid entity. CATIA V5 material contains the actual physical properties of the material as well as a specific visual representation. When a material has been added, the solid object takes on all of the characteristics of that material. The following steps will show you how to add material to an existing solid.

18.1 Select **Part1** from the top of the **Specification Tree** as shown in Figure 2.56. Make sure it is highlighted red indicating that it has been selected. Notice that the part will also highlight. **Part1** has to be highlighted so that CATIA V5 knows what to apply the material to.

Figure 2.56

18.2 Select the **Apply Materials** tool  from the bottom of the screen.

18.3 The **Library** window will appear as shown in Figure 2.57. The **Library** window contains all of the materials available in CATIA V5. Notice that the materials are divided into categories. Each category has a tab at the top of the window. Each tab can be selected to access all of the materials under that particular category.

18.4 Select the **Metal** tab. This will bring up the metals library.

18.5 Select the **Aluminum** material located in the upper left corner of the window. The **Aluminum** label will highlight indicating it has been selected.

18.6 Select the **OK** button located at the bottom of the window. The **Aluminum** material will be added to the "**T Shaped Extrusion**." An **Aluminum.1** branch will be added to the **Specification Tree**.

Figure 2.57

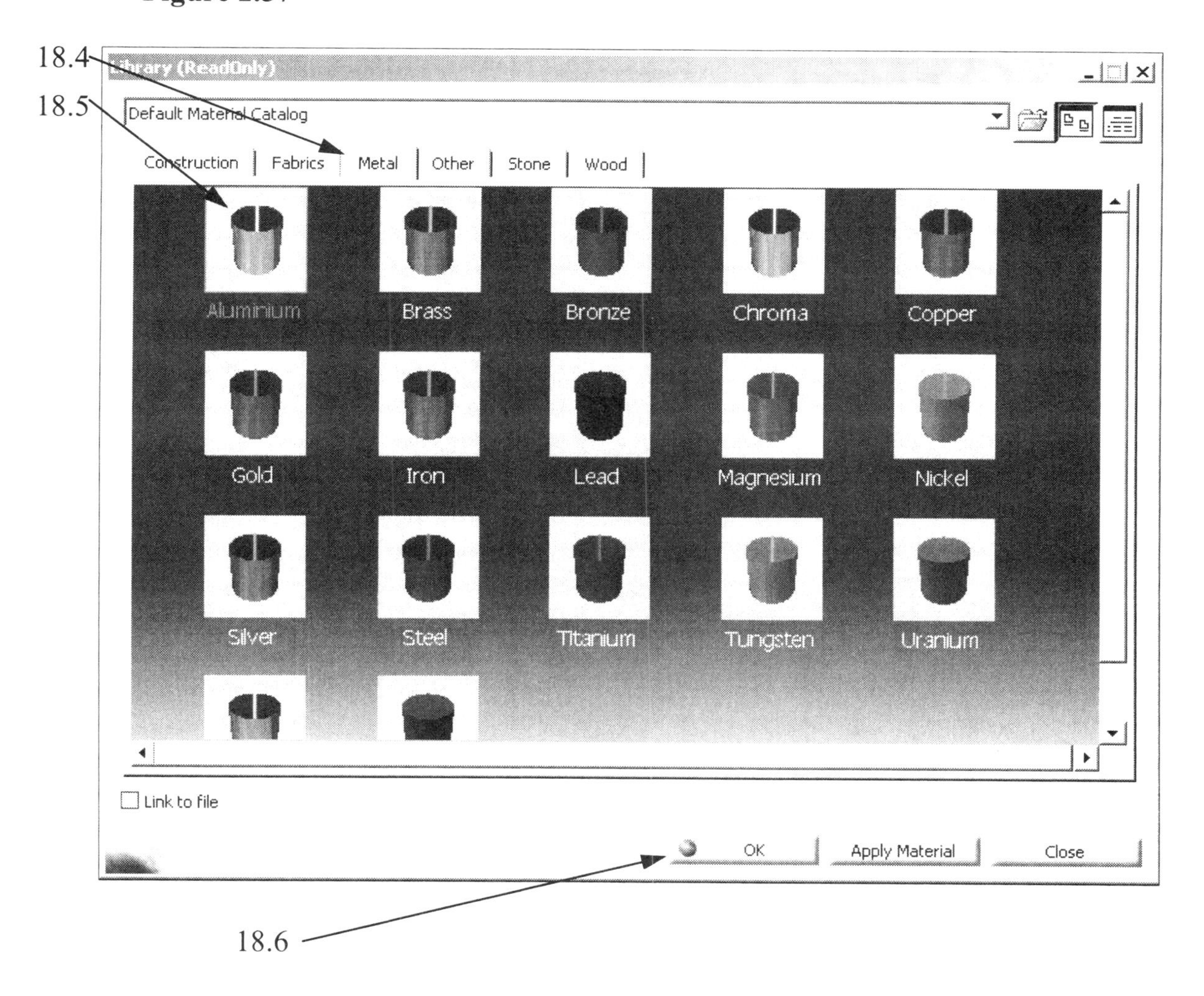

18.7 Right now the material only shows up in the tree. To view the material applied to the part, select the down arrow next to the **View Mode** tool at the bottom of the screen as shown in Figure 2.58.

18.8 When the options appear, select the **Applies Customized View Paramaters** tool [?].

18.9 This will bring up the **Custom View Modes** window. Select the **Materials** box at the bottom of the window as shown in Figure 2.59.

18.10 Select the **OK** button. The **Material** will now be applied to the **Part1** of the "**T Shaped Extrusion**." For all of the other **View Mode** options, reference the Introduction Lesson.

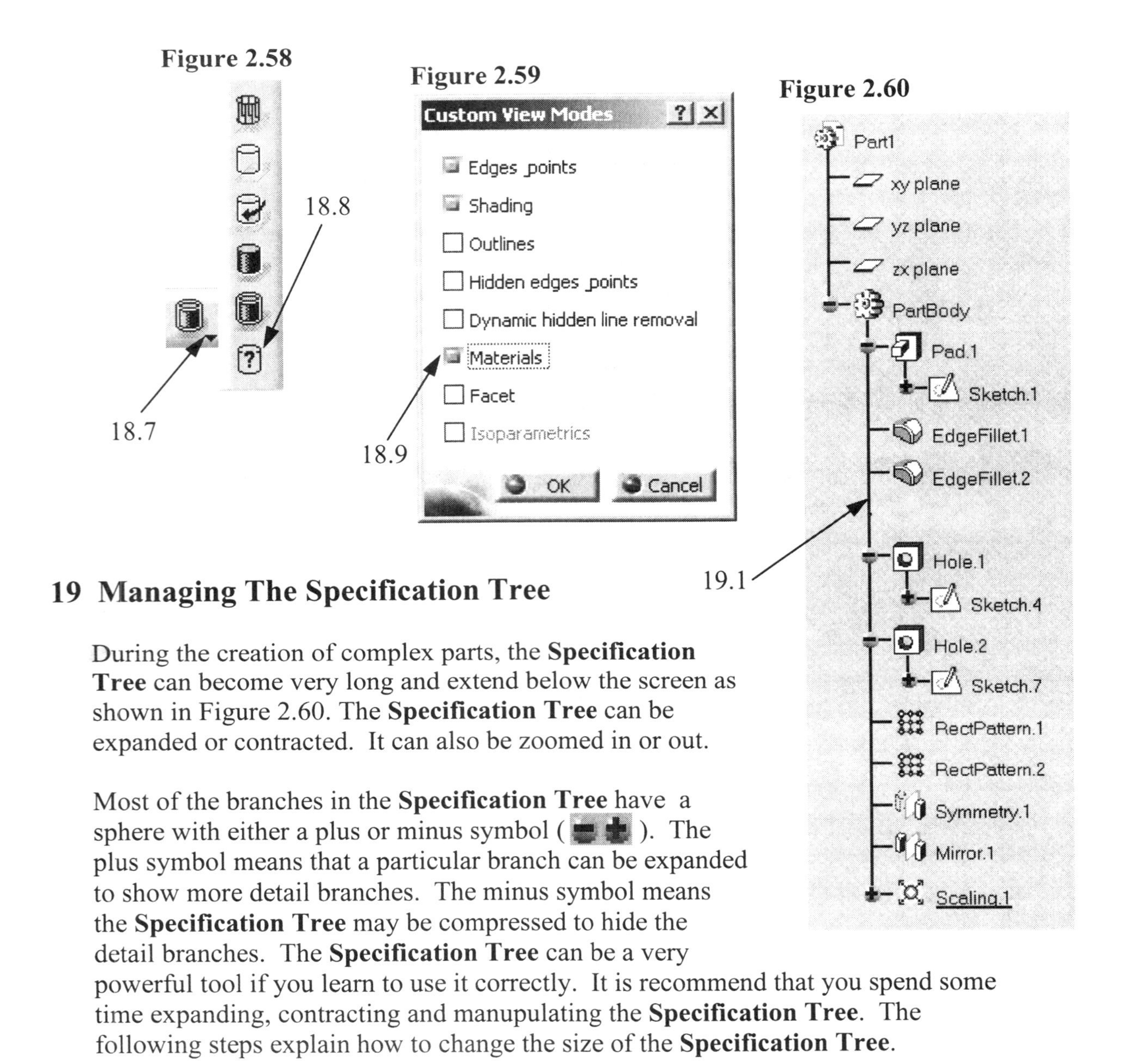

Figure 2.58

Figure 2.59

Figure 2.60

19 Managing The Specification Tree

During the creation of complex parts, the **Specification Tree** can become very long and extend below the screen as shown in Figure 2.60. The **Specification Tree** can be expanded or contracted. It can also be zoomed in or out.

Most of the branches in the **Specification Tree** have a sphere with either a plus or minus symbol ( ). The plus symbol means that a particular branch can be expanded to show more detail branches. The minus symbol means the **Specification Tree** may be compressed to hide the detail branches. The **Specification Tree** can be a very powerful tool if you learn to use it correctly. It is recommend that you spend some time expanding, contracting and manupulating the **Specification Tree**. The following steps explain how to change the size of the **Specification Tree**.

19.1 Select any one of the lines that connect the branches of the **Specification Tree** as shown in Figure 2.60. The part geometry on the screen will be dimmed to show that the **Specification Tree** is the only thing that can be manipulated.

19.2 Select the + **Magnifying Glass** tool to **Zoom** up (enlarge) the **Specification Tree**. Select the – **Magnifying Glass** tool to zoom out (shrink) the **Specification Tree**. Reference Figure 2.61 for the **Zoom** tools located on the bottom tool bar.

Figure 2.61

19.3 The **Specification Tree** can also be moved to a different location on the screen. Make sure the **Specification Tree** is still selected, the part geometry is dimmed. Select the **Move** tool located on the tool bar at the bottom of the screen or just grab a branch line, then drag and drop it where you want it located.

19.4 To unselect the **Specification Tree** so you can go back to working on the part geometry, select anywhere on the line that connects the branches of the **Specification Tree**. When the part is no longer dimmed, all of the tools will be back to normal allowing you to manipulate the part geometry instead of the **Specification Tree**.

19.5 If the **Specification Tree** extends beyond the screen length, you can use the **Move** tool or move your mouse over the tree and select the middle mouse button to activate a scrool bar for the **Specification Tree**. You can move the scroll bar with your mouse. If your middle mouse button has a roller, you can scroll up and down the **Specification Tree** with the scroll button.

20 Customizing The Specification Tree

It is easy to keep track of the **CATPart** entities for simple designs such as **Sketches**, **Pads**, and **Patterns**. But for more complex designs and assemblies, it is critical to keep all of the design information readily available. The "**T Shaped Extrusion**" required only one **Sketch** and CATIA V5 automatically named it **Sketch.1** as shown in the **Specification Tree**. Modifying a parameter and/or **Constraint** in the sketch would be easy, because it would be obvious which sketch the parameters are in. If your design had five sketches, it would be a bit more complicated. If someone else had to modify your design, how would they know what information was in which sketch? Renaming **Sketch.1** to "**L Shaped Sketch**" would help determine the information contained it that sketch. It is suggested that you name the sketch "**L Shaped Sketch**" not **T**, because the sketch profile is still in the shape of an **L**. Remember; it is the **Mirror.1** branch that completes the "**T shaped Extrusion**." CATIA V5 allows you to rename the branches of the **Specification Tree** so it has meaningful information. It is not necessary to rename every branch, use good judgment in what to rename. Renaming particular branches of the **Specification Tree** will be critical in the following lessons. Customize your **Specification Tree** by completing the following steps.

20.1 First you will rename the base of the **Specification Tree**. Select **Part1** so it is highlighted.

20.2 With your mouse over the highlighted **Part1**, push the right mouse button. This will bring up the **Edit** window.

20.3 Select the **Properties** option. Make sure you have the correct entity selected. You will change the **Properties** of the selected element only. This will bring up the **Properties** window for the **Part1** entity.

20.4 Select the **Product** tab as shown in Figure 2.62. The **Part Number** box has **Part1** in it by default. Type in "**T Shaped Extrusion**" in the place of **Part1**.

20.5 Select **OK**. Check that the base of the **Specification Tree** is now named "**T Shaped Extrusion**." This step is important, especially when you start inserting multiple **Parts** into an assembly.

Figure 2.62

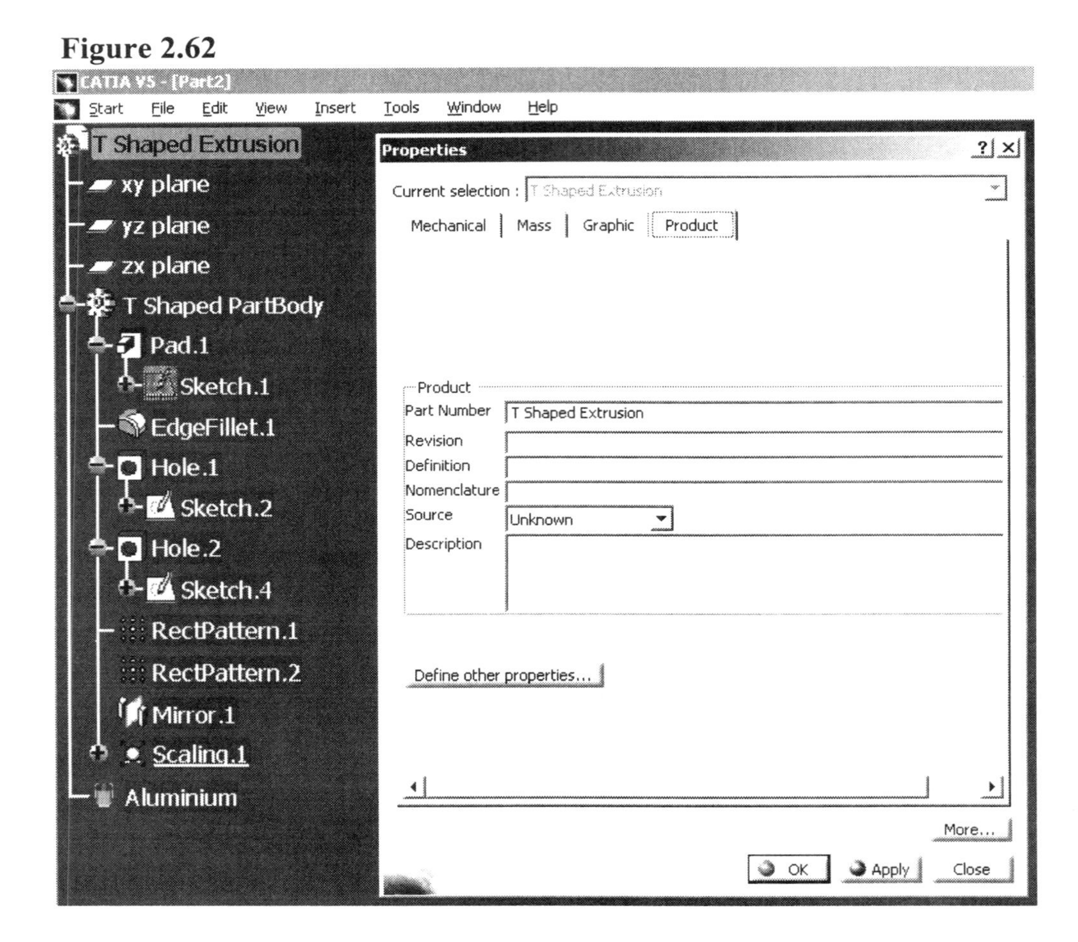

20.6 The next branch you need to rename is the **PartBody** branch. Use the same process as described in the previous steps to rename the **PartBody** branch. Rename the **PartBody** branch to "**T Shaped PartBody**." There are some differences in the **Properties** window. The **Feature Name** box is under the **Feature Properties** tab instead of the **Product** tab. The box you modify is under **Feature Name** instead of **Part Number**.

20.7 Rename **Sketch.1** to "**L Shaped Sketch**" following the same process as described in Step 20.6

If your **Specification Tree** looks similar to what is shown in Figure 2.62 you are ready for the next step. Customizing the **Specification Tree** will become more important as your designs become more complex.

21 Reviewing The Design Process Using The Specification Tree

CATIA V5 supplies you with a tool to review the processes used to create a CATPart. This tool automatically takes you through each step of the design process. The following steps will take you through a review of the "**T Shaped Extrusion**" you created.

21.1 From the **Specification Tree** select the **T Shaped PartBody** branch that you want to review.

21.2 From the top tool bar menu, select the **Edit** option. This will bring up the **Edit** pop-up window.

21.3 Select the **Scan or Define In Work Object** option located near the bottom of the window. This will bring up a **Scan** window as shown in Figure 2.63.

21.4 From the **Scan** window, select the **Forward** button. This will take you to the first step in creating the "**T Shaped Extrusion**," **Pad.1**.

21.5 Select the **Forward** button again. This will take you to the second design step where **EdgeFillet.1** was created.

21.6 Select the **Forward** button again. This will take you to the third design step where **EdgeFillet.2** was created.

21.7 Continue through the design steps by clicking the **Forward** button. This will take you through **Hole.1**, **Hole.2**, **RectPattern.1**, **RectPattern.2**, **Translate.1**, **Symmetry.1**, **Mirror.1** and **Scaling.1**.

21.8 When you get to **Scaling.1,** you will see the finished part which was the last design entity created.

21.9 You have the option of going backwards, by selecting the **Backward** button.

NOTE: If you exit out of the **Scan** window in the middle of the review process CATIA V5 will show the part as it existed at that point in the design process. In other words, it will not show the finished part. Don't panic; you haven't lost the finished part. Bring the **Scan** window back up and select the **Last Feature** button. Then select the **Exit** button.

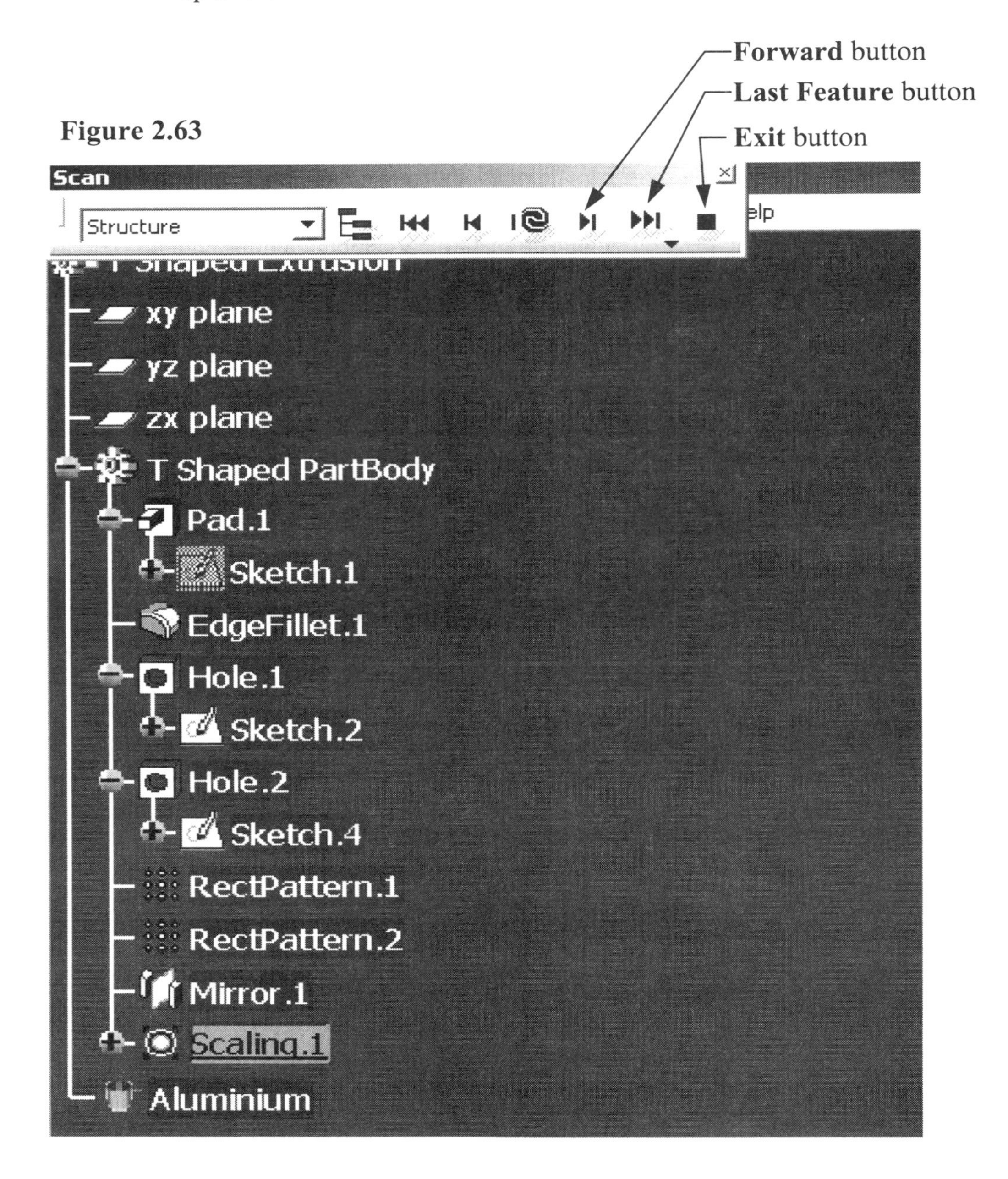

Figure 2.63

22 Save The Completed "T Shaped Extrusion" Before It Gets Away

To save a CATIA V5 work session, complete the following steps.

22.1 Verify that all operations are complete and the "**T Shaped Extrusion**" is the way you want it to be saved. For the following steps reference Figure 1.45 and 1.46 in Lesson 1.

22.2 Select **File** from the top tool bar.

22.3 Select the **Save As** option.

22.4 In the **Save In** box, select the directory you want the part to be saved to.

22.5 Type in the **File Name**; for this lesson save your part as the "**T Shaped Extrusion**." The extension "**CATPart**" will be added automatically.

22.6 If everything is the way you want it in the **Save As** window, select **Save**.

NOTE: Remember the file name and the directory you saved your file to; you will need to open it back up and use it in Lesson 3 and Lesson 4.

23 Exiting CATIA V5

To exit CATIA V5, complete the following steps.

23.1 Make sure you saved the part. If you have made any unsaved changes to the part, CATIA V5 will prompt you to save when exiting.

23.2 Select **File** from the top tool bar as shown in Figure 2.64.

23.3 Select **Exit.**

23.4 Since CATIA V5 is Windows compliant, another way to exit is to select the **Close** tool at the very top right of the screen. Both methods of exiting CATIA V5 will put you back at the **Windows NT Desktop**.

Figure 2.64

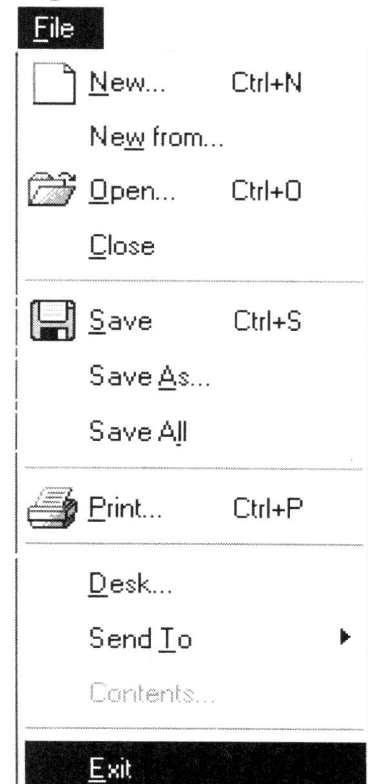

Lesson 2 Summary

The **Part Design Work Bench** gives the designer a variety of powerful, time saving tools to work with. The **Part Design Work Bench** is a natural transition from the **Sketcher Work Bench**. It is important to be able to differentiate between **Sketcher** entities and **Part Design** entities.

Lesson 2 Review

After completing this lesson you should be able to answer the questions and explain the concepts listed below.

1. List two different methods of deleting an entity.
2. What three things must be selected to create a hole that is accurately located in a part?
3. How would you modify the diameter of a hole that is already created?
4. How would you modify the location of a hole that is already created?
5. CATIA V5 gives you a choice of five different types of holes. List all five.
6. CATIA V5 gives you several methods to rotate a part (changing the parts relationship to the 0,0,0 point). List two of them.
7. **T** or **F** Applying **Material** to your part automatically updates the visual representation of the part.
8. How can the **Specification Tree** be expanded to show additional branches?
9. What key on the keyboard do you hold down to multi-select two lines at the same time?
10. The **Rectangular Pattern** tool is found in which tool bar?
11. What color (by default) does CATIA V5 shade the part when it is in the process of updating?
12. What two things must be specified in order to use the **Rotate** command from the **Transformation Features** tool bar?
13. Name the five tool bars discussed in this lesson.
14. When creating a chamfer, what are the two different mode options available in the **Chamfer Definition** box?
15. How do you make the part dim so you can manipulate the **Specification Tree**?
16. What lets you know there are hidden tool bars at the bottom right side of the screen?

17. What tool would you use to move an object from one location to another?

18. **T** or **F** **Constraints** can be used in the **Sketcher Work Bench** only.

19. Why is it important to rename the specific branches in the **Specification Tree**?

20. Explain what the **Compass** can be used for.

Lesson 2 Practice Exercises

Now that your CATIA V5 toolbox has more tools in it, put them to use on the following practice exercises. The shapes are simple and can be completed in one sketch. The dimensions represent the constraints you are to use in the **Sketcher Work Bench** prior to extruding them in the **Part Design Work Bench**. This lesson does not have any suggested steps, but there are helpful hints for each of the practice exercises.

Each exercise has a suggested name to use when saving it; this is critical, because some of these exercises are used later on in this workbook. Good Luck!

1 This first exercise will get you more familiar with the **Chamfer** and **Hole** commands. Save this part as "**Lesson 2 Exercise 1.CATPart**."

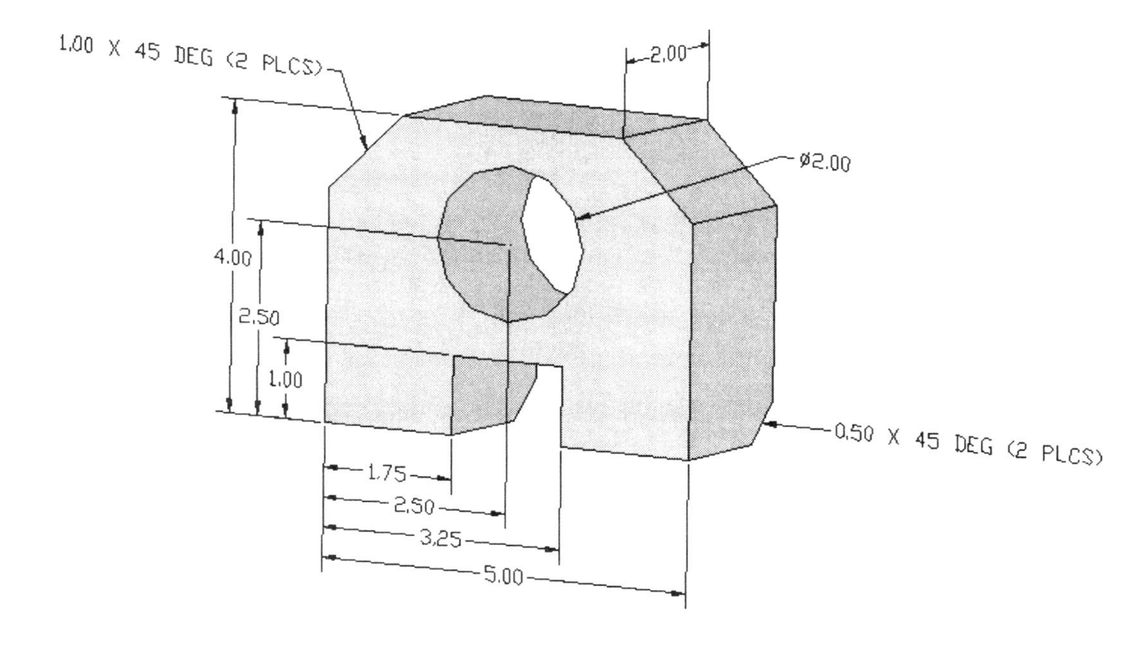

HINT: Sketch the part with square edges. Add the two big chamfers to the part after it has been extruded into a part using the **Chamfer** tool. Use the **Hole** tool to add the hole in the part.

2 This part is pretty simple. Save this part as "**Lesson 2 Exercise 2.CATPart**."

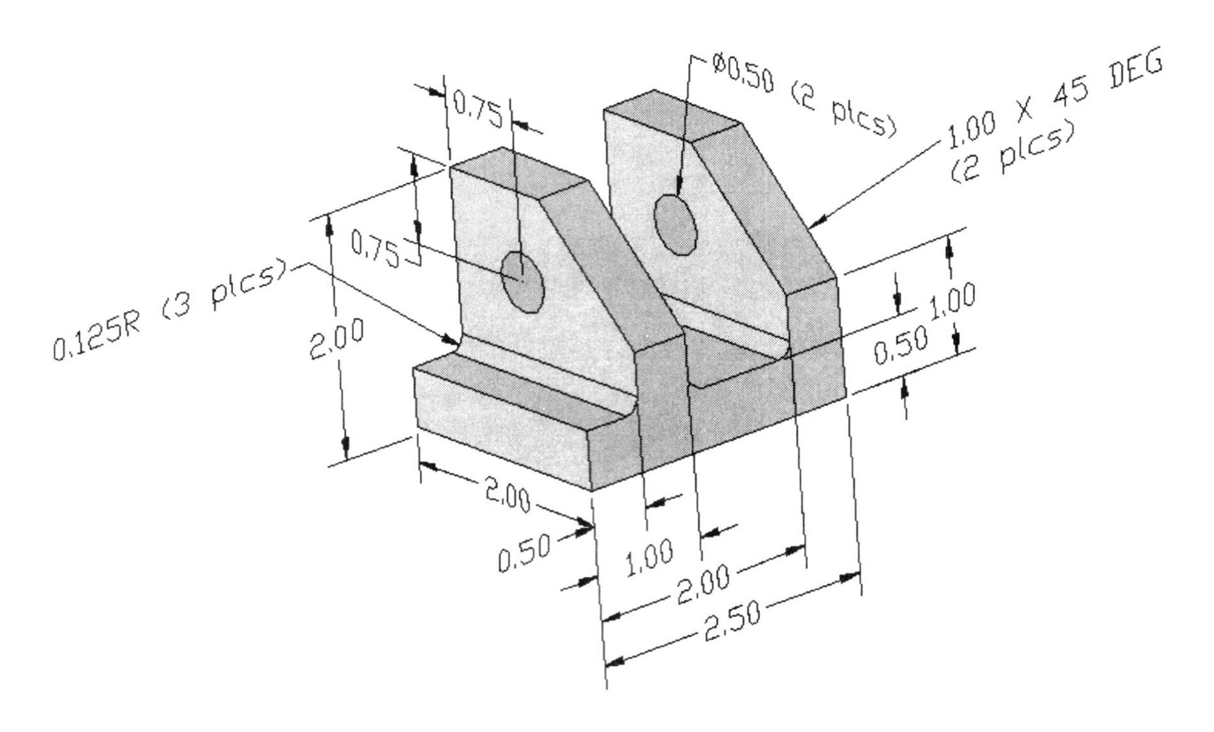

HINT: The two holes can be put in at the same time. Use the **Up To Last** option in the **Extension** tab as shown in Figure 2.19.

3 This part provides good practice with the **Fillet** and **Hole** tools. Save this part as "**Lesson 2 Exercise 3.CATPart**."

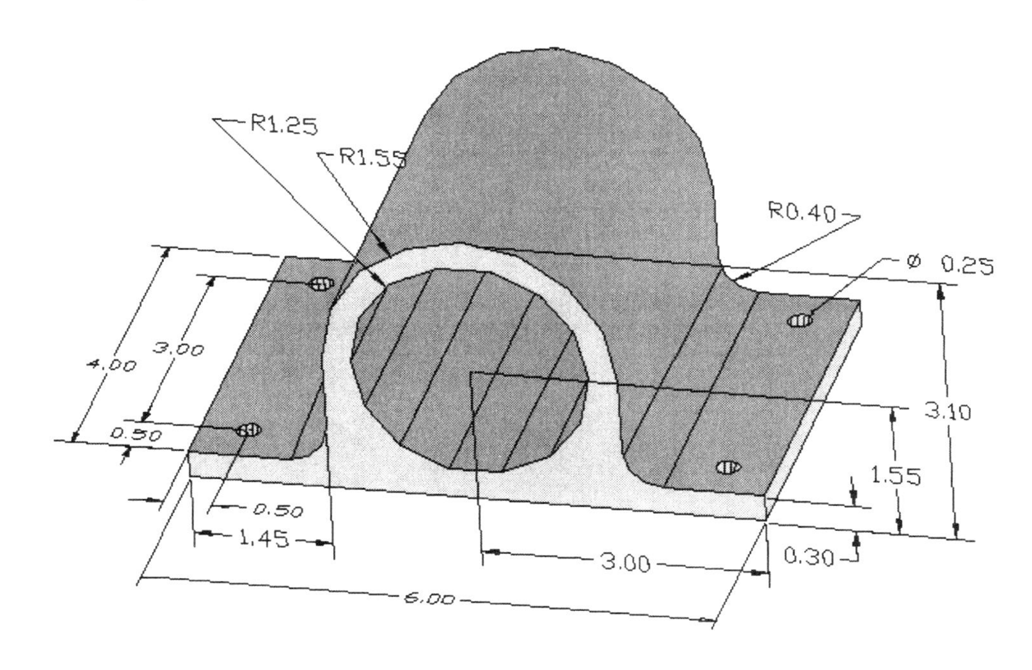

HINT: Use the **Sketcher Work Bench** to create the part using three rectangles. Add the radii in the **Part Design Work Bench** using the **Fillet** tool. Don't forget, you need to use the **Pad** tool in the **Part Design Work Bench** prior to using the **Fillet** tool. The 1.55 radius can be created using two fillets of the same radius.

4 This part will give you more practice with the **Fillet** and **Hole** tools. Save this part as "**Lesson 2 Exercise 4.CATPart**."

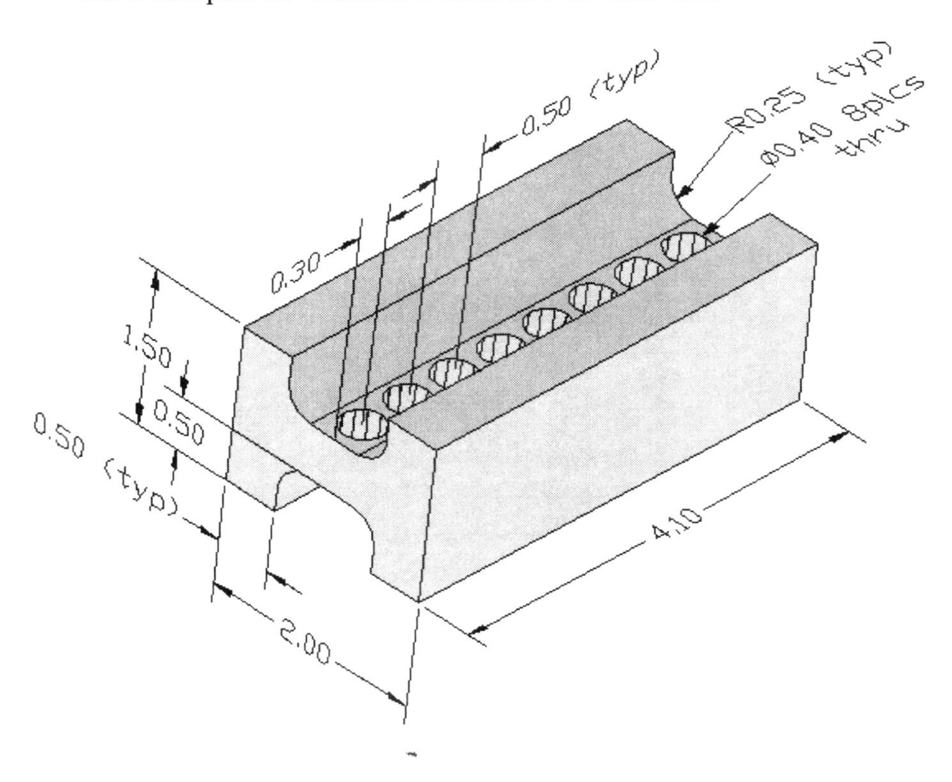

HINT: For some practice using the **Mirror** tool, create one half of the I-beam in the **Sketcher Work Bench**. Extrude it using the **Pad** tool in the **Part Design Work Bench**. Use the **Mirror** tool to create the other half. Also, create the first hole in the channel and use the **Rectangular Pattern** tool to create the rest of the holes.

Introduction To Creating Sheets And Views Using The Drafting Work Bench

In this lesson you will create a production drawing of the "**T Shaped Extrusion**" created in Lesson 1 and Lesson 2. The Lesson 3 format is the same as Lesson 1 and Lesson 2 in that it is not a comprehensive lesson on the **Drafting Work Bench**. It is not the intent of this lesson to be a comprehensive reference, but a basic step-by-step for the most common tools and functions in the CATIA V5 **Drafting Work Bench**.

Lesson 3 Objectives

This lesson will introduce you to the basics of creating sheets and views using the **Drafting Work Bench**. Lesson 4 covers the text and dimensioning found in the **Drafting Work Bench**.

This lesson covers the following areas:

- How to select the **Drafting Work Bench**
- The **Drafting Work Bench** layout
- How to customize the **Drafting Work Bench**
- How to create **Sheets**
- How to create **Views**
- How to modify an existing view
- How to move a view from one sheet to another
- How to create **Detail**, **Section** and **Auxiliary Views**
- How to **Save** the newly created sheets (drawings) "***.CATDrawing**"
- How to **Print** the newly created sheets (drawings)

Drafting Work Bench Tool Bars 

There are thirteen standard tool bars found in the **Drafting Work Bench.** Five of the thirteen tool bars are shown on the following pages; the other eight will be covered in Lesson 4. The individual tools found in each of the five tool bars are also shown and include the tool name and a brief definition. This chapter will not cover every tool found in the **Drafting Work Bench**.

The **Style** Tool Bar

Allows the user to use default drafting standards or customize the drafting standards.

The **Select** Tool Bar

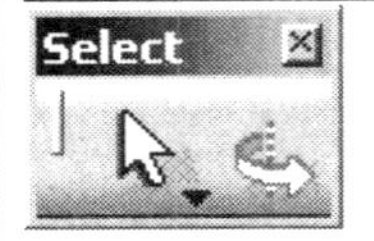

Reference menu 5 of the Introduction for complete list of tools.

Tool Bar	Tool Name	Tool Definition
	Select	Select was covered in the previous lessons.
	Free Rotation	A toggle option from the **Select** tool.

The **Drawing** Tool Bar

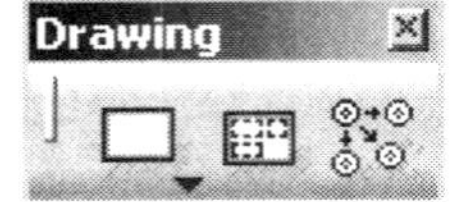

Tool Bar	Tool Name	Tool Definition
	Instantiate 2D Component	Creates a 2D component instance.
	New View	Creates a new view. The user defines the view required.
	Sheets Tool Bar	
	New Sheet	Creates a new sheet.
	New Detail Sheet	Creates a new detail sheet.

The **Views** Tool Bar

Tool Bar	Tool Name	Tool Definition
	Projections Tool Bar	
	Front View	Creates a front view.
	Unfolded View	Creates an unfolded view.
	View From 3D	Extracts a view defined in 3D.
	Projection View	Creates a projection view defined by the user.
	Auxiliary View	Creates an auxiliary view defined by the user.
	Isometric View	Creates an isometric view based on the plane the profile was created on.
	Advanced Front View	Creates a front view with advanced options.
	Sections Tool Bar	
	Offset Section View	Creates a section view defined by the user.
	Aligned Section View	Creates an aligned view defined by the user.
	Offset Section Cut	Creates an offset section cut defined by the user.
	Aligned Section Cut	Creates an aligned section cut defined by the user.
	Details Tool Bar	
	Detail View	Creates a detail view defined by a circle callout.
	Detail View Profile	Creates a detail view defined by a sketch.
	Quick Detail View	Creates a quick detail view defined by a circle.
	Quick Detail View Profile	Creates a quick detail view defined by a sketch.
	Clippings Tool Bar	
	Clipping View	Creates a clipping view defined by a circle.
	Clipping View Profile	Creates a clipping view defined by a sketch.

	Break View Tool Bar	
	Broken View	Removes a defined length of a part.
	Breakout View	Removes a defined section of the solid so internal geometry can be shown.
	Wizard Tool Bar	
	View Creation Wizard	A wizard that helps the user define the views required on the sheet.
	Front, Top and Left	Creates a view configuration based on front, top and left using the current projection standard.
	Front, Bottom and Right	Creates a view configuration based on front, bottom and right using the current projection standard.
	All Views	Creates a view configuration based on all projected views using the current projection standard.

The **Dress Up** Tool Bar

Tool Bar	Tool Name	Tool Definition
	Area Fill	Creates a fill area customized by the user.
	Axis and Threads Tool Bar	
	Center Line	Creates a center line.
	Center Line With Reference	Creates a center line with a reference.
	Thread	Creates a thread.
	Thread With Reference	Creates a thread with a reference.
	Axis Line	Creates an axis line.
	Axis Line and Center Line	Creates axis lines and center lines on circular outlines.

Steps To Creating Sheets And Views Using The Drafting Work Bench

CATIA V5 uses the files created in the **Part Design Work Bench** to create drawings. As previously stated, the files created in the **Part Design Work Bench** are saved with the "**CATPart**" extension. The **Drafting Work Bench** uses the file created in the **Part Design Work Bench,** but is saved as a separate file. The **Drafting Work Bench** files have a "**CATDrawing**" extension. Both the "**CATPart**" and "**CATDrawing**" files are linked together. To pull up the "**CATDrawing**" file, CATIA V5 must be able to locate the related "**CATPart**" file. If CATIA V5 cannot find the linked file, you will get an error. If you move a "**CATPart**" file to a different directory, the "**CATDrawing**" file linked to it will not be able to locate the "**CATPart**" file. You will have to update the directory where the "**CATPart**" exists.

This lesson takes you through one of the more common methods of accessing, creating and manipulating a part ("**CATPart**" file) into a drafting or drawing format ("**CATDrawing**" file). There are multiple methods of creating a drawing. This lesson will introduce you to one method and several of the tools. You are encouraged to explore other methods to find out what works best for your requirements.

1 Start CATIA V5

If you need to review this step refer to Lesson 1 Step 1.

2 Select The Part Design Work Bench

Yes, this lesson covers the **Drafting Work Bench,** but it is easier to specify what **"CATPart"** you are bringing into the **Drafting Work Bench** by making it active in the **Part Design Work Bench**.

3 Open The "T Shaped Extrusion.CATPart" Document

Figure 3.1

T Shaped Extrusion
xy plane
yz plane
zx plane
Select here

Open the "**T Shaped Extrusion.CATPart**." The "**T Shaped Extrusion**" started in Lesson 1 as the "**L Shaped Extrusion**." In Lesson 2 the "**L Shaped Extrusion**" was mirrored into a T shaped extrusion; thus, at the completion of Lesson 2 the part was saved as the "**T Shaped Extrusion.CATPart**." Once your "**T Shaped Extrusion**" part is on the screen, select the very top branch labeled "**T Shaped Extrusion**" in the **Specification Tree** as shown in Figure 3.1. The default **Part1** label was revised in Lesson 2 to "**T Shaped Extrusion**." Make sure "**T Shaped Extrusion"** is highlighted.

4 Select The Drafting Work Bench

This step uses the "**T Shaped Extrusion.CATPart**" to create a production drawing using the **Drafting Work Bench**. With a few simple menu selections, CATIA V5 will automatically produce an orthographic layout of your part. The steps are defined below.

4.1 Using one of the methods described in Lesson 1 select the **Drafting Work Bench**.

Figure 3.2

Drafting work Bench

4.2 Selecting the **Drafting Work Bench** will bring up the **New Drawing Creation** pop-up window shown in Figure 3.3.

The **Drafting Work Bench** has many options. This step will show you how to create a drawing relatively quickly.

Figure 3.3

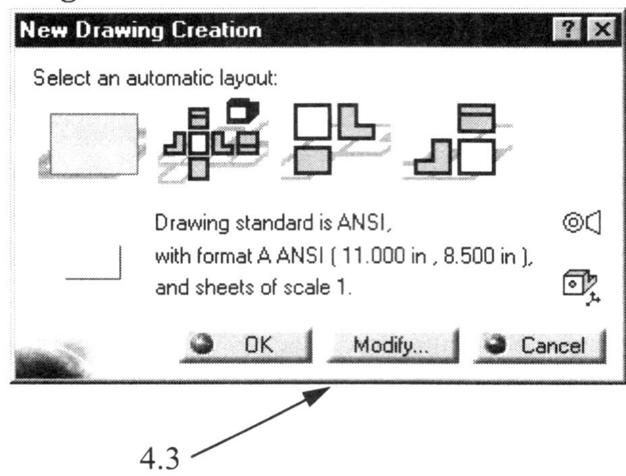

4.3 Select the **Modify** button in the **New Drawing Creation** window as shown in Figure 3.3.

4.4 This selection will bring up an additional pop-up window, the **New Drawing** window as shown in Figure 3.4.

Figure 3.4

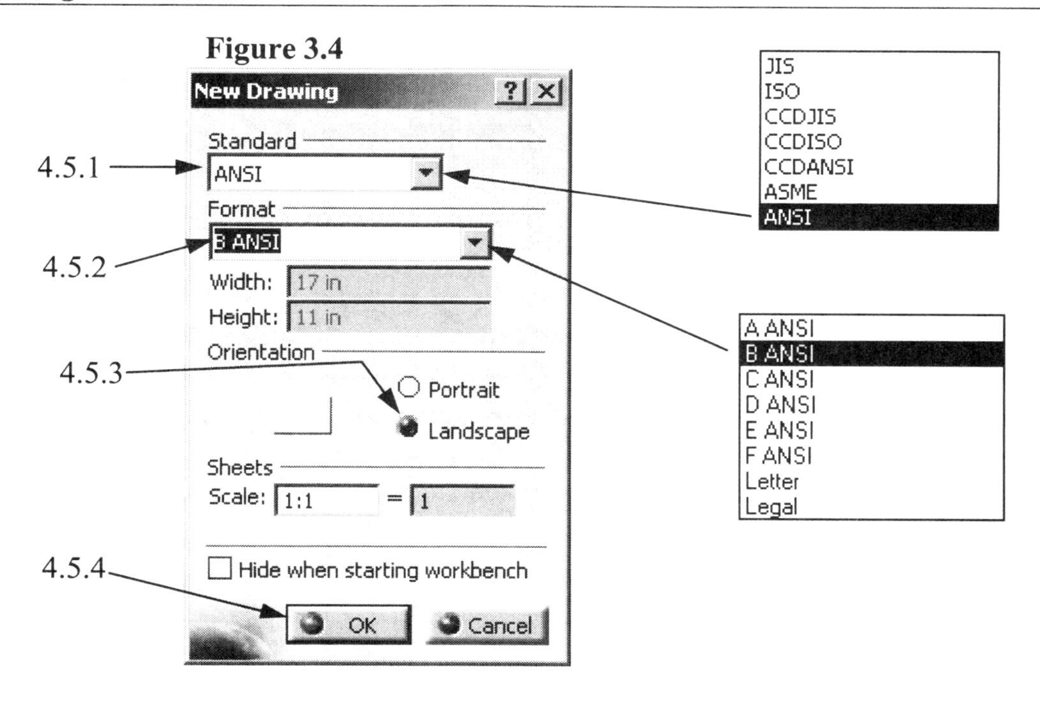

4.5 Select the arrow under the **Standard** option. This will give you several options for the standard you want to use for your drawing.

4.5.1 For this lesson select "**ANSI**," as shown in Figure 3.4.

4.5.2 For the **Format** box, select "**B ANSI**." The size of "**B ANSI**" will show up in the **Width** and **Height** boxes.

4.5.3 Set the **Orientation** to "**Landscape**" and the **Scale Of Sheets** to "**1**."

4.5.4 When all of the options are defined, select the **OK** button.

4.6 At this point, the **New Drawing** window will disappear leaving the **New Drawing Creation** window as shown in Figure 3.5. The information displayed in the window should be updated to show the options you just defined, such as the **Standard**, **Format**, and **Orientation**. This window gives you four drawing layout options. For this lesson, select the third option: **Front**, **Bottom** and **Right**.

Figure 3.5

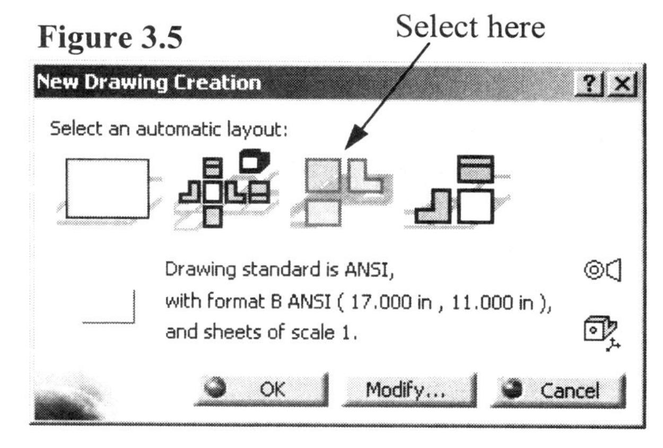

4.7 Select **OK**. The pop-up window will disappear and the CATIA V5 drafting screen will appear. The "**T Shaped Extrusion**" part will appear in the three views you just defined: **Front**, **Bottom** and **Right**. Reference Figure 3.6 for the CATIA V5 drafting screen.

Figure 3.6

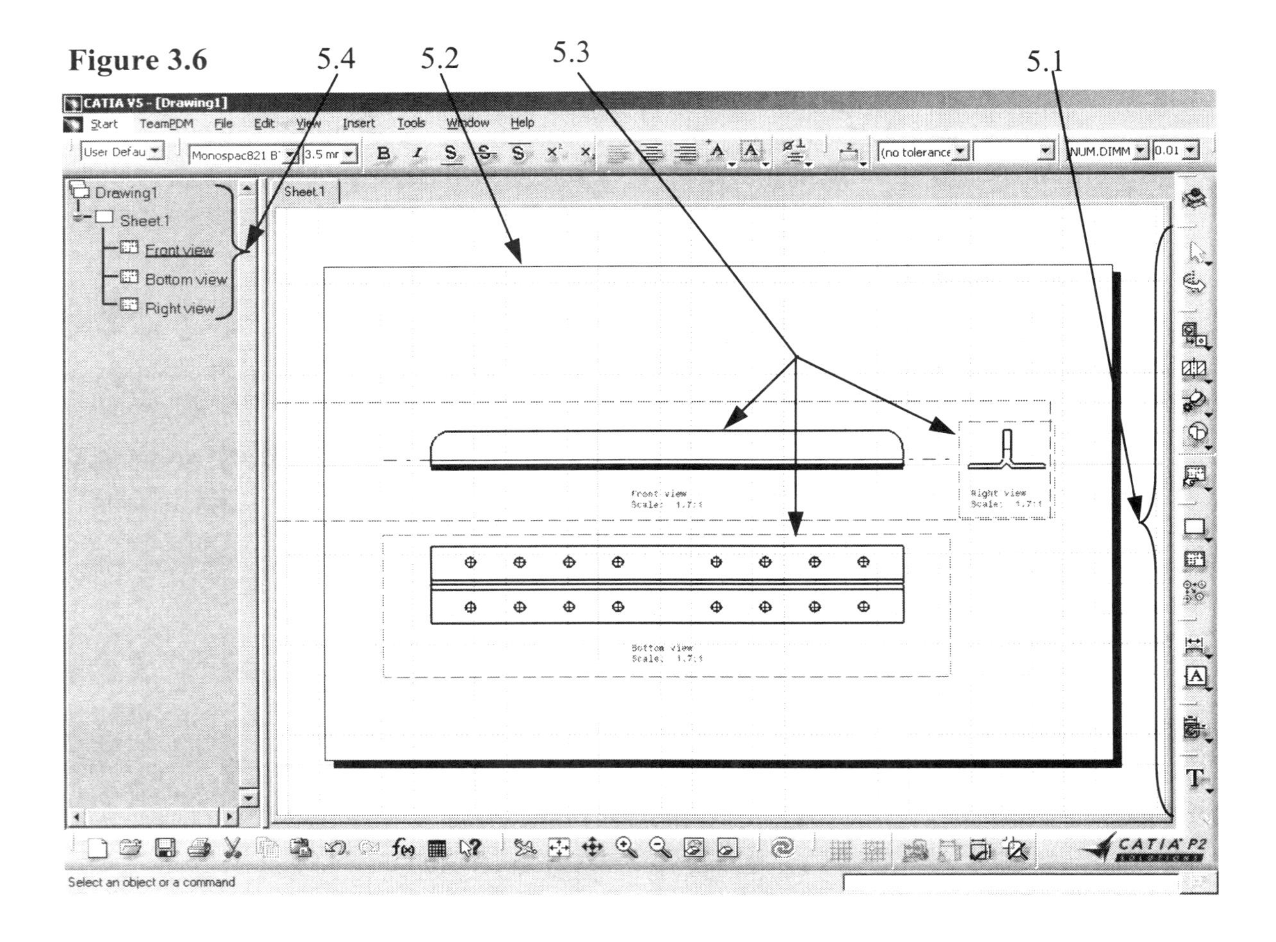

5 The Drafting Work Bench Layout

One of the first things you should notice is that the screen looks significantly different than the **Sketcher** and **Part Design Work Benches**. The **Drafting Work Bench** is truly limited to two dimensions. All of the rotation and other 3D capabilities are gone. Your 3D part is now being represented in orthographic projection, the foundation on which drafting is based.

5.1 **Tool Bars**: By default there are thirteen tool bars in the CATIA V5 **Drafting Work Bench**. Thirteen tool bars is too many to possibly show on one side bar. By default, CATIA V5 adds a tool bar across the top of the screen. This is still not enough room to represent all of the tool bars. Notice, both the side tool bar and top tool bar have the double arrow symbol signifying that there are additional tool bars avalaible, but not visible. To help you get familar with all of the tool bars, pull all thirteen tool bars out on the screen so they are visible and available for selection.

5.2 **Sheet.1**: CATIA V5 automatically created **Sheet.1** for you. The size was specified by you in the **New Drawing Creation** window. You selected "**B ANSI**" which is 11″ × 17″. The sheet is represented by the shadowed box in Figure 3.6.

5.3 **View Layout**: You determined the view layout in the **New Drawing Creation** window. You selected the third choice, which included the **Front**, **Bottom** and **Right** views. By default, the views include the **View Name** and **View Scale**. You will be shown how to modify this later in this lesson.

5.4 **Drafting Work Bench Specification Tree**: CATIA V5 creates a **Specification Tree** for the **Drafting Work Bench**. The **Specification Tree** is separate from the **Part Design Specification Tree,** but has similar qualities and can be manipulated (modified) in the same manner.

5.5 Modifiying the **Drafting Work Bench Standards**: As in the **Sketcher** and **Part Design Work Benches**, the standards are quick and easy to change.

The only action required by you in this step is to pull all thirteen tool bars out on the screen so they are visible and can be easily selected. If you have completed this task you can continue on to Step 6.

6 Customizing The Default Values

If you are going to change anything about the **Drafting Work Bench** defaults, now would be the best time. Customizing the **Drafting Work Bench** is similar to the customizations made in Lesson 1 and Lesson 2. The steps are detailed below.

6.1 Go to the top tool bar (pull down menu) and click on **Tools**, **Options** as shown in Figure 3.7. This brings up the **Options** window with **File Tab** options on the right side of the screen and **File Type** options on the left. From the options on the left, select **Drafting**. The tabbed options on the right change accordingly.

Figure 3.7

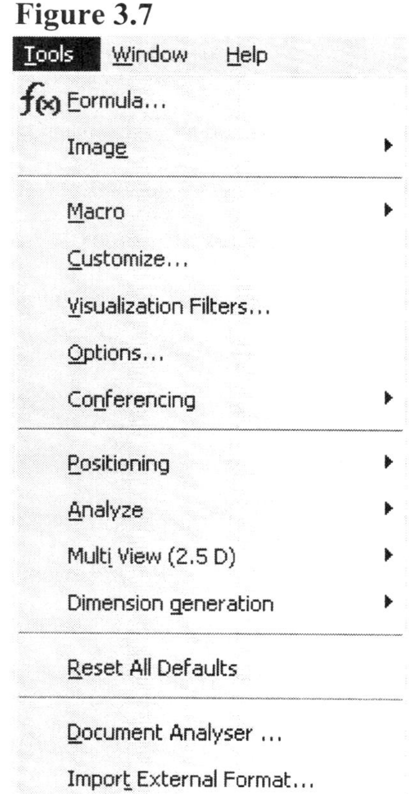

6.2 Under **Drafting,** there are six tabs. They are **General**, **Layout**, **View**, **Generation**, **Geometry**, **Dimension**, **Manipulators**, **Anotation and Dress-Up** and **Administration**. The following is a brief explanation of the most frequently used tabs.

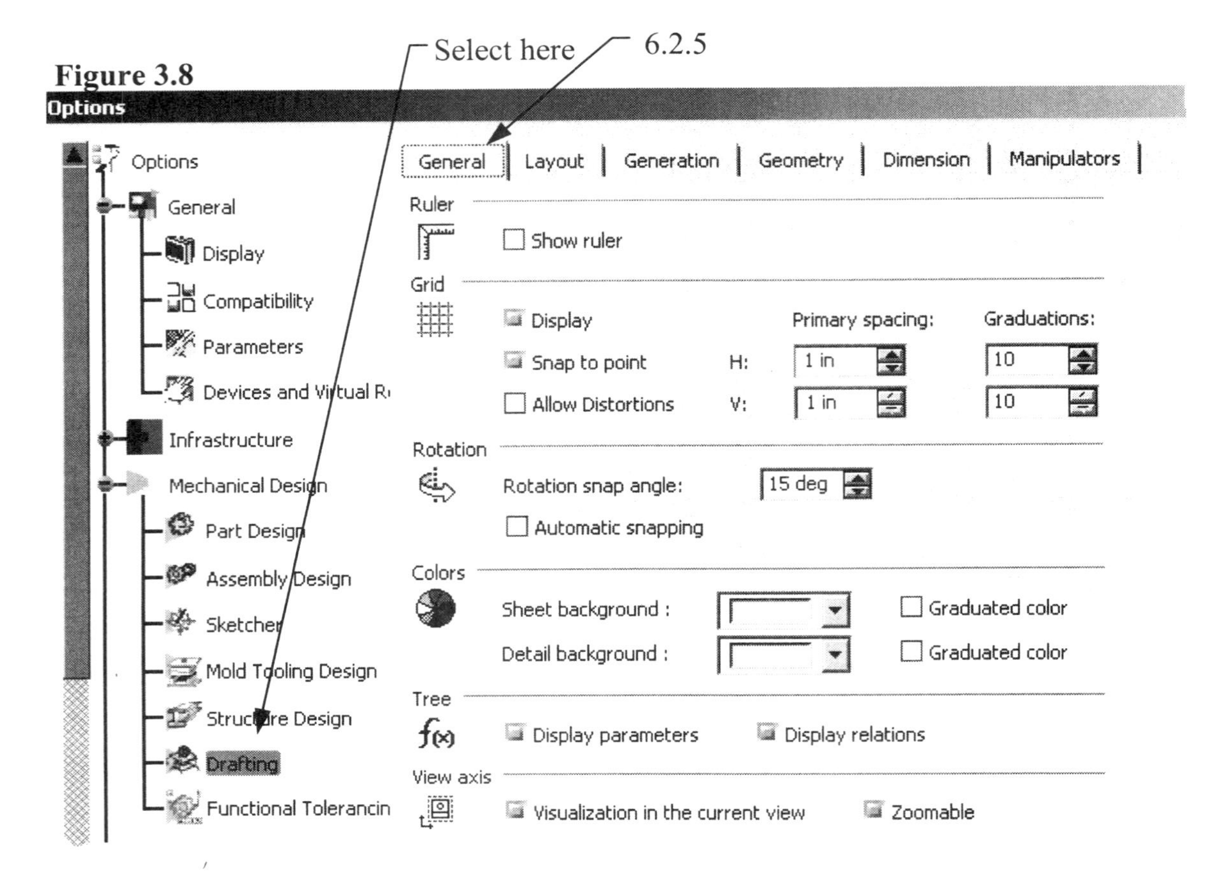

Figure 3.8

6.2.1 **General**: Figure 3.8 shows the different sections of the **General** tab and the variables they control. The controls are similar to the **Sketcher** and **Part Design** controls. For this step set the **Grid primary Spacing** value to **1″** and the **Graduations** to **10**. Turn on **Grid Display** and **Snap To Point**.

Figure 3.9

General | Layout | Generation | Geometry | Dimension

View creation

View name

Scaling factor

View frame

Propagation of broken and breakout specfications

New Sheet

Copy background view

Source sheet: First sheet

Other drawing

6.2.2 **Layout**: Figure 3.9 displays the toggle controls in the **Layout** tab. This tab controls the default setup of the drawing views. For this step set up your **Layout** the same as shown in Figure 3.9.

6.2.3 **View**: Figure 3.10 displays the options on the **View** tab. The **View** tab controls which entities are shown on your drawing.

6.2.4 **Generation**: The toggle control for the **Generation** tab is displayed in Figure 3.11. Set up your **Generation** tab the same as shown in Figure 3.11.

Figure 3.10

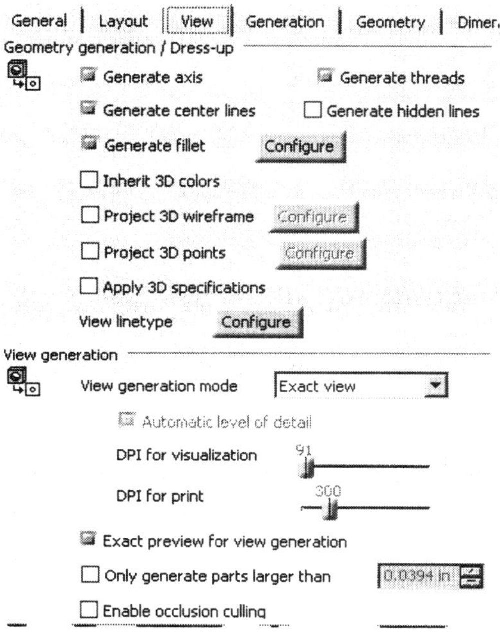

Figure 3.11

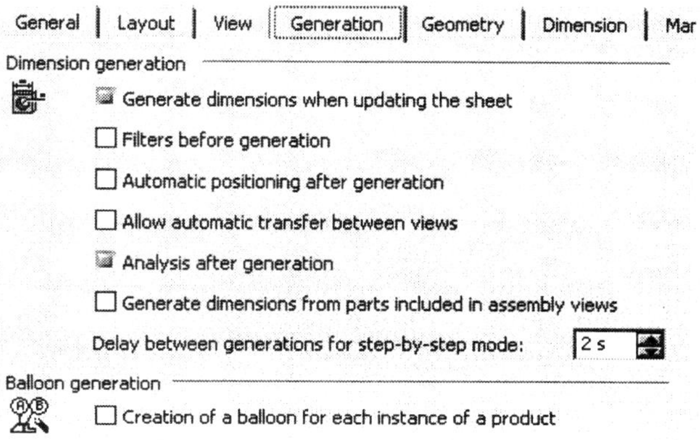

6.2.5 **Geometry**: This tab controls how the geomentry created in the **Drafting Work Bench** will be displayed. Some of the toggle controls also appear in the bottom tool bar such as **Constraint Visualization**, reference Figure 3.12 and Figure 3.6. Set up your **Geomtry** tab the same as is shown in Figure 3.12.

6.2.6 **Dimension**: This tab controls how the dimensions are created and visualized. There are additional windows with more controls. This tab is not covered in detail but you now know where the controls are. As you create dimensions you are encouraged to test the different controls. For this step set up the tab as shown in Figure 3.13.

Figure 3.12

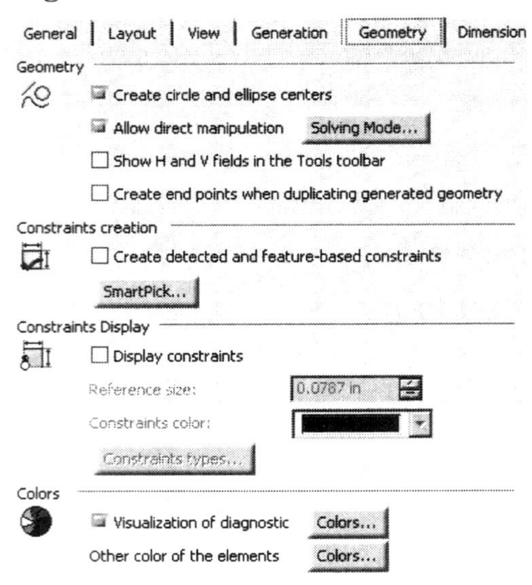

Figure 3.13

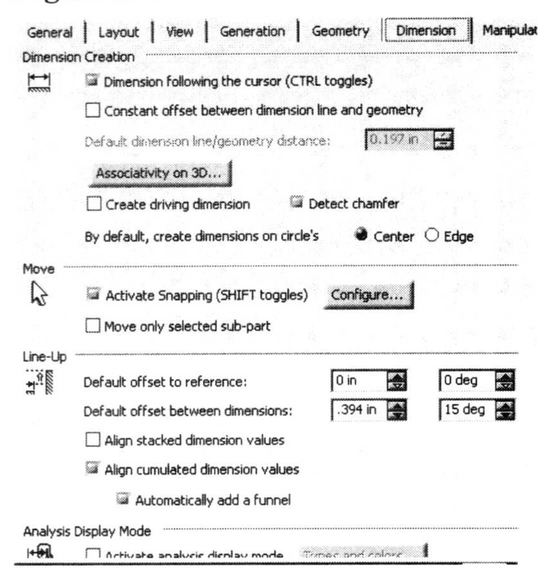

6.2.7 **Manipulators**: This tab controls the tools that allow you to modify the dimensions. For this step set up the tab as shown in Figure 3.14.

6.2.8 **Annotation and Dress-Up**: This tab provides tools for configuring annotation and snap controls. Reference Figure 3.15.

6.2.9 **Administration:** The **Administration** tab allows you to control the administration of the drawing. Toggles on this tab can prevent the manipulation of the backgroundand standards. Reference Figure 3.16.

Figure 3.14

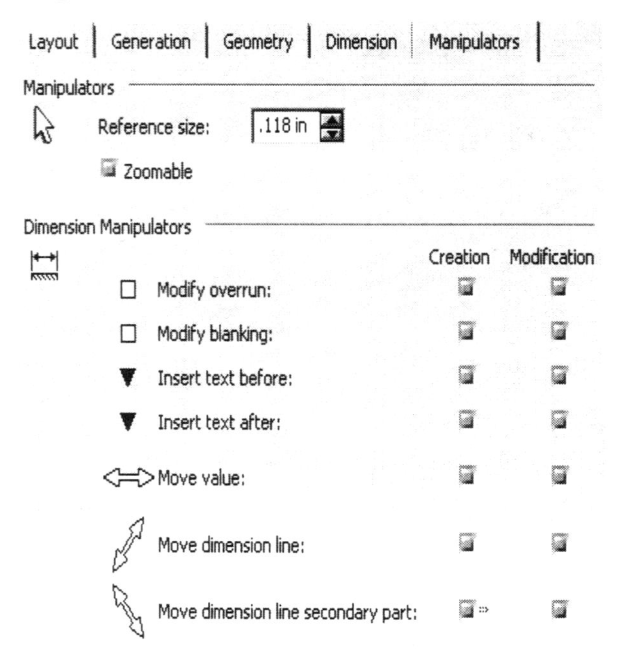

Figure 3.15

Geometry | Dimension | Manipulators | Annotation and Dress-Up | Ac

Annotation Creation

Apply snapping to

Text

Text with leader

Geometrical tolerance

Move

Activate snapping (SHIFT toggles) Configure...

2D Component Creation

Create with a constant size

Balloon Creation

3D associativity Numbering

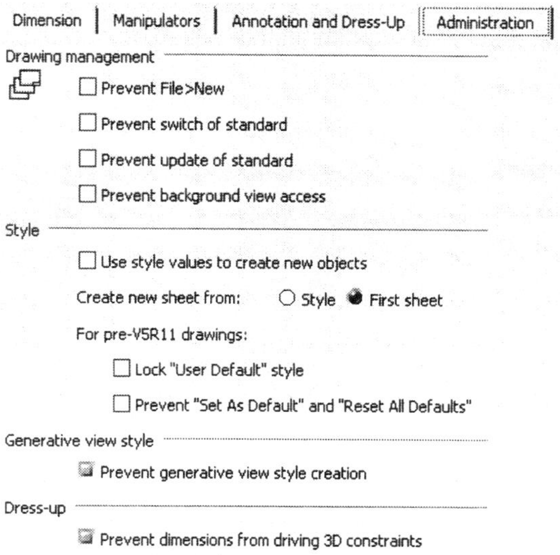

Dimension | Manipulators | Annotation and Dress-Up | Administration

Drawing management

- [ ] Prevent File>New
- [ ] Prevent switch of standard
- [ ] Prevent update of standard
- [ ] Prevent background view access

Style

- [ ] Use style values to create new objects

Create new sheet from: ○ Style ● First sheet

For pre-V5R11 drawings:

- [ ] Lock "User Default" style
- [ ] Prevent "Set As Default" and "Reset All Defaults"

Generative view style

- Prevent generative view style creation

Dress-up

- Prevent dimensions from driving 3D constraints

6.3 You are encouraged to try some of the other modifications. When the changes have been made, select the **OK** button at the bottom of the **Options** window. The changes you specified will be displayed, such as the **Grid** and **Graduations**.

7 Creating A New Sheet

The orthographic layout of the "**T Shaped Extrusion**" on **Sheet.1** does not leave much room for any additional views. To add an additional view, you will need to create a new sheet. To do this, complete the following steps.

7.1 Creating new and/or additional sheets for your drawing is very simple. First you have to be in the **Drafting Work Bench**.

7.2 In the **Drafting Work Bench,** select the **New Sheet** tool. This will create a new sheet every time you click on the tool. For this lesson create **Sheet.2**.

Figure 3.17

New sheet

Sheet.1 Sheet.2

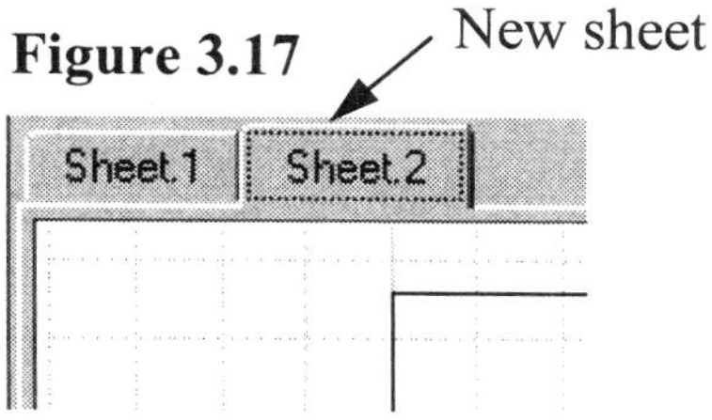

7.3 There are two ways you can verify that a new sheet has been created. One way is, at the top left of the **Drafting Work Bench** screen there are **Sheet** tabs representing each existing sheet as shown in Figure 3.17. The tabs will be labeled **Sheet.1** and **Sheet.2** in successive order. The second method of verifying a new sheet was created, is by checking the **Drafting Specification Tree.** The **Sheets** will also show up as branches as shown in Figure 3.18. The branches will be labeled the same as the tabs. The newly created sheet will become the active sheet. If you want to go back to the previous sheet, click on the file tab representing the desired sheet number. The sheet can also be selected from the **Drafting Specification Tree**.

Figure 3.18

7.4 Each newly created sheet will, by default, have the sheet size and drafting standard defined when you first enter the **Drafting Work Bench**.

7.5 No new views have been defined for the newly created **Sheet.2**. As the views are defined, the **Drafting Specification Tree** will also include the defined views.

The sheets are as easy to delete as they are to create. You can select the sheet from the **Specification Tree**, then select the **Cut** tool at the bottom of the screen. Be careful, you can delete a sheet even if it has existing views defined.

8 Creating A New View

Now that you have created a new sheet, you can create additioinal views as required and/or desired. There are several different ways to add new views to the newly created sheet; two of the most common methods are listed below.

NOTE: It is important that you know what view and orientation of the part is best for what you want to accomplish. To fully visualize what you are creating in the **Drafting Work Bench** and selecting from the **Part Design Work Bench**, split the screen horizontally so that half of the screen shows the **Drafting Work Bench** and the other half displays the **Part Design Work Bench**. To split the screen horizontally, select the pull down window labled **Window** at the top center of the screen. Select **Tile Horizontally** as shown in Figure 3.19. This selection will split all of the active windows horizontally on the screen. You will be required to select a plane or surface from the **Part Design Work Bench**. The plane or surface you select will be applied to the **Drafting Work Bench**. So, before proceeding, be sure you have the **Part Design Work Bench** and the **Drafting Work Bench** windows split on the screen as shown in Figure 3.20. This is a common window function.

Figure 3.19

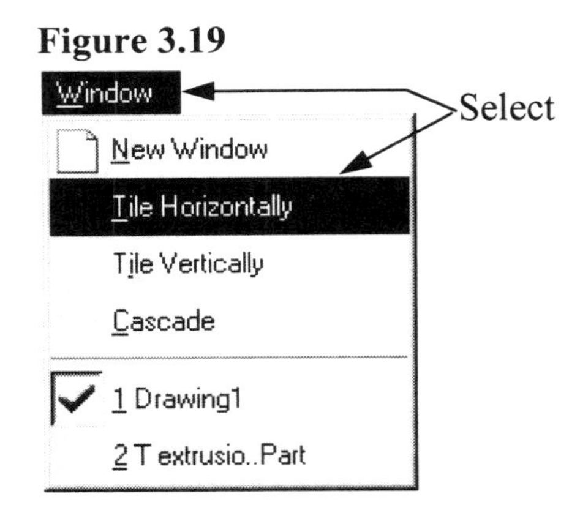

Figure 3.20 (Windows tiled horizontally)

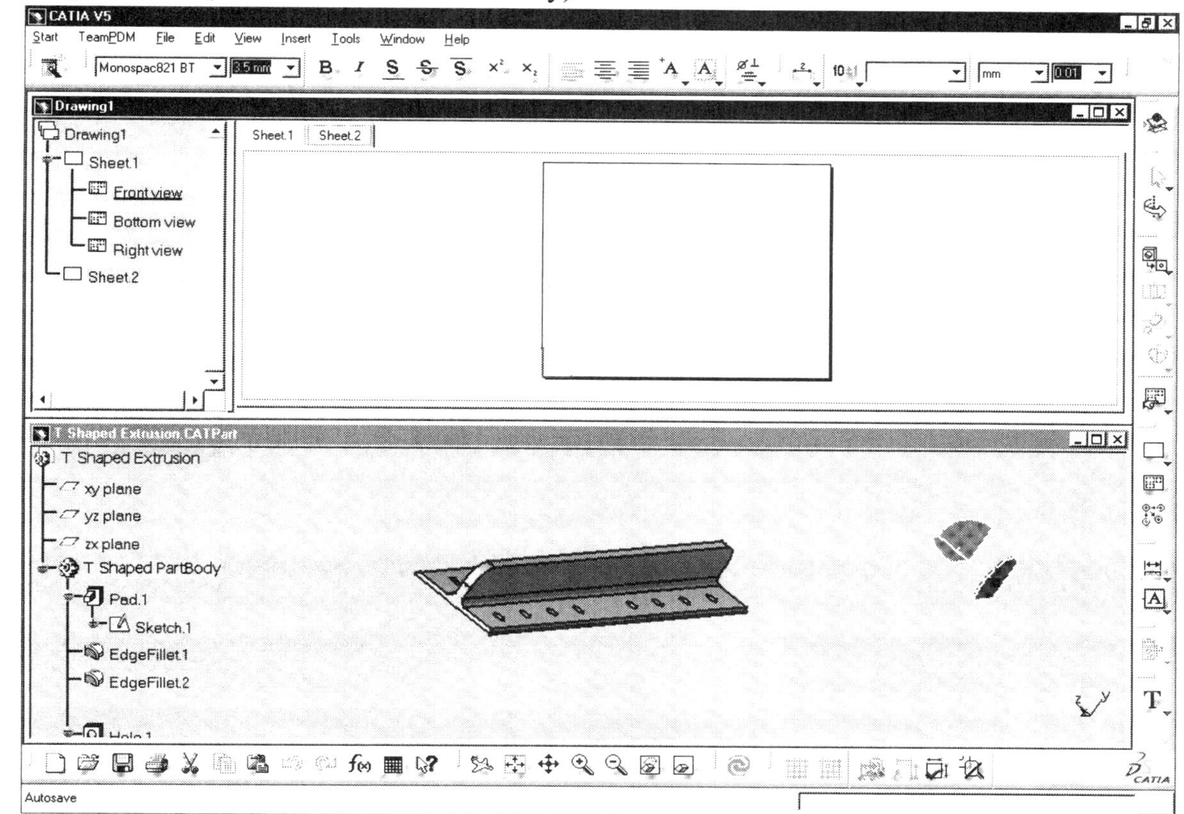

8.1 Select the **View Creation Wizard** tool from the **Views** tool bar.

8.2 This will bring up a pop-up window labeled "**View Wizard (Step ½): Predefined Configurations**." The window allows you, the user, to define the specific orthographic projections and/or isomectric views. Select the first choice (top left) "**Configuration 1 using the 3rd angle projection method**." This option will place three orthographic views on the pop-up window. At this point, you can modify the placement of the views by selecting the individual views and dragging them to the new location. Once you have the placement of the three views similar to Figure 3.21, select the **Next** button at the bottom of the pop-up window.

Figure 3.21

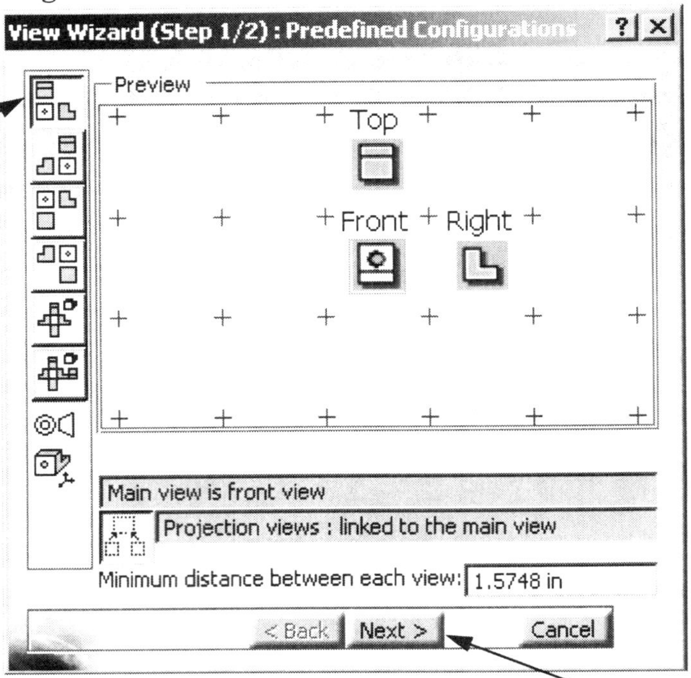

View Wizard (Step 1/2) : Predefined Configurations
Preview
Top
Front
Right
Main view is front view
Projection views : linked to the main view
Minimum distance between each view:
1.5748 in
< Back
Next >
Cancel
Select here first
Select here second

8.3 The next window allows you to select additional views (Figure 3.22). In this step you will want to add an isometric view (bottom left choice). Select the **Isometric View** button and then select the location you want the isometric view to be presented. For this lesson, place the isometric view at the top right of the existing views.

Figure 3.22

View Wizard (Step 2/2) : Arranging the Configuration
Preview
Top
Isometric
Front
Right
Main view is front view
Projection views : linked to the main view
Minimum distance between each view:
1.5748 in
< Back
Finish
Cancel
Place view here
Isometric view
When done, select here (8.4)

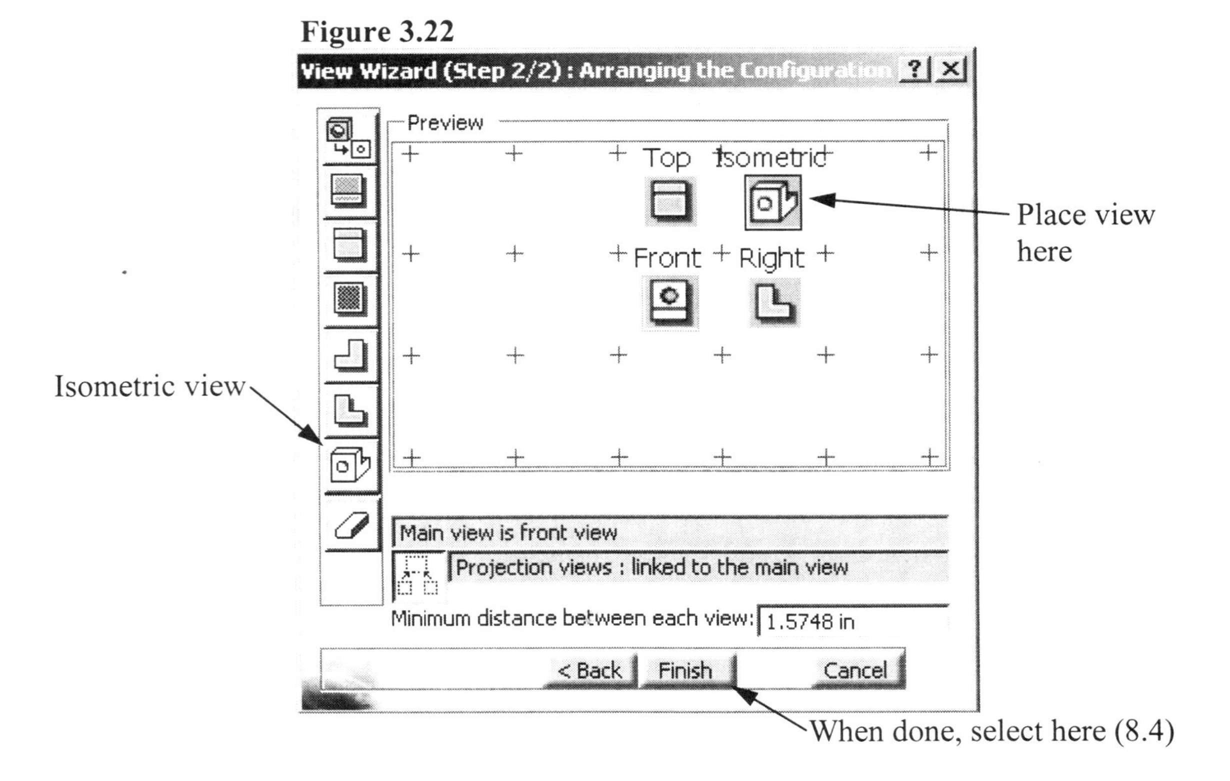

8.4 When you have the view layout as you desire, or similar to Figure 3.22, select the **Finish** button at the bottom middle of the pop-up window.

8.5 CATIA V5 will then prompt you to "**Select a reference plane on a 3D geometry**." The plane or surface you are being prompted for is refering to the part in the **Part Design Work Bench**. For this lesson, select the surface shown in Figure 3.23.

Figure 3.23

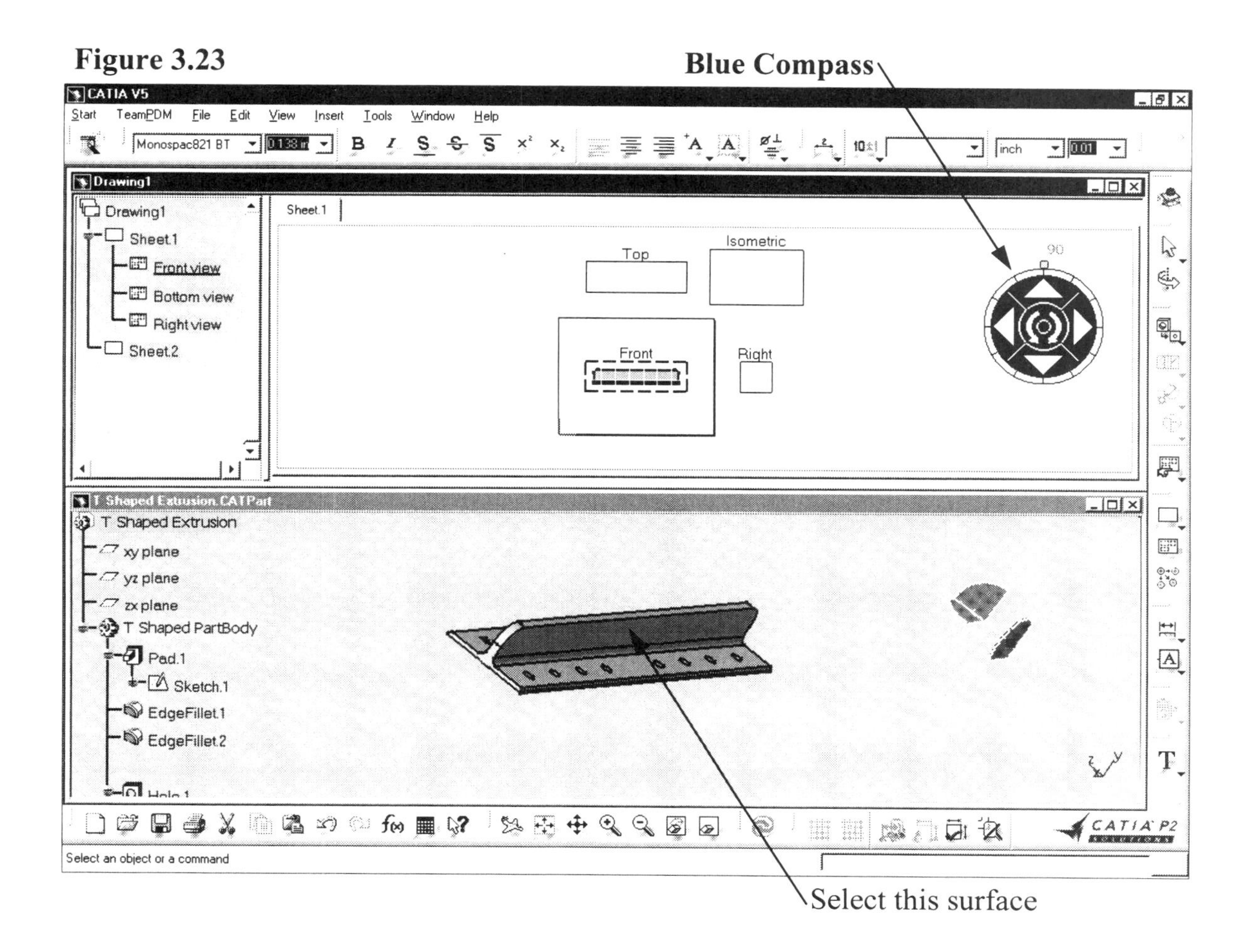

8.6 After you select the surface, four green rectanglar frames will appear in the **Drafting Work Bench** screen. The rectanglar frames will be labeled with the views you defined in the Steps 8.2 and 8.3. The part will be displayed in the **Front View**, with the surface you selected in Step 8.5 as the front view, normal to the viewer as shown in Figure 3.23. If you drag your mouse over the other views, you will notice that the part appears in the view the mouse is over. If you move the mouse over the **Top View,** the part will appear in that view, oriented as the **Top View**.

8.7 You should also notice a **Blue Compass** in the upper right hand corner of the **Drafting Work Bench** shown in Figure 3.23. The **Blue Compass** allows you to modify the views prior to completing the view orientation. The four arrows pointing 90 degrees apart will rotate the part 90 degrees in the rotational direction you select. In other words, currently the length of the "**T Shaped Extrusion**" is shown in the **Front View**, if you selected the arrow pointing to the right on the **Blue Compass,** the part would rotate 90 degrees. The orientation would change and the length of the part would be shown in the **Right View**. The half circled arrows will not change the part orientation/view relationship, but will rotate the part in the view every 15 degrees. Notice the green ball on the outside of the **Blue Compass** rotates around indicating the degree of rotation. The 15 degree rotation is the standard default. This should be enough to get you started in creating the views you need. The best way to find out what these tools are really capable of is to test and try them, don't be afraid of making a mistake.

Figure 3.24

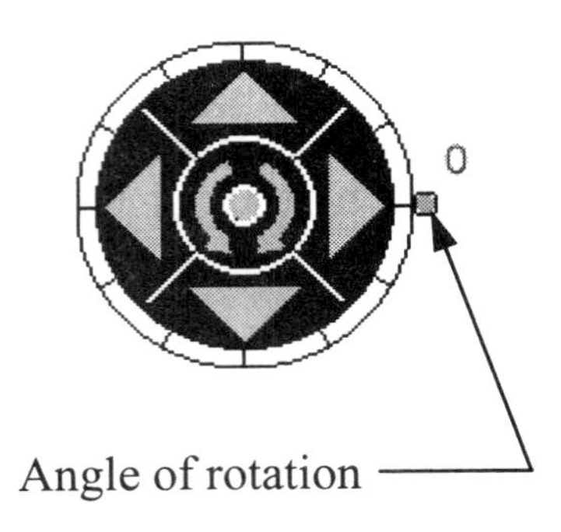

8.8 Once you have the views oriented the way they are shown in Figure 3.25, select anywhere on the sheet to lock and complete the view orientation. Notice that the **Blue Compass** disappears and the part shows up on all of the views in their correct orientation.

Figure 3.25

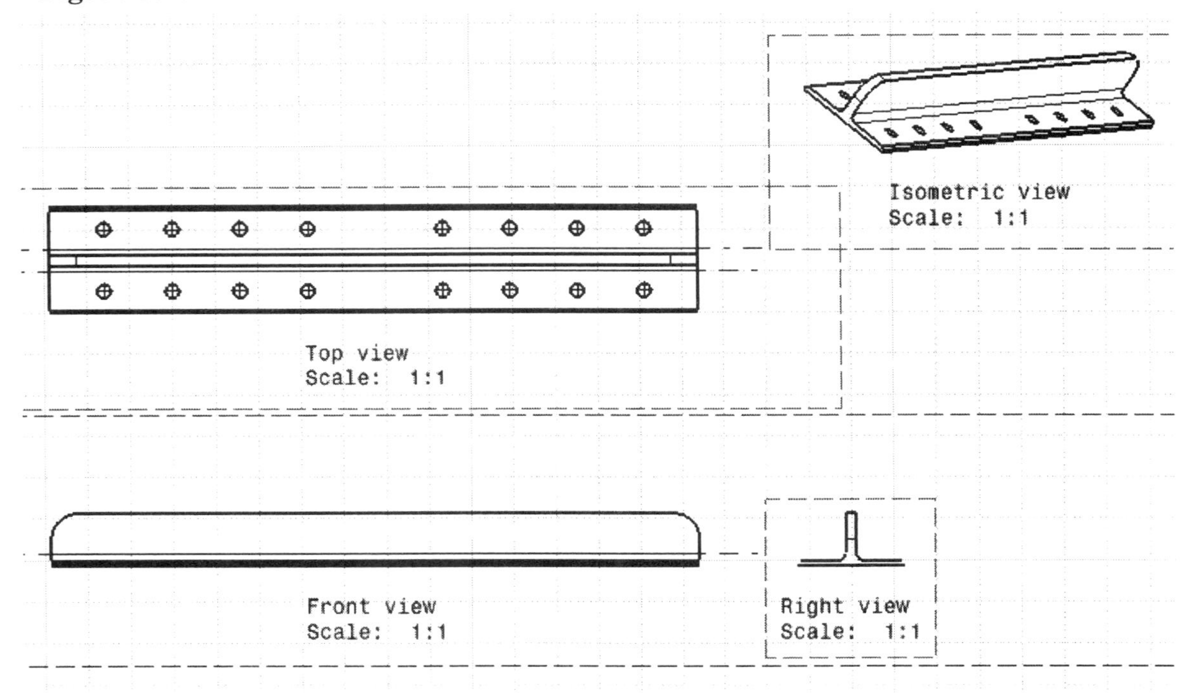

8.9 If you were testing the **Blue Compass** capabilities and oriented your part in such a way that you could not get back to where you started, delete what you have and start at Step 8.1 again. Starting at Step 8.1 may sound like a lot of work, but the steps are quick and easy. You should be able to recreate the views again in a matter of a few seconds. For this lesson make sure your view and part orientation look similar to the ones shown in Figure 3.25.

NOTE: The default incremental angle of rotation on the **Blue Compass** can be modified by holding the mouse over the green ball and pressing the right mouse button. This brings up a window that allows you select **Free Hand Rotation**, **Incremental Hand Rotation** and **Set Increment**.

9 Modifying An Existing View

Now that you have been briefly introduced to the process of creating a view, you will need to know how to modify a view. This step will cover modification of a view, meaning changing the view properties, appearance, location, scale, moving to a different sheet, copying, pasting and deleting. Being able to modify an existing view is a powerful and important aspect of creating a drawing using the **Drafting Work Bench**. These steps show you how to modify a view.

9.1 **Modifing View Properties**.

9.1.8 Select the view to be modified. The view has to be the active view. The active view will be underlined in the **Drafting Specification Tree** as shown in Figure 3.26.

Figure 3.26

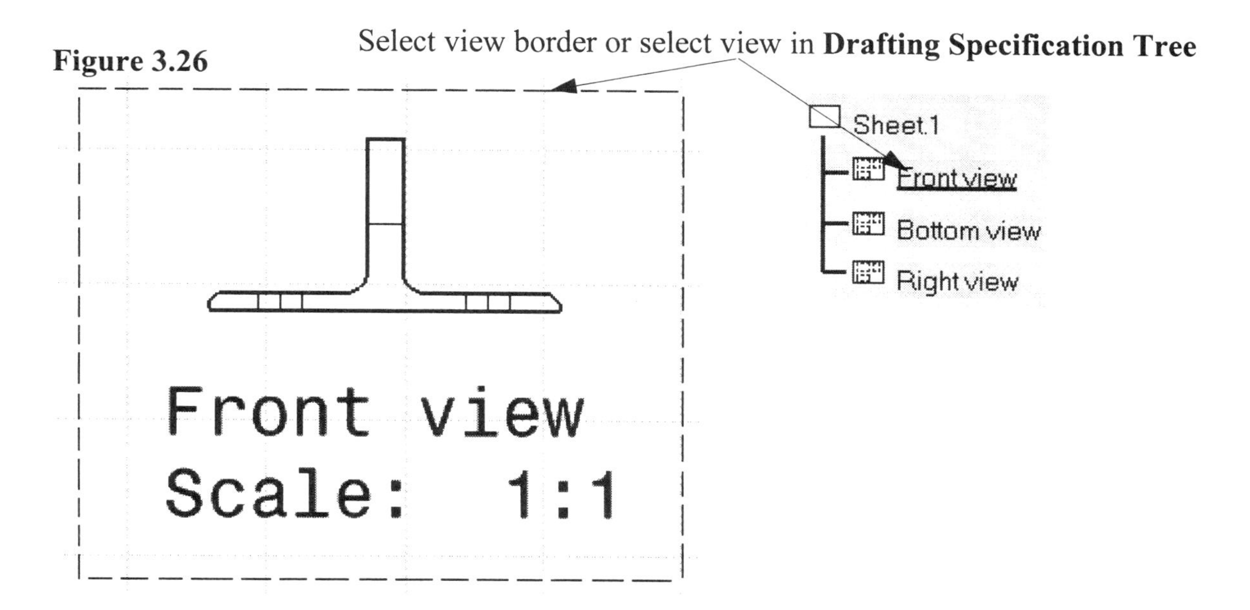

9.1.9 With the mouse over the selected view border, push the right mouse button. This will bring up the **Properties** pop-up window. Reference Figure 3.27 for all of the properties options. No changes are necessary at this time.

Figure 3.27

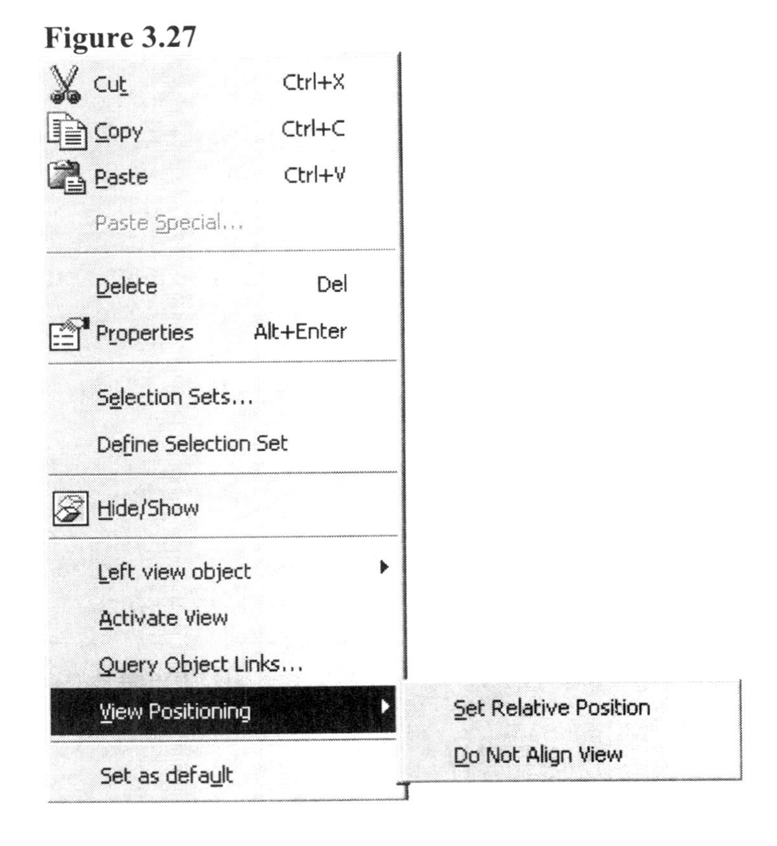

9.2 **Deleting a view from a sheet.**

9.2.1 Select the view so it is highlighted. This is done by moving the mouse over the view border and selecting it.

9.2.2 With the view highlighted, select the **Scissors** tool in the bottom left corner of the screen. You can also use the **Edit, Cut** or **Delete** options (**Windows** functions).

9.2.3 The view you selected will be deleted, as well as every entity attached to it such as text, 2D geometry and dimensions.

9.3 **Moving a view around on a sheet.**

9.3.1 Select the view so its border is highlighted.

9.3.2 Select the view border with the left mouse button and hold the mouse button down. This will hold the view. Drag the mouse to the location you want the view. When you get the view where you want it, release the mouse button and the view will stay in that location.

9.3.3 If the view was created as an **Auxiliary View** , the view will be locked at the angle the view was projected from. **Alignment** to a view means you can move the view closer or farther away, but you can not move the view to either side; it will stay in the same direction it was projected from. You can override the orthographic projected alignment by selecting **View Positioning** found in the **Properties** window. Then select the **Do Not Align View** option. This will allow you to move the view anywhere you want. If you want to re-align the view, complete the same steps, except select **Align View**.

9.4 **Moving a view from one sheet to another sheet**. This process is the same for moving a view and/or copying a view.

9.4.1 Go to the sheet with the view to be moved.

9.4.2 Select the view to be moved.

9.4.3 Use one of the multiple window functions to copy the view to the **Windows** clip board: **Ctrl** **C** , **Edit/Copy** or the **Copy** command in the **Properties** pop-up window reference Figure 3.27 and 3.28.

Figure 3.28

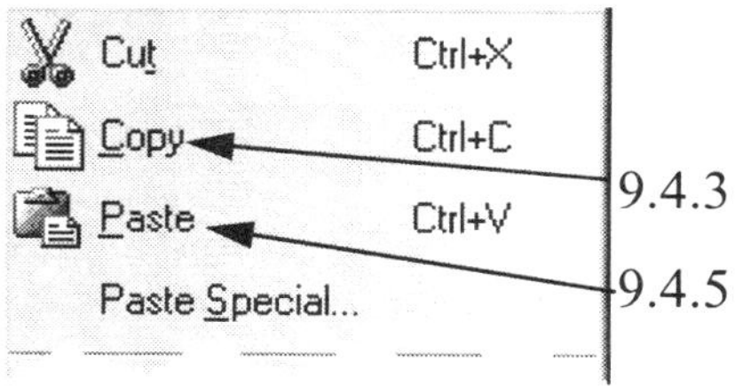

Cut Ctrl+X
Copy Ctrl+C 9.4.3
Paste Ctrl+V 9.4.5
Paste Special...

9.4.4 Go to the sheet where the view is being moved to.

9.4.5 Use one of the multiple **Window** functions to paste from the windows clip board: **Ctrl** **V**, **Edit/Paste** or the **Paste** command in the **Properties** pop-up window reference Figure 3.27 and 3.28.

9.4.6 The view will appear on the new sheet. You may have to relocate the view within the new sheet (Step 9.3).

9.4.7 The **Drafting Specification Tree** gives you one other method of moving a view from one sheet to another. In the **Drafting Specification Tree,** select the view you want to move and drag and drop it onto the sheet you want to move it to.

9.5 **Scaling a view**. To modify the scale of a view you need to go to Step 10.6.

9.6 **Changing the size of the view window**. As you create dimensions and/or drafting geometry, the view window will expand to encompass the new geometry. CATIA V5 adjusts the view frame automatically. The frame can be larger than the sheet as long as the geometry is contained within the sheet boundary. The views and geometry do not have to be inside the sheet boundary other than for printing purposes.

10 Creating A Detail View

Detail Views are used to show detailed areas of a part that are too small to see and/or dimension in a standard orthographic view. Detail views are usually scaled up several times to show the needed detail.

CATIA V5 makes creating **Detail Views** quick and easy. The steps below show you how to create a **Detail View** from an existing view. In this exercise you will create a **Detail View** of the "**T Shaped Extrusion**." The current standard orthographic views do not allow enough room for all of the dimensions required to fully define the "**T Shaped Extrusion**." To create a **Detail View** complete the following steps.

10.1 In this step you will be using the **Right View** on **Sheet.2**. Go to **Sheet.2** and activate the **Right View**. This is done by double clicking on the border of the view or by double clicking on the **Right View** in the **Drafting Specification Tree**. You can tell which view is the active view, because the border will be highlighted. Make sure the **Right View** is highlighted.

10.2 Select the **Detail View** tool .

10.3 The prompt zone will read "**NullState**." Determine and then select the center point of the **Detail View**. When selecting the center point, you may want to turn off the "**Snap To Grid**." The "**Snap To Grid**" may make it difficult to select the exact center of the **Detail View.**

For this lesson, make the center of your **Detail View,** the end of the left base leg of the "**T Shaped Extrusion**." This will allow you to dimension the chamfer on that base leg as shown in Figure 3.29. Dimensioning will be covered in the next lesson. You will need to **Zoom** in on the **Right View** to select the left side accurately.

Figure 3.29

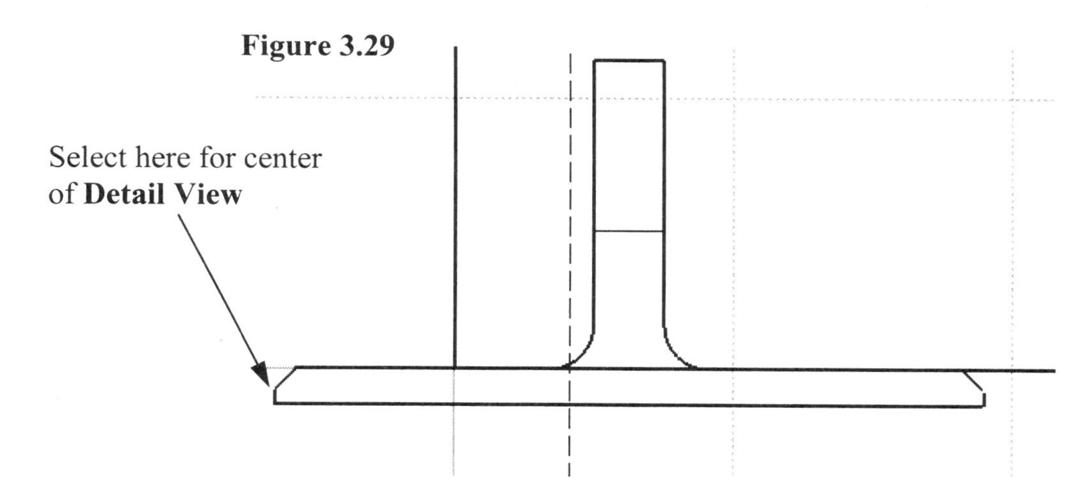

10.4 With the center of the **Detail View** selected, move the mouse away from the selected center. The distance you move away from the center determines the size of, and scope covered by, the **Detail View** . Notice the view grows as you move the mouse away from the selected center. CATIA V5 will prompt you to "**Select a point or click to define the circle radius**." When you move the mouse out, sufficient enough to show the chamfer as shown in Figure 3.30, click to set the circle. Confirm the view size by selecting the location with the mouse.

Figure 3.30

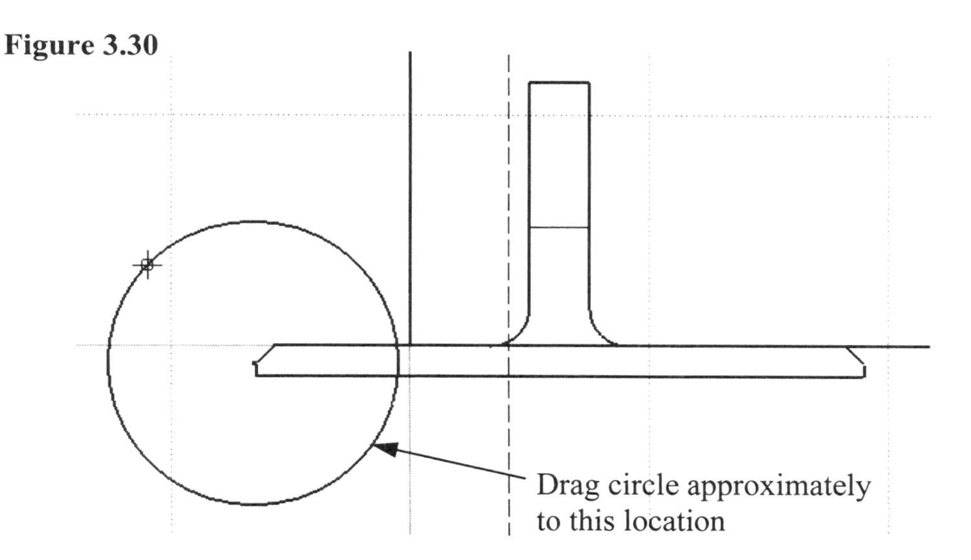

10.5 Once you have determined the size of the view by selecting the circle, CATIA V5 automatically creates the **Detail View**. Now all you have to do is determine the location of the view. You determine the location by dragging the mouse around to where you want the view located. When you get the view where you want it, select the location and the view will freeze in that location. The view will by default be created at a scale of 2 to 1. Notice CATIA V5 automatically creates a link from the standard orthographic view to the **Detail View**. By default the view is labled **View A.** Reference Figure 3.31 for the link between the **Right View** and **Detail View A**. This default lableling may be revised.

Figure 3.31

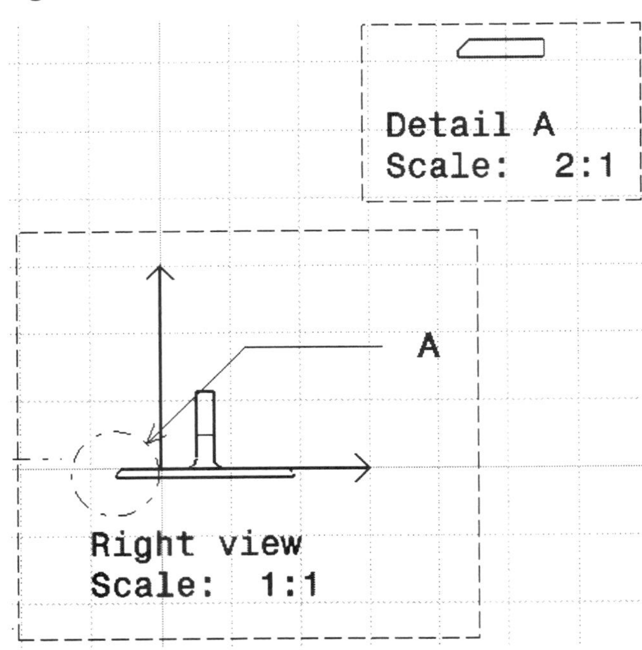

10.6 **Detail View A** will have to be scaled from the default of 2:1 to 7:1. To modify the **View Scale,** complete the following steps.

10.6.1 Make sure the **Detail View** is active, then double click on it so it is highlighted. Remember you can also select the **Detail View** from the **Drafting Specification Tree**.

10.6.2 Move the mouse so it is on top of the view border as shown in Figure 3.32. Push the right mouse button. If the mouse is on any other enity, the **Properties** window will be for that entity, not the view properties.

Figure 3.32

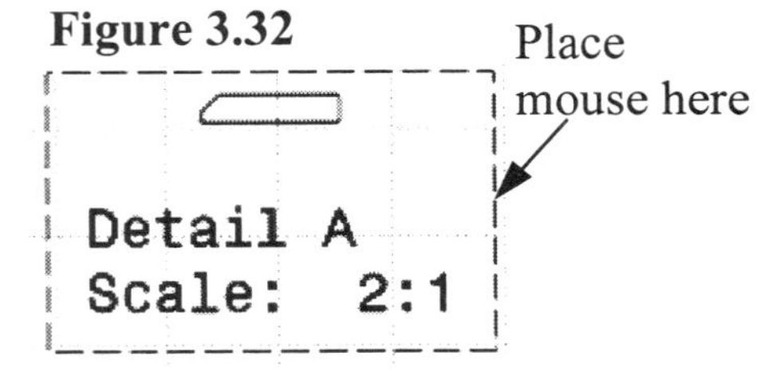

10.6.3 A pop-up window will appear. Select the **Properties** option as shown in Figure 3.33.

Figure 3.33

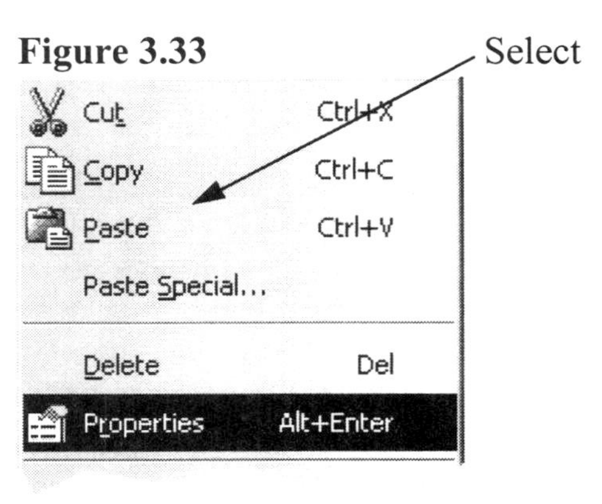

10.6.4 This will bring up all of the properties that define the **Detail View** you just created.

10.6.5 Select the **View** tab at the top left of the window. One of the variables under the **View** tab is **Parameters,** which includes the **Scale**. Type in **7** to **1** in the place of **2** to **1** as shown in Figure 3.34. Select **Apply** at the bottom right of the screen. **Apply** will allow you to preview the change before commiting to it. Notice the other view variables that can modified.

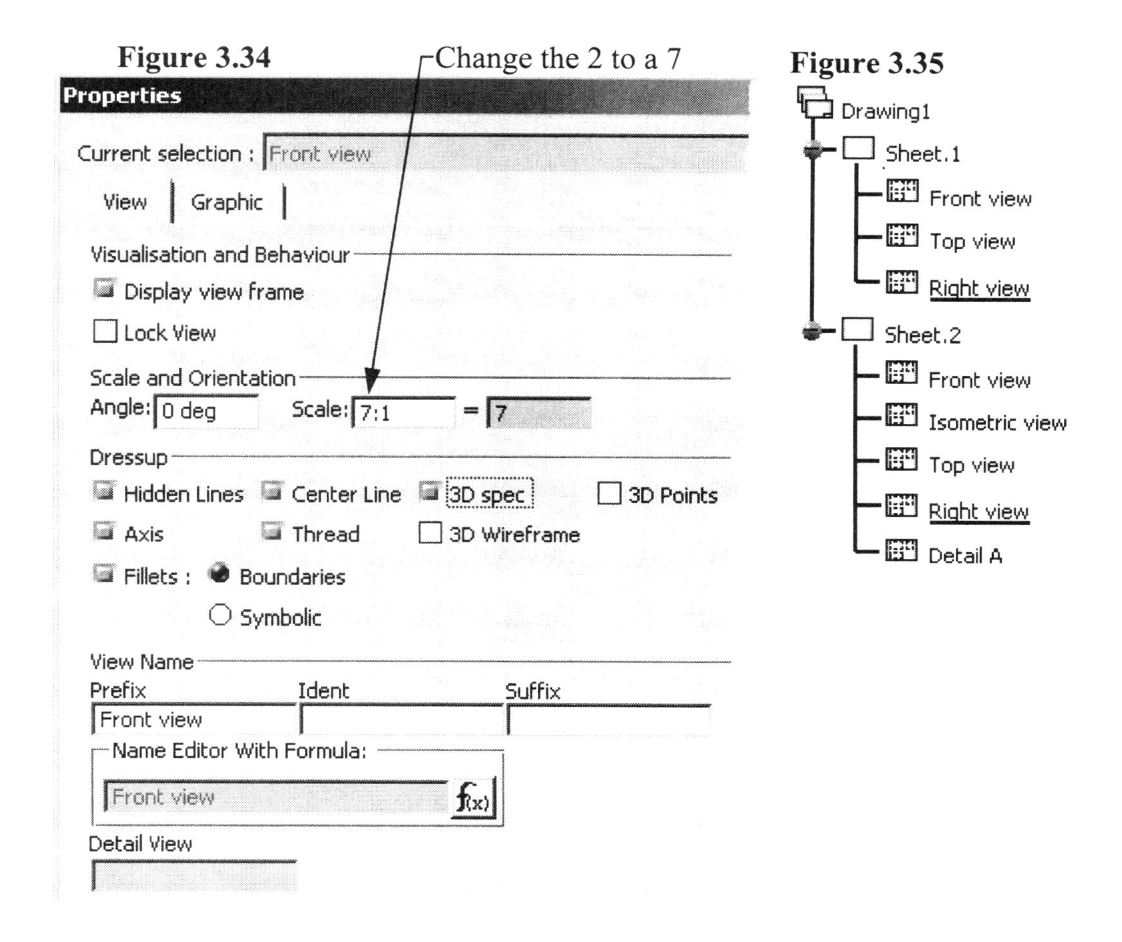

Figure 3.34

Figure 3.35

10.6.6 Select **OK**, and the modification will be completed. CATIA V5 automatically updated the scale value for you.

10.7 Notice, under **Sheet.2** in the **Drafting Specification Tree** that view **Detail A** has been added, reference Figure 3.35.

Detail View A can now be dimensioned and modified just like any other view. Congratulations! You have just created your first detail view!

11 Creating A Section View

Creation of **Section Views** has never been easier. To get the full picture of what you are creating, it is suggested that you have the screen split between the **Drafting Work Bench** and the **Part Design Work Bench**. Refer to Step 8 on splitting the screen between the two work benches.

There are four options under the **Section View** tool, they are:

1 **Aligned Section View** .
2 **Offset Section View** .
3 **Aligned Section Cut** .
4 **Offset Section Cut** .

The **View** and **Cut** are similar, except for the **Cut** option shows only what is on the cutting plane; whereas, the **View** includes the background geometry. Study the difference between the **Offset Section View** tool and the **Offset Section Cut** tool.

For this lesson you will be using the **Aligned Section View** . To create an **Aligned Section View,** complete the following steps.

11.1 Activate the **Top View**, reference Figure 3.36. The **Top View** is the view that shows the holes true size. The view is looking down on the standing leg. **The view you are going to use to create the section view must be the active view.**

11.2 Select the **Aligned Section View** tool.

Figure 3.36 Top View

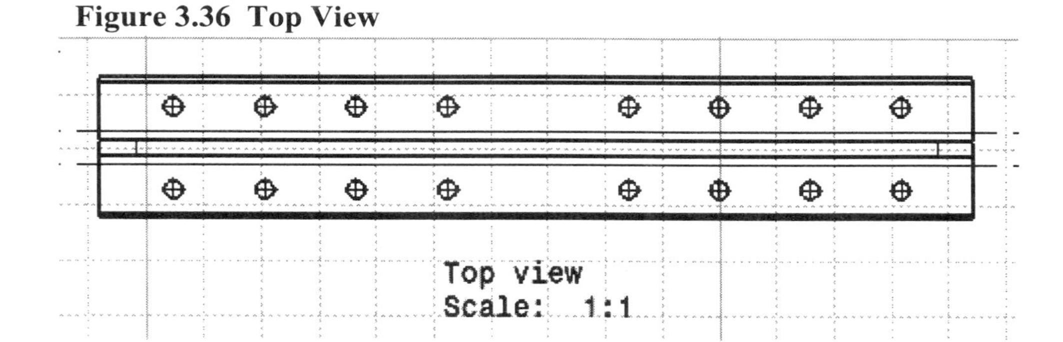

NOTE: Using the **Drafting Work Bench,** you are going to create a line that cuts across the part. This line will be represented in the **Part Design Work Bench** as a plane that cuts perpendicular through the part. As you create the line in the **Drafting Work Bench,** you can simultaniously see the cutting plane in the **Part Design Work Bench**.

After you select the **Aligned Section View** tool, CATIA V5 will prompt you with "**FirstState.**" This prompt is asking you to select the starting point for your line. Your mouse will now be accompanied by the **Autodetector** .

11.3 Move the **Autodetector** to the **Top View,** directly over the center of the hole farthest to the left of the **Top View**. Notice, as the **Autodetector** passes over the hole location, CATIA V5 creates a connecting line from the center of the hole to the **Autodetector**. This connecting line is telling you when the cutting line passes through the center of a hole. For this step you want the cutting line to pass through the center of this hole. Move the **Autodetector** about one **Primary Spacing** above the part. Reference Figure 3.37 for approximate target location. Once you have the **Autodetector** located as described, click the left mouse button once. This selection defines the starting location for the cutting plane.

Figure 3.37

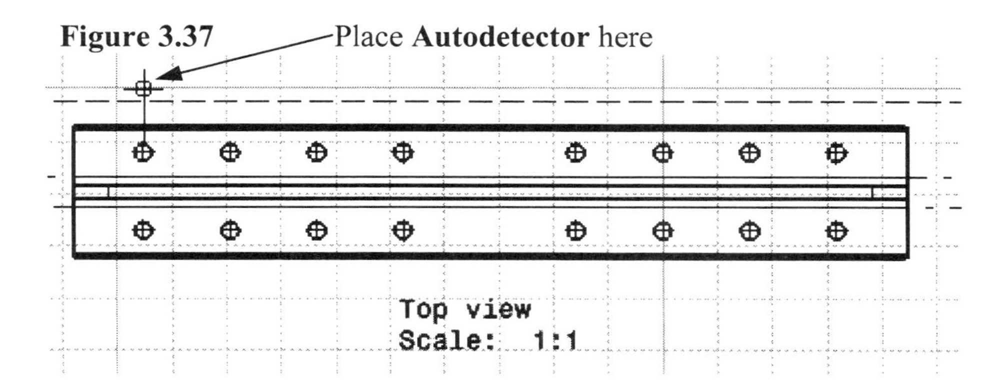

11.4 CATIA V5 will now prompt you to "**Select an object or a command.**" This prompt is asking you to select the second point of the cutting line. Move the mouse directly down through the part, the **Autodetector** will soon follow. Notice as you move the **Autodetector** through the part, you are creating a cutting line through the part in the **Drafting Work Bench**, as shown in Figure 3.38. This cutting line is visible in the **Part Design Work Bench** as a cutting plane as shown in Figure 3.39. As you move the mouse around, a visible representation is created in both the **Drafting Work Bench** (represented as a line) and the **Part Design Work Bench** (represented as a plane). The cutting line in the **Drafting Work Bench** also shows the angle of the cutting plane. For this lesson, pull the line down through the entire part and one additional **Primary Spacing** below the part. Keep the angle at 90 degrees as shown in Figure 3.38.

11.5 CATIA V5 allows you to create additional cutting planes by continuing the process of clicking and moving. For this lesson, only one cutting plane is required. Move the **Autodetector** to the location described in Step 11.4 as shown in Figure 3.38 and double click the mouse. Double clicking the mouse will end the process of creating the **Section View**.

Figure 3.38

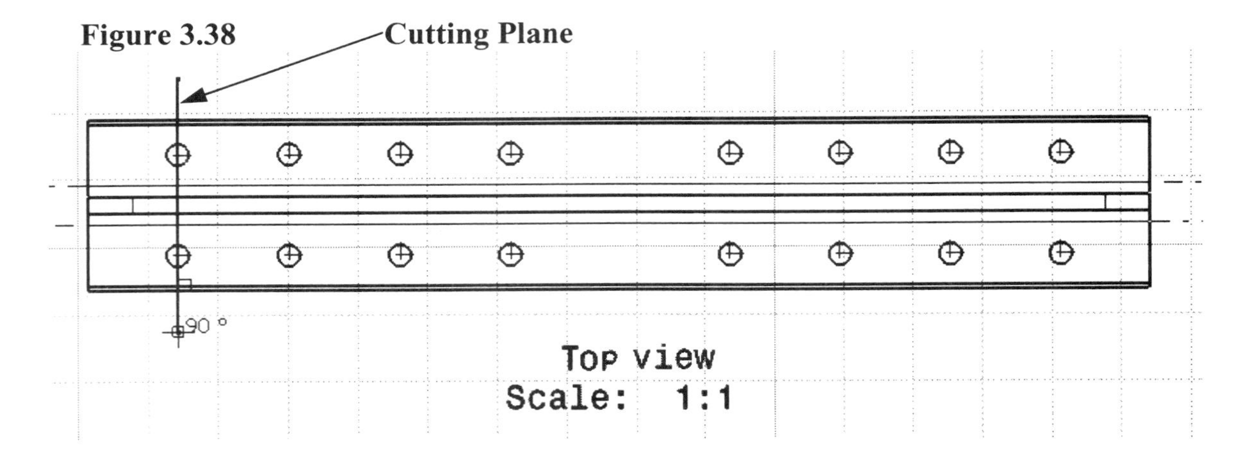

Figure 3.39

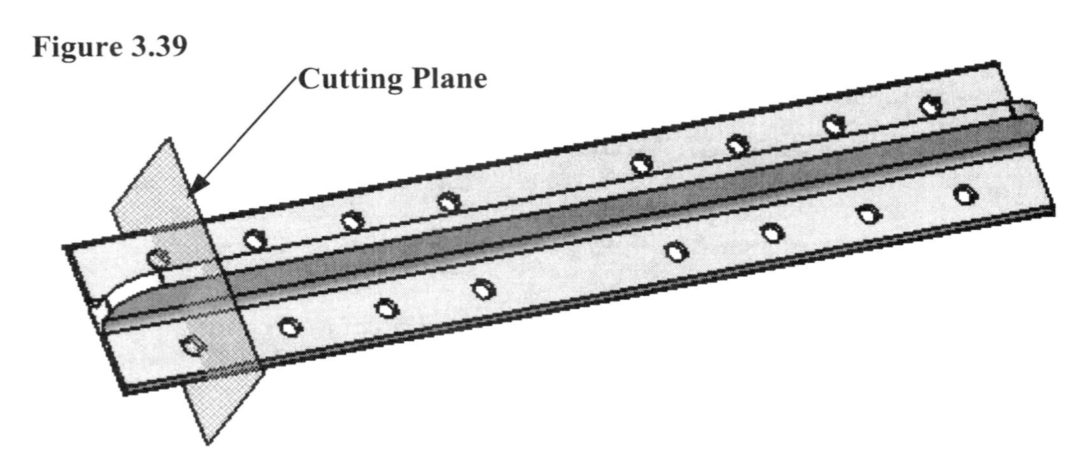

11.6 The **Aligned Section View** is created but not yet located. The **Aligned Section View** will be framed in a green border and will follow your mouse movement.

11.7 Drag the mouse to where you want the **Aligned Section View** and select that location. The **Aligned Section View** will now be frozen in the selected location. When you select the location, the part goes away and the **Section View** remains similar to what is shown in Figure 3.40. CATIA V5 will automatically call the view **Section View B-B**. You might have to go back to a full screen view and/or **Zoom Up** on the view to see all the detail clearly. If you selected the center of the hole location, you will be able to see that the **Section View** shows the holes and their centerlines.

11.8 You can modify the location, title and scale of **Section View B-B**. You can also use the **Show/No Show** tools to modify the appearance of **Section View B-B**. To change the **Properties** of the **Section View B-B,** refer to Step 9. For this step, no properties need to be modified.

Figure 3.40

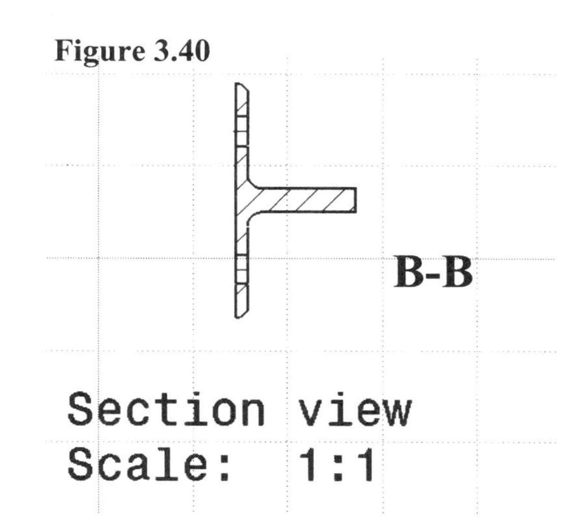

NOTE: CATIA V5 has saved you a lot of work. It has automatically created and labled the **Section View** and linked it to the **Top View**. The **Drafting Specification Tree** added the the new **Section View B-B** to the current sheet as shown in Figure 3.41. CATIA V5 has taken the responsibility of naming the views, but allows you the flexibility to modify them if so desired.

This is a very powerful drafting tool. Although this lesson does not go into any more detail on this tool, you now have an introductory knowledge of how to apply it to your drawings. You can now expand that basic knowledge by applying it to different parts. The **Practice Exercises** at the end of this lesson will give you that opportunity.

Figure 3.41

Drawing 1
Sheet.1
Front view
Top view
Right view
Sheet.2
Front view
Isometric view
Top view
Right view
Detail A
Section view B-B

12 Creating An Auxiliary View

Orthographic projection can usually show enough part detail to show all dimensions and surfaces "**True Size**." When standard orthographic views can not show all of the required detail, auxiliary views are used to show the required detail. CATIA V5 has made the creation of auxiliary views quick and easy. This step will take you through creating an auxiliary view of the "**T Shaped Extrusion**." The standard orthographic views show all of the dimensions required on this part, except for the chamfer on the base leg. To show the chamfer surface "**True Size**," an **Auxiliary View** is required. Complete the following steps to create an **Auxiliary View**.

12.1 Activate the **Right View**, the view shown in Figure 3.25. The **Right View** shows the edge of the chamfer the best. It is the edge of the chamfer that you will use to project the **Auxiliary View** from.

12.2 Select the **Auxiliary View** tool . This tool is found in the **Views** tool bar. It is not the default tool, so you may have to select the little arrow at the bottom to access the **Auxiliary View** tool.

12.3 CATIA V5 will prompt you to "**Sketch a projection plane**." This prompt is asking you to draw a line in the active view. The line will be used to project the new view 90 degrees from the line. What you need to accomplish in this step is to show the chamfered surface "**True Size**" in your **Auxiliary View**. To do this, you must create a line that is parallel to the chamfered edge in the **Right View**.

12.4 Move your mouse to the **Right View**, the **Autodetector** will follow. Notice, as you move your mouse past the chamfered edge, the edge highlights. This signifies that you are parallel with that line. When the chamfer edge is highlighted, select the point as shown in Figure 3.42. This will be the starting point for your projection line.

Figure 3.42

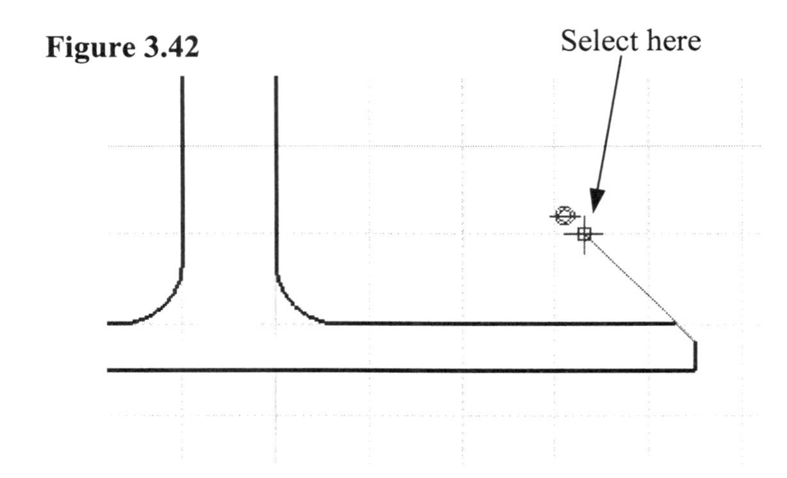

12.5 CATIA V5 will prompt you for a "**Second_State**." This means, select the second or ending point for your projection line. If you drag the mouse along (parallel to) the chamfered edge, the chamfered edge will remain highlighted. This signifies that you are still parallel to the chamfered edge line. Drag it parallel and past the part as shown in Figure 3.43, then select the point.

Figure 3.43

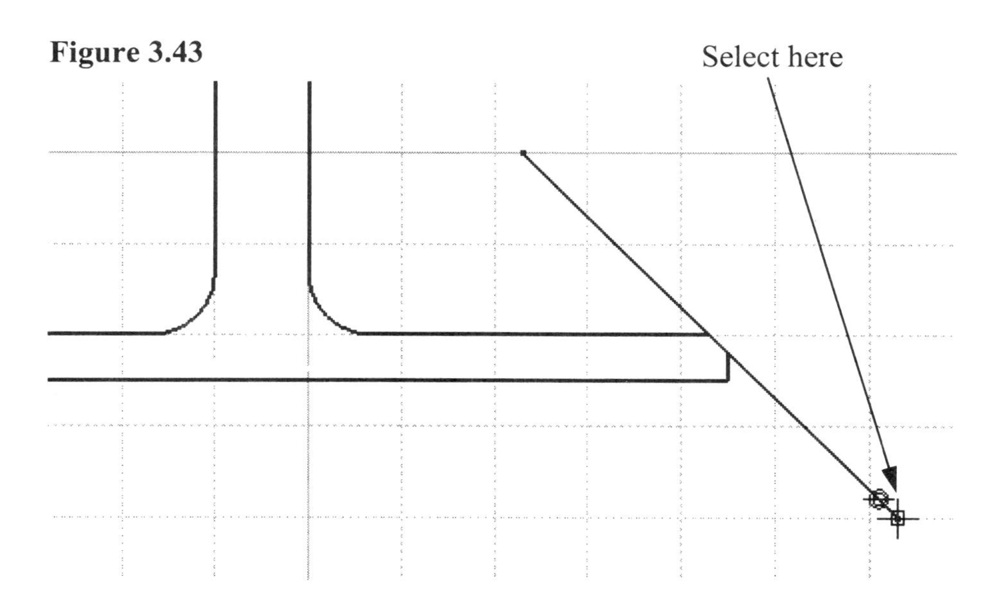

12.6 CATIA V5 will prompt you to "**Click the view position**." CATIA V5 has created the **Auxiliary View,** but is asking you where to place it. Drag the mouse 90 degrees away from the projection line you created in Steps 12.3 thru 12.5. As you drag the mouse away, you will notice the new **Auxiliary View** trailing the mouse movement. Move the mouse to an open spot on your sheet and select that location as shown in Figure 3.44. The new **Auxiliary View** will be locked in that location. The view location can be modified, but location is limited. Since the view was created 90 degrees from the projection line, the view is locked into the line of projection. You can move the view away from the projection line, but not from side to side.

Figure 3.44

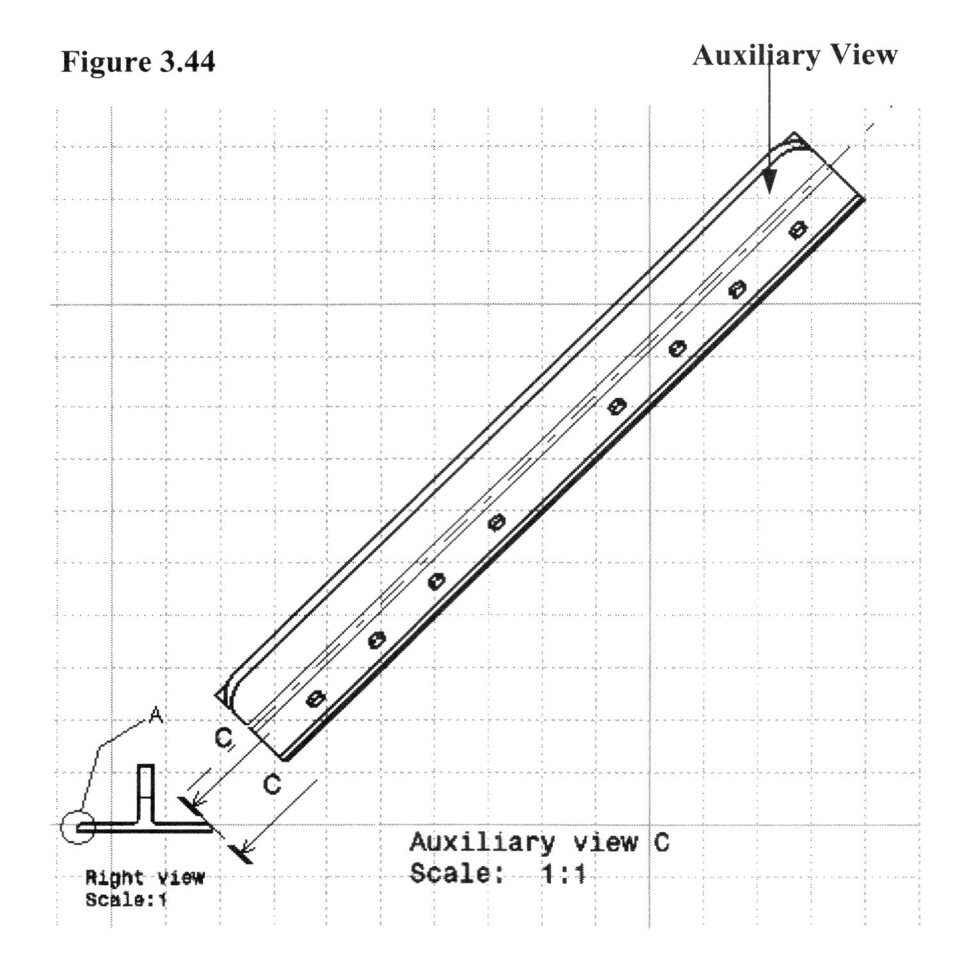

12.7 Once you lock in the view location, the view name and scale appear as shown in Figure 3.44. An automatic link is made to the **Auxiliary View** and the view it was projected from. CATIA V5 creates the projection lines and reference to the **Auxiliary View**.

12.8 The newly created **Auxiliary View** is very long and takes up a lot of room on the sheet. If a small portion of the view will show all the information needed, you can clip the view. The **Clipping** tool allows you to show only the portion of the view you select. Select the **Clipping View Profile** tool located in the **View** tool bar. All you need to do is draw a frame around the part of the view you want to be visible. The portion of the view outside of the frame will not be visible. The frame is created using successive lines, then double clicking to end. For this lesson, use the **Clipping View Profile** tool to hide the overall length of the "**T Shaped Extrusion**," reference Figure 3.45 and 3.46.

Figure 3.45

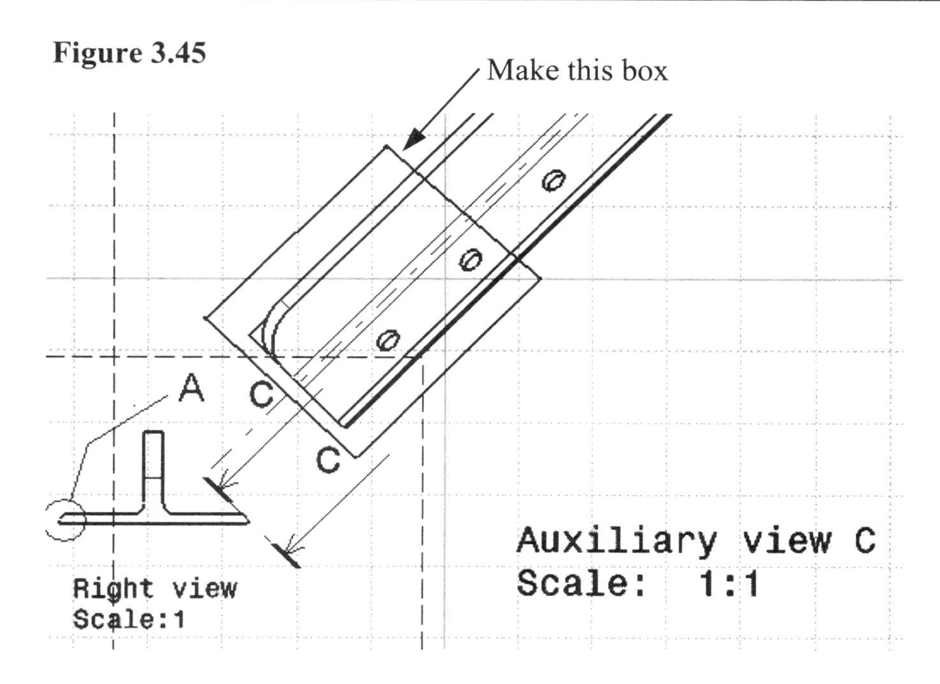

Figure 3.46 (after **Clipping** has been performed)

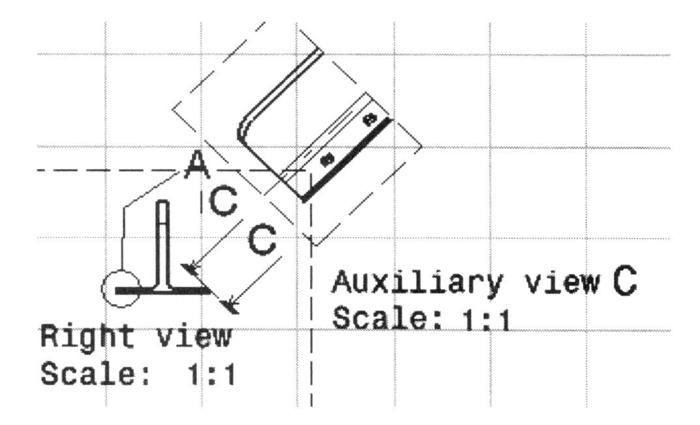

The combination of the tools available give you a limitless possiblity of view creation and manipulation. This lesson was a brief introduction to the possiblities. This lesson did not cover all of the tools available. The other tools are similar, so if you get the idea of what you need to do, you can easily and quickly learn the other tools.

13 Saving Your Newly Created Drawing

This process is basically identical to saving a file covered in Lesson 1. The only difference is that it will be saved with the name you give it followed by the extension of "**CATDrawing**." This "**CATDrawing**" file will be linked to the "**CATPart**" file it used to create the views. Next time you bring up this "**CATDrawing**" file, it will go to the same directory looking for the "**CATPart**" file. If the "**CATPart**" file has been deleted or moved to another directory, you will get a warning that it can not find the "**CATPart**" file required to complete the CATIA Session. You will have the option to bring up just the "**CATDrawing**" file, but the part will not be loaded. If the file was moved to another directory, you will have the opportunity to type in the new directory so the file can be linked.

For this lesson, save your new drawing as "**T Shaped Extrusion.CATDrawing**."

14 Printing The Newly Created Sheets

Printing the final product is very similar to almost any other **Windows** printing process. To print **Sheet.1** and/or print **Sheet.2**, complete the following steps.

14.1 Before you print, go to the the **Page Setup** window found in the **File** pull down menu. The **Page Setup** window shown in Figure 3.47, allows you to change the **Format** of your sheet, including the standard size (**Width & Height**). It allows you change the **Orientation** to **Landscape** or **Protrait**. If you have not changed your units to inches in the **Tools**, **Options**, **General** menu, the **Width** and **Height** will be in mm. The last option is to modify the **Background**. If you have made the appropriate changes along the way, this should be just a verification process. Reference Figure 3.47 for the layout of the **Page Setup** window.

Figure 3.47

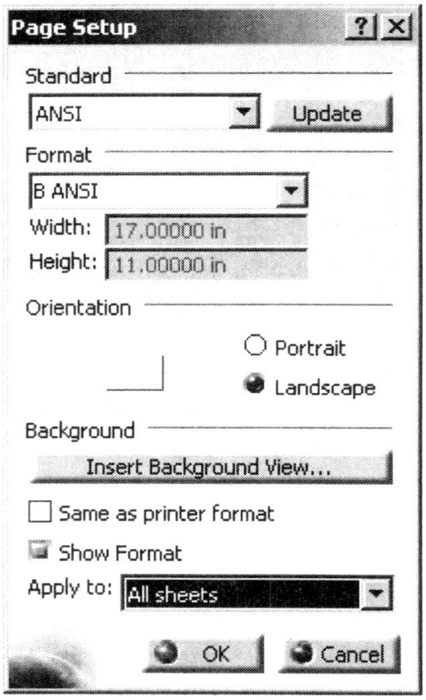

14.2 Select the **File**, **Print Preview** option. The **Print Preview** window will pop up. The window will give you a good idea of what your print will look like before you commit it to paper. If you are satisfied with the what is shown on the **Print Preview** window, select the **Print** tool to print.

14.3 If you already know that your sheet is set up the way you want it, you can skip Step 2 and select the **Print** option. This will bring up the **Print** pop-up window. The **Print** window offers all the print options any **Windows** application does; **Printer**, **Print Range**, **Copies, Preview** and **Options** buttons. For the **Print** window layout and options, reference Figure 3.48. You were instructed to selected the "**B ANSI**" sheet size which is 11″ × 17″. If you have a printer/plotter, that will plot that size, you are good to continue. If you have a standard 8.5″ × 11″ printer, you will have to select the "**Fit toPage**" option. This will print the 11″ × 17″ on 8.5″ × 11″; hence, the print will not be to scale.

Figure 3.48

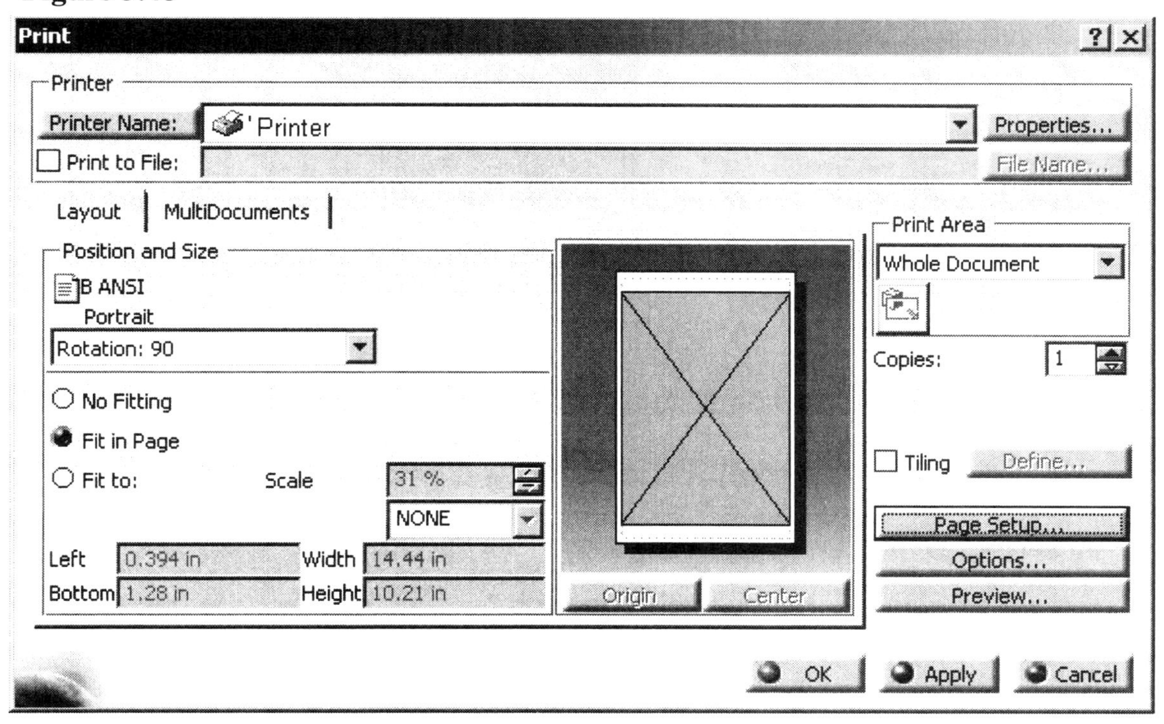

14.4 If everything is set up the way you want it, select the **OK** button at the bottom of the **Print** window. A click of the **OK** button and your masterpiece should be on its way!

This lesson does not require you to print any of the sheets you created. However, it would be a good idea to try printing now. Future lessons will refer back to this step when printing is required.

Lesson 3 Summary

Creating a drawing is no small task. CATIA V5 has gone a long way in simplifying and automating the process, where possible. The best thing CATIA V5 has done is to make the process easier to learn. This lesson supplies the introductory tools required to start producing drawings immediately. You now have enough knowledge to explore and learn the more advanced options.

Lesson 3 Review

After completing this lesson you should be able to answer the questions and explain the concepts listed below.

1. A file created in the **Drafting Work Bench** has what type of file extension?
2. What are the similarities between the **Drafting Work Bench** and the **Sketcher Work Bench**?
3. **T** or **F** A line created in the **Drafting Work Bench** is a three-dimensional entity.
4. List at least three types of **Views** CATIA V5 creates automatically for you.
5. **T** or **F** The **Drafting Work Bench** has a **Specification Tree.**
6. **T** or **F** CATIA V5 allows a maximum of 5 views per sheet.
7. **T** or **F** CATIA V5 allows you to transfer a view from one sheet to another.
8. **T** or **F** By default, all of the tool bars are visible when entering into the **Drafting Work Bench**.
9. How do you toggle the **Snap to Grid** on and off?
10. How do you modify the **Primary Spacing** of the grid?
11. **T** or **F** You have to use the **View Creation Wizard** to create a sheet.
12. What tool bar contains the following tools: **Centerline**, **Centerline With Reference**, **Thread**, **Thread With Reference** and **Axis Line**?
13. **T** or **F** CATIA V5 allows you to create an unlimited amount of projected views.
14. What is the function of the **View Creation Wizard**?
15. **T** or **F** When creating a **Section View**, CATIA V5 helps you visualize it by creating a cutting plane through the part in both the **Drafting Work Bench** and the **Part Design Work Bench**.
16. How do you activate the **Properties** window for a view?

17. **T** or **F** CATIA V5 saves the drafting sheets created in the **Drafting Work Bench** and parts created in the **Part Design Work Bench** in the same file with the same extension.

18. Does the **Drafting Work Bench** have tools to create drafting geometry, such as 2-D lines, arc, circles etc.?

19. Explain how you can modify an **Auxiliary View** so you can move it anywhere on the sheet and not just in the direction of the created projection (hint: **Alignment**).

20. How do you split the screen so you can see the **CATDrawing** screen and the **Part Design** screen at the same time?

Lesson 3 Practice Exercises

The practice exercises for this lesson require you to use the "**CATPart**" files you created back in Lesson 1 and extruded in Lesson 2. Each practice part has its own list of special requirements. The requirements are a minimum, so you have to use the tools learned in this lesson. This workbook is not meant to teach any drafting standards; although, good practices are strongly encouraged. It is up to you to layout the drawings as to the best of your experience and/or education will allow. Remember text and dimensions will be covered in Lesson 4. The file extension used in the **Drafting Work Bench** is "**CATDrawing**." Good Luck!

1 Use the "**Lesson 2 Exercise 1.CATPart**" to create a drawing with the following sheet and view requirements.

1.1 Create a drawing with three standard orthographic views of this part.
1.2 Put the drawing on **ANSI C** sheet.
1.3 Create a second sheet.
1.4 Create an **Aligned Section View**. Run the section view vertically through the center of the hole.
1.5 Create an **Auxiliary View** off of the angled surface noted in the drawing below.
1.6 Create a **Detail View** of the chamfer shown on the drawing below. Scale the detail view 4:1.
1.7 Save the drawing as "**Lesson 3 Exercise 1. CATDrawing**."

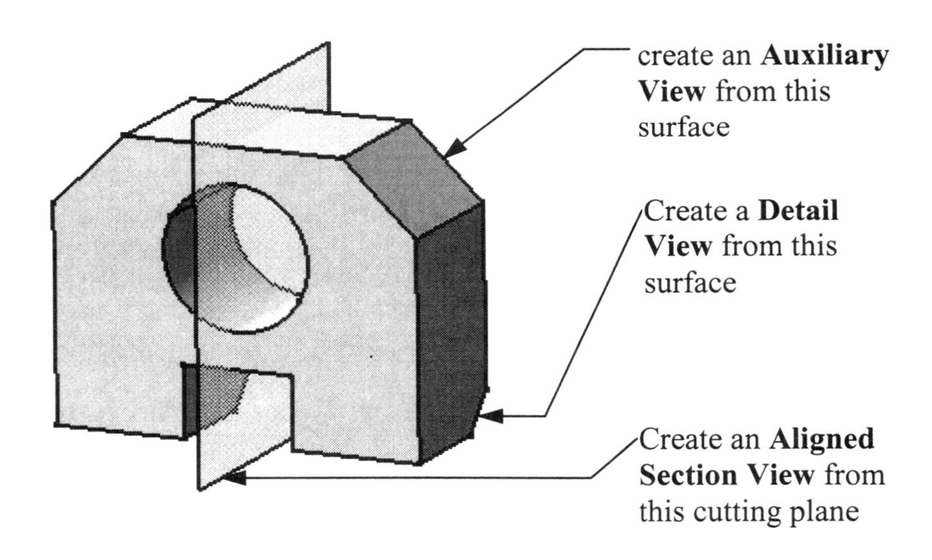

2 Use the "**Lesson 2 Exercise 2.CATPart**" to create a drawing with the following sheet and view requirements.

2.1 Create a drawing with three standard orthographic views of this part.
2.2 Put the drawing on **ANSI B** sheet.
2.3 Create a second sheet.
2.4 Create an **Aligned Section View**. Run the section view vertically through the center of the holes.
2.5 Create an **Aligned Section Cut**. Run the section cut vertically through the center of the holes.
2.6 Create an **Auxiliary View** off of the angled surface noted in the drawing below.
2.7 Save the drawing as "**Lesson 3 Exercise 2.CATDrawing**."

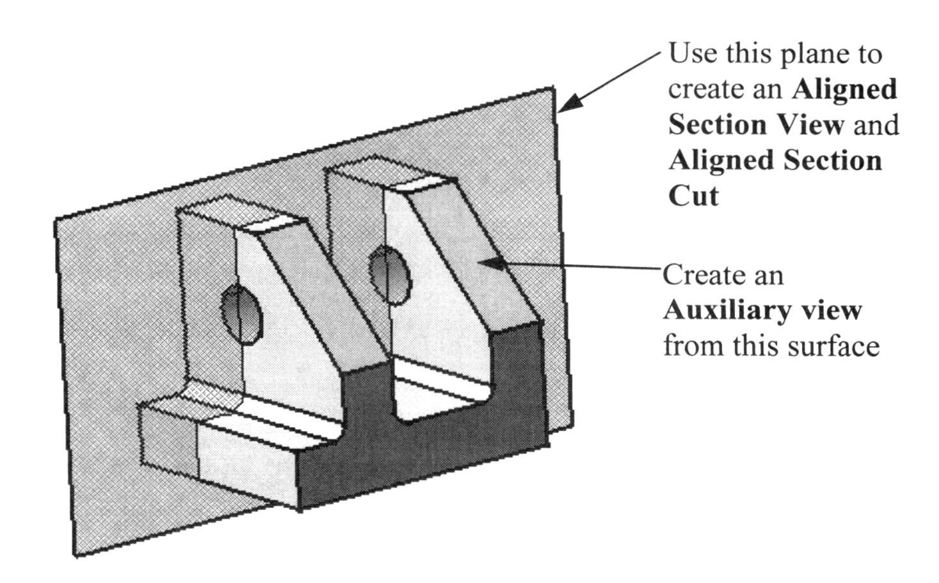

3 Use the "**Lesson 2 Exercise 3.CATPart**" to create a drawing with the following sheet and view requirements.

3.1 Create a drawing with three standard orthographic views of this part.
3.2 Put the drawing on **ANSI C** sheet.
3.3 Create a second sheet.
3.4 Create an **Aligned Section View**. Run the section view vertically through the center of the hole.
3.5 Save this part as "**Lesson 3 Exercise 3. CATDrawing**."

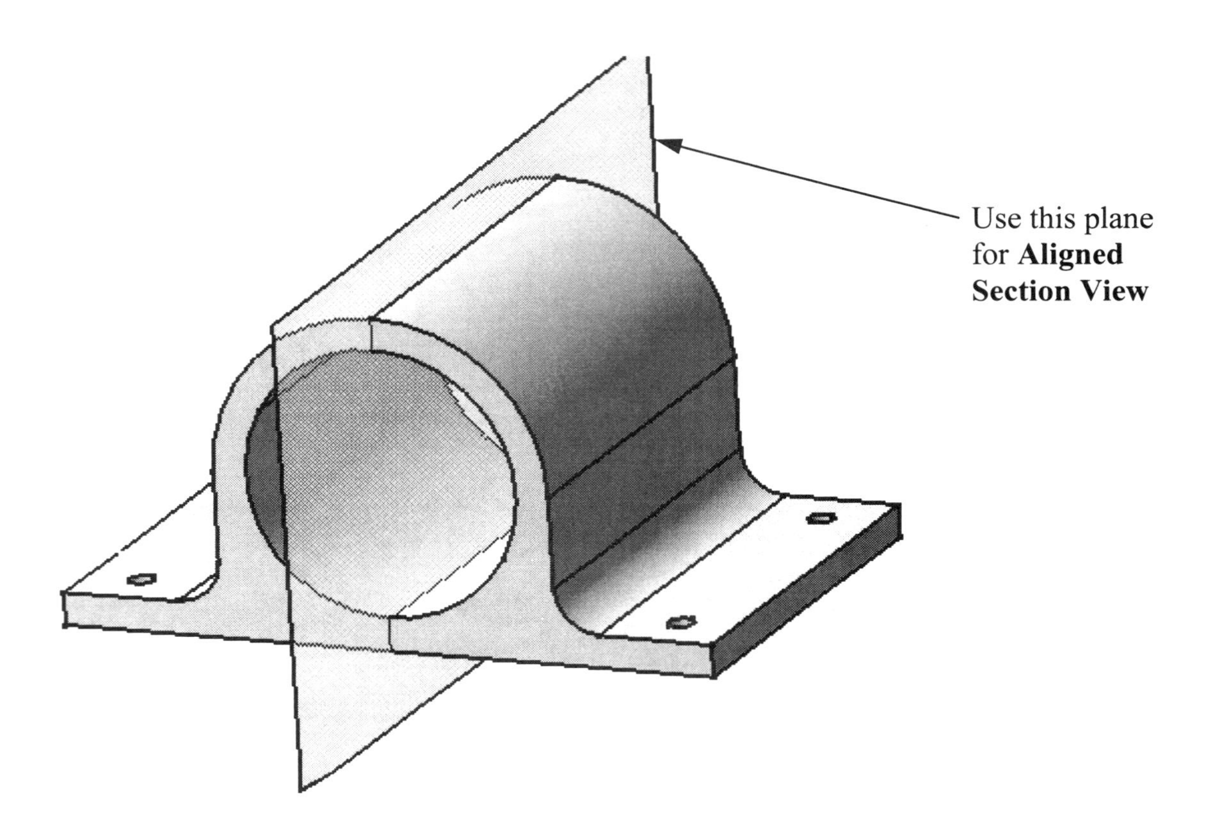

4 Use the "**Lesson 2 Exercise 4.CATPart**" to create a drawing with the following sheet and view requirements.

4.1 Create a drawing with three standard orthographic views of this part.
4.2 Put the drawing on **ANSI A** sheet.
4.3 Create a second sheet.
4.4 Create an **Aligned Section View**. Run the section view vertically through the center of the holes.
4.5 Create an **Aligned Section Cut**. Run the section cut vertically through the center of the holes.
4.6 Save this part as "**Lesson 3 Exercise 4. CATDrawing**."

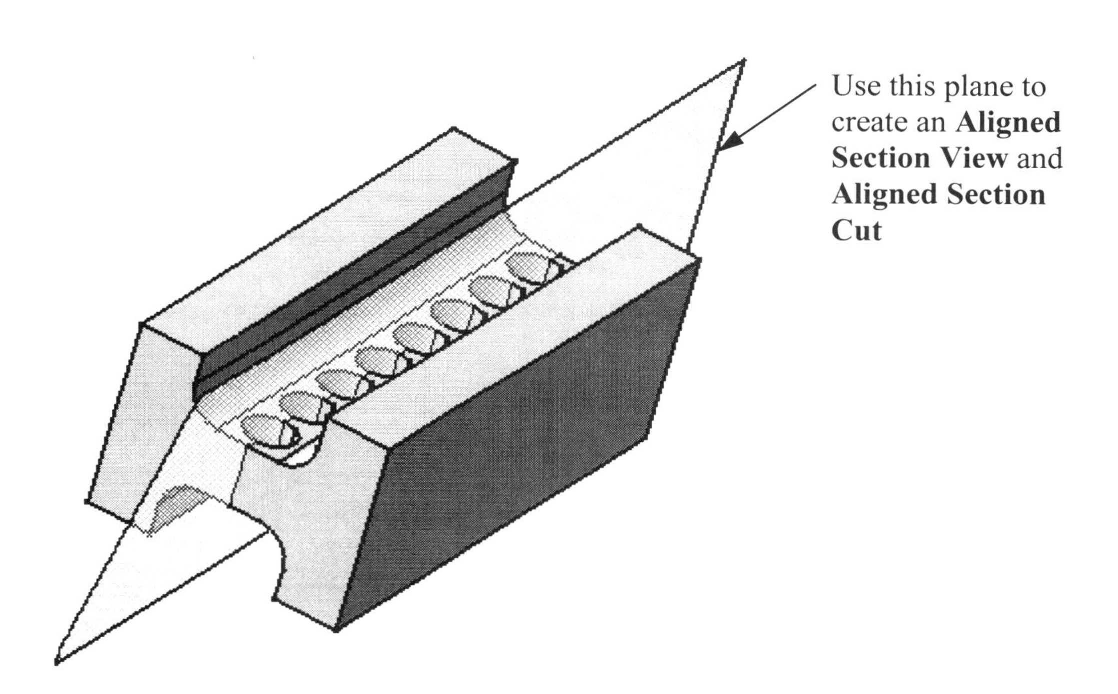

NOTES:

Introduction To Creating Text And Dimensions Using The Drafting Work Bench

This lesson continues from Lesson 3 with the **Drafting Work Bench**. You will use the **Sheets** and **Views** of the "**T Shaped Extrusion.CATDrawing**" created in Lesson 3. This lesson will instruct you on how to add the text and dimensions to the views created in Lesson 3. Like Lesson 3, this is not a comprehensive lesson on the **Drafting Work Bench**.

Figure 4.1

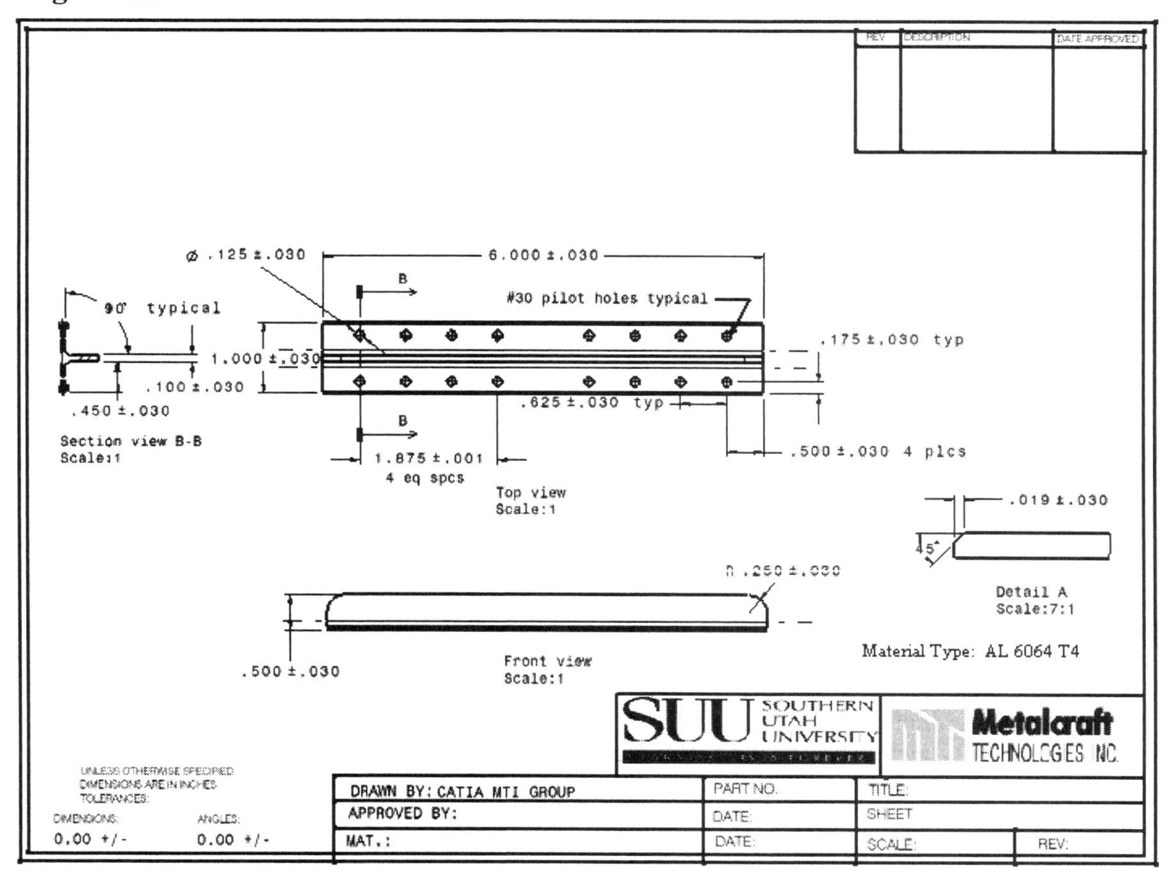

Lesson 4 Objectives

The objective of this lesson is to supply you with enough instruction that you can complete a production drawing of the "**T Shaped Extrusion**" as shown in Figure 4.1. You should be able to apply this newfound knowledge to the exercise parts and any other part that requires a production drawing. This chapter will not cover every tool found in the **Drafting Work Bench**. It is not the intent of this lesson to be a comprehensive reference, but a basic step-by-step for the most common tools and functions in the CATIA V5 **Drafting Work Bench**.

The tools and processes covered in this lesson are listed below.

- Additional **Drafting Work Bench** tools
- Review of lesson 3
- How to create and modify **Text**
- How to create and modify **Leaders**
- How to create and modify **Dimensions**
- How to create a **Border** and **Title Block**
- How to insert the **Bill Of Materials**
- How to insert **Picture Files**
- Review how to **Save** and **Print** the completed drawing sheets (**CATDrawing**)

Drafting Work Bench Tool Bars 

There are thirteen standard tool bars found in the **Drafting Work Bench**. Five of the thirteen tool bars were covered in Lesson 3. The remaining tool bars are covered in this lesson. The individual tools found in each of the five tool bars are also shown and include the tool name and a brief definition.

The **Geometry Creation** Tool Bar

This tool bar is the same as the one used in the **Sketcher Work Bench** for this reason all the options are not covered in this section. The entities created in this work bench are 2D drafting entities only.

Tool Bar	Tool Name	Tool Definition
	Point by Clicking	Creates a point.
	Line	Creates a line.
	Circle	Creates a circle.

	Profile	Creates profile.
	Spline	Creates a spline.

The **Geometry Modification** Tool Bar

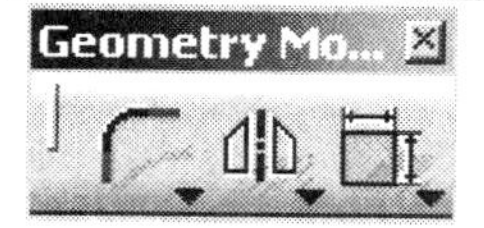

This tool bar is the same as the one used in the **Sketcher Work Bench** for this reason all the options are not covered in this section. The entities created in this work bench are 2D drafting entities only.

Tool Bar	Tool Name	Tool Definition
	Corner	Provides seven different methods to modify drafting geometry; Corner, Chamfer, Trim, Break, Quick Trim, Close and Complement.
	Symmetry	Provides several different methods of translating and duplicating drafting geometry.
	Geometrical Constraint	Provides a method of constraining drafting geometry

The **Graphic Properties** Tool Bar

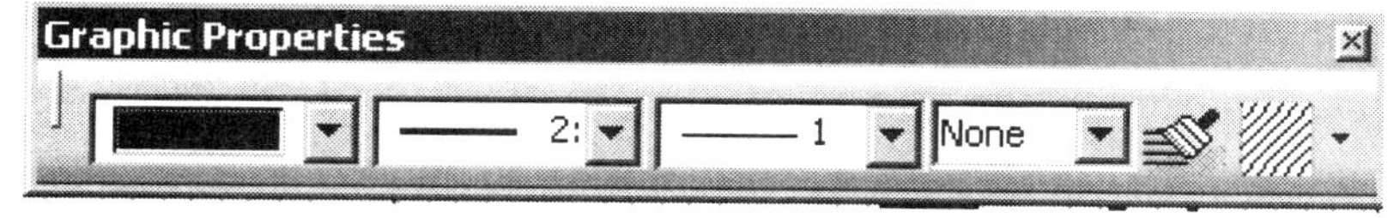

This tool bar is located in the top tool area.

Tool Bar	Tool Name	Tool Definition
	Colors	Allows the user to specify a color to drafting geometry.
	Line Thickness	Allows the user to select from various line thicknesses for the drafting geometry.
	Line Type	Allows the user to specify a line type.
	Layers/Filters	Applies layers and filters to the drafting geometry.
	Copy Object Format	Copies the format of one object to other selected objects.
	Pattern	Applies cross hatching patterns to specified areas.

The **Tools** Tool Bar

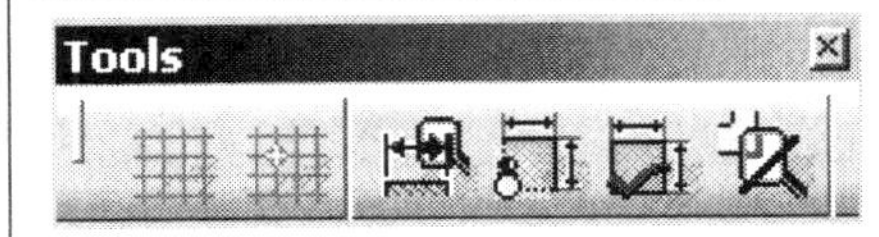

Tool Bar	Tool Name	Tool Definition
	Sketcher Grid	Toggles the grid on or off in the **Drafting Work Bench**.
	Snap To Point	Toggles the Snap to Point on or off.
	Analysis Display Mode	Displays the various types of dimensions in different and customizable colors.
	Show Constraints	Shows 2D constraints in the active view.
	Create Detected Constraints	Creates 2D constraints in the active view.
	Filter Generated Elements	Applies a virtual filter on generated elements.

The **Dimensioning** Tool Bar

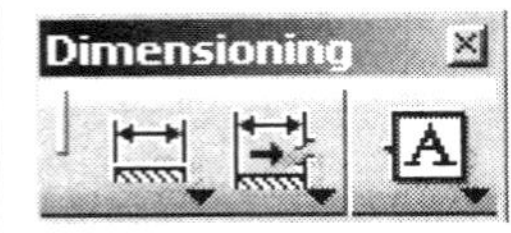

Tool Bar	Tool Name	Tool Definition
Dimensions Tool Bar		
	Dimensions	Creates dimensions.
	Cumulated Dimensions	Creates cumulated dimensions.
	Stacked Dimensions	Creates stacked dimensions.
	Length / Distance Dimensions	Creates length and distance dimensions.
	Angle Dimensions	Creates angled dimension.
	Radius Dimension	Creates a radius dimension.
	Diameter Dimension	Creates a diameter dimension.

	Chamfer Dimension	Creates a chamfer dimension.
	Thread Dimension	Creates a thread dimension in both top and/or side views.
	Coordinate Dimension	Creates coordinate dimensions.
	Hole Dimension Table	Creates a hole dimension table.
	Coordinate Dimension Table	Creates a coordinate dimension table.
Extension Line Tool Bar		
	Create Interruption	Creates an interruption of the selected extension line.
	Remove Interruption	Removes an extension line interruption.
	Remove All Interruptions	Removes all extension line interruptions.
Tolerance Tool Bar		
	Datum Feature	Creates a datum feature.
	Geometric Tolerance	Creates geometric tolerance

The **Generation** Tool Bar

Tool Bar	Tool Name	Tool Definition
Generating Dimensions Tool Bar		
	Generating Dimensions	Generates dimensions from 3D (**Part Design** and **Assembly Design Work Benches**).
	Generating Dimensions step by step	Generates dimensions from 3D step-by-step.
	Generate Balloons	Generates balloons in the active view.

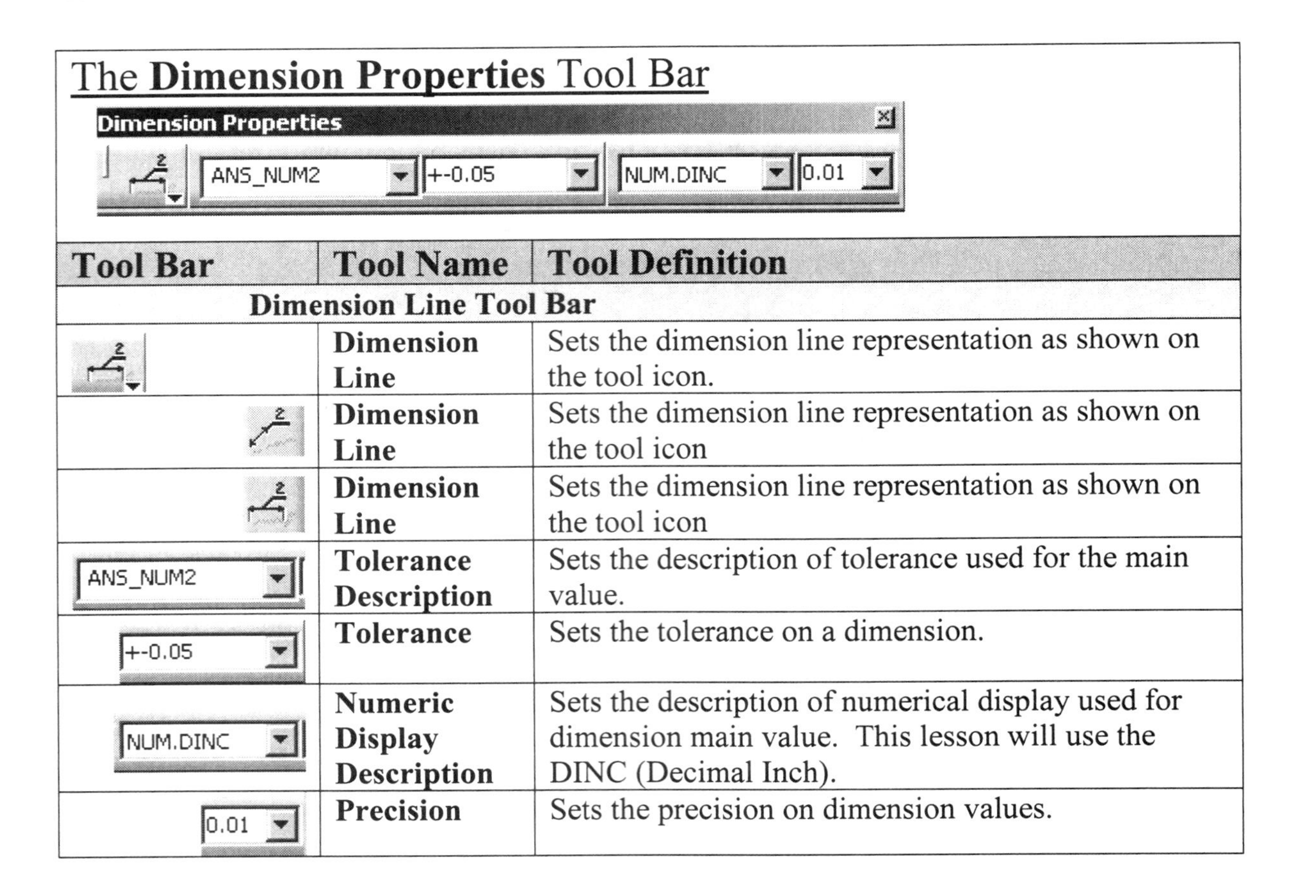

The **Dimension Properties** Tool Bar

Tool Bar	Tool Name	Tool Definition
	Dimension Line Tool Bar	
	Dimension Line	Sets the dimension line representation as shown on the tool icon.
	Dimension Line	Sets the dimension line representation as shown on the tool icon
	Dimension Line	Sets the dimension line representation as shown on the tool icon
ANS_NUM2	**Tolerance Description**	Sets the description of tolerance used for the main value.
+-0.05	**Tolerance**	Sets the tolerance on a dimension.
NUM.DINC	**Numeric Display Description**	Sets the description of numerical display used for dimension main value. This lesson will use the DINC (Decimal Inch).
0.01	**Precision**	Sets the precision on dimension values.

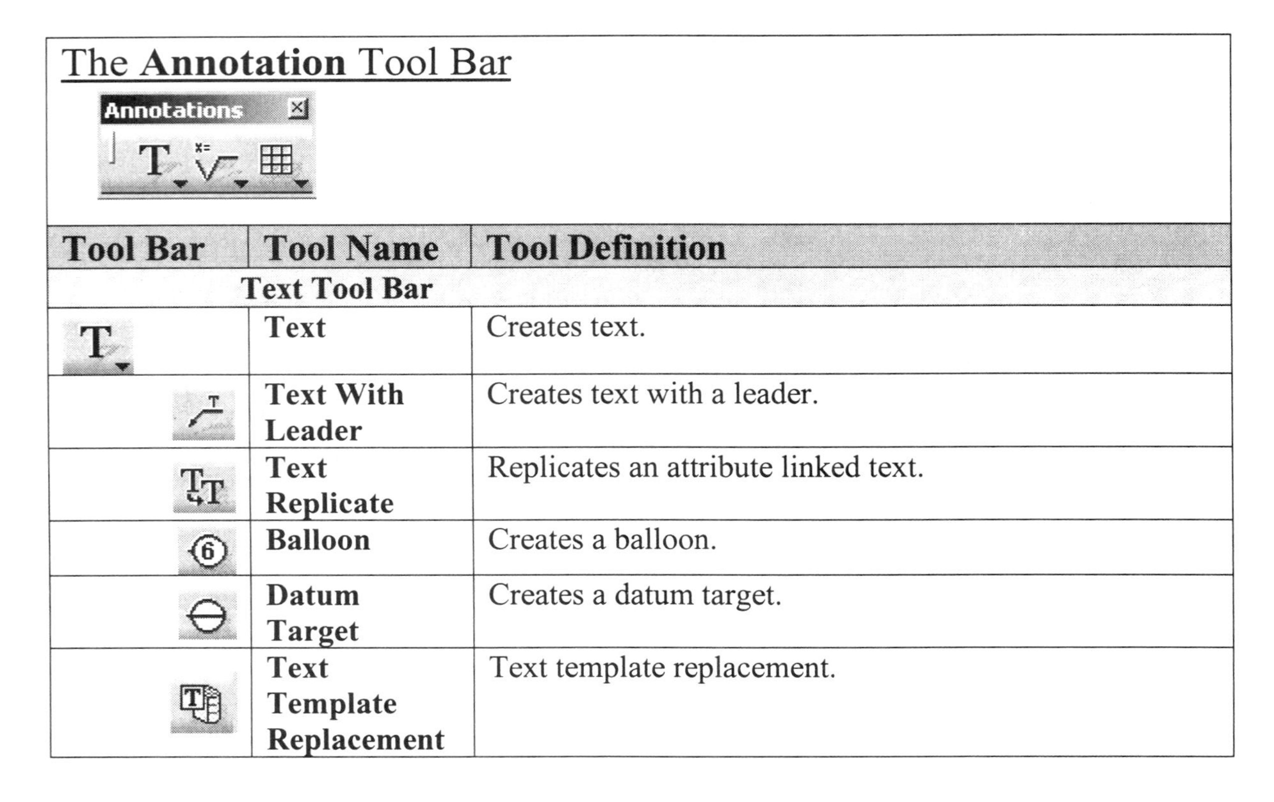

The **Annotation** Tool Bar

Tool Bar	Tool Name	Tool Definition
	Text Tool Bar	
	Text	Creates text.
	Text With Leader	Creates text with a leader.
	Text Replicate	Replicates an attribute linked text.
	Balloon	Creates a balloon.
	Datum Target	Creates a datum target.
	Text Template Replacement	Text template replacement.

Symbols Tool Bar		
	Roughness Symbol	Creates a roughness symbol.
	Weld Symbol	Creates a weld symbol.
	Weld	Creates a weld.
Table Tool Bar		
	Table	Creates a table.
	Table from CSV	Creates a table from a CSV file.

The **Text Properties** Tool Bar

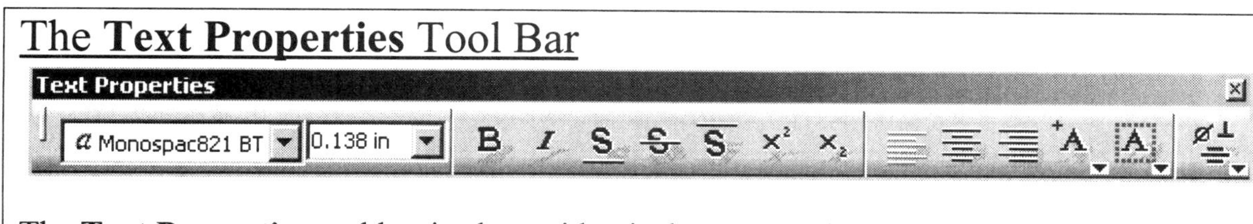

The **Text Properties** tool bar is almost identical to most Microsoft text products and therefore will not be covered in detail.

Adding Text And Dimensions To Drawings Using The Drafting Work Bench

CATIA V5 uses the files created in the **Part Design Work Bench** to create drawings. As previously stated, the files created in the **Part Design Work Bench** are saved with the "**CATPart**" extension. The **Drafting Work Bench** uses the file created in the **Part Design Work Bench**, but is saved as a separate file. The **Drafting Work Bench** files have a **"CATDrawing"** extension. Both the "**CATPart**" and "**CATDrawing**" files are linked together. To pull up the "**CATDrawing**" file, CATIA V5 must be able to locate the related "**CATPart**" file. If CATIA V5 cannot find the linked file, you will get an error. If you move a "**CATPart**" file to a different directory, the "**CATDrawing**" file linked to it will not be able to locate the **"CATPart"** file. You will have to update the directory where the "**CATPart**" exists. This can be done using the **Edit**, **Links** tools. This lesson takes you through the steps of accessing and adding text and dimensions to your existing **"T Shaped Extrusion.CATDrawing"** file. There are many different methods and standards to dimension a drawing. This lesson concentrates on how to create, place and modify text and dimensions. This lesson does not attempt to teach any one dimensioning standard or style. You are encouraged to explore other methods; find out what works best for your requirements.

1 Start CATIA V5

If you need to review this step, refer to Lesson 1.

2 Select The Drafting Work Bench

In this lesson you do not need to open the **Part Design Work Bench** because you already made the "**CATPart**" to "**CATDrawing**" link in Lesson 3. Select the **Drafting Work Bench**. The **New Drawing** pop-up window will appear. Select the **Cancel** button since you will be working on an existing drawing.

3 Open The "T Shaped Extrusion. CATDrawing" Document

This lesson uses the "**T Shaped Extrusion.CATDrawing**" you created in Lesson 3. Open the "**T Shaped Extrusion.CATDrawing**."

4 The Drafting Work Bench Layout

The **Drafting Work Bench** layout was covered in Lesson 3. It is mentioned in this lesson, because it would be a good idea to review the layout, where everything is and how to access all of the tool bars. You are encouraged to go back to Lesson 3 and review the **Drafting Work Bench** layout.

Since you have opened an existing "**CATDrawing**" file, your layout will be a bit different than the **Drafting Work Bench** without any sheets or views created. The following steps cover the same items as Step 5 in Lesson 3, except the information is specific to the opened "**T Shaped Extrusion.CATDrawing**."

4.1 **Tool Bars**: Lesson 3 covered six of the thirteen **Drafting Work Bench** tool bars. This lesson will cover the remaining tool bars. Pull all of the tool bars up on the screen so they are visible and pickable. It would be beneficial to preview all the tools you will be using in this lesson. If you have any detailed questions on the tool bars, refer back to Lesson 3.

4.2 **Sheets**: When you opened the "**T Shaped Extrusion.CATDrawing**," a minimum of two sheets should appear on the sheet tab at the top left of the screen. If you created any additional sheets in Lesson 3, they should also be represented on a tab.

4.3 **View Layout**: You determined the view layouts in Lesson 3 when you created the drawing and added the additional views.

4.4 **Drafting Work Bench Specification Tree**: All of the sheets and views you created in Lesson 3 will be represented in the **Drafting Specification Tree**.

4.5 **Modifiying the Drafting Work Bench Standards**: As in the **Sketcher** and **Part Design Work Benches** the standards are quick and easy to change.

The only action required by you in this step, was to review the tool bars from Lesson 3 and preview the tool bars for this lesson. If you have completed this task, you can continue on to Step 5.

5 Customizing The Default Values

Customizing the default values was covered in detail in Lesson 3. This is a critical step, so it was left in this lesson as a reminder. You need to think about them before you get started. If you have any questions on how to modify any of the default values, refer back to Lesson 3 Step 6. If you do not have any questions on this subject, continue on to the next step.

6 Creating And Modifying Text

CATIA V5 does not have all of the functions of a true word processor, but it is a vast improvement from past CAD system attempts to incorperate text functionality. True **Windows NT** functionality gives CATIA V5 several ways to create and modify text. If you are familiar with **Windows** functions, the **Text** tool bar is almost idenitical. If you look at the top tool bar, you should be able to recognize many similar tools such as **Font Style**, **Font Size**, **Bold**, **Italics**, **Under Line**, **Align Right**, **Align Center, Align Left** and so on.

The **Annotation** tool bar has two options. The first one is the **Text** tool which has four options, they are:

1 The **Text** tool . This is the tool used to create text.

2 The **Text With Leader** .

3 The third option is **Balloon** . **Balloon** places a balloon around a drawing annotation such as a part number.

4 The last option is **Datum Target** .

The other option on the **Annotation** tool bar is the **Bold Symbol** tool. The **Bold Symbol** tool has three options, they are:

1 The **Roughness Symbol** .

2 The **Welding Symbol** .

3 The **Weld Symbol** .

This lesson will not be using any of the last three options, but you might want to know them for furture projects.

To add text to your "**T Shaped Extrusion.CATDrawing**," complete the following steps.

6.1 Text will be attached (linked) to the view that is active at the time the text is created. You must activate the view you want the text to be attached to. If the view is deleted, the attached text will be deleted with it. For this step you will be required to put text in the **Top View**, so activate that view. Select the **Top View** by selecting the border of the **Top View** or by selecting the **Top View** in the **Drafting Specification Tree**.

6.1.1 Select the **Text** tool in the **Annotation** tool bar.

6.1.2 CATIA V5 will prompt you to "**Indicate The Text Anchor Point**." Move the mouse to the location in the active view where you want the text created and select that location. This is where the text will be anchored (located). For this step, select a location just under and to the left of the "**View Title**" and "**Scale Value**" as shown in Figure 4.2. The text can be easily relocated later if you change your mind.

Figure 4.2

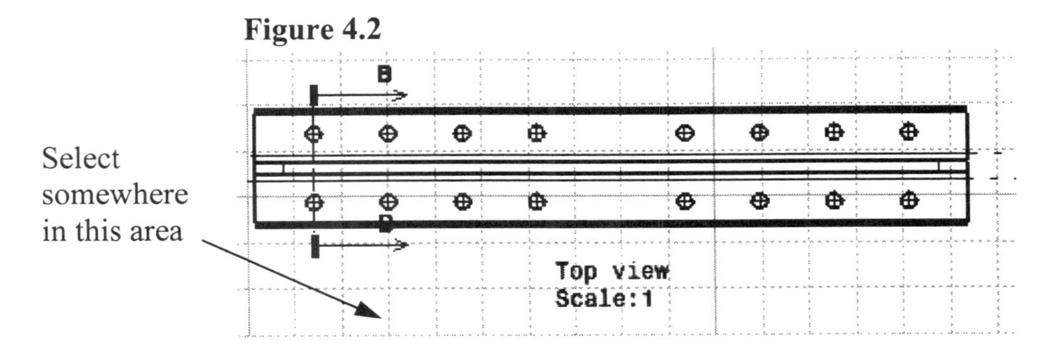

6.1.3 A **Text Editor** pop-up window will appear as shown in Figure 4.3. As the name indicates, this is the text editor. For this step, you will be adding a note specifying the material type. In the **Text Editor** window, type in the following: "**Material Type is 6064 Aluminum in the T6 Condition**."

Figure 4.3

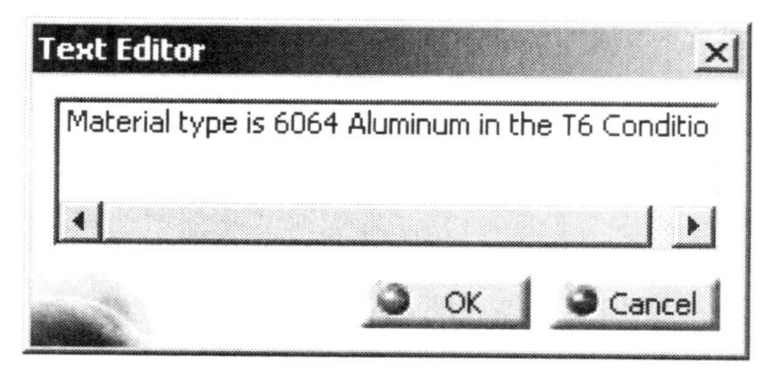

6.1.4 When you've entered the text as shown in Figure 4.3, select the **OK** button. The **Text Editor** window will disappear and the text will show up where you indicated in Step 6.1.2. Your **Top View** should now look similar to the view shown in Figure 4.4.

6.1.5 You may need to move the text around a little if it isn't similarly located as shown in Figure 4.4. The newly created text is created within a text window. The text window can be moved by selecting it and dragging it around to where you want it. This is a **Windows** drag and drop operation.

6.1.6 You can change the size of the text window by selecting one of the window handles (box) and moving the mouse accordingly, reference Figure 4.5 .

Figure 4.4

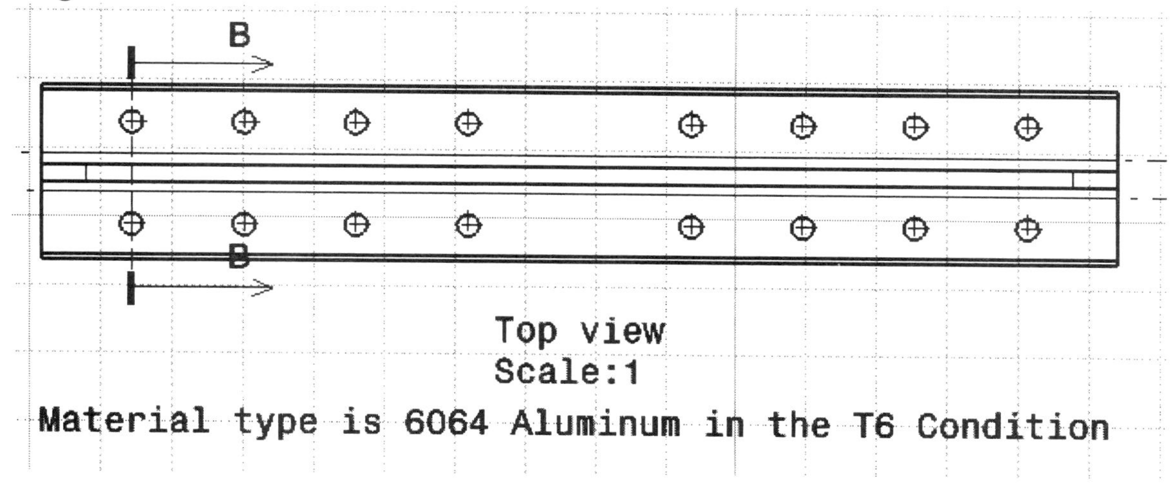

Figure 4.5

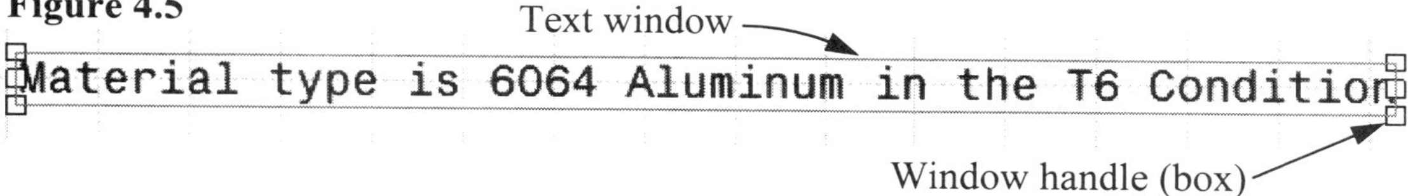

6.1.7 The text can be edited by double clicking on the text. Double clicking on the text brings up the **Text Editor** pop-up window with the text you just selected in the window. You can delete, add and/or modify the text as you would in any other **Windows** text editor. When you have completed all of the text modifications, select the **OK** button to apply the modifications.

6.1.8 To change the properties of the text, move the mouse over the text window as shown in Figure 4.6. The text window will highlight. While the mouse is over the text window, right click the mouse. From the pop-up window that appears, select the **Properties** option. This will bring up the **Properties** pop-up window as shown in Figure 4.7.

Figure 4.6

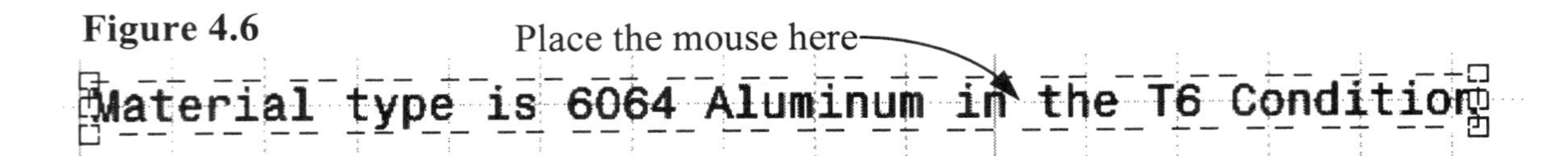

The **Properties** window has three main tabs, they are:

1 **Font**, reference Figure 4.7 .

2 **Text**, reference Figure 4.8 .

3 **Graphic**, reference Figure 4.9 .

Figure 4.7

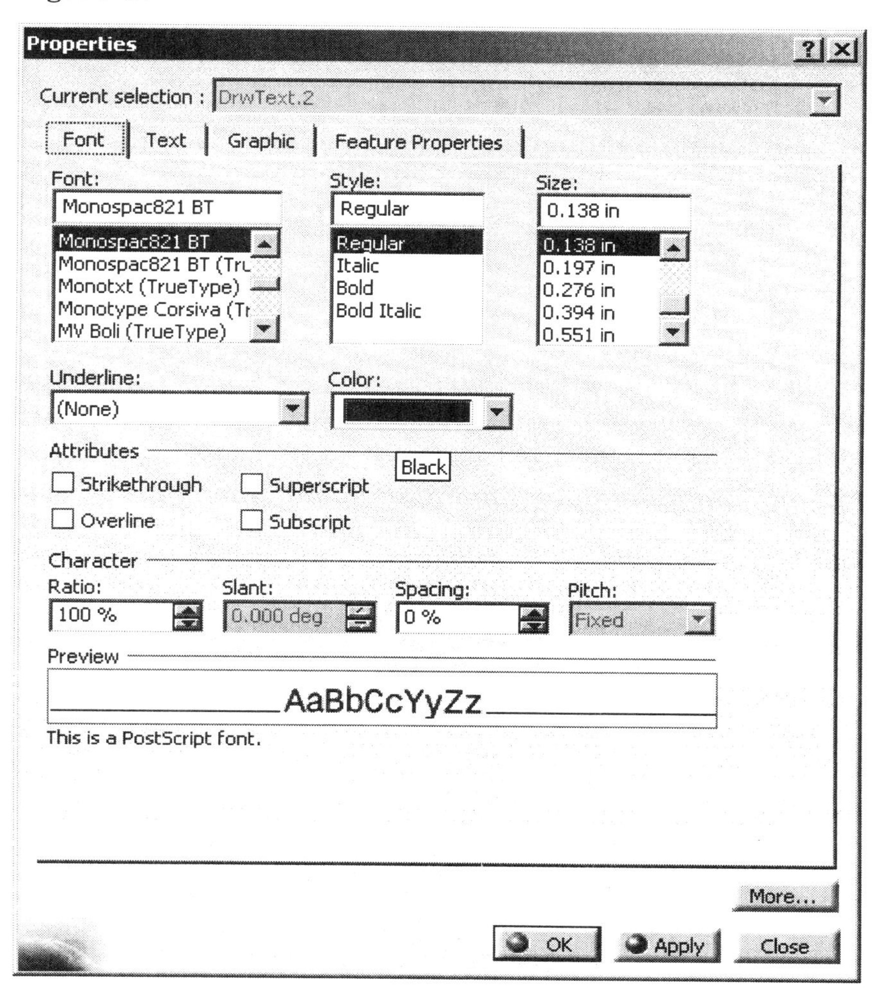
Properties
Current selection : DrwText.2
Font
Text
Graphic
Feature Properties
Font:
Monospac821 BT
Monospac821 BT
Monospac821 BT (Tru
Monotxt (TrueType)
Monotype Corsiva (Tr
MV Boli (TrueType)
Style:
Regular
Regular
Italic
Bold
Bold Italic
Size:
0.138 in
0.138 in
0.197 in
0.276 in
0.394 in
0.551 in
Underline:
(None)
Color:
Black
Attributes
Strikethrough
Superscript
Overline
Subscript
Character
Ratio:
100 %
Slant:
0.000 deg
Spacing:
0 %
Pitch:
Fixed
Preview
AaBbCcYyZz
This is a PostScript font.
More...
OK
Apply
Close

Figure 4.8

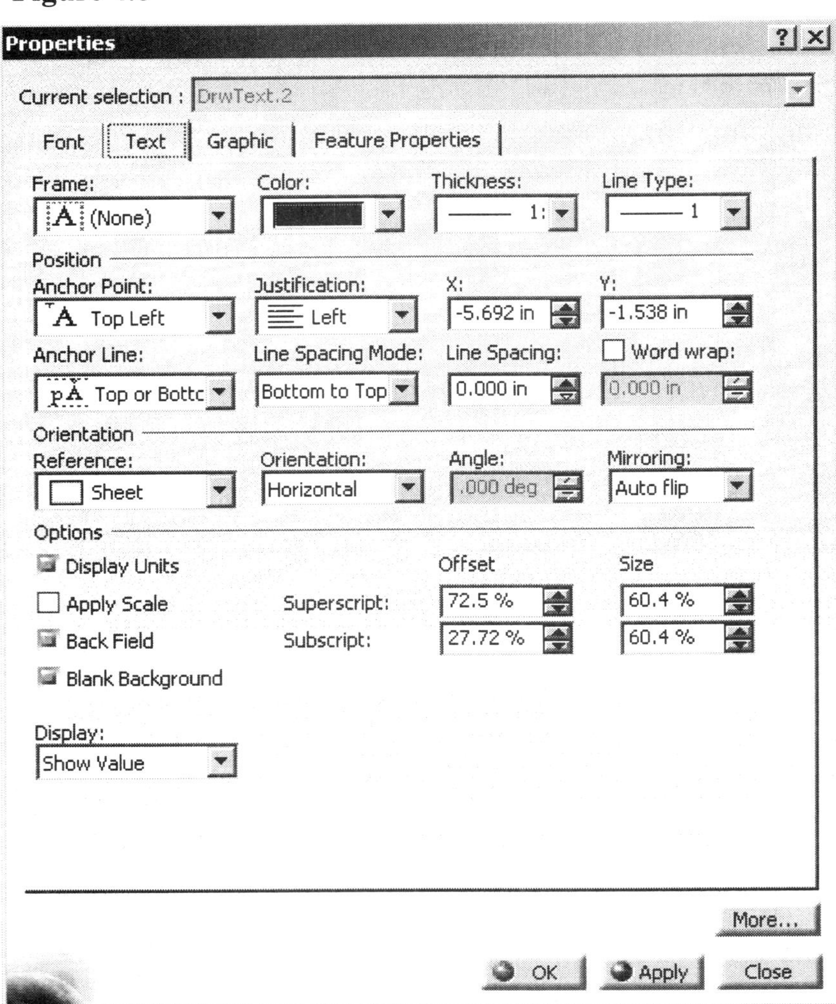
Properties
Current selection : DrwText.2
Font
Text
Graphic
Feature Properties
Frame:
(None)
Color:
Thickness:
Line Type:
Position
Anchor Point:
Top Left
Justification:
Left
X:
-5.692 in
Y:
-1.538 in
Anchor Line:
Top or Bottc
Line Spacing Mode:
Bottom to Top
Line Spacing:
0.000 in
Word wrap:
0.000 in
Orientation
Reference:
Sheet
Orientation:
Horizontal
Angle:
.000 deg
Mirroring:
Auto flip
Options
Display Units
Apply Scale
Back Field
Blank Background
Offset
Size
Superscript:
72.5 %
60.4 %
Subscript:
27.72 %
60.4 %
Display:
Show Value
More...
OK
Apply
Close

Figure 4.9

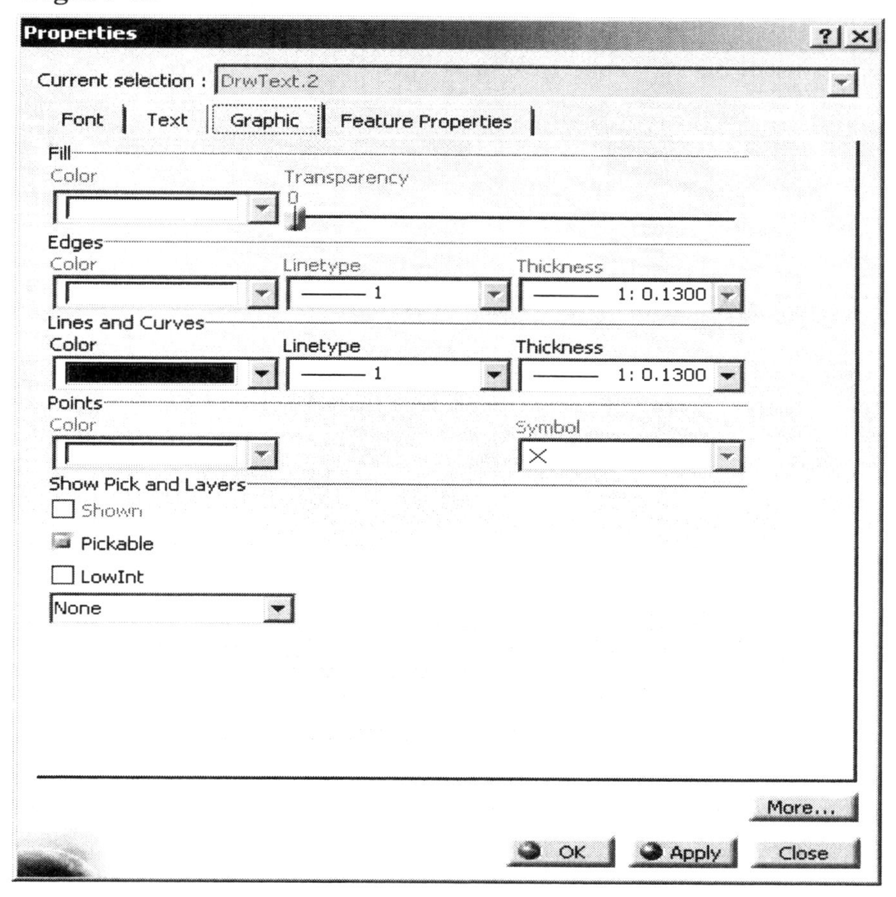

Once all the changes have been made, select **Apply** to verify that the changes are indeed what you expected. If the applied changes are as you expected, select **OK** and the changes will become permanent.

Make the following modifications to the text you created in Step 6.1.3.

1 **Edit** the text so the material type is in the "**T4**" condition instead of "**T6**." Reference Step 6.1.7.

2 Change the **Font** to **ROM3**. Reference Step 6.1.8 under the **Font** tab.

3 Change the **Style** to **Bold**. Reference Step 6.1.8 under the **Font** tab.

4 Change the **Size** to **.15**. Reference Step 6.1.8 under the **Font** tab.

5 Change the the **Color** to **Green**. Reference Step 6.1.8 under the **Font** tab.

6 Change the **Justification** to **Center**. Reference Step 6.1.8 under the **Text** tab.

After completing the changes listed above, the **Top View** should look similar to the one in Figure 4.10.

Figure 4.10

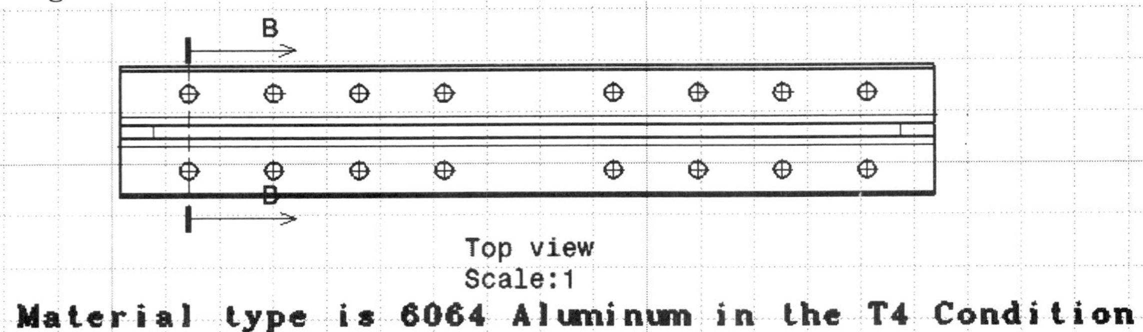

7 Creating And Modifying Leaders

This step is actually a continuation of Step 6. The process of creating a leader is the same as creating text, with one additional step. The following explain the process of creating **Leaders**.

The **Text With Leader** tool is found in the **Annotation** tool bar, it is a sub option under the **Text** tool. All of the information in Step 6 applies to the **Text With Leader** tool, except that the first point selected is the starting point of the leader. CATIA V5 will prompt you to "**Select Element Or Indicate The Leader Anchor Point**." The second point is actually the "**Text Anchor Point**." Once you have selected the leader anchor point and text anchor point, the process is identical to the **Text** tool. The text created with the **Text With Leader** tool can be modified the same as the text created with the **Text** tool, covered in Step 6. The leader anchor point and text anchor point can also be modified. Select the text and leader so they are highlighed. Once the leader text is highlighted, handles on the leader and text box show up. You can select and drag each handle to the location you want.

For this step create a leader anchored to the top right hole in the **Top View**. Anchor the text a little above and to the right of the part. In the **Text Editor** window, type "**#30 Pilot Holes Typical**." The **Top View** should look similar to Figure 4.11.

Figure 4.11

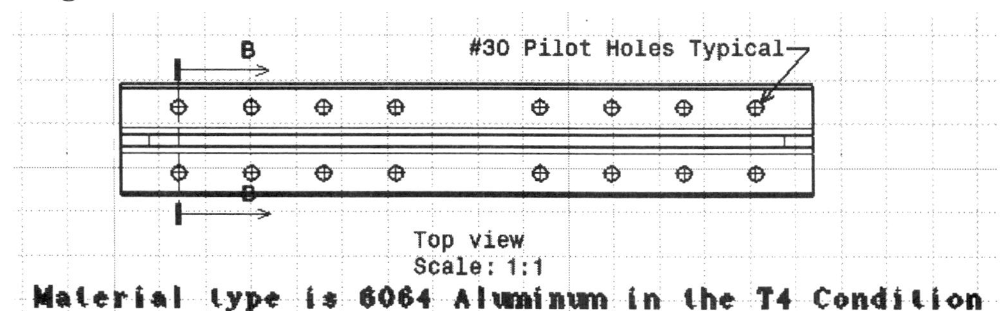

8 Creating And Modifying Dimensions

CATIA V5 has simplified the dimensioning process and at the same time added some powerful tools. CATIA V5 knows what kind of dimensions to create by the different entities you select. There are several different ways of creating dimensions.

This step concentrates on the **Dimensioning** tool bar. The tools in the **Dimensioning** tool bar are listed below.

1 The **Dimension** tool .

2 The **Datum Feature** tool .

3 The **Coordinate Dimension** tool .

4 The **Geometric Tolerance** tool .

This step will show you how to create several of the basic types of dimensions using the **Dimensioning** tool. This step also will show you how to modify the dimensions created.

Another more advanced method of creating dimensions is using constraints created in the **Sketcher** and **Part Design Work Benches**. The **Constraint** method is a powerful concept, but also one that takes some up front planning and time to get a handle on. Even though this book does not cover the process, for your reference, the tools are **Generating Dimensions** and **Generating Dimensions Step By Step,** both found in the **Generation** tool bar.

There are two main things that affect how your dimensions look and are created. The first thing is your dimension standards. Select the **Tools**, **Options** as shown in Figure 4.12. This will bring up the **Options** window. Select the **Drafting** branch, you may have to expand the **Mechanical Design** branch to access the **Drafting** branch. Select the **Dimension** tab. This will bring up the dimension options, your window should look similar to the one shown in Figure 4.13. This step does not require you to change any of the standards, just be aware what options are available.

Figure 4.12

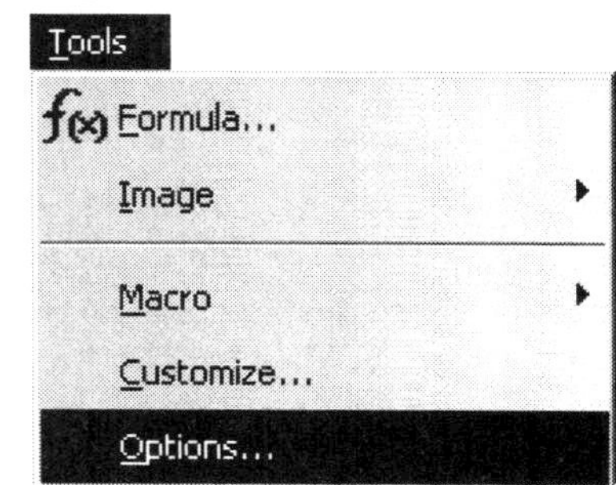

Figure 4.13

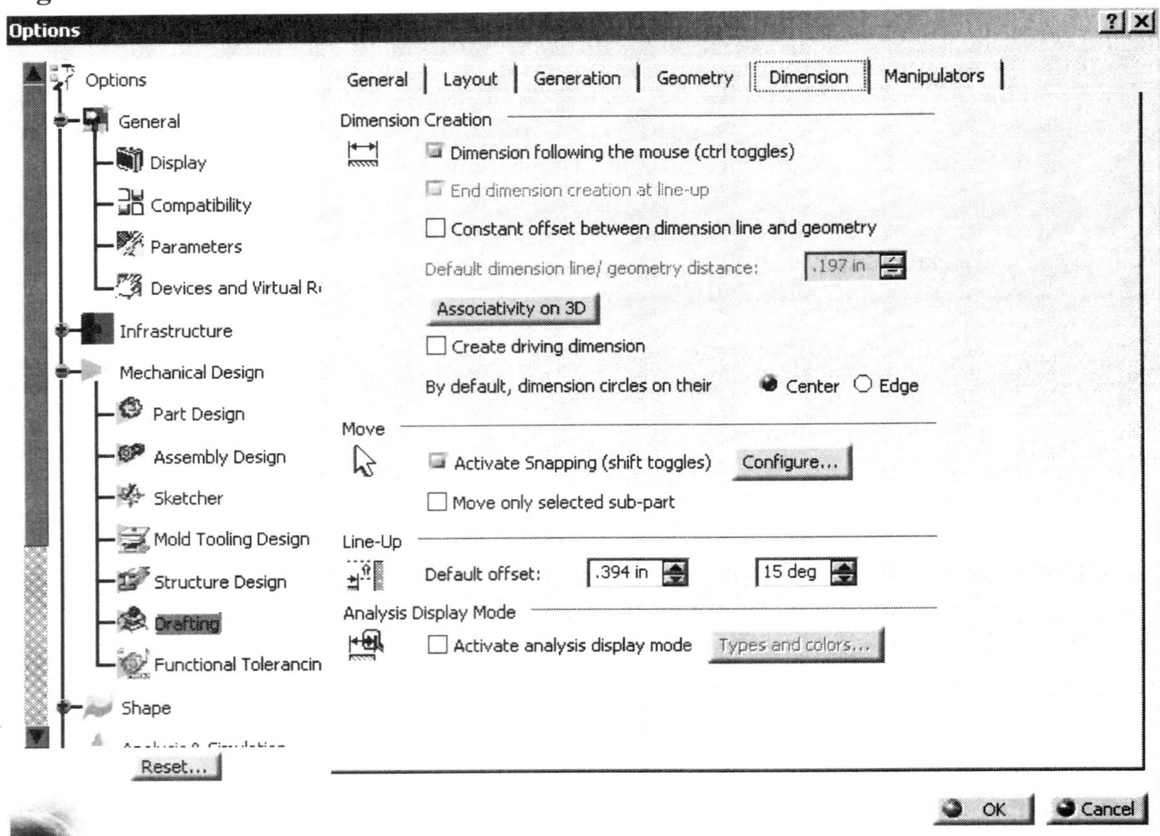

The second critical thing in creating dimensions is located in the top right tool bar, the **Dimension Properties** tool bar. This is where you control the **Dimension Line**, **Tolerance Type**, **Tolerance**, **Unit** and **Percision**, reference Figure 4.14. There is one more thing that should be considered when creating dimensions, the units the "**CATPart**" file was created in. CATIA V5 will allow you to create the "**CATPart**" in one unit and dimension it in another. Be sure you have the units set the way you want. For this step and lesson, everything is in inches.

Figure 4.14

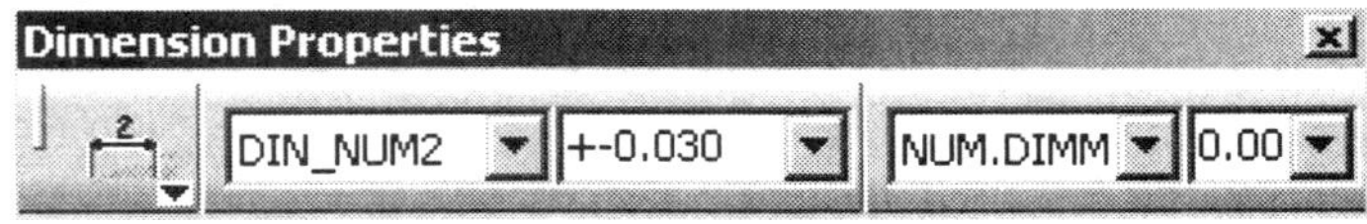

The following step will show you how to create and modify several of the basic dimension types.

8.1 Before you actually create a dimension, set the **Dimension Properties** to the following specifications.

8.1.1 Set the **Dimension Line** to where the value is between the dimension lines (reference Figure 4.15).

8.1.2 Set the **Tolerance Description** to **DIN_NUM2** (reference Figure 4.15).

8.1.3 Set the **Tolerance** to **+-.030** (reference Figure 4.15).

8.1.4 Set the **Numeric Display Description** to **NUM.DIMM** (reference figure 4.15).

8.1.5 Set the **Precision** to three places to the right of the decimal (.001) (reference Figure 4.15).

Figure 4.15

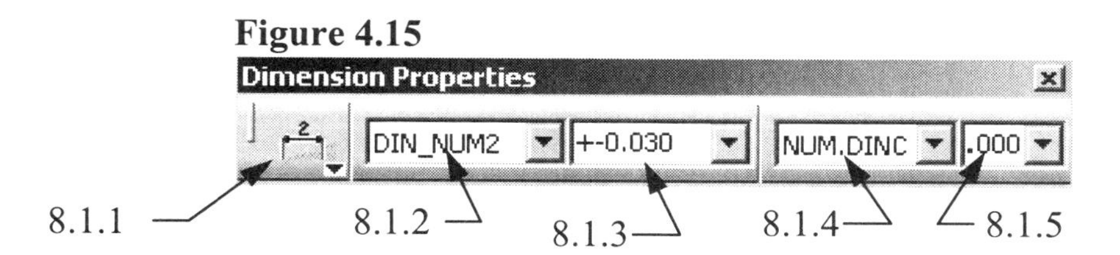

Now you are ready to create some dimensions. Follow the steps listed below.

8.2 Select the **Dimension** tool in the **Dimensioning** tool bar. CATIA V5 will prompt you to "**Select The First Element For Dimension Creation**." If you select only one element, it will create a **Dimension** for the one element. For example, if you select one edge of the part, you will get the length of that one edge. If you select a circle you will get a diameter **Dimension**. In this step you are going to create the overall length of the extrusion in the **Top View**.

8.3 Select the left most edge of the "**T Shaped Extrusion**" in the **Top View**. A dimension length of the edge selected will be created; ignore it! Reference Figure 4.16.

Figure 4.16

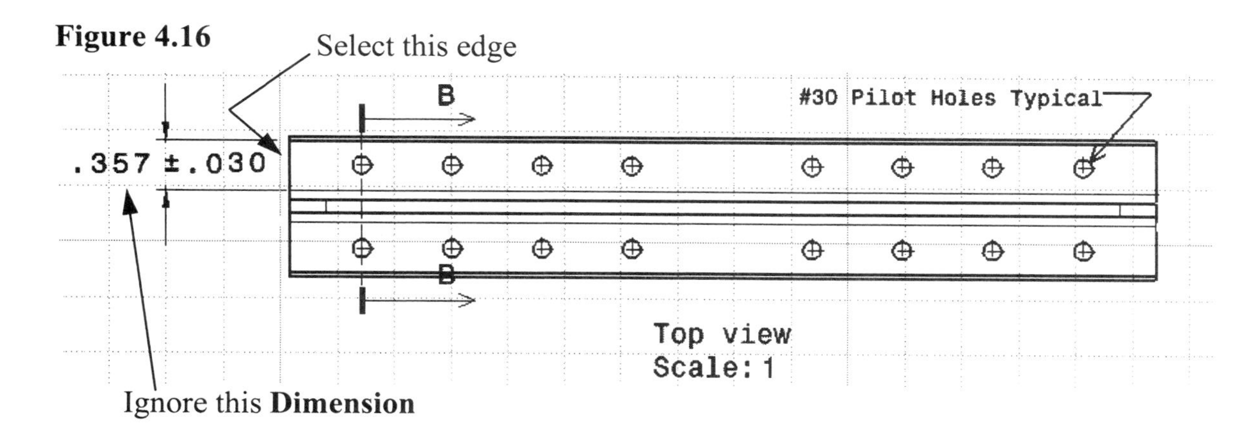

8.4 Select the right most edge of the "**T Shaped Extrusion**" in the **Top View**. The length **Dimension** created in Step 8.3 will be replaced with a distance **Dimension**

between the two ends of the part. Your newly created **Dimension** should look similar to the one shown in Figure 4.17.

Figure 4.17

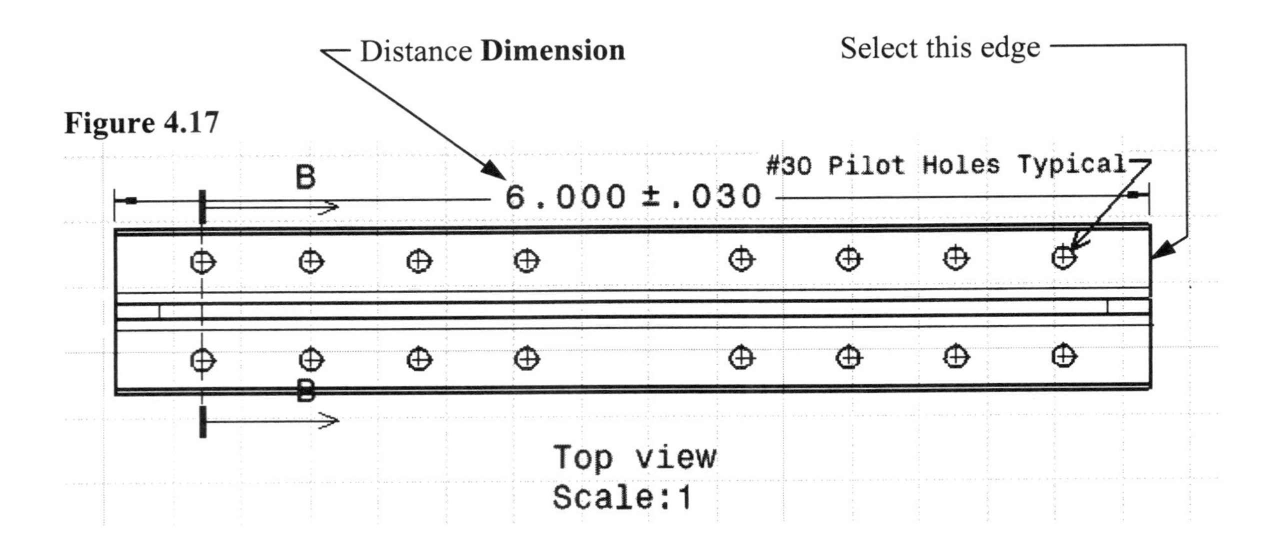

8.5 The newly created **Dimension** can be modified by completing the following steps.

8.5.1 Select the **Dimension** so it is highlighted.

8.5.2 Select the part of the **Dimension** that you want to change and drag it to the new location. For example, the dimension created in Figure 4.17 is too close to the part; select the dimension line to modify it's location.

8.5.3 Drag the **Dimension** line to a position similar to the one shown in Figure 4.19.

Figure 4.18

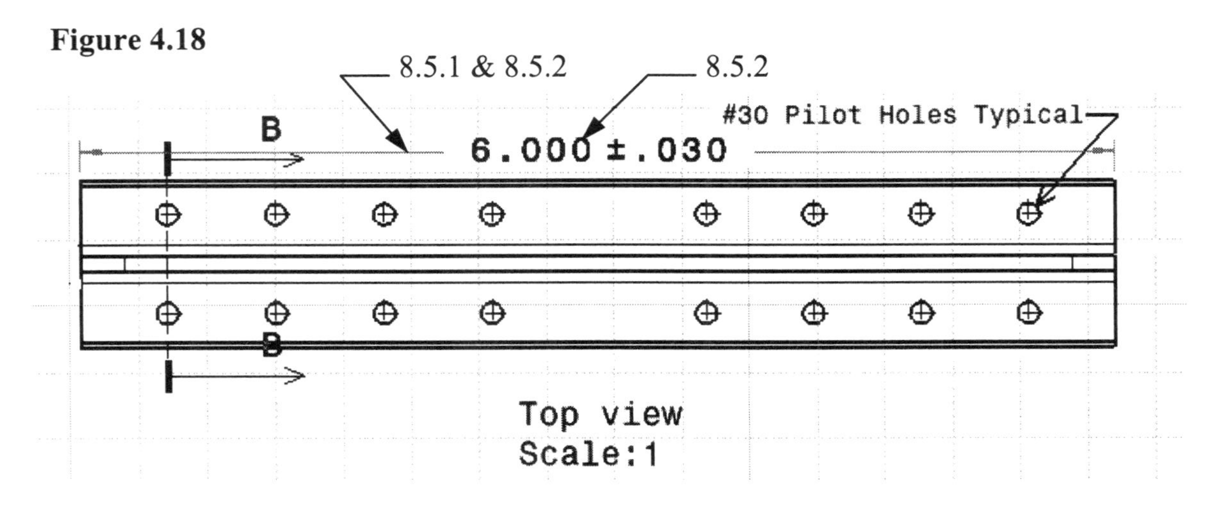

Figure 4.19

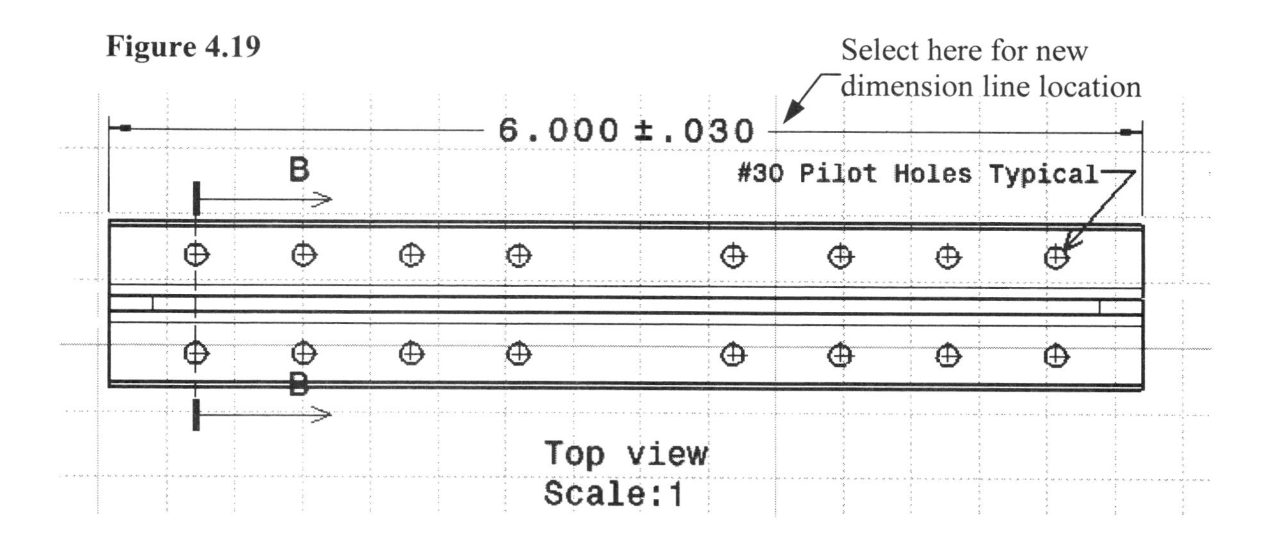

8.6 In this step you will create a radius dimension using the **Front View**. There are several different ways you can accomplish this. The most common method would be to use the **Dimension** tool bar. An alternative method is shown in Figure 4.20. Select **Insert** from the top pull-down menu. Select **Dimensioning, Dimensions** and **Radius Dimensions.** Then select the radius on the right side of the standing leg of the "**T Shaped Extrusion**" in the **Front View** as shown in Figure 4.21. Indicate the location of the radius dimensions. You can right click on the radius dimension to change the **Properties**. The **Properties** window allows you to modify the **Dimension Text**, **Font**, **Text** and **Graphic**.

Figure 4.20

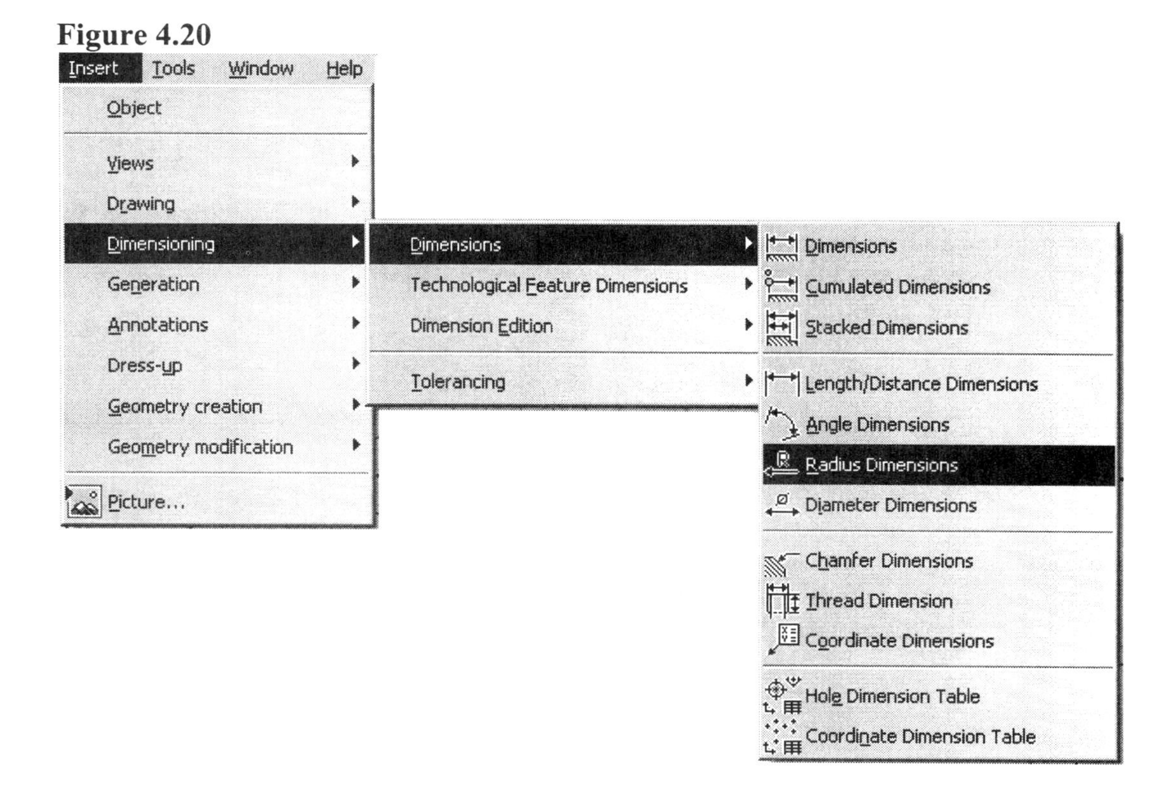

8.7 This time select the **Diameter Dimension** tool located in the **Dimension** tool bar. Select the top left hole in the **Top View**. A diameter **Dimension** will be created. The arrows will be pointing in toward the center of the circle. Selecting one of the arrows will reverse the direction of the arrows to where they point out away from the center of the circle. The hole is too small to clearly show the arrows this way, so switch them back by selecting the arrows again, reference Figure 4.22.

Figure 4.21

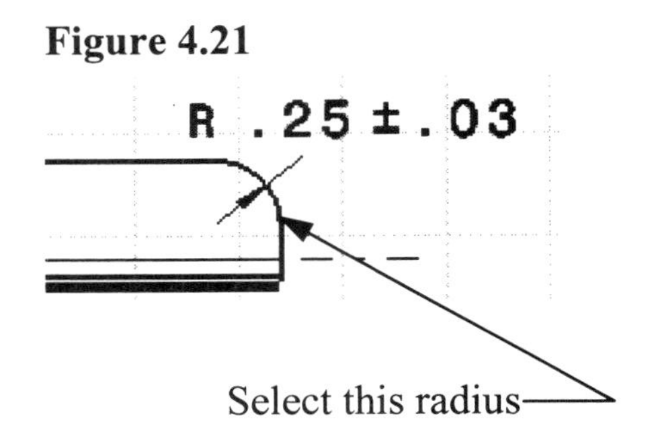

Figure 4.22

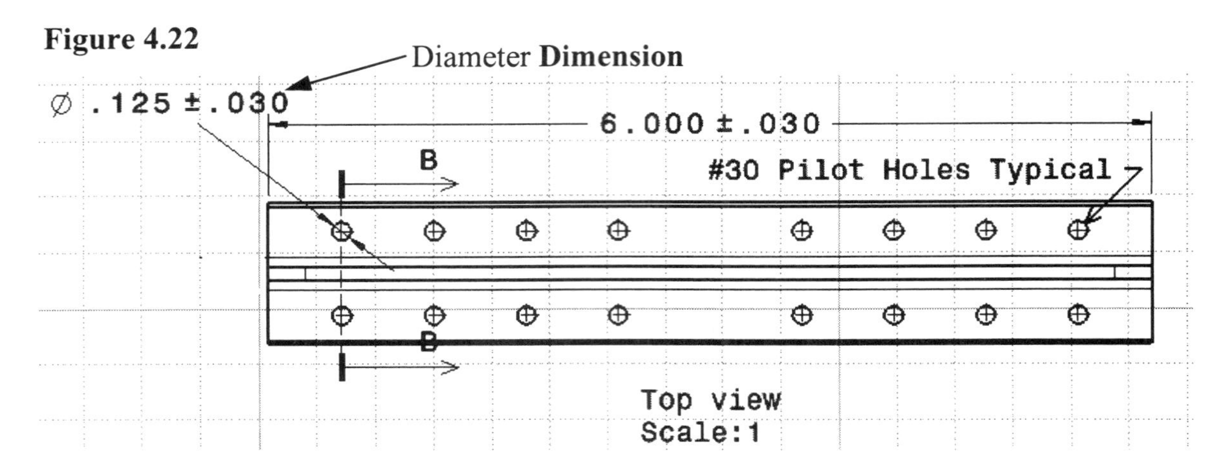

8.8 CATIA V5 knows to put the degree symbol in a **Angled Dimension**, reference the **Dimension Properties** tool bar at the top right of the screen; the **Unit** defaults to degrees. The one catch to this, is that CATIA V5 does not change the **Tolerance** or the **Precision**. In most cases, the **Tolerance** and **Precision** are different than the length and/or distance **Dimension**. The **Tolerance** and **Precision** can be modified after creation. In this step you will create an **Angle Dimension** using the **Right View**.

8.8.1 Change the **Dimension** defaults prior to creating a **Dimension**. In the **Dimension Properties** tool bar, select the **Tolerance** arrow so all of the sub options show up. Select the **+-0.030** , reference Figure 4.23.

Figure 4.23

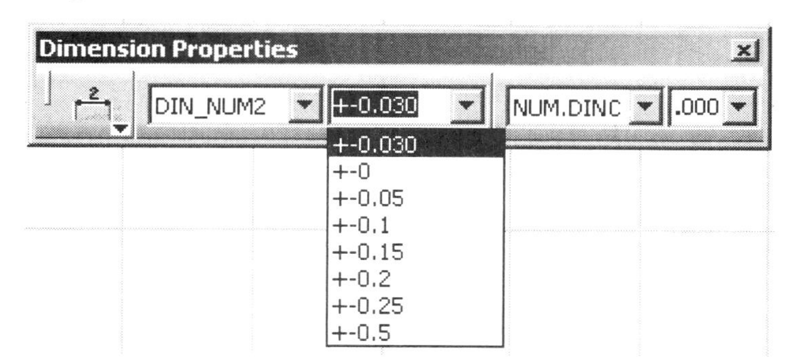

8.8.2 **Zoom Up** on the **Section View B-B** so the chamfer can be clearly identified and pickable, similar to what is shown in Figure 4.24.

8.8.3 From the **Dimension** tool bar select the **Angle** tool.

8.8.4 Select the right side of the standing leg of the **"T Shaped Extrusion"** in the **Section View B-B**. Figure 4.24 points out the location of selection 1.

Figure 4.24

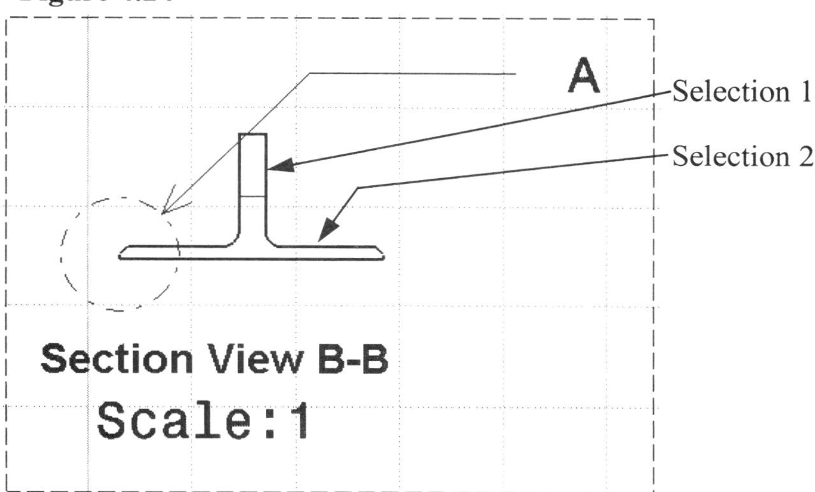

8.8.5 Select the top right side of the horizontial leg of the "**T Shaped Extrusion**" in the **Section View B-B**. Figure 4.24 points out the location of selection 2. This will create the **Angled Dimension**. If the **Dimensions** bunch too close together don't worry, a few quick mouse clicks and the **Dimension** can be modified just the way you want it. The **Dimension** created should read 90 degrees with no tolerance as shown in Figure 4.25.

Figure 4.25 (Angled Dimension as created by default)

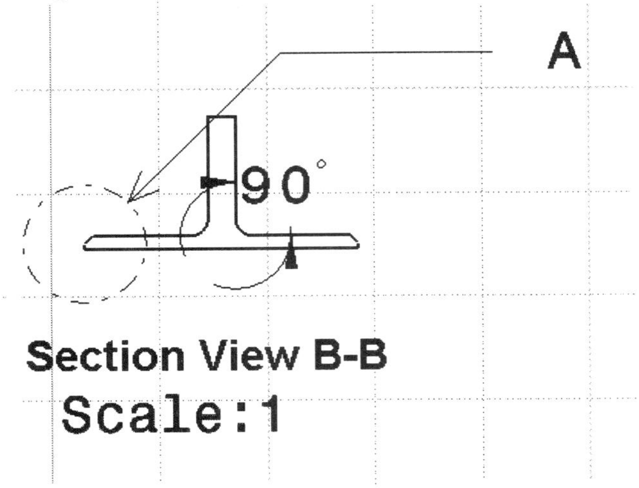

Figure 4.26

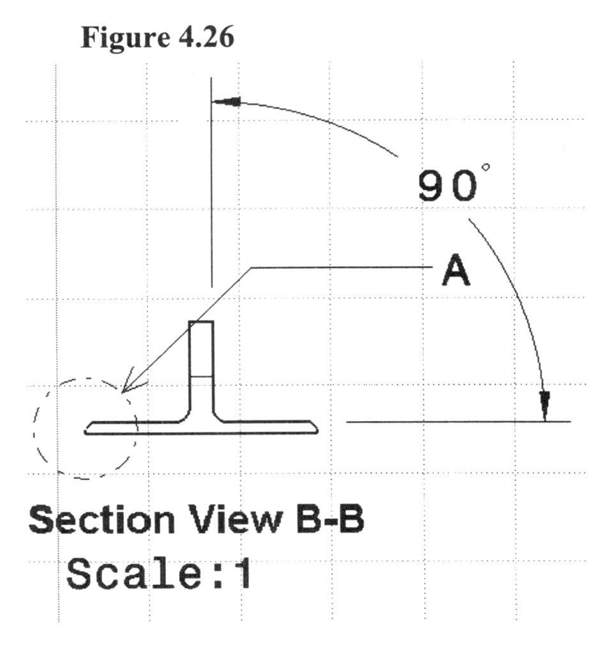

8.8.6 To move the **Dimension,** select the **Dimension** arrow (arc) and drag it to the upper right quadrant of the view.

8.8.7 To move the text, select the text and drag it in either direction along the dimension arrow (arc). Notice the **Dimension** value is tied to the arc. If the **Dimension** value is too close to the part, you will have to reselect the arrow **Dimension** (arc) and drag it farther away from the part, similar to what is shown in Figure 4.26.

As with the other **Dimensions**, you can add pre fix/suffix text to the dimensions. This is done by selecting the text and then bringing up the **Properties** window. The **Dimension Text** tab has a text box for prefix and suffix text. For this step, add the suffix text that reads "**Typical**" as shown in Figures 4.27.

Figure 4.27

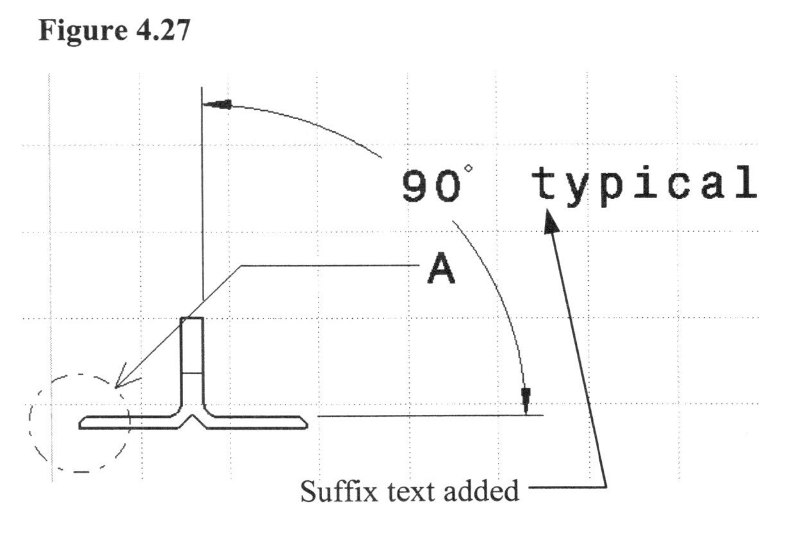

8.9 The preceeding steps introduced you to creating the basic types of **Dimensions**. Your assignment now is to complete the **Dimensions** required to complete the drawing of the "**T Shaped Extrusion**." You now have the basic tools to **Dimension** the drawing. Good luck! Don't worry, you're not totally abandoned. Figure 4.31 is meant to be a guide and/or a suggestion as to what your completed dimensioned drawing should look like.

9 Creating A Border/Title Block Using The Geometery Creation Tool Bar

One of the main components of a production drawing is the title block. Title blocks can vary from simplistic to complex depending on how much information you are required to present. This lesson will give you the basic tools to create a title block. How complex the title block is, is up to you. The tools and processes used to create the title block are almost identical to the tools in the **Sketcher Work Bench**. If you have any detail questions on how to use the **Profile**, **Line**, **Trim**, **Corner** and even the **Constraints** tools, refer back to Lesson 1 on the **Sketcher Work Bench**. Figure 4.31 is an example of an average title block.

The title block can be saved and imported to any other drawings, so you won't have to recreate a title block for every drawing. CATIA V5 allows you to import graphic files into your title block, so you can import your school and/or company logo as a graphics file. The **SUU** (Southern Utah University) and **MTI** (Metalcraft Technologies, Inc.) logo on the title block in Figure 4.31 are .tif files imported into the title block. To create a title block, complete the following steps.

9.1 Make sure all of your views fit on the sheet. This may require you to move a few of the views around (Lesson 3 Step 9). You may also discover that you do not have enough room on the current sheet, so you may have to create a new sheet (covered in Lesson 3 Step 7).

9.2 Turn on the **Grid Lines**, if not already on. It is the tool at the bottom right of the screen. The tool toggles the grid lines **On** and **Off**. If the **Primary Spacings** on the grid lines do not balance with sheet outlines, modify the grid spacing (Lesson 3 Step 6.2.1).

9.3 Turn on the **Snap To Grid** if not already on. It is the tool at the bottom right of the screen. The tool toggles the **Snap To Grid On** and **Off**.

9.4 Select the **Edit**, **Background** option located at the bottom of the pull down window as shown in Figure 4.28. This will turn off all of the current views making it so that they can not be modified. The title block option in CATIA V5 can only be used in the **Background** mode.

9.5 Select the **Insert** pull down window from the top of the screen as shown in Figure 4.29.

9.6 From the **Insert** pull down window, select the **Drawing** option and then the **Frame And Title Block** option. This will bring up the **Insert Frame and Title Block** window, shown in Figure 4.30. From inside this window sample title blocks can be inserted into the drawing. These sample title blocks will automatically update to fit your drawing sheet so, select the **OK** button.

Figure 4.28

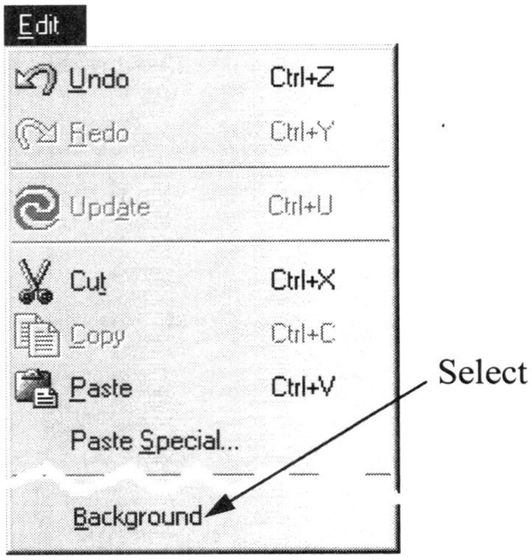

Edit
Undo Ctrl+Z
Redo Ctrl+Y
Update Ctrl+U
Cut Ctrl+X
Copy Ctrl+C
Paste Ctrl+V
Paste Special...
Background
Select

Figure 4.29

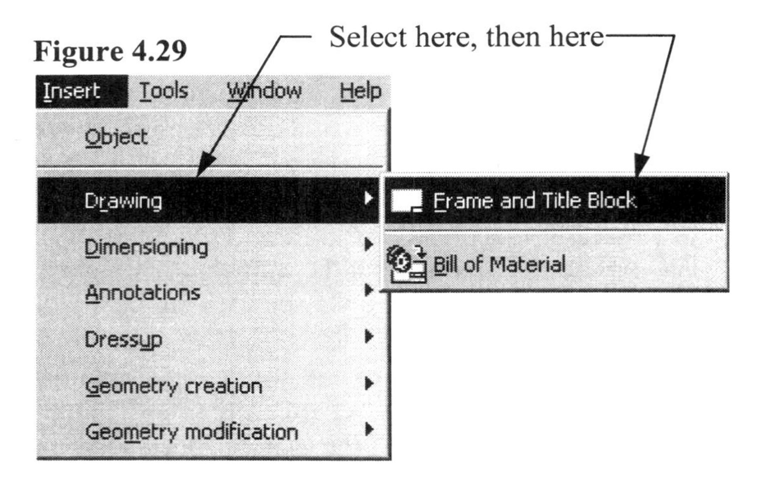

Figure 4.30

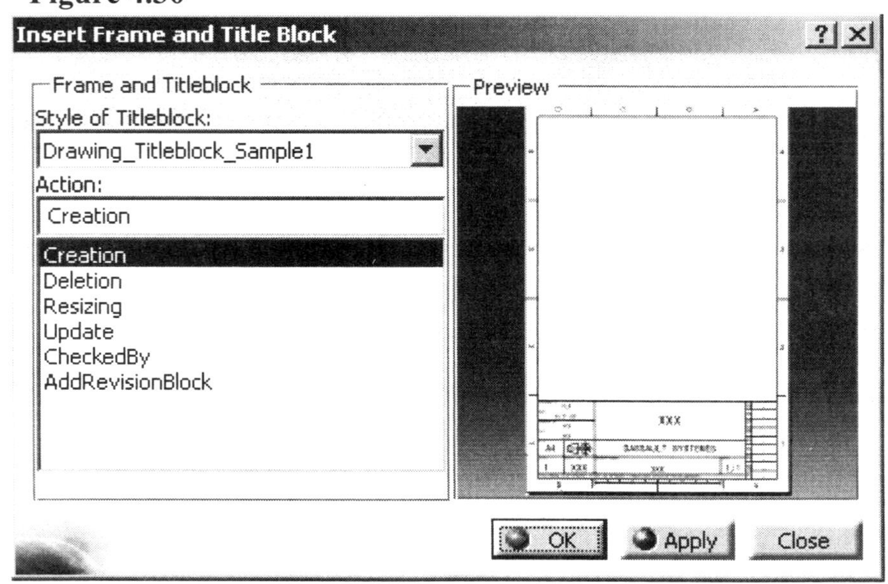

9.7 The title block has been inserted into the drawing with default information and formatting. Move the lines of the title block around until they are in the locations you want. To move them, select the line and drag it to the location you want. Change the information in the title block to fit your personal needs. Figure 4.31 gives you an example of what the title block could look like. Your title block doesn't have to look exactly like the one shown in Figure 4.31.

Figure 4.31

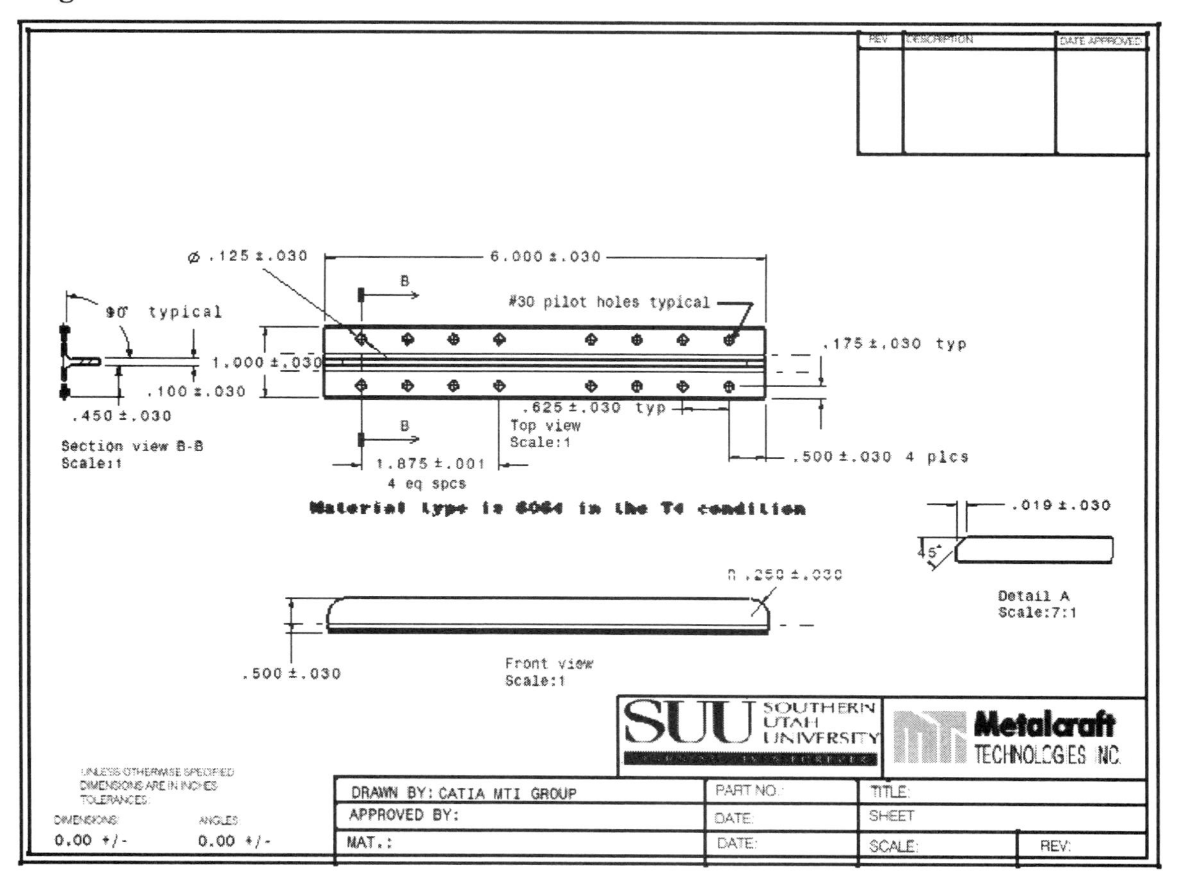

9.8 The default line weight for the title block is too thin. To make the border line thicker, complete the following.

9.8.1 Bring the **Properties** window up for one of the lines in the title block you just created, reference Figure 4.32

9.8.2 Select the **Graphic** tab.

9.8.3 Go down to the **Lines And Curves** section.

9.8.4 Select the arrow to the right of the **Thickness** option. Selecting the arrow will show you the different line weight options available to you.

9.8.5 Select the line that is one weight thicker than the default line.

9.8.6 Select **Apply** to preview the change.

9.8.7 Select **OK** to accept the change.

9.8.8 Repeat Steps 9.8.1 thru 9.8.7 to change the line thickness of the remaining lines of the title block.

9.8.9 Now, verify that the border line is thicker.

9.9 Add the **Text** to the title block where appropriate. Remember, Figure 4.31 is for your reference. Reference Step 6 of this lesson on how to use the **Text** tool. This is a good opportunity to use your creative side. This step does not have any exact requirement, other than to complete the title block.

NOTE: The **Geometry** tool bar can be used to create additonal lines and other draw entities that your generic title block does not have. These commands work similar to the **Sketcher Work Bench** explained in Lesson 1.

Figure 4.32

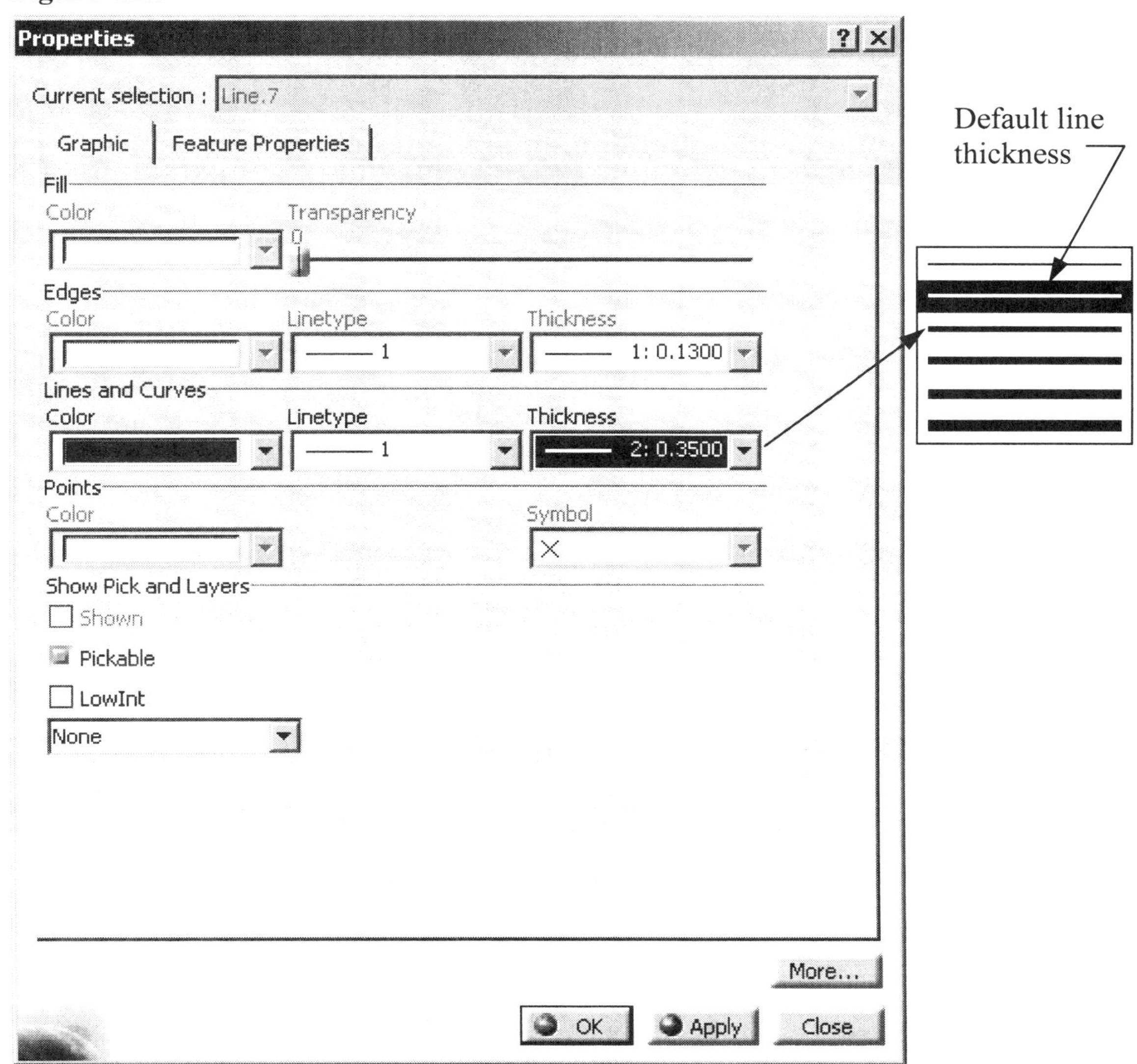

10 Inserting Bill Of Materials Into The Title Block

CATIA V5 allows you to automatically add the **Bill of Materials** to the title block as Figure 4.29 shows. The only catch is, it requires a "***.CATProduct**" file. So far you have created "***.CATPart**" and "**.CATDrawing**" files. Lesson 5 and 6 cover "***.CATProducts**," the lessons also explain how to add the **Bill of Materials** to your drawings. This is a powerfull and time saveing tool and is worth mentioning, but for detailed instructions, reference Lessons 5 and 6.

11 Inserting Picture Files Into The Title Block

CATIA V5 allows you to customize your **Title Blocks** by inserting your choice of picture files. Figure 4.34 gives you a sample of all the options you have for importing into the **Title Block**. The following steps detail the process of adding a picture file to your title block. This can be any picture file you want, but you need to supply it. This step uses a sample file for demonstation purposes only.

11.1 Things you need to know before you start.

11.1.1 Make sure you know what kind of graphics file you are going to import.

11.1.2 Make sure the size and quality is what you expect.

11.1.3 Make sure you know where the file is located and what the name of the file is.

11.2 Select the **Insert** tool from the top pull down menu as shown in Figure 4.33.

11.3 Select the **Object** option. This will bring up an **Insert Object** window as shown in Figure 4.34.

Figure 4.33

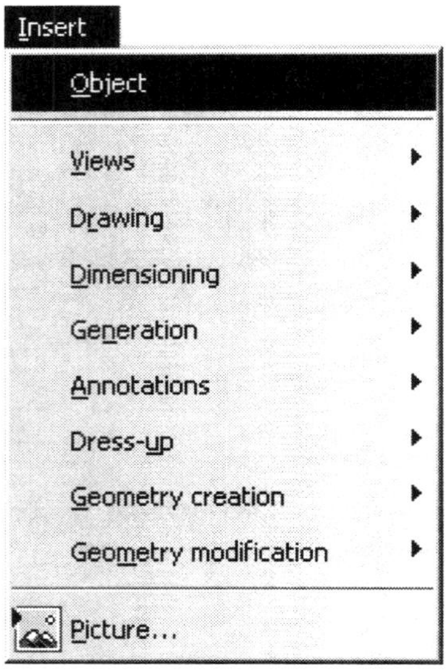

Insert
Object
Views
Drawing
Dimensioning
Generation
Annotations
Dress-up
Geometry creation
Geometry modification
Picture...

Figure 4.34

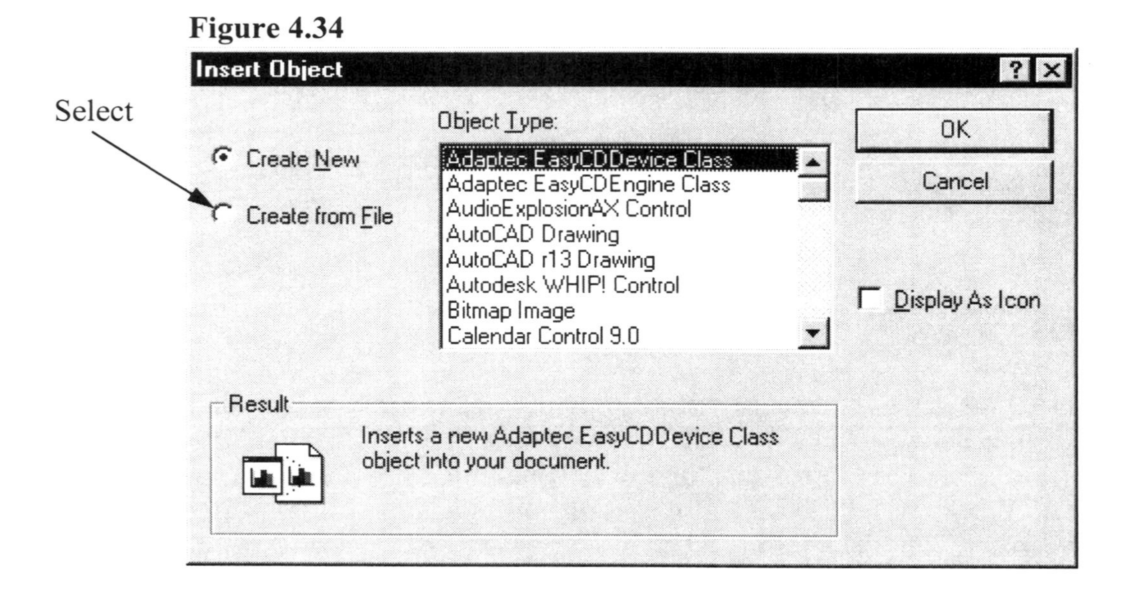

11.4 From the **Insert Object** window select the **Create From File** option as shown in Figure 4.34.

11.5 From the same window, select the **Browse** option. This will bring up the **Standard Windows Browse** window which allows you to search through the directories for the specific picture file.

Figure 4.35

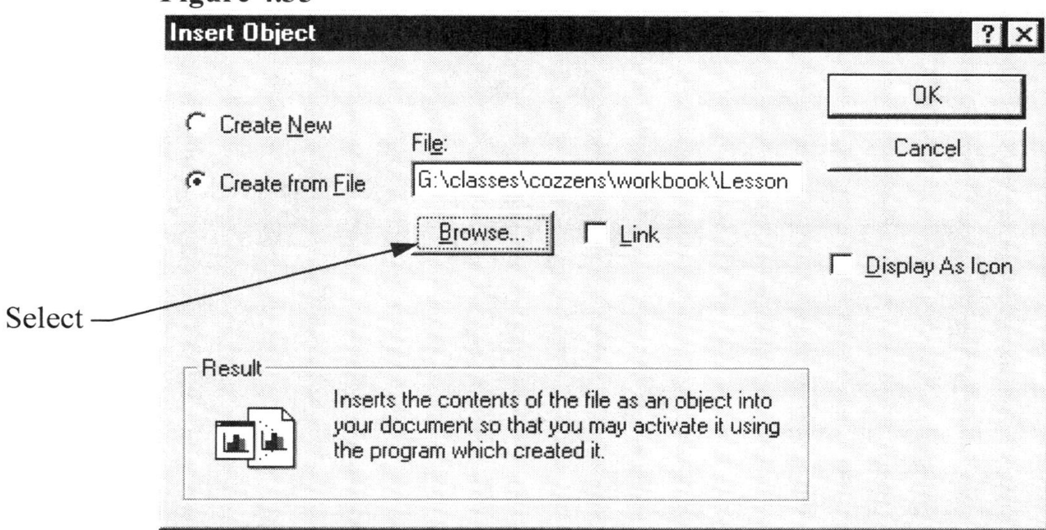

11.6 Find the graphics file you want to insert and select it.

NOTE: If you want the picture to be displayed make sure the "**Display As Icon**" is not selected, reference Figure 4.35.

11.7 Select **OK**.

11.8 The graphic will be placed on the (0,0) point on the active sheet.

11.9 To move the graphic to the proper location, use the drag and drop process.

11.10 If your School/Business icon is placed on your title block similar to what is shown in Figure 4.31 you are ready to move on.

12 Saving Your Updated Drawing

This process is basically identical to saving a file covered in Lesson 1. The only difference is that it will be saved with the name you give it followed by the extension of "**CATDrawing**" as explained in Lesson 3.

13 Printing The Finished Drawings

This process is covered in detail in Lesson 3 Step 14. Remember to make sure all of the veiws fit on the sheets. You can use the **Print Preview** window to preview the print and make the necassary adjustments. Figure 4.36 shows the "**T Shaped Extrusion**" in the **Print Preview** window. If the views do not fit within the sheet boundaries, review Lesson 3 Step 9 for detailed information on how to move views around. For this lesson print out all of the drafting sheets you created.

Figure 4.36

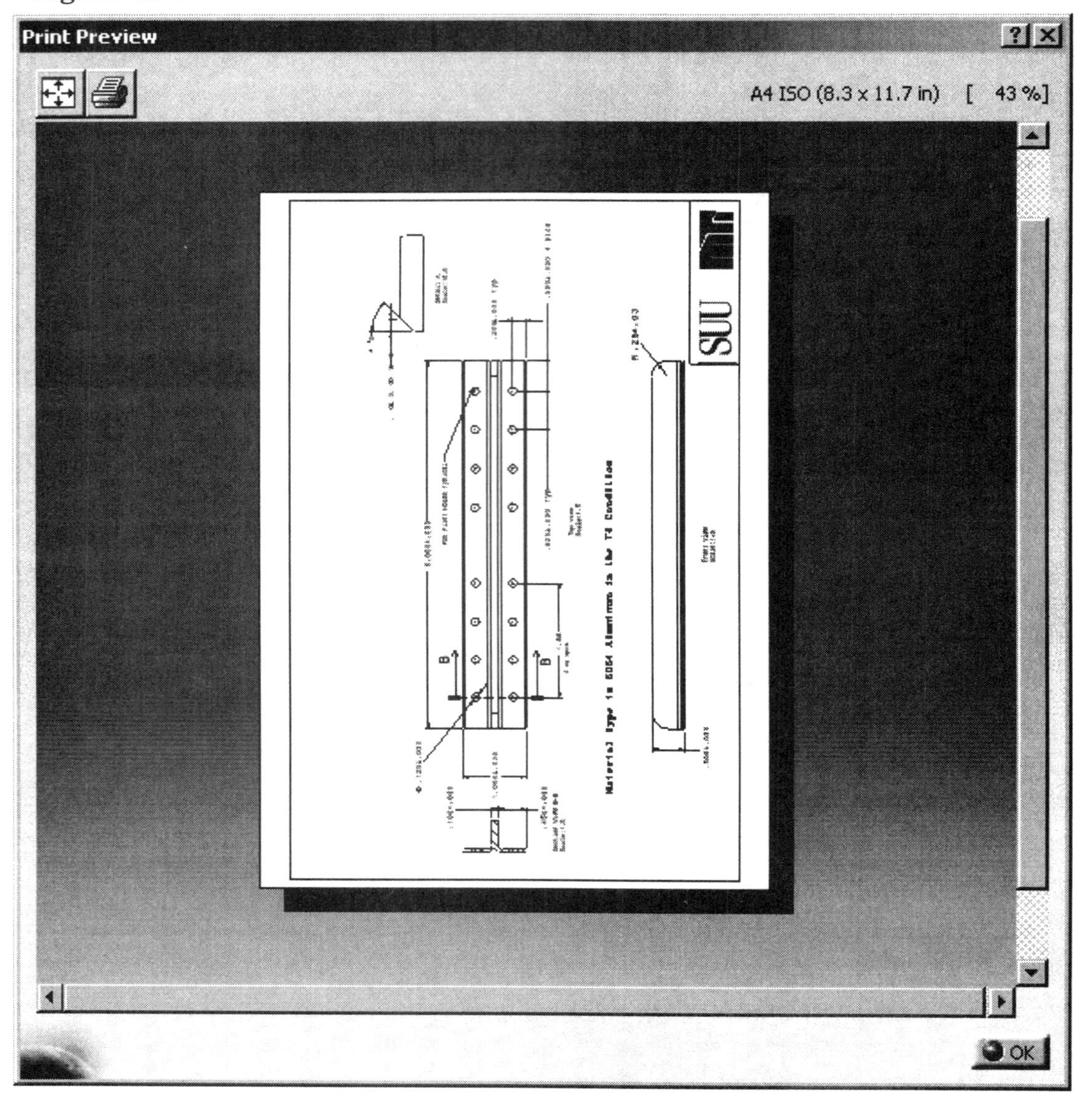

Lesson 4 Summary

Creating a complete drawing is no small task. CATIA V5 has come a long way in simplifying and automating the process where possible. One of the best things CATIA V5 does is to make the process of learning quicker and easier. This lesson supplies the introductory tools required to start producing drawings almost immediately. You now have enough knowledge to explore and learn the more advanced options on your own.

Lesson 4 Review

After completing this lesson, you should be able to answer the questions and explain the concepts listed below.

1. A file created in the **Drafting Work Bench** has what type of file extension?

2. **T** or **F** When you create a **CATDrawing** from a **CATPart**, CATIA V5 makes a link between the two different files.

3. **T** or **F** The **Drafting Work Bench** has some similarities to the **Sketcher Work Bench**.

4. Justify your answer to question 3.

5. List at least three types of views that CATIA V5 automatically creates for you.

6. **T** or **F** CATIA V5 does not allow dimensions to be modified. You must delete the dimension and start over.

7. **T** or **F** It is possible to generate dimensions from **Constraints** created in the **Sketcher Work Bench**.

8. **T** or **F** To create a radius dimension, you use the same **Dimension** tool you would use to create an angled dimension.

9. Does the **Drafting Work Bench** have a **Specification Tree**?

10. **T** or **F** By default, all of the tool bars are visible when entering the **Drafting Work Bench.**

11. How do you toggle the **Snap To Grid**, **On** and **Off**?

12. How do you modify the **Primary Spacing** and **Graduations** of the drafting grid?

13. What tool bar contains the following tools: **Centerline**, **Centerline With Reference**, **Thread**, **Thread With Reference** and **Axis Line**?

14. **T** or **F** Modifying CATIA V5 text is as simple as double clicking on the text to be modified to bring up the **Text Editor** window, making the changes to the text and then selecting **OK** to apply the changes.

15. **T** or **F** Dimensions created in a view will show up on the **Drafting Specification Tree**.

16. What tool bar controls the dimension tolerance?

    a. **Dimensioning**
    b. **Graphic properties**
    c. **Dimension Properties**
    d. **Generation**
    e. None of the above

17. What must be done to a sheet prior to inserting a CATIA V5 title block?

18. **T** or **F** Once the title block is inserted you can go back and modify the views.

19. What tools are required to insert a picture file into your title block?

    a. **Insert**
    b. **Object**
    c. **Import Picture**
    d. Both b and c
    e. Both a and b

20. **T** or **F** CATIA V5 allows two-dimensional entities to be created in the **Drafting Work Bench** to help supplement the three-dimensional views.

Lesson 4 Practice Exercises

The practice exercises for this lesson require you to use the "**CATDrawing**" files you created Lesson 3. Complete the drawings started in Lesson 3 using the **Text** and **Dimensioning** tools. Create a border/title block for each of drawings. It is not the intent of this book to teach any drawing and/or drafting standard; therefore, the standard and style is up to you.

Remember the file extension used in the **Drafting Work Bench** is **CATDrawing**. Good Luck!

1 Open "**Lesson 3 Exercise 1.CATDrawing**." Add the required text, dimensioning and border/title block information to make the sheets a completed production drawing. Insert a picture file of your school and/or company logo. Save the drawing as "**Lesson 4 Exercise 1. CATDrawing**."

2 Open "**Lesson 3 Exercise 2.CATDrawing**." Add the required text, dimensioning and border/title block information to make the sheets a completed production drawing. Save the drawing as "**Lesson 4 Exercise 2. CATDrawing**."

3 Open "**Lesson 3 Exercise 3.CATDrawing**." Add the required text, dimensioning and border/title block information to make the sheets a completed production drawing. Save the drawing as "**Lesson 4 Exercise 3. CATDrawing**."

4 Open "**Lesson 3 Exercise 4.CATDrawing**." Add the required text, dimensioning and border/title block information to make the sheets a completed production drawing. Save the drawing as "**Lesson 4 Exercise 4. CATDrawing**."

Introduction To Creating Complex And Multiple Sketch Parts

Product Structure

In Lesson 1 you created a sketch and in Lesson 2 you extruded the sketch into a three-dimensional part. Lesson 1 taught you that an unclosed sketch could not be extruded. This statement is still true. In this lesson you will learn how to create multiple sketches that make one part. This means that you will go into the **Sketcher Work Bench** several times to create separate sketches. The separate sketches will be extruded in the **Part Design Work Bench** to form one part. The **Specification Tree** will show only one part **(Part.1)**. The difference is, **Part.1** will be made up of several sketch branches. Figure 5.1 shows you the first two multiple sketch parts you will create in this lesson. Figure 5.2 shows the **Specification Tree** for **Part.1**. Notice the multiple sketches are all under the **PartBody** branch, which make up the **Part.1 Specification Tree**.

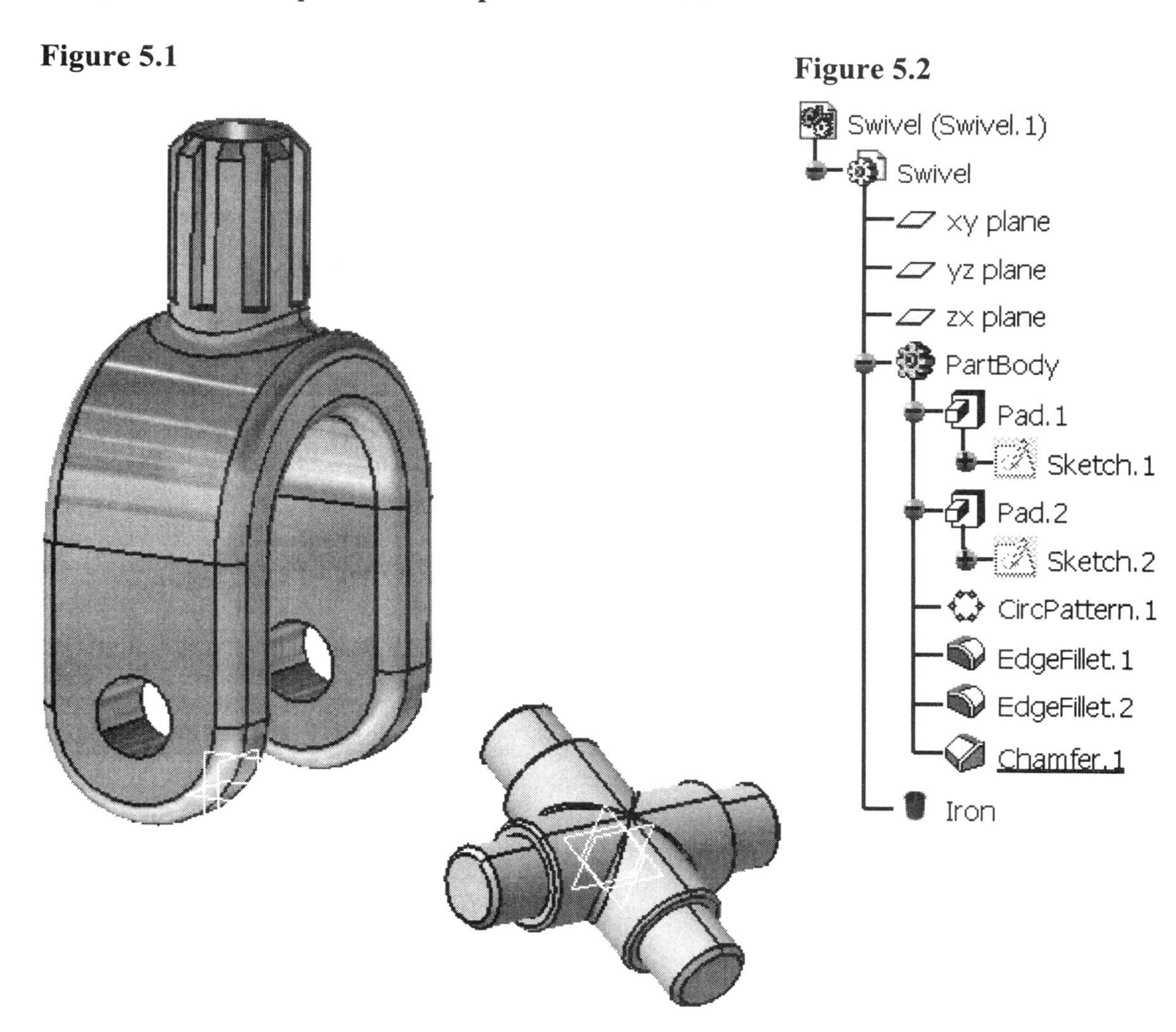

Figure 5.1

Figure 5.2

The first part of this lesson will teach you the tools needed to create the complex and multiple sketch parts. The second part of this lesson gives step-by-step instructions on creating three different parts. The first part will be the **Swivel**. The **Swivel** will be created using multiple sketches. The second part is the **Top U-Joint**. It will also be created using multiple sketches. The third part is the **Bottom U-Joint**. You will use **Boolean Geometry** to create this part.

The three parts created in this lesson will be used in Lesson 6, "**Introduction To Creating An Assembly Using The Assembly Design Work Bench**."

Lesson 5 Objectives

The objective of this lesson is to:

- Re-enforce what was learned in Lessons 1 and 2.

- Show different approaches to creating parts using the **Sketcher** and **Part Design Work Benches**. This will be accomplished by creating a part that requires multiple sketches.

- Create the detail parts for Lesson 6, "**Introduction To Creating An Assembly Using The Assembly Design Work Bench**."

- Introduce you to some of the tools found in the **Product Structure Work Bench**.

- Use **Boolean Geometry** to create parts (yes **Boolean Geometry** still exists).

The step-by-step instructions will not be as detailed as in Lesson 1 and 2. It will be assumed the tools used in the first two lessons are now common knowledge. If they are not, you are encouraged to go back and review the subjects you have questions on.

Tools Used For Complex & Multiple Sketch Parts

For the most part, the tools used in this lesson are the same tools used in Lesson 1 and Lesson 2. If you have questions on any of the **Part Design Work Bench** and/or the **Sketcher Work Bench**, review Lesson 1 and Lesson 2.

This lesson does use a new work bench, the **Product Structure Work Bench**. The **Product Structure Work Bench** and the associated tools are shown next with a brief explanation. The **Product Structure Work Bench** is very helpful when you start to deal with multiple **CATPart**s, as you will see in this lesson.

The **Body** and **Boolean Geometry** tools are also shown, as you will be using them later in this lesson.

The **Product Structure Work Bench** has several tool bars. This lesson uses only one of the tool bars. The tool bar is labeled "**Product Structure**" tool bar. Only four of the tools are used or discussed in this lesson, they are the tools with the * beside them. The same tools can be accessed from the **Insert** pull down window when the **Product Structure Work Bench** is active. Note that the **Insert** pull down window will be different depending on what work bench is active (in use). Additional **Product Structure Work Bench** tools will be discussed in Lesson 6.

The **Filtered Selection** Tool Bar

Tool Bar	Tool Name	Tool Definition
	Products Selection	Allows you to select a specific product.

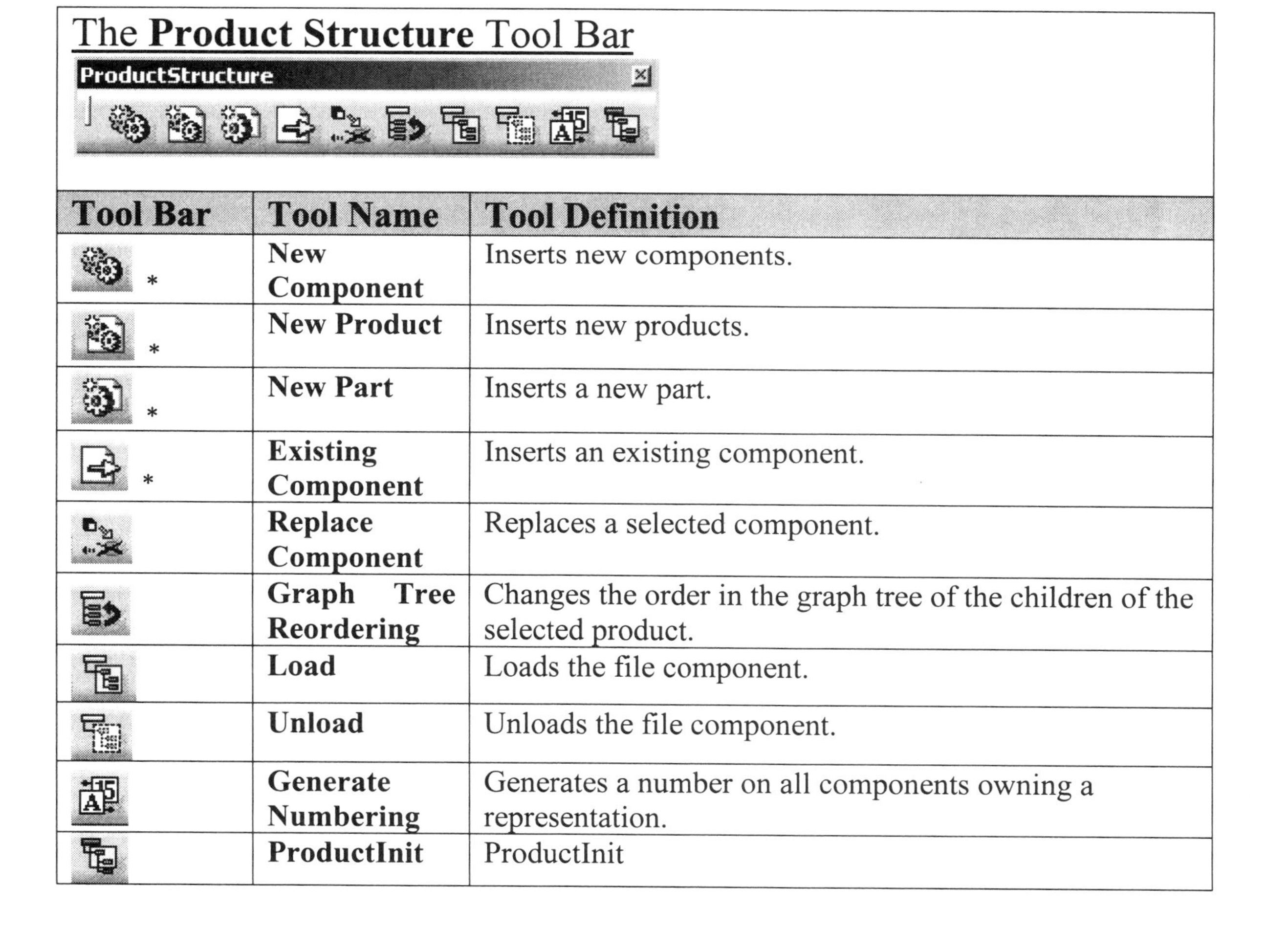

The **Product Structure** Tool Bar

Tool Bar	Tool Name	Tool Definition
*	**New Component**	Inserts new components.
*	**New Product**	Inserts new products.
*	**New Part**	Inserts a new part.
*	**Existing Component**	Inserts an existing component.
	Replace Component	Replaces a selected component.
	Graph Tree Reordering	Changes the order in the graph tree of the children of the selected product.
	Load	Loads the file component.
	Unload	Unloads the file component.
	Generate Numbering	Generates a number on all components owning a representation.
	ProductInit	ProductInit

The **Representation** Tool Bar

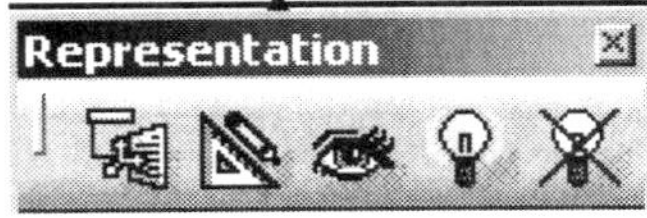

Tool Bar	Tool Name	Tool Definition
	Manage Representations	Manages the representation of the selected component.
	Design Mode	Switches the selected component to design mode.
	Visualization Mode	Switches the selected component to visualization mode.
	Activate Node	Activates a shape representation of the selected component.
	Deactivate Node	Deactivates a shape representation of the selected component.

The **Enovia VPM** Tool Bar

Tool Bar	Tool Name	Tool Definition
	No Connection	Opens a connection to a specific database.

The **Resource** Tool Bar

Tool Bar	Tool Name	Tool Definition
	Create a Resource	Creates a resource from an ASMPRODUCT selected.
	Activate a Resource	Activates an existing resource.
	Deactivate A Resource	Deactivates an existing resource.

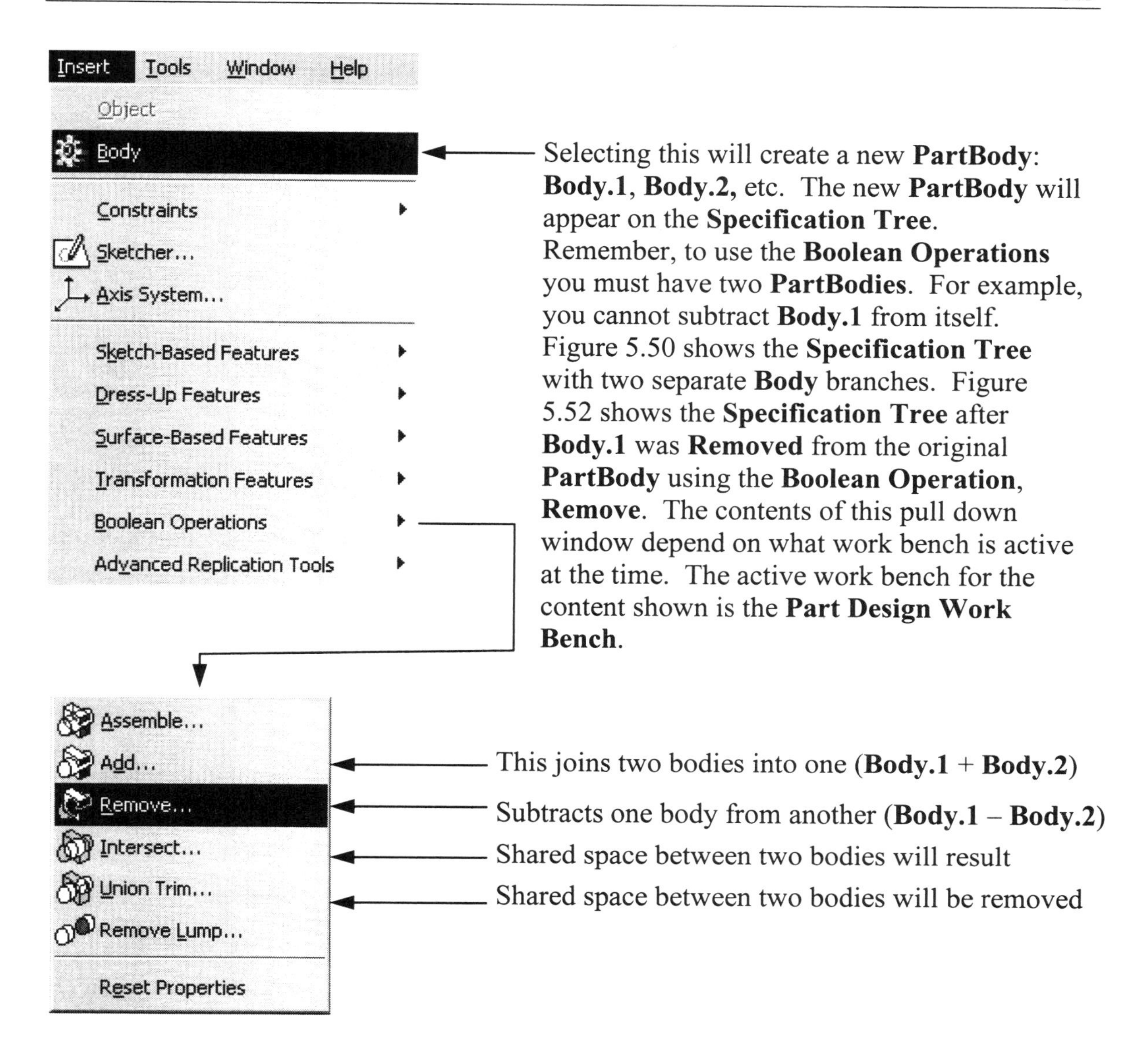
Insert
Tools
Window
Help
Object
Body
Constraints
Sketcher...
Axis System...
Sketch-Based Features
Dress-Up Features
Surface-Based Features
Transformation Features
Boolean Operations
Advanced Replication Tools
Selecting this will create a new PartBody: Body.1, Body.2, etc. The new PartBody will appear on the Specification Tree. Remember, to use the Boolean Operations you must have two PartBodies. For example, you cannot subtract Body.1 from itself. Figure 5.50 shows the Specification Tree with two separate Body branches. Figure 5.52 shows the Specification Tree after Body.1 was Removed from the original PartBody using the Boolean Operation, Remove. The contents of this pull down window depend on what work bench is active at the time. The active work bench for the content shown is the Part Design Work Bench.
Assemble...
Add...
Remove...
Intersect...
Union Trim...
Remove Lump...
Reset Properties
This joins two bodies into one (Body.1 + Body.2)
Subtracts one body from another (Body.1 – Body.2)
Shared space between two bodies will result
Shared space between two bodies will be removed

Steps To Creating Complex & Multiple Sketch Parts

Figure 5.3

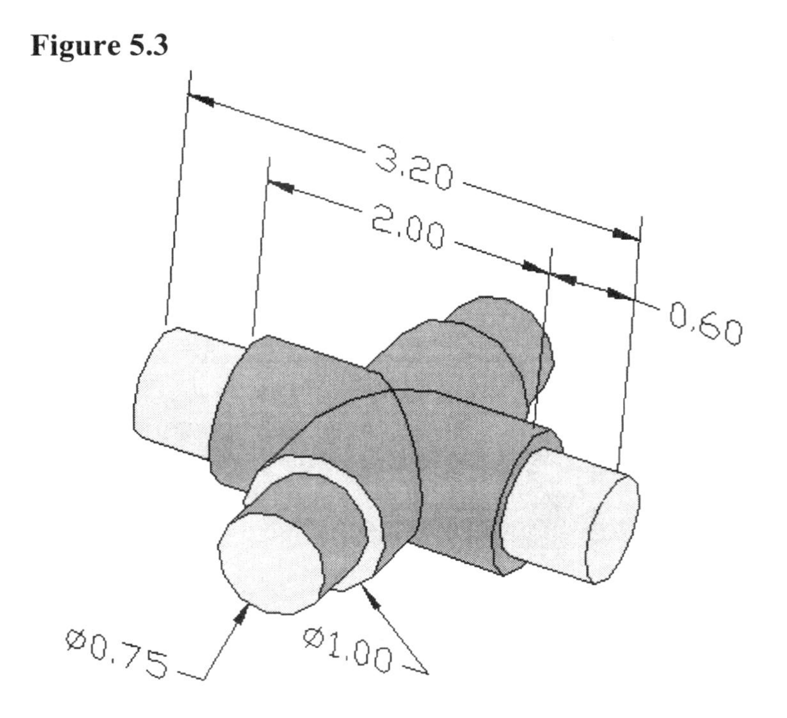

1 Creating The "Swivel" Using Multiple Sketches

There are multiple methods to creating a **CATPart**. A few of them have been used and/or explained in the earlier lessons. This lesson will show you a few more ways of accomplishing the same thing. Figure 5.3 shows the dimensions that will be used to create the "**Swivel**." Complete the following steps to create the "**Swivel**" as shown in Figure 5.3.

1.1 Start CATIA V5 (the obvious!)

1.2 Verify that the default **Properties** are set the way you want them such as **Units**, **Primary Spacing** and **Graduations**. For this step, set the **Units** to inches, the **Primary Spacing** to **1** inch and the **Graduations** to **10**.

1.3 By default CATIA V5 will bring up the **Product Structure Work Bench**. If your CATIA V5 default has been changed select the **Product Structure Work Bench**. Select **Product1** from the **Specification Tree**. With the **Product1** branch highlighted select the **New Part** tool from the **Product Structure** tool bar as shown in Figure 5.4. This will bring up the **New Part** window as shown in Figure 5.5. It is important that the **Product1** branch is selected otherwise the **Part Number** window will not pop up.

Figure 5.4

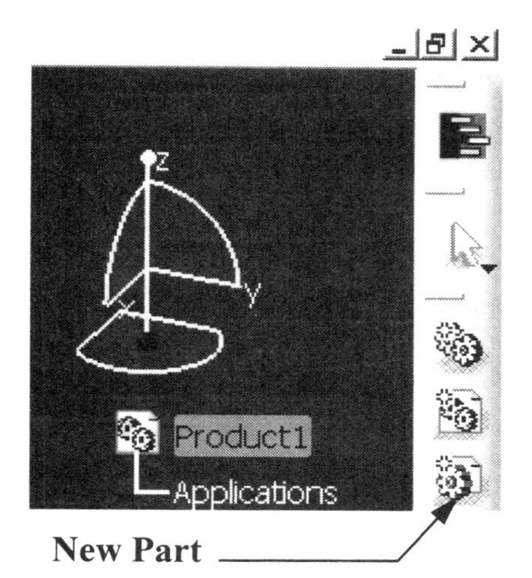

New Part

Figure 5.5

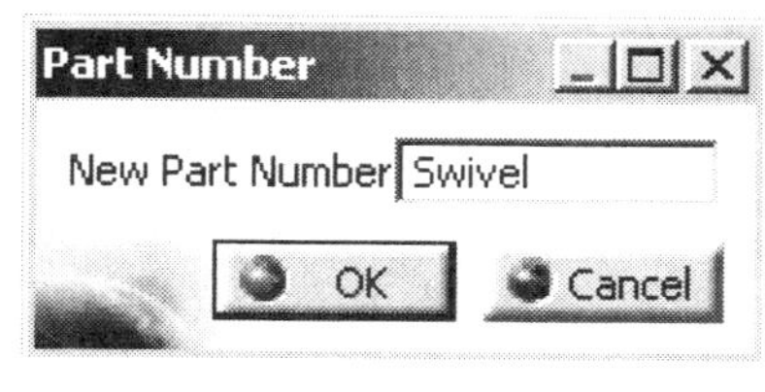

Figure 5.6

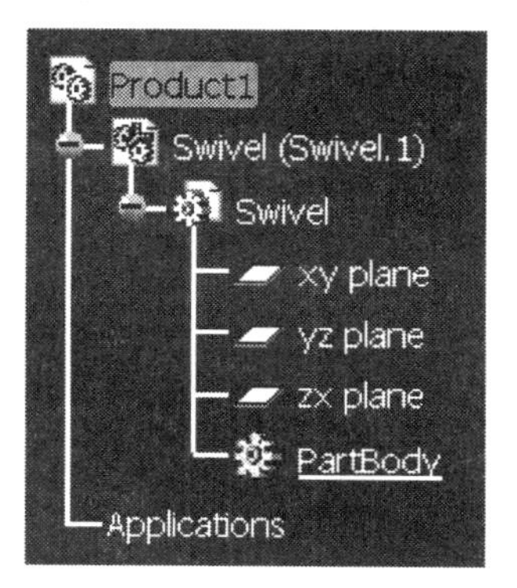

1.4 For the **New Part Number** box type in "**Swivel**." This method allows you to name the part as you create it rather than modifying the name after the part is created as you did in Lessons 1 and 2.

1.5 Select **OK** to create the **Swivel.CATPart**. Notice that a **Swivel (Swivel.1)** branch was just created in the **Specification Tree**. If you expand the **Swivel (Swivel.1)** branch you will have a **Specification Tree** similar to what is shown in Figure 5.6.

1.6 Now you need to enter the **Part Design Work Bench**. There are several different methods you can use to accomplish this. One method is to expand the **Swivel (Swivel.1)** branch as shown in Figure 5.6 and then double click on the next branch labeled "**Swivel**." This will bring up the **Part Design Work Bench**. The second method is to expand the **Swivel (Swivel.1)** branch and select (highlight) the next branch, **Swivel**, then select the **Part Design Work Bench** from either the **Start** pull down menu or the **Welcome To CATIA V5** window. Reference Lesson 1 about the **Welcome To CATIA V5** window.

1.7 Enter the **Sketcher Work Bench** using the **ZX Plane** as shown in Figure 5.7.

1.8 Create a **Circle** at the coordinates (0,0) with a radius of **.5** inches as shown in Figure 5.8.

Figure 5.7

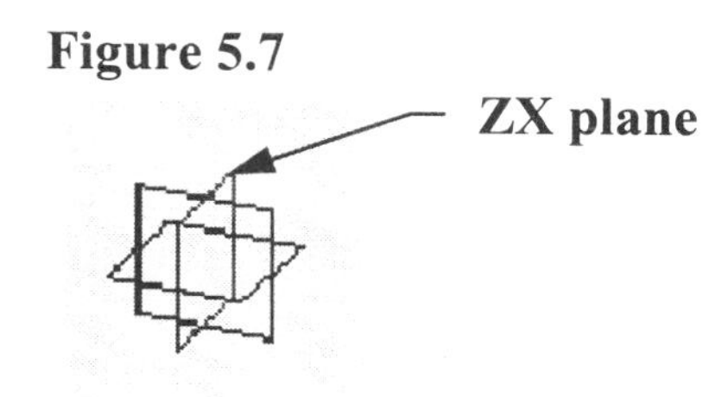

ZX plane

Figure 5.8

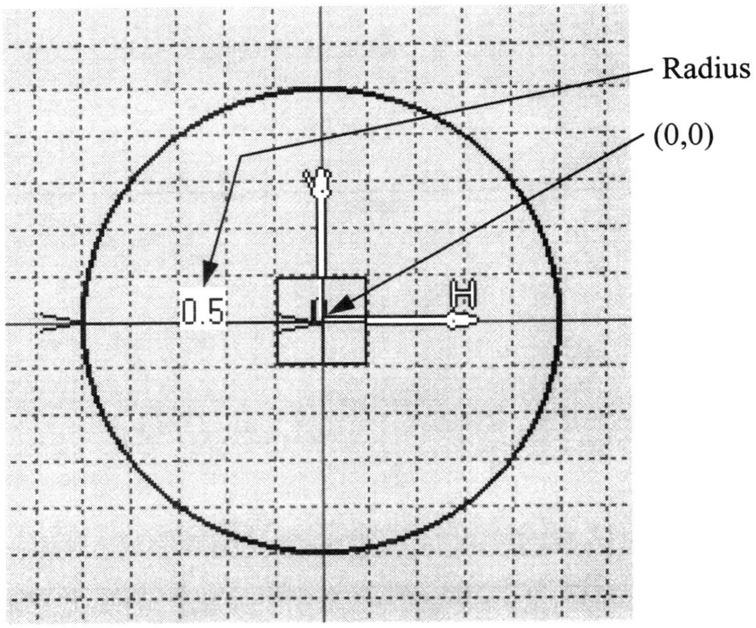

Radius

(0,0)

0.5

1.9 Exit the **Sketcher Work Bench**. Remember, this will put you back into the **Part Design Work Bench**.

1.10 Select the **Pad** tool.

1.11 When the **Pad Definition** window appears, as in Figure 5.9, select the **More** button. This will expand the **Pad Definition** window to show the **Second Limit** box. Figure 5.10 shows the **Pad Definition** window expanded to include the **Second Limit**.

Figure 5.9

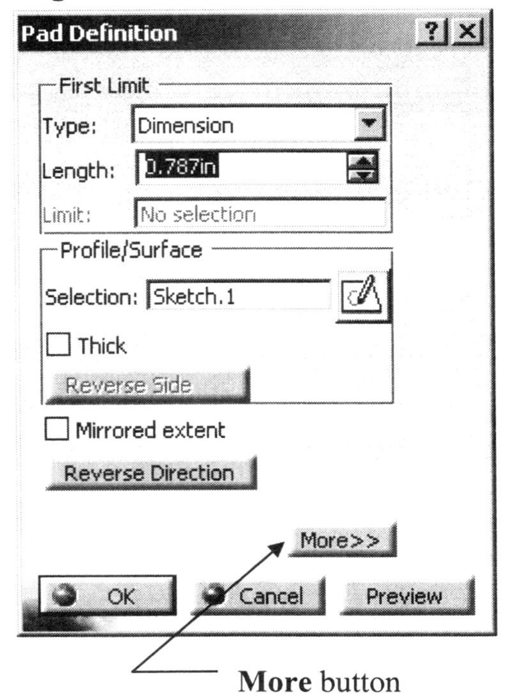

Figure 5.10 (Pad Definition **window extended)**

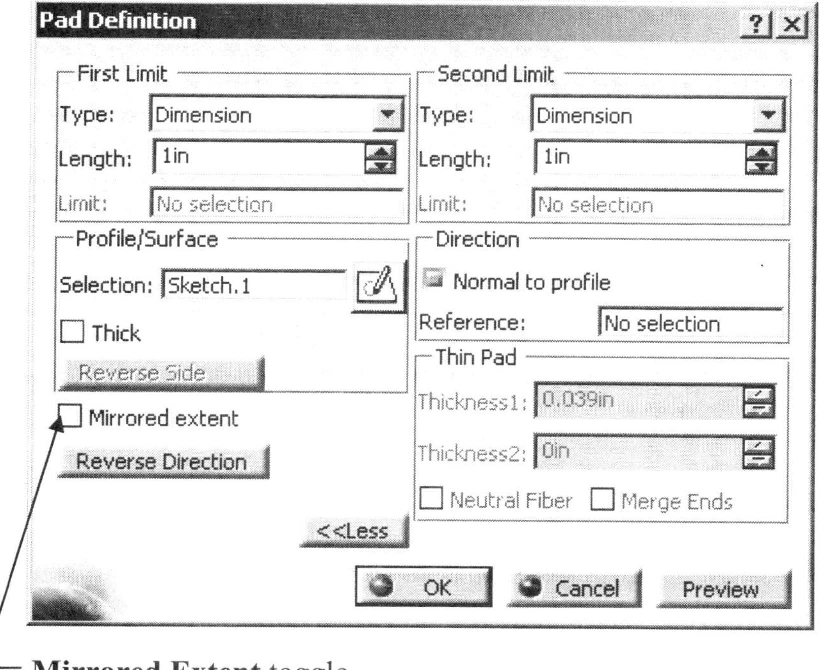

Mirrored Extent toggle

1.12 In the **First Limit** area, enter "**1in**" for the **Length** box. Leave the **Type** box set at "**Dimension**," as shown in Figure 5.10.

1.13 In the **Second Limit** area, enter "**1in**" for the **Length** box, as shown in Figure 5.10.

1.14 In the **Profile** section, the **Selection** box should show **Sketch.1**.

1.15 Select the **OK** button. Notice that the circle has been extruded 1 inch in both directions, reference Figure 5.11.

1.16 Select the **ZX Plane,** reference Figure 5.7. Enter the **Sketcher Work Bench**. Notice Step 1.7 created **Sketch.1**. This step creates an additional sketch, **Sketch.2**.

Figure 5.11

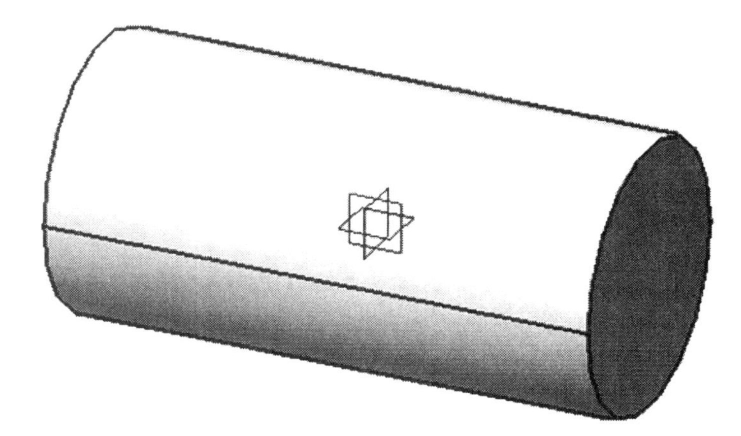

1.17 Create another **Circle** at the coordinates (0,0) with a radius of **.375** as shown in Figure 5.12.

1.18 Exit the **Sketcher Work Bench.** Select the **Pad** tool.

1.19 Enter "**1.6**" as the **First Limit Length**. Instead of selecting the **More** button and entering a length for the **Second Limit Length**, select the **Mirrored Extent** button, reference Figure 5.10. As long as the extrusion length is symmetrical you can use this button.

1.20 Select the **OK** button. The part should look similar the part shown in Figure 5.13.

Figure 5.12

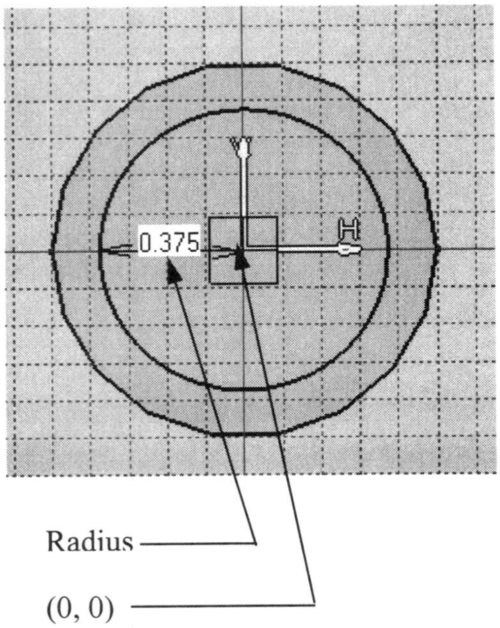

0.375

Radius

(0, 0)

Figure 5.13

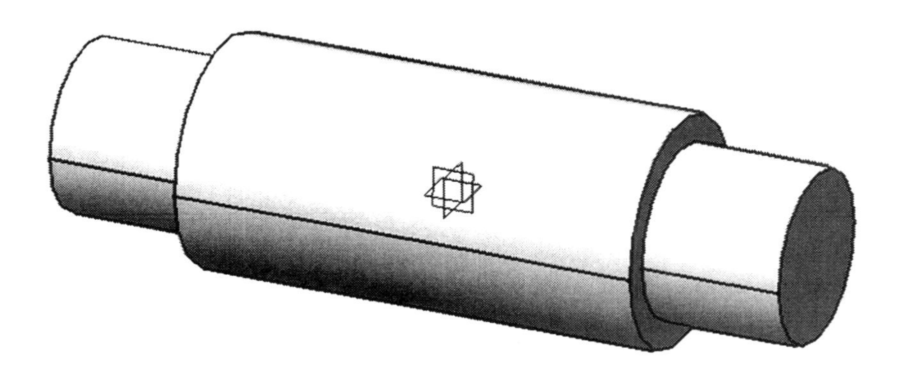

1.21 You now need to create or duplicate **Pad.1** and **Pad.2** in the **YZ Plane.** You could create two new sketches and extrude them using the **Pad** tool, but that would be more work and time consuming than necessary. The **Circular Pattern** tool will allow you to use the geometry you already created. You will want to apply the **Circular Pattern** to both **Pad.1** and **Pad.2** in the same operation. For this step hold the **Ctrl** button down as you select **Pad.1** and **Pad.2** from the **Specification Tree**, this is the Windows multi-select process. Reference Figure 5.14.

Figure 5.14

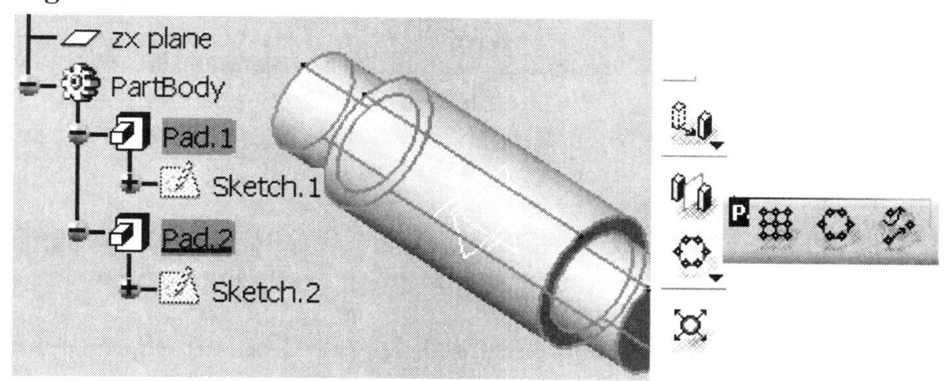

1.22 With **Pad.1** and **Pad.2** highlighted select the **Circular Pattern** tool. It is found in the **Transformation Features** tool bar. You may have to expand the **Rectangular Pattern** tool to locate it, as shown in Figure 5.14.

1.23 This will bring up the **Circular Pattern Definition** window as shown in Figure 5.15. Fill in the boxes as shown in Figure 5.15. Make sure the **Angular Spacing** is **90deg** and **Pad.1** and **Pad.2** show up in the **Object To Pattern** box.

1.24 Select the **Reference Element** box, and then select the **YZ Plane** from the **Specification Tree**. This will bring up a **Feature Definition Error** window.

Figure 5.15

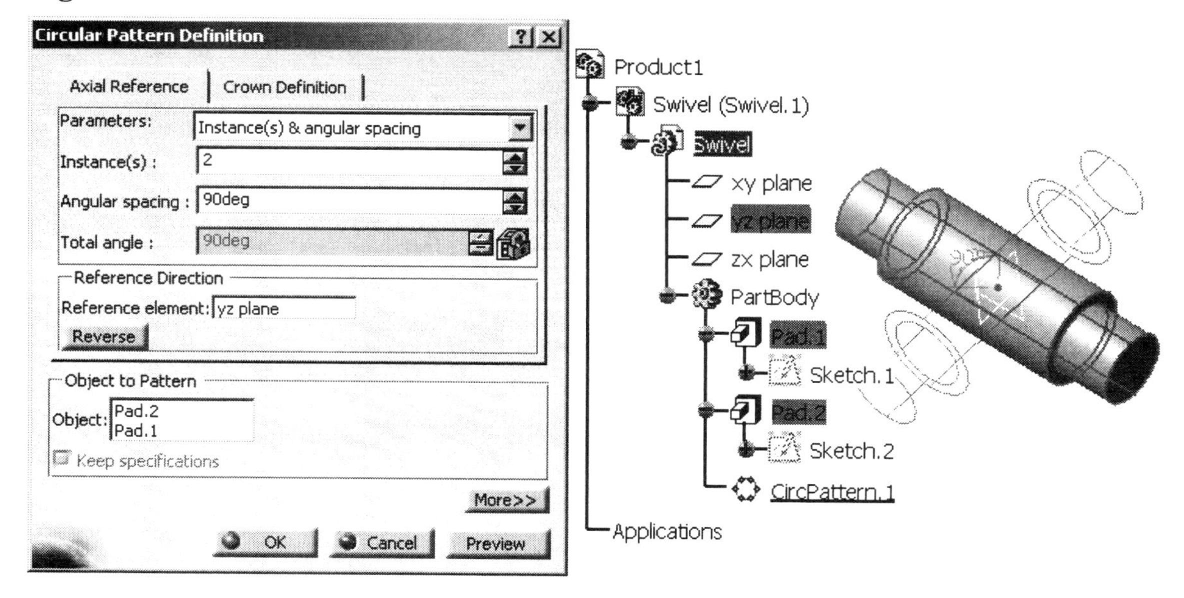

1.25 In the **Feature Definition Error** window, select **OK**. Your screen should look similar to what is shown in Figure 5.15. CATIA V5 is looking for some entity to rotate the pattern around. You could have created a line and/or used other existing entities to define the rotation, but no other elements exist at this point. You get a warning when using the **XY Plane** but it still lets you use it, notice it shows up in the **Reference Element** box.

1.26 In the **Circular Pattern Definition** window select the **OK** button. At this point your part should look similar to the one shown in Figure 5.16.

1.27 Create a .04 radius fillet on the outer edges of **Pad.1** (the large cylinder), reference Figure 5.16. You can create the fillet on all four edges and add only one **Fillet** branch to the **Specification Tree** by using the Windows multi-select method (Select the edges while you hold down the **Ctrl** key).

1.28 Create a .06 radius fillet at the intersection of the **Pad.1** (large cylinder), reference Figure 5.16. Use the Windows multi-select method so you only create one **Fillet** branch to the **Specification Tree**.

1.29 Create (.060 X 45 deg.) **Chamfers** on the ends of all four smaller pins as shown in Figure 5.16. Again, use the Windows multi-select method.

Figure 5.16

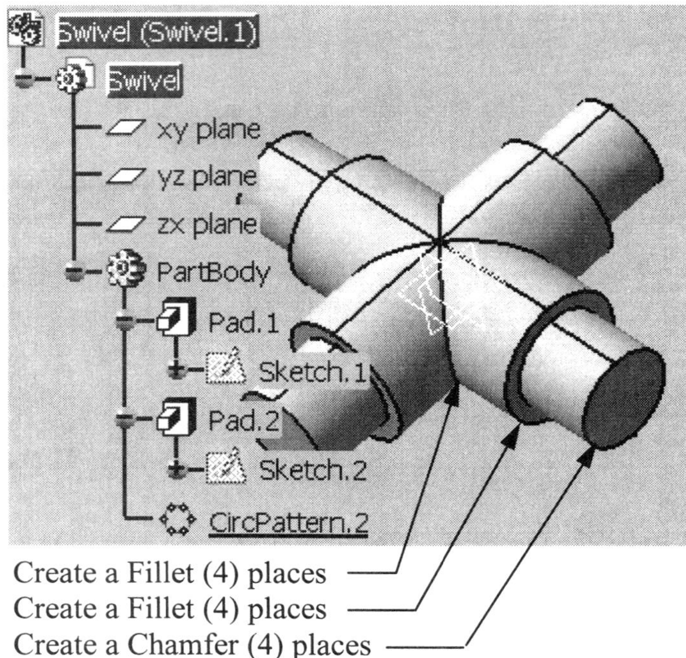

Figure 5.17

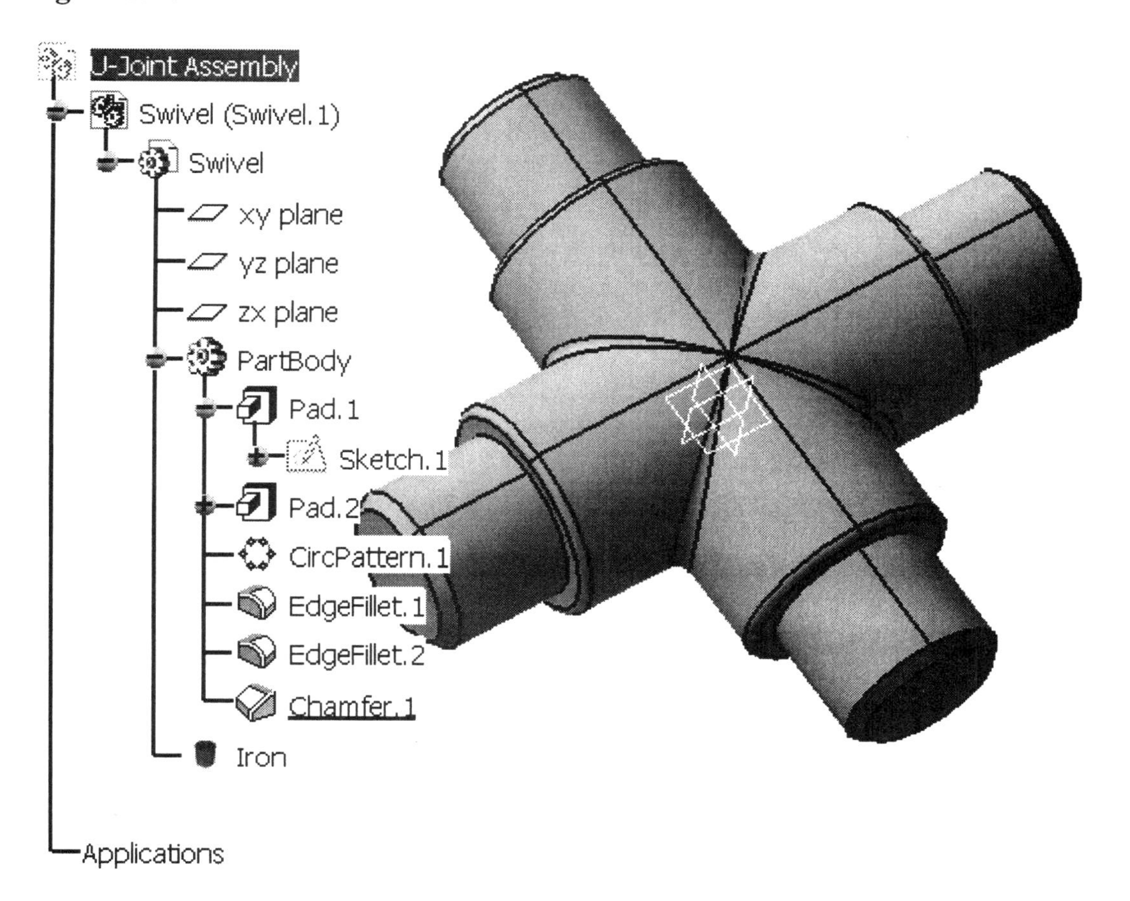

1.30 Select **Product1** from the base of the **Specification Tree**. Rename **Product1** to **U-Joint Assembly**. Reference Lesson 2 Step 20 on how to rename **Specification Tree** branches; the process is the same for **Products**, **Parts**, **PartBodys** and all other **Specification Tree** branches.

1.31 Apply **Iron** material to the "**Swivel**" part. Remember you must select (highlight) the **Part** you are applying the material to. You must also apply the **Material Visualization**. Reference Lesson 2 Step 18 for detailed instructions.

1.32 If your "**Swivel**" looks similar to what is shown in Figure 5.17, you are ready to continue to the next step.

1.33 To add a new part to the "**U-Joint Assembly**," double click on "**U-Joint Assembly**" on the **Specification Tree**. This will take you out of the **Part Design Work Bench** and bring up the **Product Structure Work Bench**.

1.34 With the "**U-Joint Assembly**" selected (highlighted), select the **New Part** tool . This will bring up the **Part Number** window, the same as in Step 1.3 and Figure 5.5.

1.35 Name the new part "**Top U-Joint**," then select **OK**. Notice the new part "**Top U-Joint**" part will show up on the **Specification Tree**.

1.36 A **New Part: Origin Point** window will come up. Select to **NO** keep the same **Part Origin**.

1.37 Save the "**U-Joint Assembly**." The file will be saved as a **CATProduct**, not a **CATPart**. If you saved the file prior to adding the new part the file would have been saved as a **CATPart**. A **CATProduct** is two or more **CATParts**. Lesson 6 will go into detail on how existing **CATParts** are brought into a **CATProduct**.

2 Creating The "Top U-Joint" Using Multiple Sketches

The "**Top U-Joint**" is the second of three parts you will create in this lesson, reference Figure 5.18. You will create the "**Top U-Joint**" using the same method you did in Step 1, using multiple sketches. There are many possible methods of creating this part. Step 1 is a suggested method. At this point, you should be capable of creating this part on your own. You are encouraged to try your own method, maybe even use the following steps as helpful hints. General steps to create this part are provided below if you are not ready to attempt this on your own. Since this part will be used in Lesson 6, **Assembly Design Work Bench**, it is critical that the dimensions shown in Figure 5.18 are used.

Figure 5.18

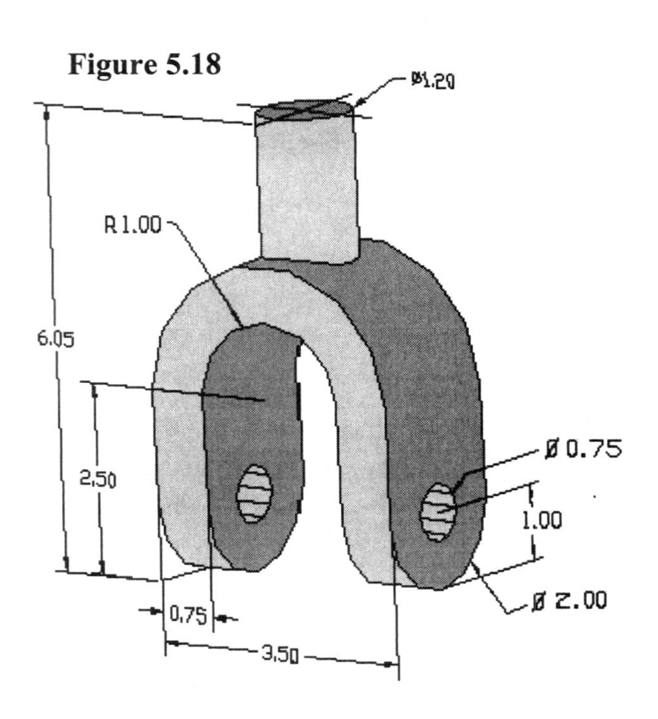

2.1 The "**Top U-Joint"** part is a new part and completely separate from the "**Swivel**" part you just created. Since the "**Top U-Joint**" part is a new part, you will need to double click on the "**Top U-Joint**" branch of the **Specification Tree**. This will bring up the **Part Design Work Bench** for the selected part, the "**Top U-Joint**."

2.2 Select the **YZ Plane**, shown in Figure 5.19.

2.3 Enter the **Sketcher Work Bench**.

Figure 5.19

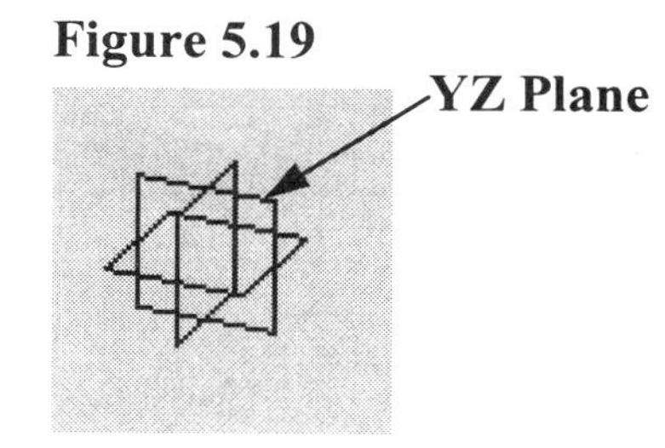

2.4 Sketch the profile shown in Figure 5.20. There are many different ways this sketch could be created and most of them would be correct. This step is not limiting you to one particular method as long as you get the desired result. The method used in Figure 5.20 is a sample or suggested method. (Hint: the part is symmetrical). Complete the following steps to create the sketch as shown in Figure 5.20. Reference Figure 5.20 for the dimensions.

2.4.1 Create a rough sketch using the **Profile** tool.

2.4.2 Constrain the sketch and adjust the dimensions by double clicking the constraints and using the **Constraint Definition** window.

2.4.3 If your sketch looks similar to the one shown in Figure 5.20 you are ready to move on to the next step.

Figure 5.20

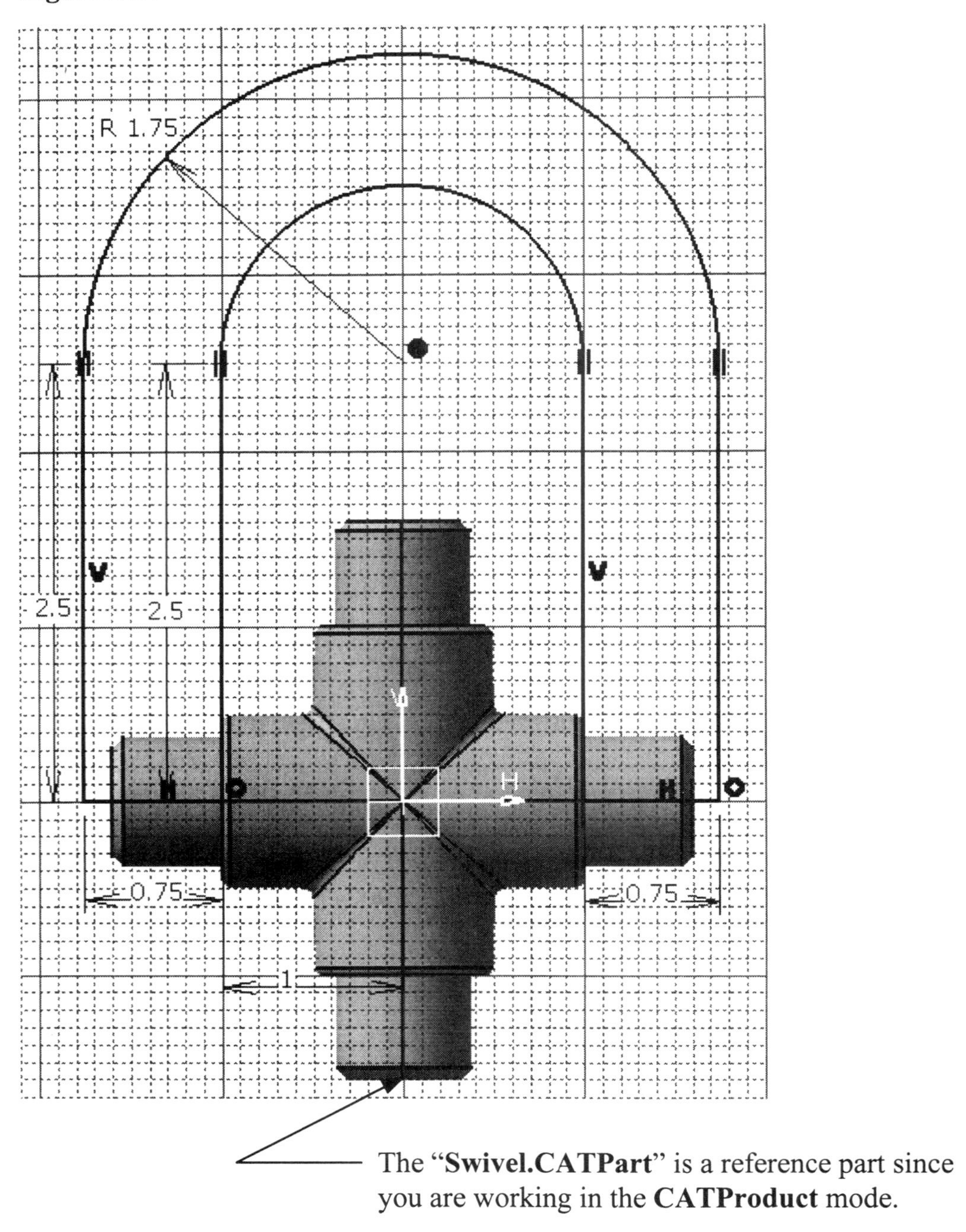

The "**Swivel.CATPart**" is a reference part since you are working in the **CATProduct** mode.

2.5 Exit the **Sketcher Work Bench**.

2.6 Select the **Swivel (Swivel.1)** branch of the **Specification Tree**.

2.7 With the **Swivel (Swivel.1)** branch highlighted select the **Hide/Show** tool at the bottom of the screen. This will place the **Swivel** in **Hide** mode. It will still be there it is just moved from view. Notice the **Swivel (Swivel.1)** branch is dimmed; this signifies that the branch is in **Hide** mode.

Figure 5.21

Swivel.CATPart is in **Hide** mode.

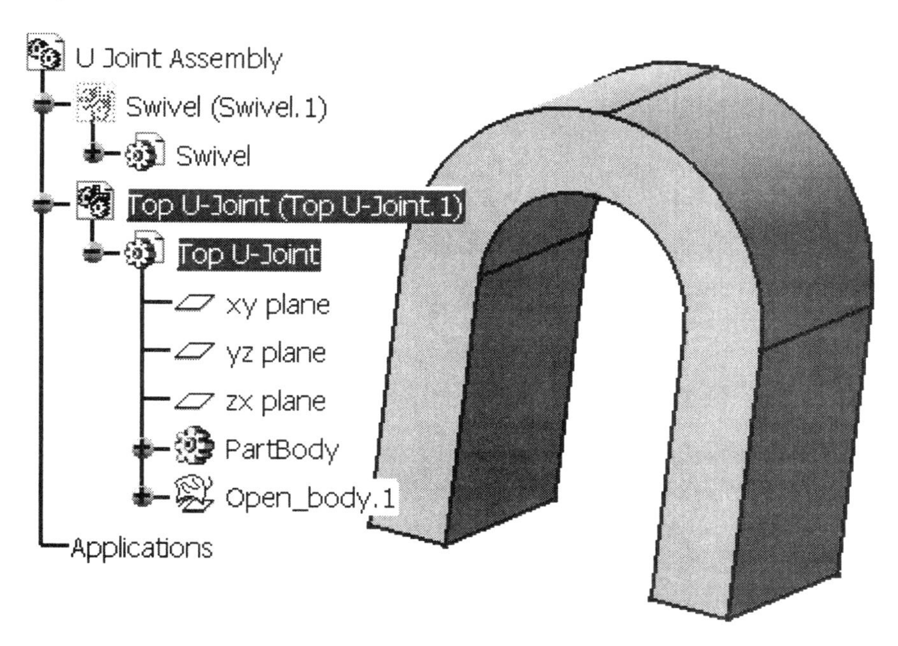

2.8 Use the **Pad** tool to extrude the profile 1 inch, then select the **Mirrored Extent** option so the extrusion will be a total of 2 inches thick but centered on the **YZ Plane**. Your newly created pad should look similar to what is shown in Figure 5.21.

2.9 The next step is to create a **2″** diameter rounded edge on the bottom of both legs. This process can be simplified by using the **Tritangent Fillet** tool. By default, the **Tritangent Fillet** tool will be a sub option under the **Edge Fillet** tool. To select the **Tritangent Fillet** tool, select the arrow next to the **Edge Fillet** tool. This will display the sub options available, reference Figure 5.22.

Figure 5.22

2.10 Select the **Tritangent Fillet** tool. This will bring up the **Tritangent Fillet Definition** window as shown in Figure 5.23. The first box is the **Faces To Fillet** box. This box allows you to select two faces to be joined with a fillet. The second box is **Face To Remove**. This box allows you to select the face that will be removed and replaced with the fillet.

Figure 5.23

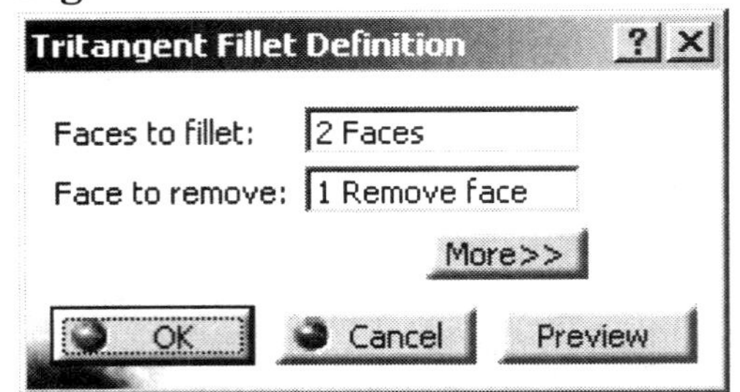

Figure 5.24

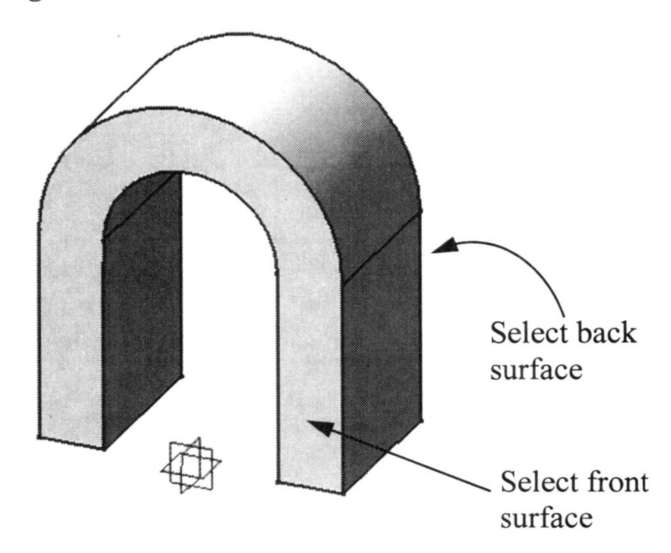

2.11 Select the front and back surfaces as the **Faces to Fillet** on the "**Top U-Joint**," as shown in Figure 5.24.

2.12 Select the bottom surface of the right leg. This surface joins the front and back surfaces. It is the **Face to Remove**, as shown in Figure 5.25.

Figure 5.25 Figure 5.26

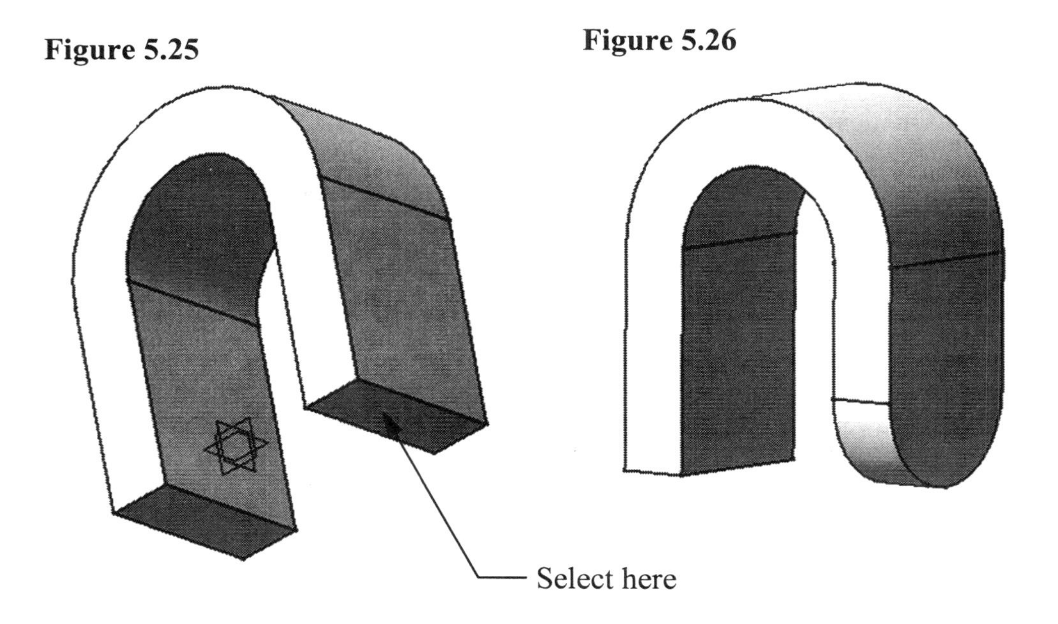

2.13 Select the **OK** button. The bottom surface selected will be removed and replaced with a radius. The radius will be the same size as the length of the surface it replaced, reference Figure 5.26.

2.14 Repeat Steps 2.9 thru 2.13 to create the radius on the other leg. With the radii created on both legs, the part should look similar to the one shown in Figure 5.27.

Figure 5.27

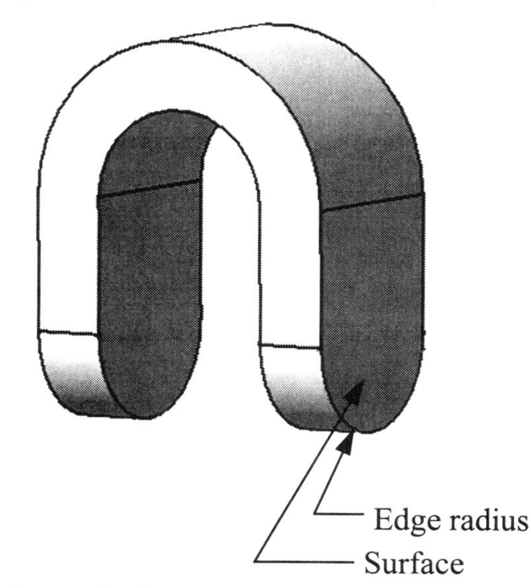

2.15 To put the **.75** inch diameter hole in the two legs of the part, multi select the edge and surface shown in Figure 5.27. If you have questions on this process, reference Lesson 2.

2.16 Select the **Hole** tool. In the **Hole Definition** window, enter ".**75**" inches for the **Diameter** box and select "**Up To Last**" as the **Hole Extension**. This will create the hole in both legs at the same time.

2.17 Select **OK** to create the hole. Your part should look similar to the one in Figure 5.28.

Figure 5.28

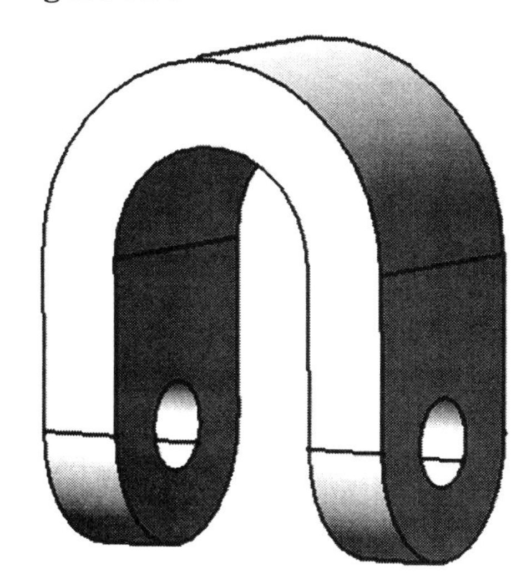

NOTE: Selecting the edge of the radius before selecting the **Hole** tool will create the hole at the center of that radius as shown in Figure 5.27. Selecting an arc is the same as multi selecting two edges to locate the hole.

2.18 The next step is to create the shaft on the top of the "**Top U-Joint**." To accomplish this you will need to create a **Plane** that will represent the top of the shaft. This **Plane** is where you will create the sketch for the shaft. Since the **Plane** tool is not a default tool in the **Part Design Work Bench**, you will have to add it to the tool bar or enter the **Generative Shape Design Work Bench**. Since the **Generative Shape Design Work Bench** has not been covered, complete the following steps to add the **Plane** tool to the **Part Design Work Bench**.

Figure 5.29

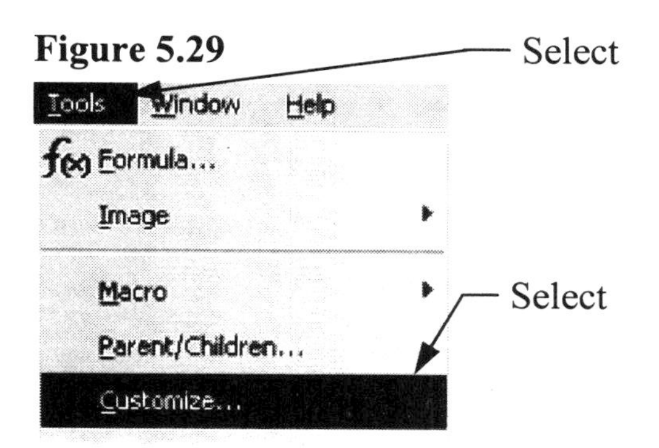

2.18.1 Select the **Tools, Customize** options from the top pull down menu as shown in Figure 5.29.

Figure 5.30

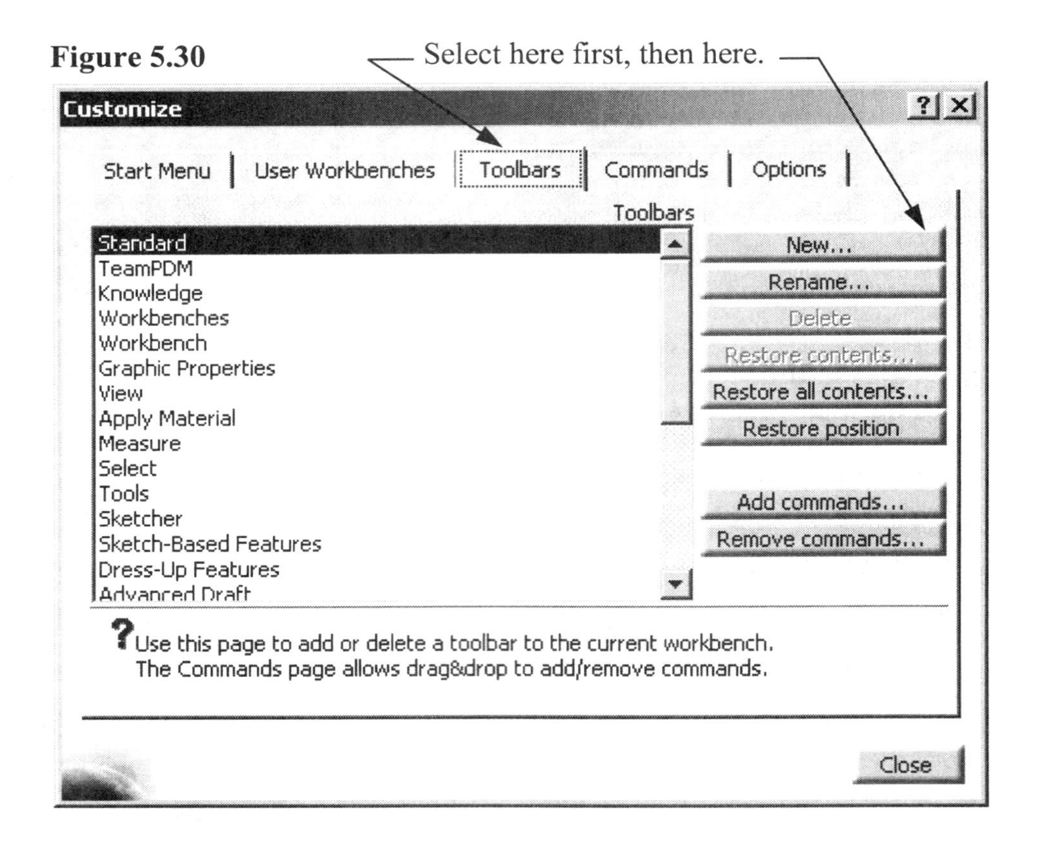

Select here first, then here.
Customize
Start Menu
User Workbenches
Toolbars
Commands
Options
Toolbars
Standard
TeamPDM
Knowledge
Workbenches
Workbench
Graphic Properties
View
Apply Material
Measure
Select
Tools
Sketcher
Sketch-Based Features
Dress-Up Features
Advanced Draft
New...
Rename...
Delete
Restore contents...
Restore all contents...
Restore position
Add commands...
Remove commands...
Use this page to add or delete a toolbar to the current workbench.
The Commands page allows drag&drop to add/remove commands.
Close

Figure 5.31

Select
Select
New Toolbar
Toolbar Name:
WireFrame
Workbenches:
FreeStyle
Shape Design Workbench
Sheet Metal Design
Hybrid Part Workbench
Circuit Board Design
Toolbars:
ShapeModification
ShapeSurfaces
ShapeCurves
WireFrame
ShapeView
ShapeCloud
ShapeAnalysis
OK
Cancel

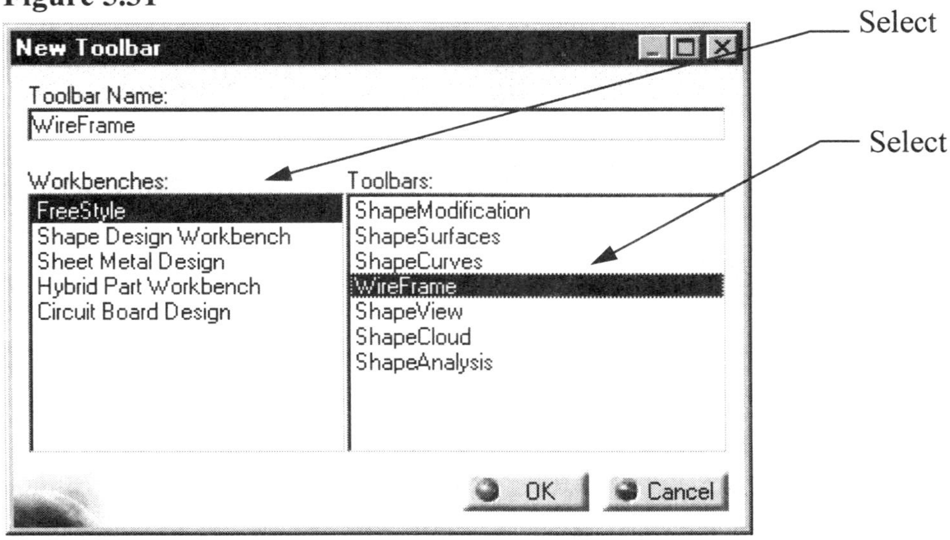

2.18.2 When the **Customize** window appears on the screen, select the **Toolbars** tab and then select the **New** button located on the right side of the window as shown in Figure 5.30.

2.18.3 When the **New Toolbar** window appears, select **FreeStyle** as the **Work Bench** and **WireFrame** as the **Toolbar** as shown in Figure 5.31.

2.18.4 The **WireFrame** tool bar that contains the **Plane** tool will appear on the screen as shown in Figure 5.32. If a different tool bar appeared, select the "**X**" in the top right corner of the incorrect tool bar to remove it and repeat Steps 2.18.1 thru 2.18.3.

Figure 5.32

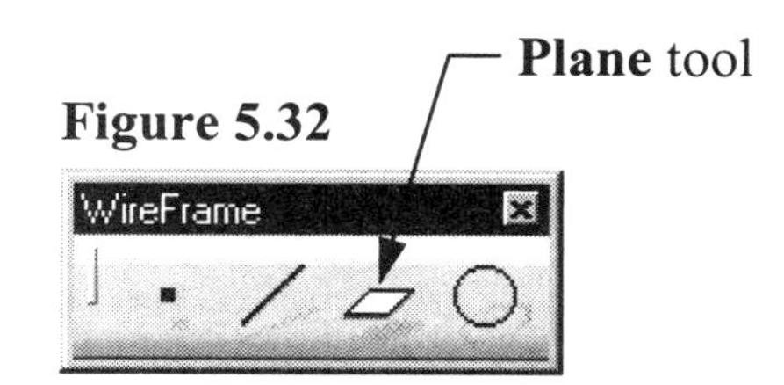

Plane tool

2.18.5 Select the **OK** button in the **New Toolbar** window to accept the new tool bar. Select the **Close** button in the **Customize** window to remove it from the screen. You can now move the **WireFrame** tool bar to a convenient location on the screen. This was the long method of accessing the **Plane** tool, but it also taught you how to customize your tool bars. Just because you are in the **Part Design Work Bench** does not mean you cannot have access to tools from other work benches.

2.18.6 You can access the **Plane** tool quickly by selecting the **View** pull down menu on the top of the screen. Select **Toolbars** (the first option). This will bring up a long list of tool bars. The ones that are checked are the tool bars you currently have access to on the right side of your screen. Check the tool bar labeled **Reference Elements (Extended)**. This will add the **Reference Element** tool bar to the right side of your screen. This tool bar contains the **Plane**, **Point** and **Line** tools.

2.19 Select the **Plane** tool from the **Reference Elements** tool bar just added. This will bring up the **Plane Definition** window. Set the **Plane Type** to **Offset From Plane**. Select the **XY Plane** as the **Reference**. Enter **6.05** inches as the **Offset**. Reference Figure 5.33. Select **OK** to create the plane.

2.20 Select the new **Plane**. With the new **Plane** highlighted select the **Sketcher** tool to enter the **Sketcher Work Bench**. If the part goes off the screen use the "**Fit All In**" tool to bring it back. Make sure the **PartBody** is highlighted so the sketch will be created in the **PartBody** branch and not the **Open_Body.1** branch. If the sketch is created in the **Open_Body.1** branch of the **Specification Tree** you will not be able to use the **Pad** tool to create the solid part.

Figure 5.33

New **Plane**

U-Joint Assembly
Swivel (Swivel.1)
Swivel
Top U-Joint (Top U-Joint.1)
Top U-Joint
xy plane
yz plane
zx plane
PartBody
Pad.1
TritangentFillet.1
TritangentFillet.2
Hole.1
Open_body.1
Applications

Offset= 6.05in
Move
Reference

Plane Definition
Plane type: Offset from plane
Reference: xy plane
Offset: 6.05in
Reverse Direction
Repeat object after OK
OK Apply Cancel

Figure 5.34

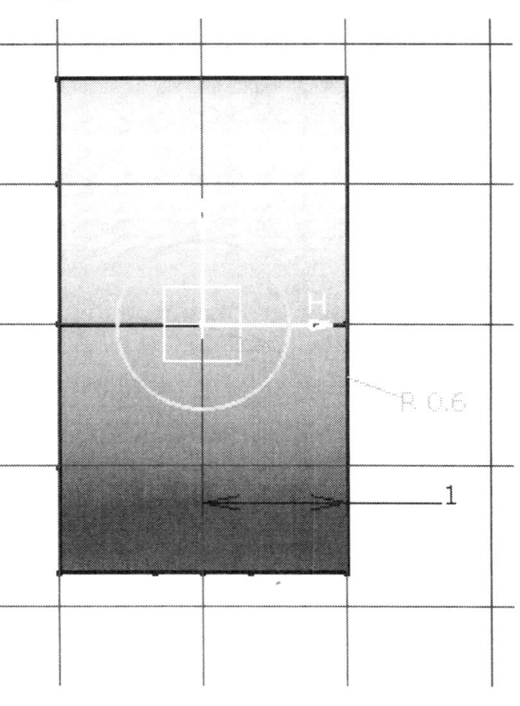

2.21 Create a **Circle** with a radius of **.6** inches, and **Constrain** it as shown in Figure 5.34.

2.22 Exit the **Sketcher Work Bench** and extrude the circle down to the top surface of the "**Top U-Joint**" as shown in Figure 5.35.

HINT: To extrude the circle down to the top of the "**Top U-Joint**" without having to enter in the exact distance, set the **Type** box to "**Up To Next.**" This will extrude the circle down until it hits the top surface of the "**Top U-Joint.**" If the extrusion arrow is pointing away from the "**Top U-Joint**" surface, select the "**Reverse Direction**" button in the **Pad Definition** window. This will invert the arrow.

2.23 Add a **.15″** × **45** deg **Chamfer** to the top edge of the shaft as shown in Figure 5.36.

Figure 5.35

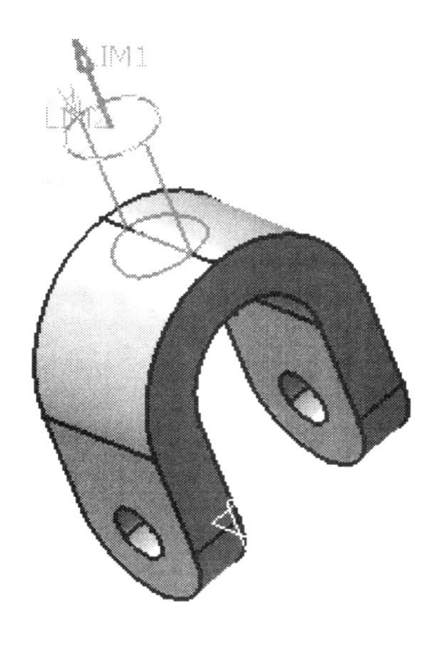

Figure 5.36

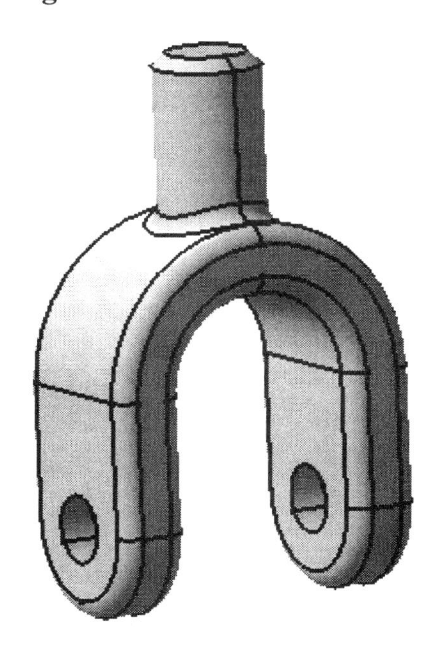

2.24 Add a **.2″** radius **Fillet** to all of the exterior edges of the solid as shown in Figure 5.36.

2.25 Add a **.2″** radius **Fillet** where the **Pad.1** (U-shaped sketch) intersects with **Pad.2** (shaft). Reference Figure 5.36.

2.26 The next few steps will take you through the process of creating the gears on the shaft. Complete the following steps to create the gears.

2.26.1 Create the sketch shown in Figure 5.37 on the **ZX Plane**. The sketch consists of one straight line.

Figure 5.37

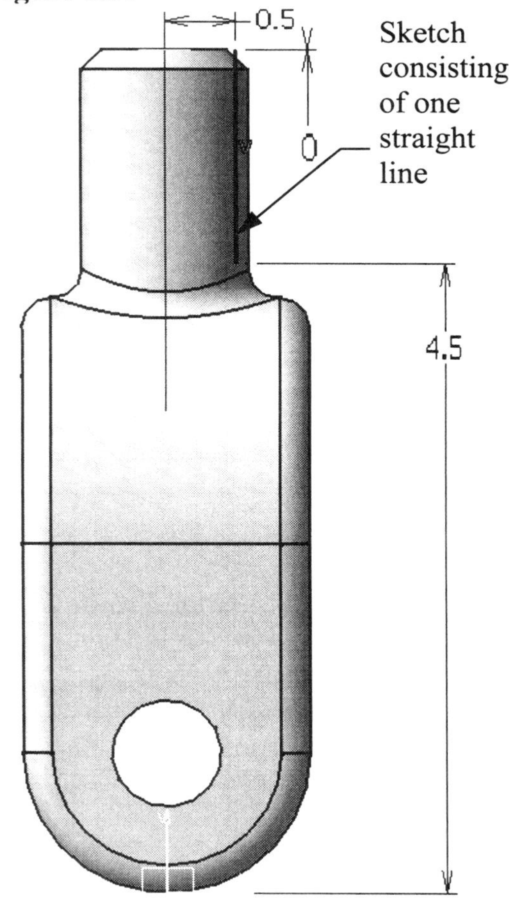

2.26.2 **Constrain** the sketch as shown in Figure 5.37. Notice the sketch is not a closed sketch. The **Slot** tool does not require a closed sketch, but it is important that you **Constrain** the sketch as shown. Future modifications will not work if the sketch is not constrained as shown in Figure 5.37.

2.26.3 Exit the **Sketcher Work Bench**.

Figure 5.38

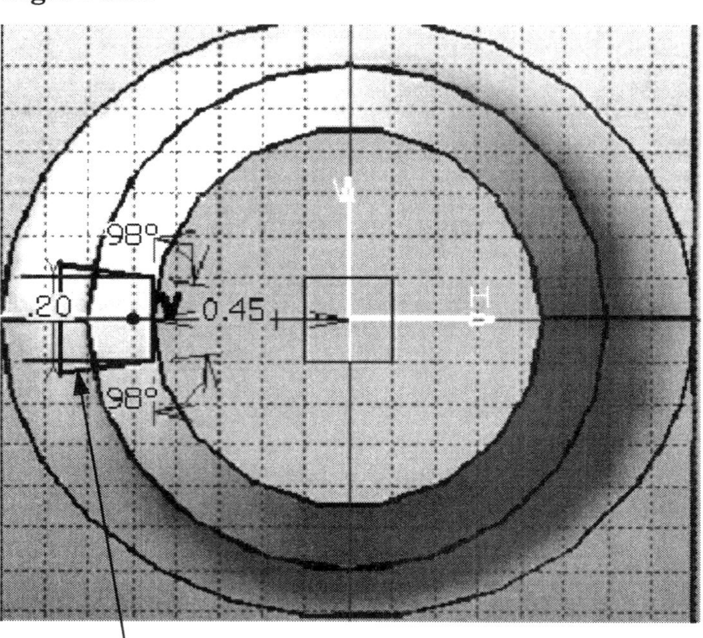

Profile sketch

2.26.4 Create another sketch; select the same **Plane** that was used to create the shaft sketch. Create the sketch that is shown in Figure 5.38.

2.26.5 **Constrain** the sketch as shown in Figure 5.38.

2.26.6 Exit the **Sketcher Work Bench**.

2.26.7 Select the **Slot** tool . This will bring up the **Slot Definition** window as shown in Figure 5.39.

2.26.8 For the **Profile** box select the sketch you created in Figure 5.38, in this case it is **Sketch.6**.

2.26.9 For the **Center Curve** box, select the sketch you created in Figure 5.37, in this case it is **Sketch.5**.

2.26.10 Select the **OK** button to create the **Slot**. Figure 5.40 represents what the completed slot should look like.

Figure 5.39

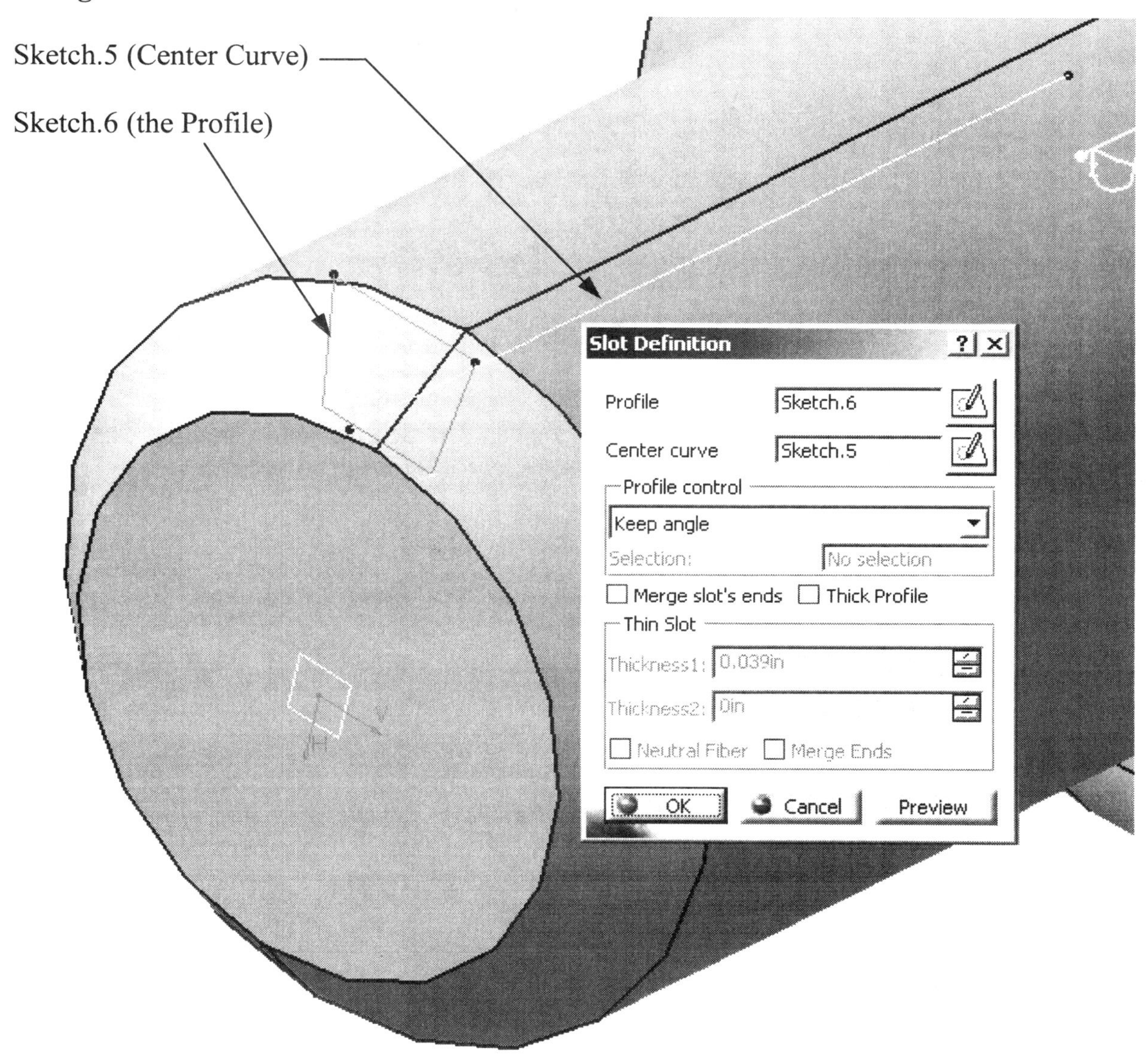

Figure 5.40 (completed slot)

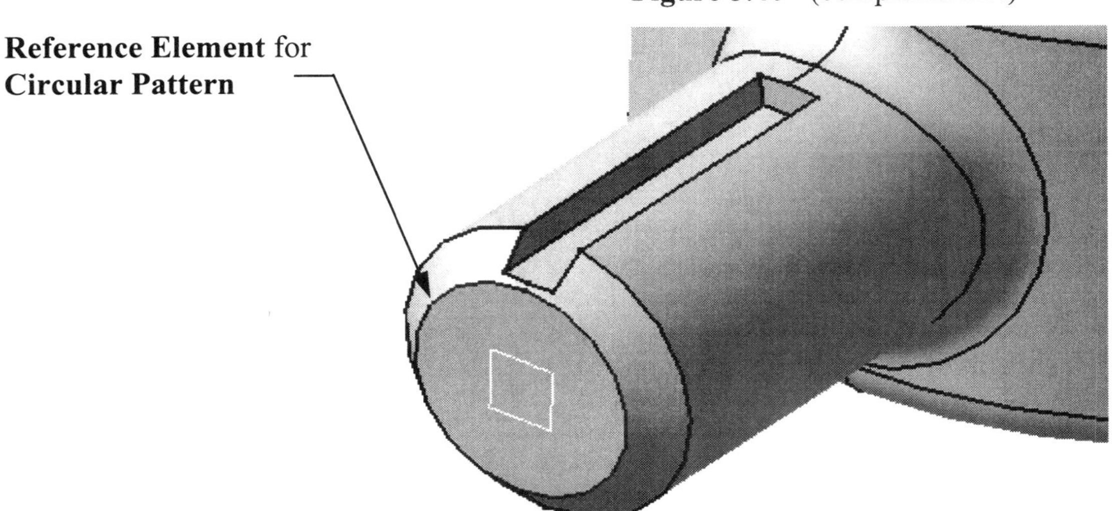

2.27 The original slot is complete. Now you can use the original slot to create a circular pattern around the shaft. The following steps detail this process.

2.27.1 Make sure that **Slot.1** is highlighted. Select the **Circular Pattern** tool. This will bring up the **Circular Pattern Definition** window as shown in Figure 5.41.

2.27.2 Select **Angular Spacing & Total Angle** for the **Parameters** box.

2.27.3 Enter **45 deg** for the **Angular Spacing** box.

2.27.4 Enter **360 deg** for the **Total Angle** box.

Figure 5.41

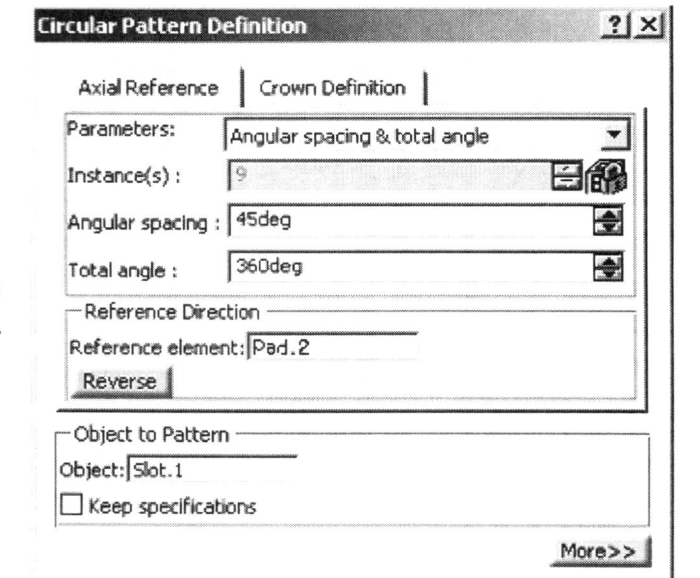

2.27.5 Select the rounded surface on the end of the shaft for the **Reference Element**. Figure 5.40 shows that it was **Pad.2**. For this particular selection

2.27.6 (**Angular Spacing & Total Angle**) you can select a circular edge to define the circular pattern. Other selections require a line or axis.

2.27.7 **Slot.1**, the slot you created in the previous steps, should show in the **Object To Pattern** box; if not, select it.

2.27.8 Select the **Preview** button. If the preview shows the slot being rotated and duplicated around the rounded edges as shown in Figure 5.42, select the **OK** button to create the **Circular Pattern**.

2.27.9 If your "**Top U-Joint**" looks similar to the one shown in Figure 5.42 you are ready to move on.

Figure 5.42

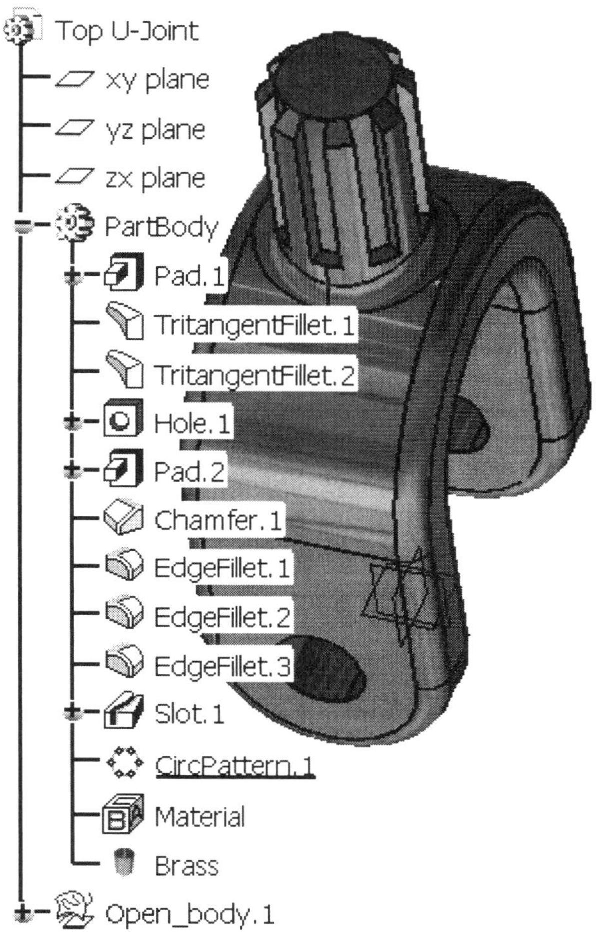

2.27 You are done adding features to the "**Top U-Joint**" but not quite finished with the part. Add "**Brass**" material to the "**Top U-Joint**." The only reason for applying brass as the material is so the part can easily be differentiated from the other parts (the "**Swivel**" and "**Bottom U-Joint**"). Remember that you must change the **Visualization** mode before the material will be applied to the part, reference Lesson 2 on **Applying Material**.

2.28 Earlier in this lesson you were instructed to put the "**Swivel**" part in **Hide** mode so it would not get in the way while you constructed the "**Top U-Joint**." Now is a good time to bring it back so you can see both parts (the "**Top U-Joint**" and "**Swivel**"). To do this, select the "**Swivel**" branch of the **Specification Tree**. Then select the **Hide/Show** tool at the bottom of the screen. This will bring the "**Swivel**" back into view

2.29 Double click on the "**U-Joint Assembly**," the base of the **Specification Tree**. Remember this will take you to the **Product Structure Work Bench**. Save the "**U-Joint Assembly**." Remember, you must be in the **Product Structure Work Bench** to save as a "**U-Joint Assembly.CATProduct**." If the "**Swivel**" part is active in the **Part Design Work Bench**, only the "**Swivel.CATPart**" will be saved.

2.30 Step 3 will take you through the process of creating the "**Bottom U-Joint**" using a different approach.

3 Creating The "Bottom U-Joint" Using Boolean Geometry

The third and last part you will create in this lesson is the "**Bottom U-Joint**," shown in Figure 5.64. The "**Bottom U-Joint**" is identical to the "**Top U-Joint**." In Industry, you would create the "**Bottom U-Joint**" as efficiently as possible, which would be by duplicating the "**Top U-Joint**." Since it is this books objective to show you how to use as many tools as possible, this step will use a different method in creating the "**Bottom U-Joint**." At the moment, this may not be the most efficient method, but it will help you be a more efficient and knowledgeable CATIA V5 user. This additional method will be recognized by anyone who has experience on the older legacy solid modeling programs. The method still has applications in the solid modeling world. The method is, creating and manipulating solids using **Boolean Geometry**. The following instructions step you through the process of creating the "**Bottom U-Joint**" using **Boolean Operations**.

3.1 Start a new part by selecting **File**, **New** from the Windows tool bar. This will bring up the **New** window as shown in Figure 5.43.

3.2 From the **New** window, select **Part**. This will create a new part with a new **Specification Tree**. The part number will depend on how many parts you have started/created since you started the session. Rename the part to "**Bottom U-Joint**." Remember, you can rename a branch by right clicking on the part branch and selecting **Properties**, reference Lesson 2.

Figure 5.43

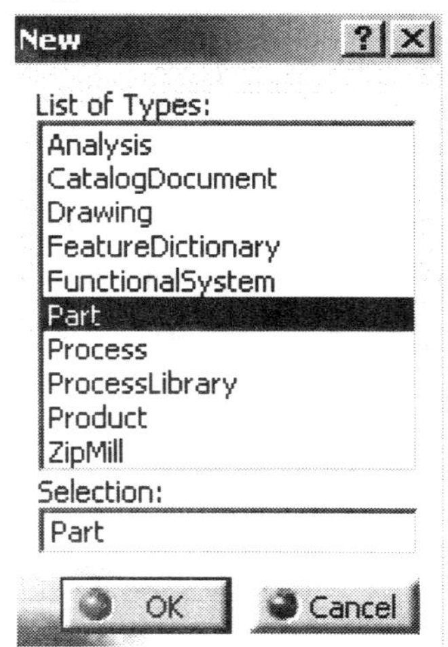

NOTE: Yes, it would have been more efficient to select **New Product** in the **Product Structure Work Bench** as you did when creating the "**Swivel**" and "**Top U-Joint**." There is a method to this madness (additional work). This process will allow you use additional tools and processes that you otherwise would not have the opportunity to experience. So, please continue as instructed.

3.3 In the **Part Design Work Bench** and the **Part Specification** branch renamed to "**Bottom U-Joint**," create a sketch in the **ZX Plane**.

3.4 Create a rectangle in the **Sketcher Work Bench** as shown in Figure 5.44.

Figure 5.44

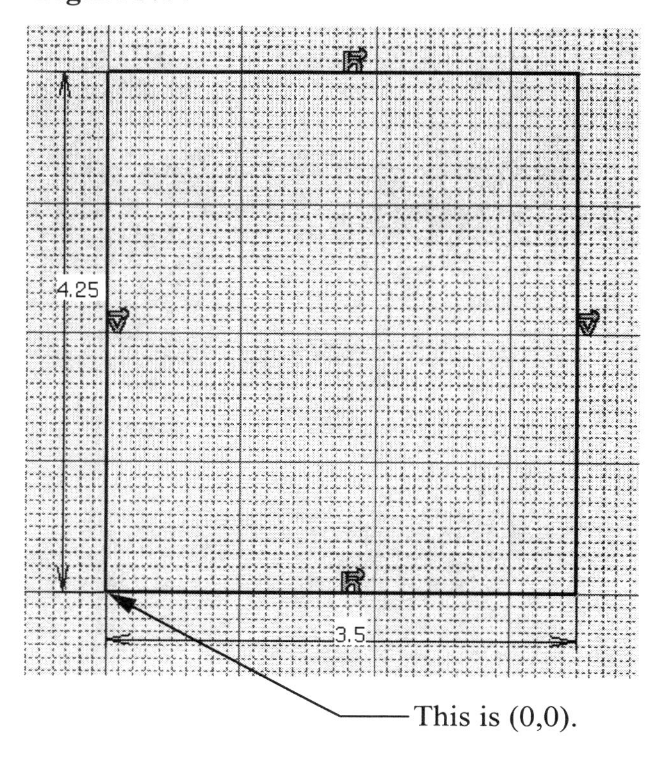

3.5 Exit the **Sketcher Work Bench**, extrude the sketch 1 inch and mirror the extent so the total thickness is 2 inches. Reference Figure 5.45.

3.6 Select the **Insert** tab from the top pull down menu as shown in Figure 5.46.

3.7 From the **Insert** window, select the **Body** option. This will insert a **Body.2** into the **Specification Tree** as shown in Figure 5.45.

3.8 Select **Body.2** in the **Specification Tree**. This will make **Body.2** the active branch.

3.9 Select the front surface of the box as shown in Figure 5.45 and enter the **Sketcher Work Bench**. The selected surface defines the sketcher plane, the same as the plane in previous lessons. The new sketch will show up on the **Specification Tree** under the **Body.2** branch.

Figure 5.45 Figure 5.46

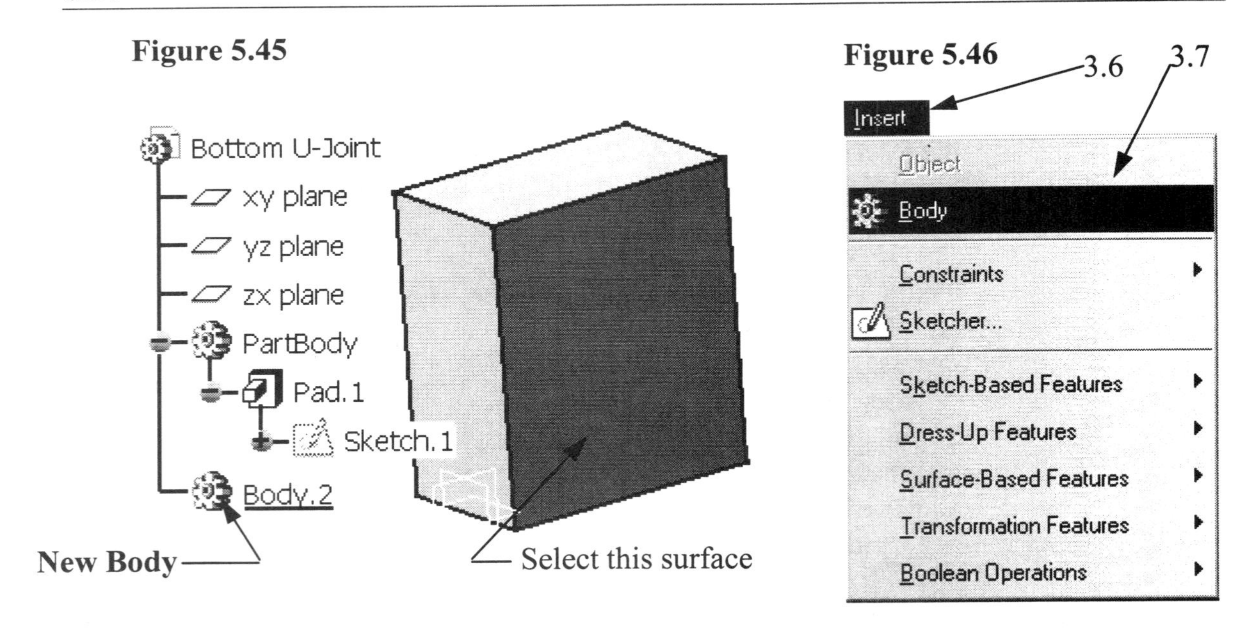

3.10 Sketch the profile shown in Figure 5.47. CATIA V5 will allow you to **Constrain** the new sketch to the edges of the existing box (**Sketch.1**) as shown in Figure 5.47.

Figure 5.47

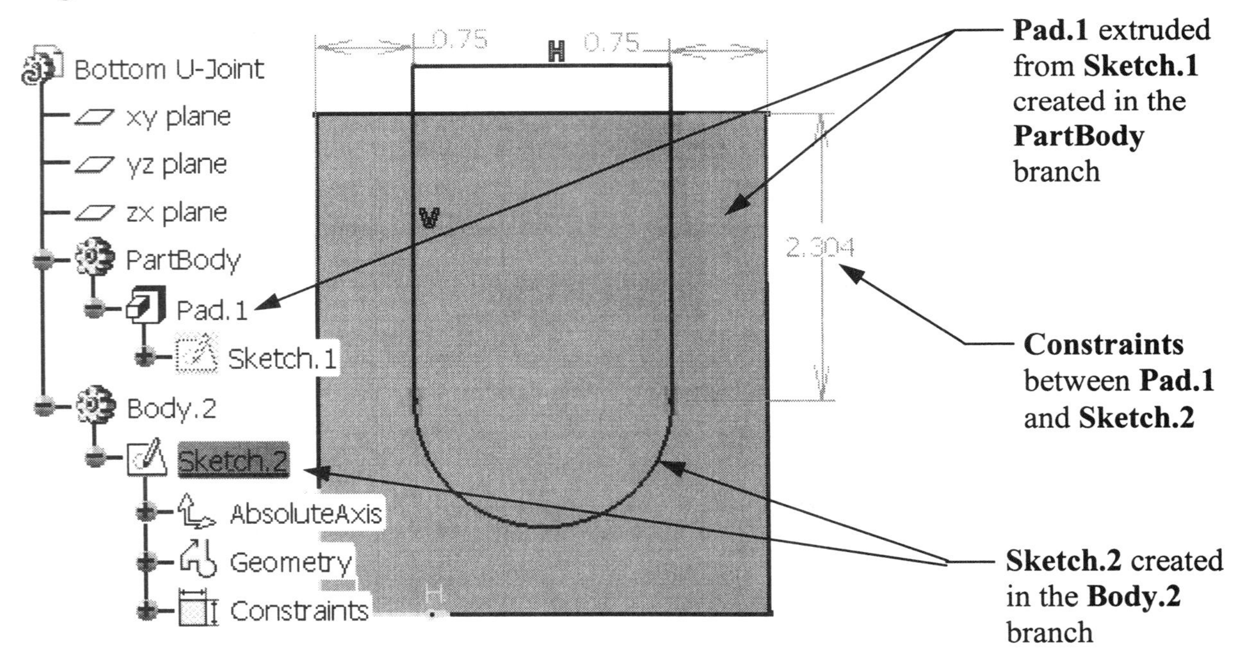

3.11 Exit the **Sketcher Work Bench** and select the **Pad** tool.

3.12 In the **Pad Definition** window, select the **More** button. Figure 5.48 shows the **Pad Definition** window expanded.

3.13 In the **First Limit** box for the **Length**, enter "**2.5in**" as shown in Figure 5.48.

3.14 In the **Second Limit** box for the **Length**, enter "**.5in**" as shown in Figure 5.48.

Figure 5.48

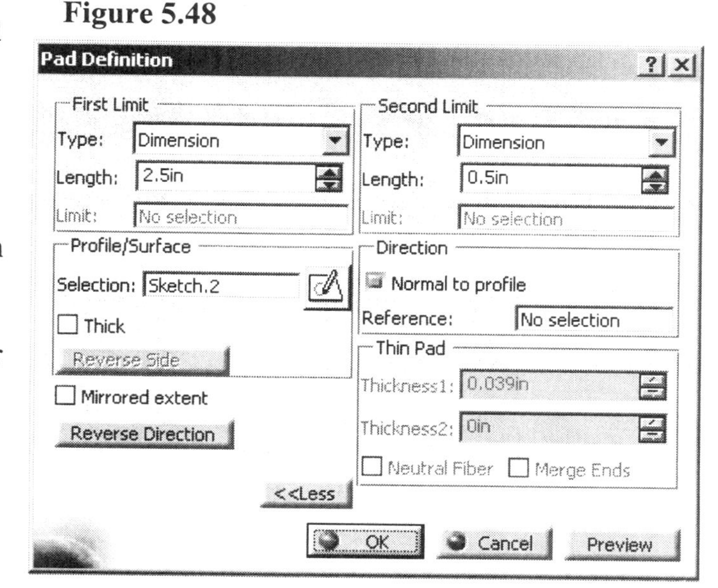

3.15 Select the **OK** button. The extruded solid should look similar to the one shown in Figure 5.49. If the solid was extruded in the wrong direction you may need to hit the **Reverse Direction** button to reverse the direction of the extrusion.

NOTE: As **Sketch.2** was being created, it was being recorded in the **Specification Tree** under the **Body.2** branch that was inserted in the **Specification Tree** in Step 3.7. Figure 5.50 shows what the **Specification Tree** should look like with the **Body.2** branch extended to show the previous steps.

Figure 5.49 **Figure 5.50**

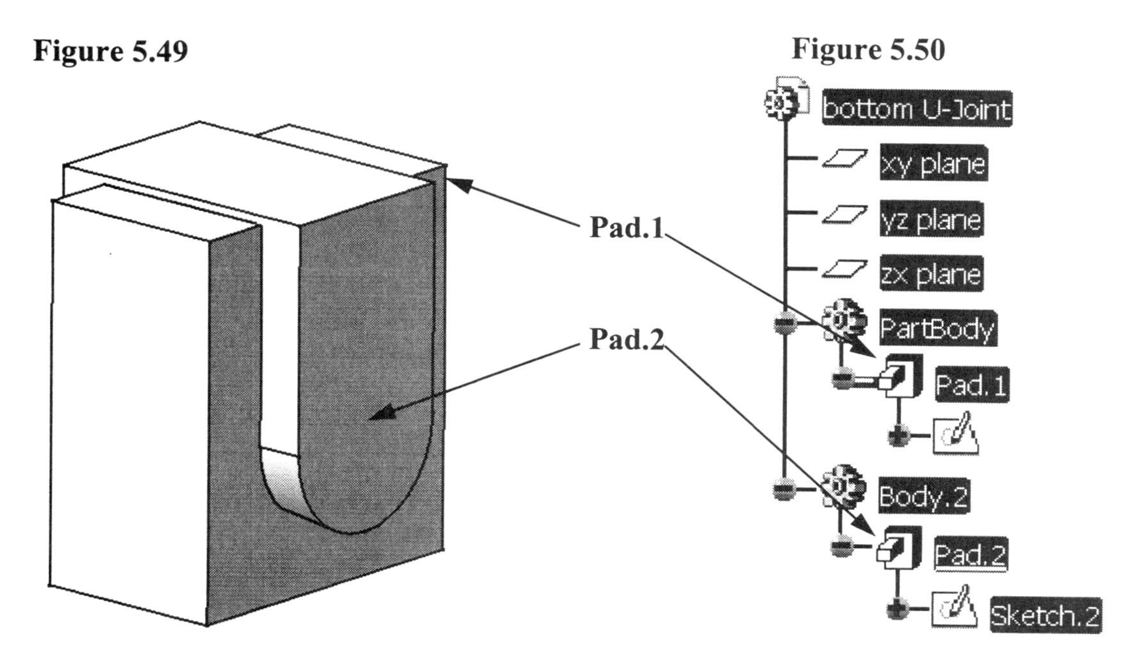

3.16 Select the branch, **Body.2**, from the **Specification Tree**. This should highlight indicating it has been selected. The corresponding solid on the screen, **Body.2**, will also highlight. If **Body.2** is not selected, the following steps will not work!

3.17 Select the **Insert** tab from the top pull down menu as shown in Figure 5.51.

3.18 Select **Boolean Operations** from the bottom of the **Insert** window as shown in Figure 5.51.

3.19 Select **Remove** from the list of **Boolean Operations** as shown in Figure 5.51. This will remove (subtract) the second pad from the first pad. Reference Figure 5.52. A **Remove.1** branch was added to the **Specification Tree**.

Figure 5.51

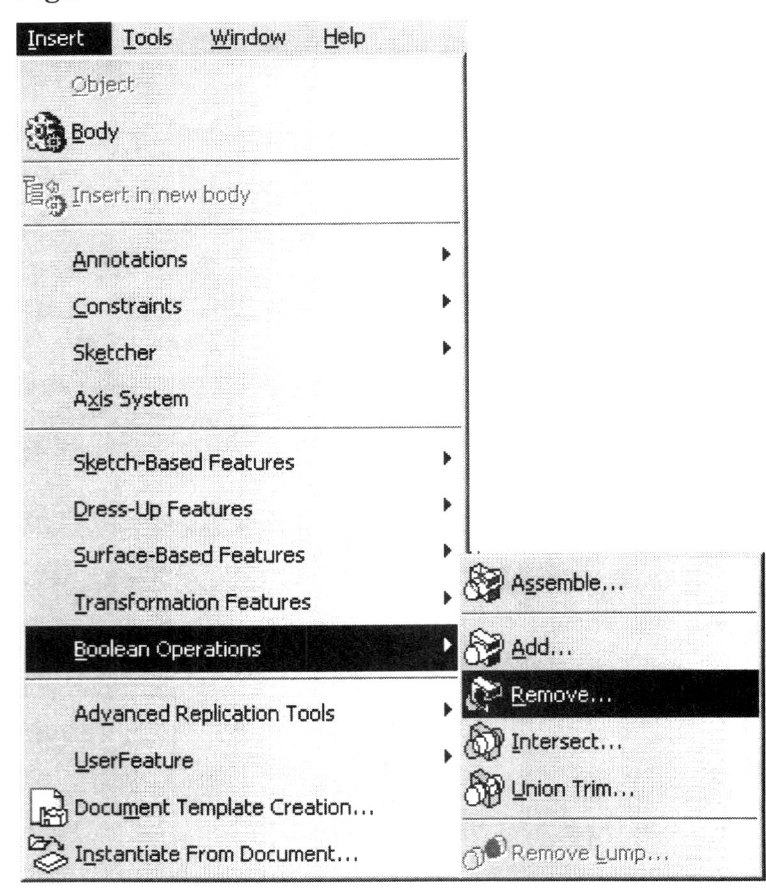

Figure 5.52

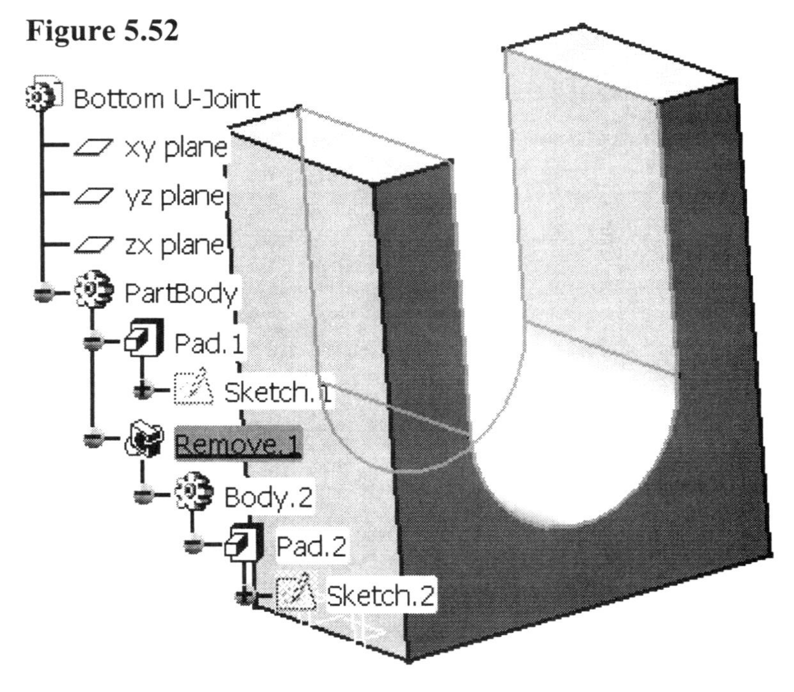

Figure 5.53

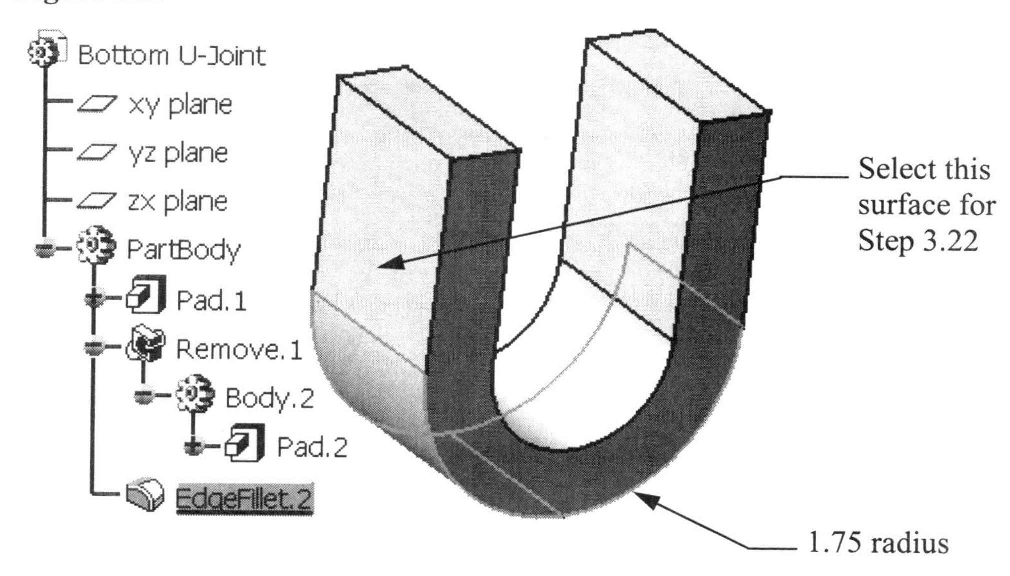

3.20 Create a **1.75** radius **Fillet** on the bottom two edges of the part. This can be accomplished with one fillet if you multi-select the two edges to be filleted, otherwise you will have to do it in two separate fillet operations. Figure 5.53 was completed in one fillet operation.

3.21 Add a **New Body** to the **Specification Tree**. Review Steps 3.6 and 3.7 on how to add a **New Body**.

3.22 Select the left side surface of the part as shown in Figure 5.53 and enter the **Sketcher Work Bench**.

Figure 5.54

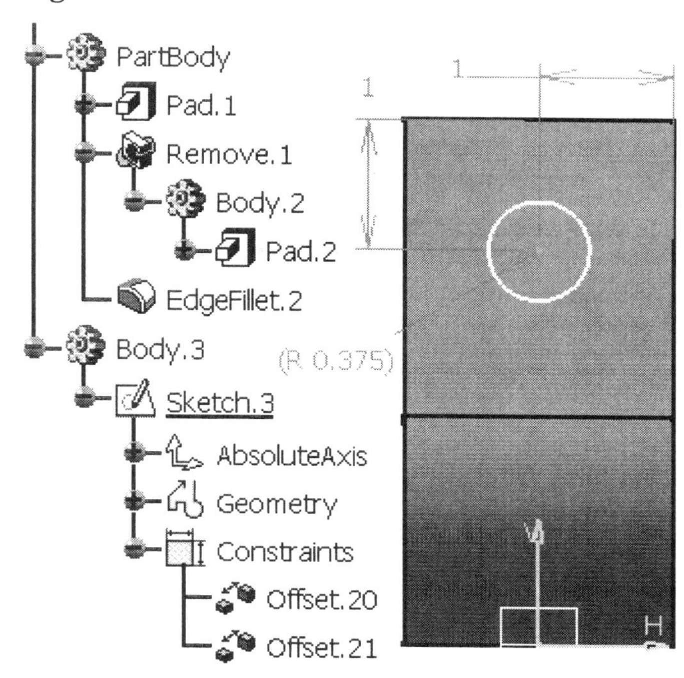

3.23 Create a **Circle** with a **.375** inch radius and **Constrain** it as shown in Figure 5.54.

3.24 Exit the **Sketcher Work Bench** and extrude the circle in both directions as shown in Figure 5.55. The length is not critical as long as it extends beyond both sides of the U-shaped **Body**.

3.25 Select the **Body.3** branch from the **Specification Tree**. Use the **Remove** tool as described in Steps 3.18 and 3.19 to remove **Body.3** (the cylinder you just created).

Figure 5.55

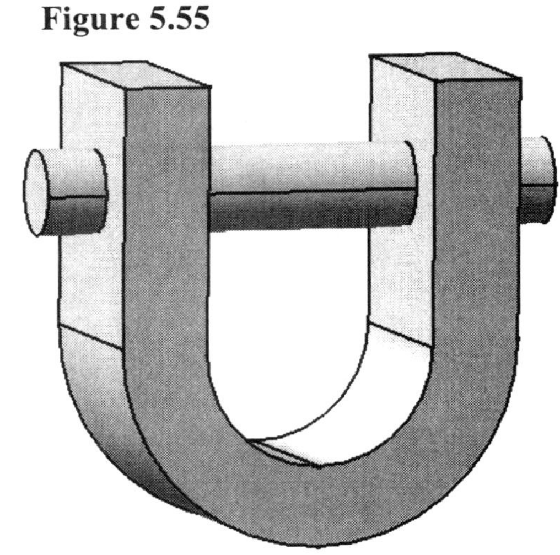

3.26 If your "**Bottom U-Joint**" looks similar to the one shown in Figure 5.56 you are ready to continue on to the next step.

Figure 5.56

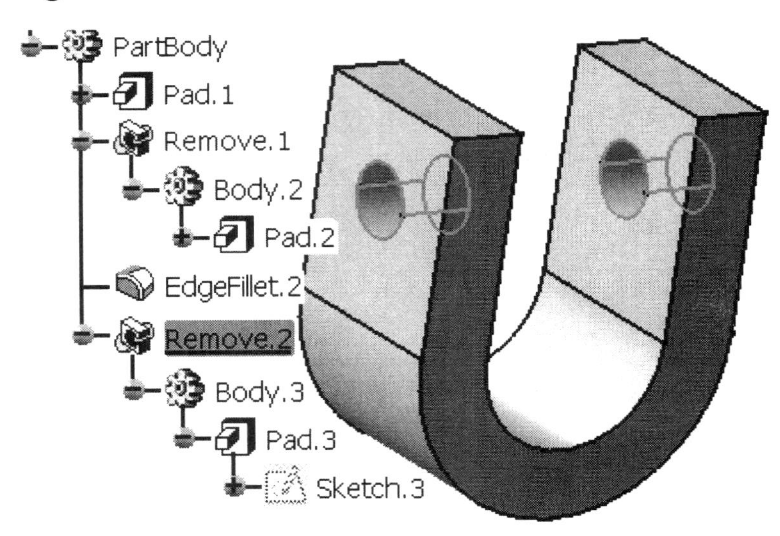

3.27 The next step in creating the "**Bottom U-Joint**" is creating the shaft. It would be simpler to create the shaft using the same method you used for the "**Top U-Joint**." This step will have you create another **Body** so you can experience the **Add** operation in the **Boolean Geometry** tool bar. This will also allow you to see the difference in how it affects the **Specification Tree.** So, for this step, create another **New Body**. The **New Body** should be **Body.4**.

Figure 5.57

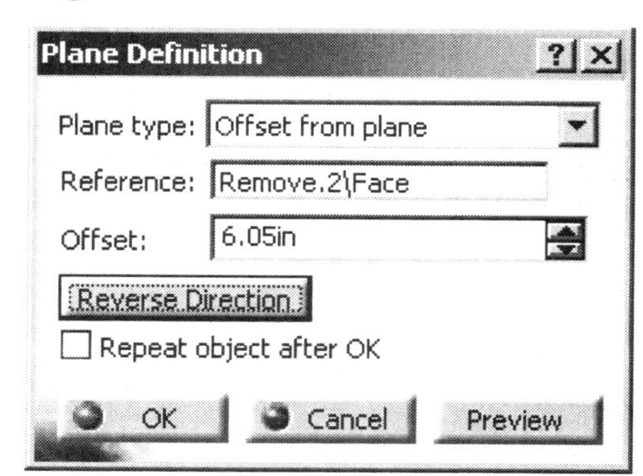

Figure 5.58

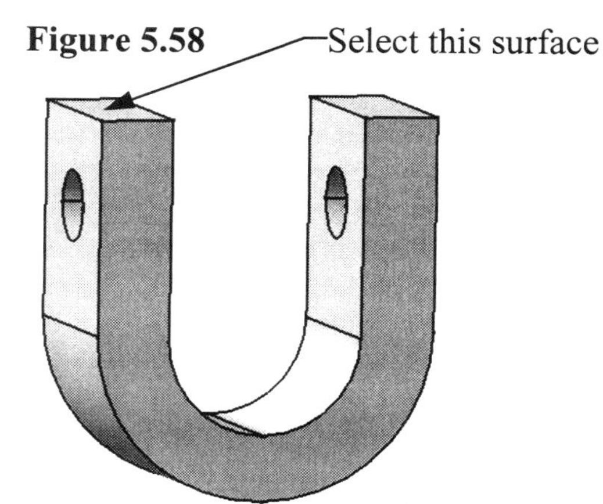

3.28 Select the **Plane** tool. When the **Plane Definition** window appears, set the **Plane Type** box to "**Offset From Plane**" as shown in Figure 5.57. If the **Plane** tool is not available, refer back to Step 2.18 to add the tool to the **Part Design Work Bench**.

3.29 Figure 5.58 shows the surface to select as the reference **Plane**.

3.30 Enter "**6.05**" in the **Offset** box as shown in Figure 5.57.

3.31 Make sure the plane is located at the opposite end of the part than the selected surface. You may need to select **Reverse Direction**. If the plane is located as shown in Figure 5.60, then select the **OK** button to create it.

3.32 Select the new **Plane** and enter the **Sketcher Work Bench.** Create a **Circle** with a **.6in** radius and **Constrain** it as shown in Figure 5.59.

Figure 5.59

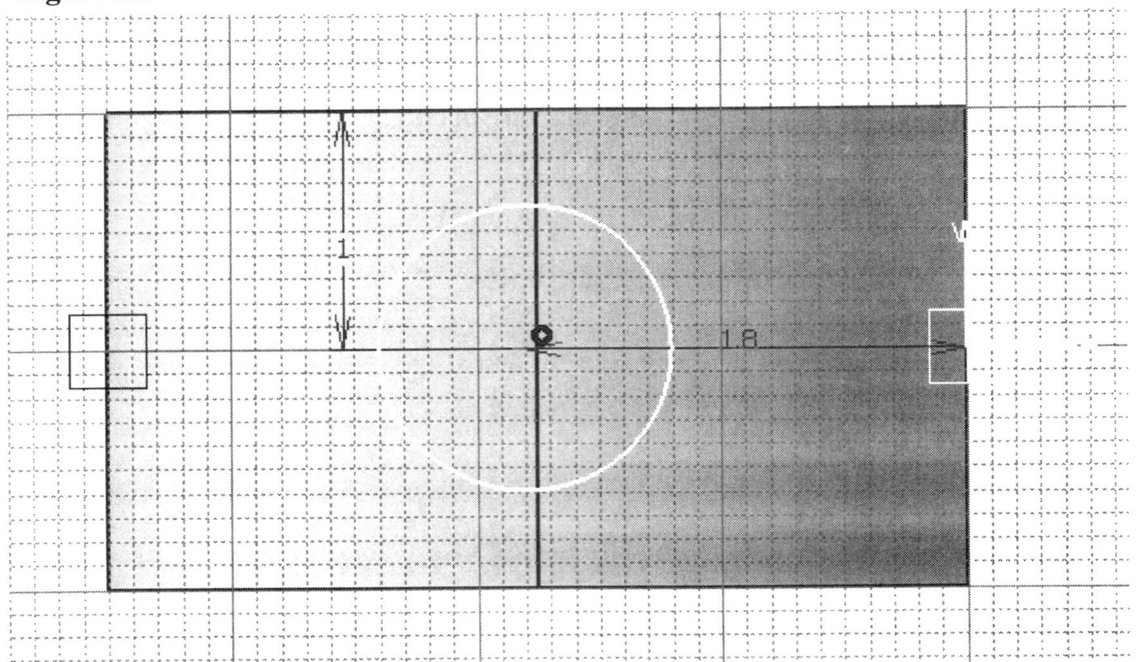

3.33 Exit the **Sketcher Work Bench** and extrude the sketch up to the bottom surface of the part using **Up To Surface** in the **Pad Definition** window. Your "**Bottom U-Joint**" should look similar to the one shown in Figure 5.60.

3.34 If you select the shaft, you will notice that only the shaft will highlight (**Body.4** in the **Specification Tree**), reference Figure 5.61. The shaft is separate from the U-shaped solid. In Step 2 you created the shaft using the same method, but as the same body. To make the "**Bottom U-Joint**" as one complete part, you must add the two solid parts (**Bodies**) together. This is done in the same menu window as Step 3.19. The only difference is you select the **Add** tool instead of the **Remove** tool. Add the two **Bodies** (**Body.1** and **Body.4**) together.

Figure 5.60

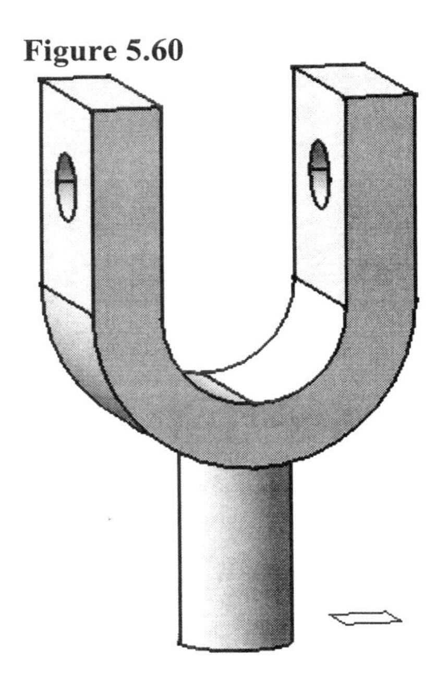

3.35 After completing the **Add** operation, your **Specification Tree** should look similar the one shown in Figure 5.62. Compare the difference between the **Specification Tree** in Figure 5.61 and Figure 5.62. Notice where the **Body.4** is located, it was added to the **PartBody** branch.

3.36 The next step is to create a **2.00″** diameter rounded edge on both legs. For this step, you will use the same tool used in Step 2, the **Tritangent Fillet** tool. Create the **Tritangent Fillet** exactly as you did in Step 2.9. For detailed instruction, refer back to Step 2.9 through 2.13.

Figure 5.61 (**Bodies** not joined)

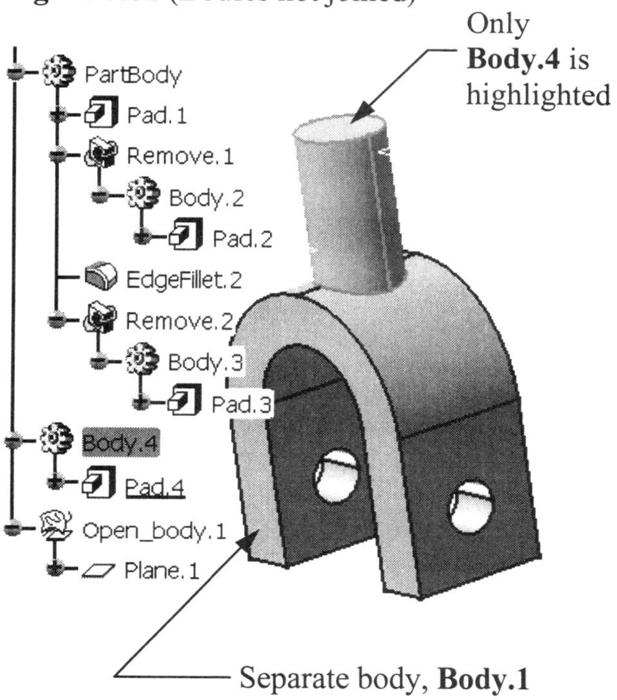

Only **Body.4** is highlighted

Separate body, **Body.1**

Figure 5.62

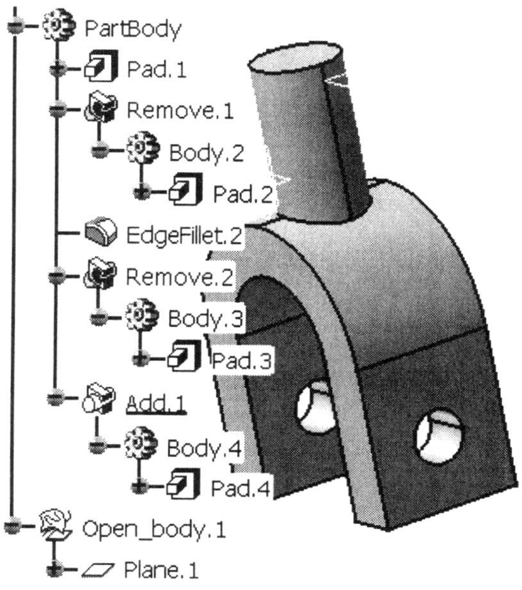

3.37 Add a **.15in Chamfer** to the top edge of the shaft as shown in Figure 5.63.

3.38 Add a **.2in** radius **Fillet** to all of the exterior edges of the solid as shown in Figure 5.63. Notice you can do this in one **Fillet** operation using the Windows multi-select, but you must be sure to select only the edges that require the fillet. Any line that is highlighted will receive a fillet. If you have trouble completing the fillet in one operation, you can use as many as you need (one edge at a time). For this step the **Specification Tree** shows that all of the fillets were created in one operation (**EdgeFillet.3**).

Figure 5.63

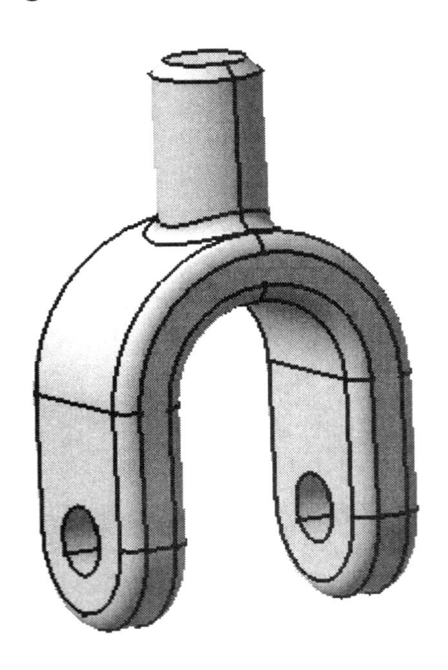

Figure 5.64

3.39 Create the gears for the shaft the same as you did for the "**Top U-Joint**," using the **Slot** tool. Refer to Steps 2.26 and 2.27 and the related figures on how to use the **Slot** tool and **Circular Pattern** tool.

3.40 In the design process you realize that the shaft is 2 inches too short. CATIA V5 gives you the tools to modify the design without requiring you to re-create any entities. CATIA

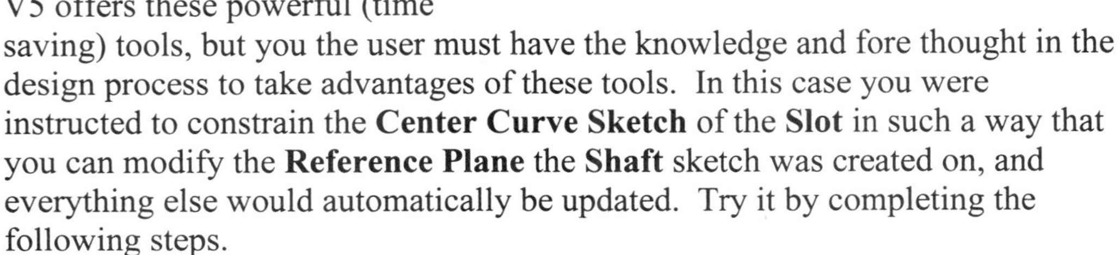

V5 offers these powerful (time saving) tools, but you the user must have the knowledge and fore thought in the design process to take advantages of these tools. In this case you were instructed to constrain the **Center Curve Sketch** of the **Slot** in such a way that you can modify the **Reference Plane** the **Shaft** sketch was created on, and everything else would automatically be updated. Try it by completing the following steps.

3.40.10 Go to the **Open Body** branch of the **Specification Tree**.

3.40.11 Expand the **Specification Tree** so the **Offset** branch is showing.

3.40.12 Double click on the **Offset** branch. This will bring up the **Edit Parameters** window.

3.40.13 The offset distance in the **Edit Parameters** window will be **6.05**. This is the value that was assigned when creating the **Plane** using the **Offset** option. Now, modify the **6.05** inches value to **8.05** inches.

3.40.14 Select the **OK** button. The length of the shaft will go from **6.05** to **8.05** inches along with the **Slot** and **Circular Pattern** that make up the gear.

3.41 One of the last requirements for "**Bottom U-Joint**" is to apply the **Titanium** material and update the **Render Style** so the material is displayed.

3.42 If your "**Bottom U-Joint**" looks similar to the one shown in Figure 5.64, save the file. Make sure you save the file as "**Bottom U-Joint.CATPart**." The file will be required in the subsequent lesson.

Lesson 5 Summary

The tools and principals covered in Lesson 1 and Lesson 2 still apply, but with the ability to create multiple (limitless) sketches, the **Part Design Work Bench** becomes even more powerful. The **Product Structure Work Bench** helps you to keep things organized as you create multiple parts. If you like to create parts using the **Boolean Geometry**, CATIA V5 allows that option. CATIA V5 allows for many ways to create and design.

Lesson 5 Review

After completing this lesson you should be able to answer the questions and explain the concepts listed below.

1. Name the two types of **Fillets** used in this lesson.
2. Explain the differences between the two types of **Fillets**.
3. What is one of the two methods used to reverse the direction of a profile being extruded (using the **Pad** tool)?
4. What button is used to expand the options in the **Pad Definition** window?
5. What are the steps required to create a **New Body**?
6. What determines the size of the **Fillet** when the **Tritangent Fillet** tool is used?
7. When you were creating the hole in the legs of the "**Top U-Joint**," why was the edge of the radius selected instead of the straight edge of the part?
8. The **Part Design Work Bench**, by default, does not contain the **Plane** tool. How can you make the **Plane** tool available for entities in the **Part Design Work Bench**?
9. **T** or **F** In the **Specification Tree**, a **Plane** falls under the **Open_Body** branch.
10. What "**Type**" option in the **Pad Definition** window is used to extrude a cylinder up to the surface without having to specify the exact distance?
11. The "**Top U-Joint**" and the "**Bottom U-Joint**" parts are identical, but were created using different methods. Which method was the quickest and easiest?
12. What two pull down menus allow you access to the **Boolean Operations**?
13. How could you tell what method was used to create the "**Top U-Joint**" part?
14. **T** or **F** To complete a **Boolean Operation**, you must have two separate **Bodies**; for example, **Body.1** and **Body.2**.
15. What two **Boolean Operations** were used in this lesson?
16. List the other **Boolean Operations**.
17. Describe the process required to subtract one body from another.

18. What method did you use to create the holes in the "**Top U-Joint.CATPart**"?

19. What method did you use to create the holes in the "**Bottom U-Joint.CATPart**"?

20. **T** or **F** The **Specification Tree** will have two separate **Body** branches after completing an **Add** operation (adding two different **Bodies** together).

Lesson 5 Practice Exercises

You now have some knowledge, and a little bit of experience, in creating complex parts that require multiple sketches. The following parts will help you improve and further develop your skill and knowledge.

1 This part is a good starter part. It requires multiple sketches, but is quite simple. If you apply what you have learned in this lesson, you should not have any problems. Save the part as "**Lesson 5 Exercise 1.CATPart**."

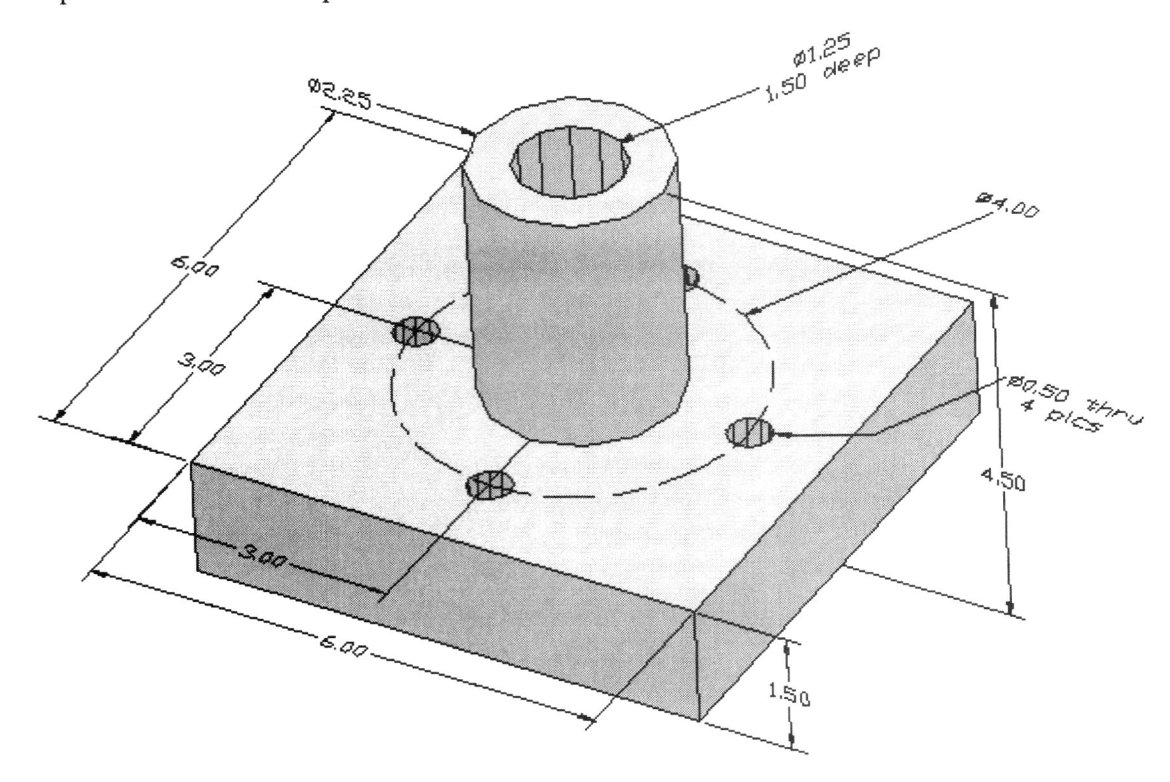

Suggested steps:

1.1 Create the box first. Create the sketch on the base plane. Extrude the box 1.50 inches.

1.2 Create the 2.25 inch diameter cylinder. You can use the base plane or the top of the new box as the sketch plane for the cylinder. The only difference it will make is how far you extrude it.

1.3 Use the **Hole** tool to hollow out the cylinder. Remember, you can select the outside radius of the cylinder to locate the hole in the center of the cylinder.

1.4 Use the **Hole** tool again to create the first of four .50 thru holes. Use two of the edges of the box to properly locate the hole. Use the first hole to create the **Pattern** of holes.

2 This part is an excellent opportunity to use both the multiple sketch and **Boolean Geometry** methods of creating a part. There are many possible methods of creating this part. Use your brief experience and your ingenuity to create this part. Remember, at this point there is no right or wrong method. Save your completed part as "**Lesson 5 Exercise 2.CATPart**."

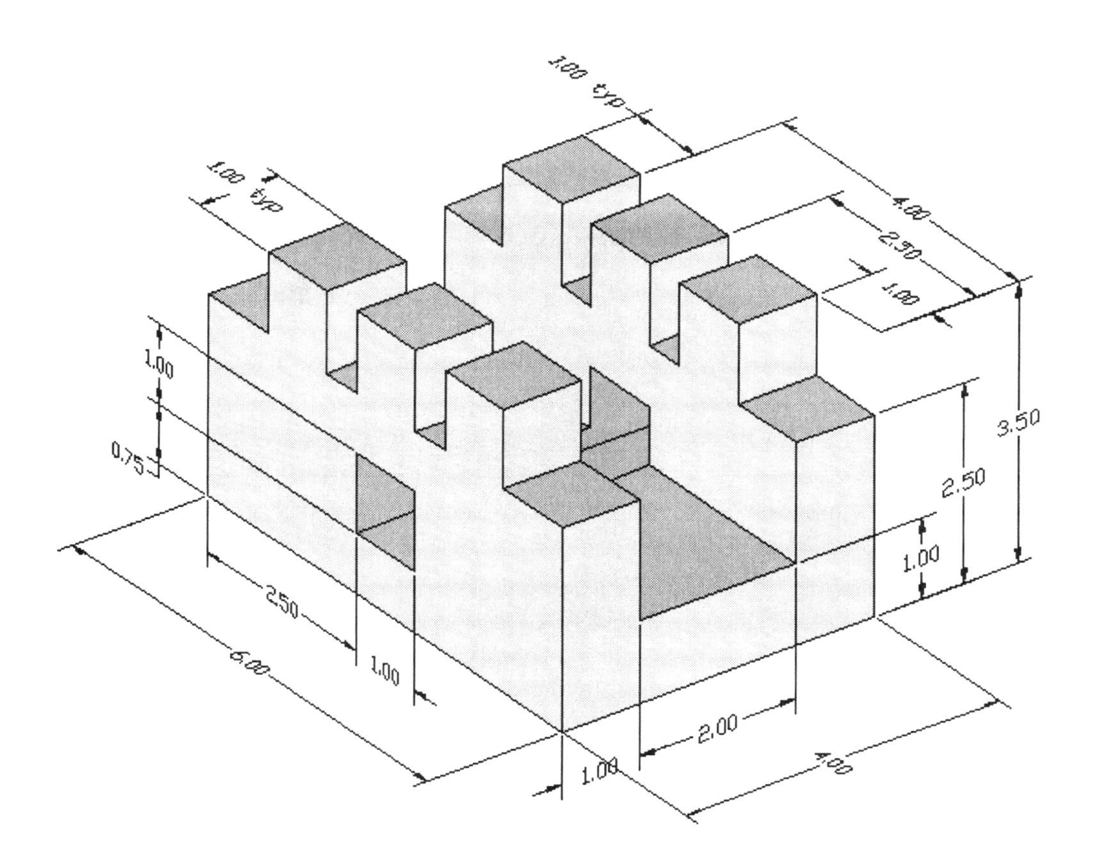

HINT: Create a base solid to begin with and then begin subtracting the different slots and channels. The only feature that requires **Boolean Geometry** subtracted out is the 1 inch square slot. Remember you need to create a **New Body** in the **Specification Tree** before you create the solid.

3 This part will give you more practice with joining and subtracting different bodies. Save the part as "**Lesson 5 Exercise 3.CATPart**."

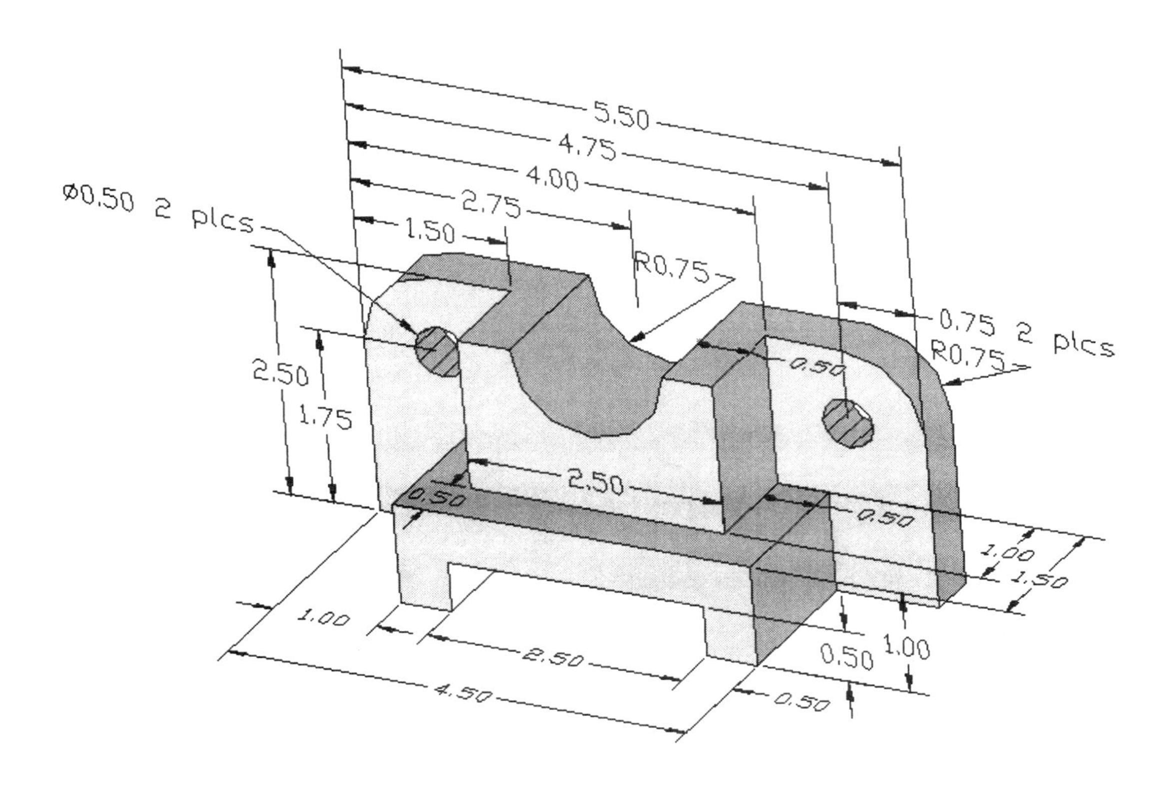

4 This part involves the same operations as the previous parts. When you are done, save the part as "**Lesson 5 Exercise 4.CATPart**."

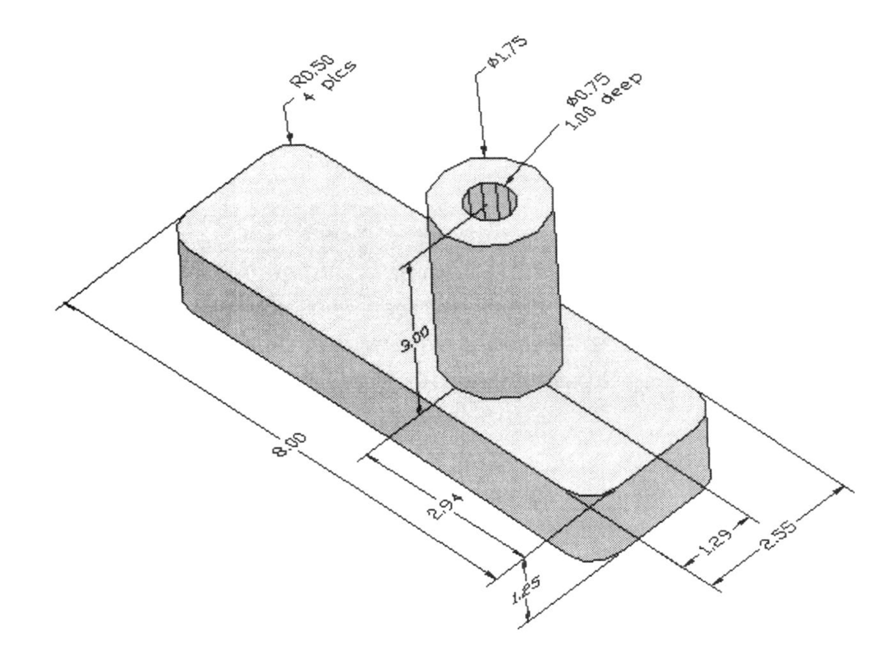

5 This part gives more practice with parts that contain multiple bodies. When you are done, save this part as "**Lesson 5 Exercise 5.CATPart**."

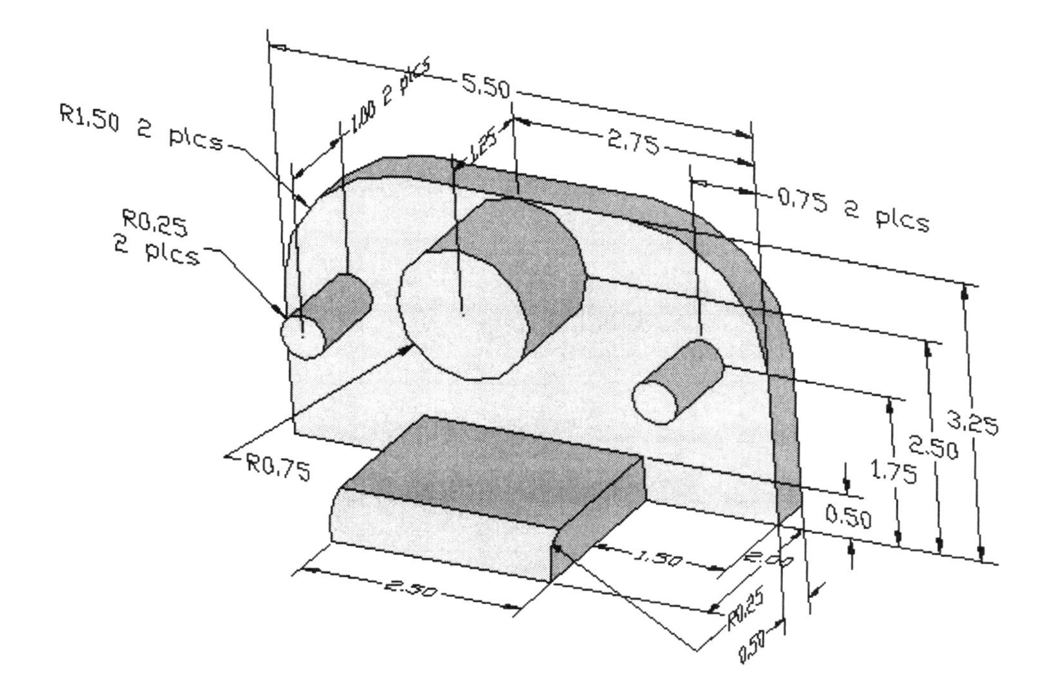

Extra Credit: If you like a challenge, this one is for you. When you are done, save the part as "**Lesson 5 Extra Credit.CATPart.**"

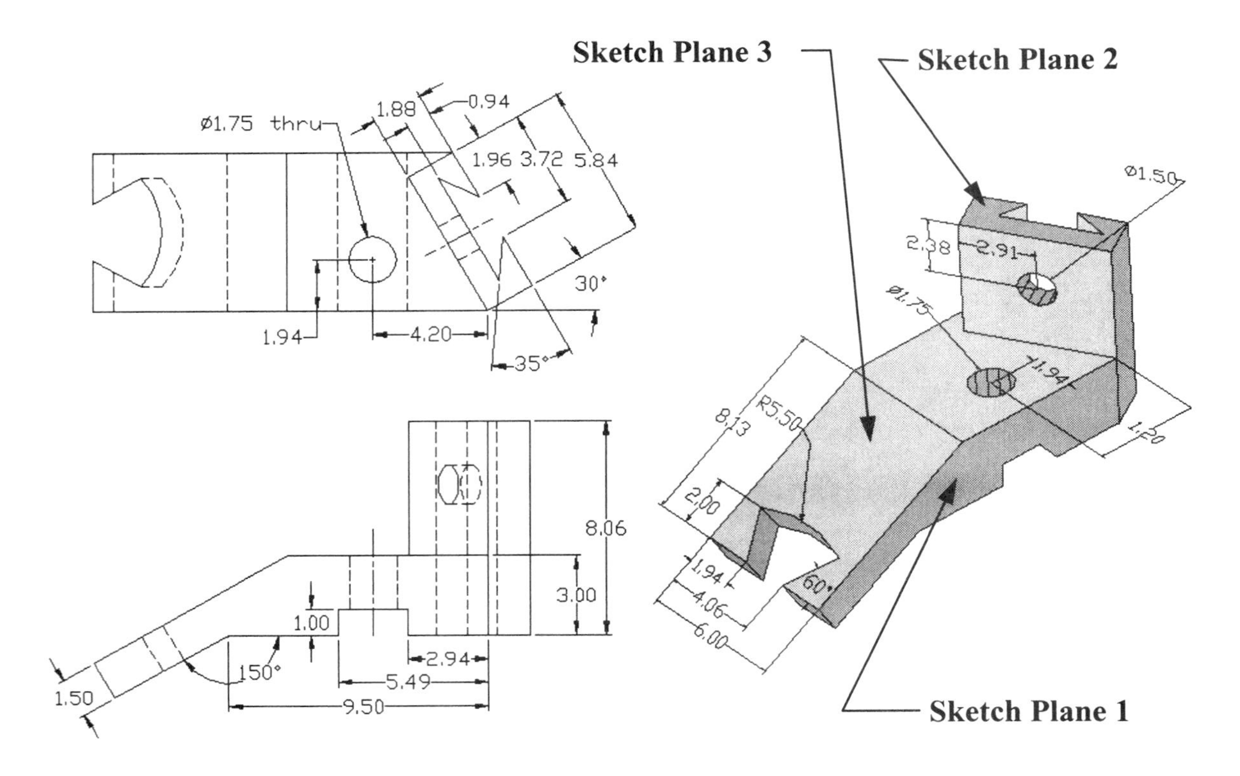

Suggested steps:

EC.1 The individual features of this part are not that difficult. What makes this part difficult is how the different features are oriented to the other features. Most of the parts used thus far have had parallel and perpendicular sketch planes. When deciding which plane to use in sketcher, select the profile that shows the most detail. In this case it would be the plane labeled **Sketch Plane 1**.

EC.2 Create the standing box profile **Sketch Plane 2** using the bottom of the base plane. Create it parallel to the base and then rotate it the 30 degrees by using the **Rotation** tool found in the **Transformation Features** tool bar.

EC.3 Create a **New Body** on the sketch plane labeled **Sketch Plane 3** and subtract it from the body created in Step EC.1.

EC.4 Create the holes using the **Hole** tool.

Lesson 6 — Assembly Design Work Bench

Introduction To Creating An Assembly Using The Assembly Design Work Bench

This lesson uses the three detail parts created in Lesson 5. The three parts are the "**Swivel**," the "**Top U-Joint**" and the "**Bottom U-Joint**." This lesson will take you through a step-by-step process of assembling the three detailed parts using the **Assembly Design Work Bench**.

The first part of this lesson will introduce you to all of the default tool bars found in the **Assembly Design Work Bench**. The second part of this lesson gives step-by-step instructions for entering the **Assembly Design Work Bench**, and inserting, moving, and assembling the detail parts in the **Assembly Design Work Bench**. Upon completion of this lesson, your assembly should look similar to the one shown in Figure 6.1. As in the previous lessons, not all of the **Assembly Design Work Bench** tools will be used and/or explained in this lesson.

Following the step-by-step instructions, there are 20 **Review Questions** to help you solidify your assembly knowledge.

The last section of this lesson has some **Practice Exercises**.

Figure 6.1

Lesson 6 Objectives

The objective of this lesson is to:

- Learn how to use some of the basic tools in the **Assembly Design Work Bench**.
- Learn how to insert existing components into an assembly.
- Learn how to move individual components around in an assembly.
- Learn how to constrain individual components in an assembly.
- Learn how to create additional components in the **Assembly Design Work Bench**.

This lesson teaches these concepts by giving step-by-step instruction followed by sample problems. Concepts covered in the previous lessons will not be covered in detail; you will be expected to know them.

Assembly Design Work Bench Tool Bars

There are twelve default tool bars in the **Assembly Design Work Bench**. The twelve tool bars are shown on the following pages. The individual tools found in each of the twelve tool bars are also shown and include the tool name and a brief definition.

As explained in the previous chapters, the arrow located at the bottom right of the tool icon indicates more than one variation of that tool is available. For a review and/or more information on how to select another variation refer back to Lesson 1.

The **Filtered Selection** Tool Bar

This is the same tool bar found in the **Product Structure Work Bench**.

Tool Bar	Tool Name	Tool Definition
	Products Selection	Allows you to select a specific product.

The **Product Structure Tools** Tool Bar

This tool bar is very similar to the **Product Structure** tool bar found in the **Product Structure Work Bench**.

Tool Bar	Tool Name	Tool Definition
	New Component	Inserts new components.
	New Product	Inserts new products.
	New Part	Inserts a new part.
	Existing Component	Inserts an existing component.
	Replace Component	Replaces a selected component.
	Graph Tree Reordering	Changes the order in the graph tree of the children of the selected product.
	Generate Numbering	Generates a number on all components owning a representation.
	ProductInit	ProductInit
	Manage Representations	Manages representations of selected the selected components.
	Multi-Instantiation Tool Bar	
	Fast Multi Instantiation	Repeat components using the parameters previously set in the **Define Multi Instantiation** command.
	Define Multi Instantiation	Repeat components as many times as you wish in the direction of your choice.

The **Move** Tool Bar

Tool Bar	Tool Name	Tool Definition
	Manipulation	Move a component by free hand translation or rotation. This is a quick move tool.
	Explode	Edits a group of objects you want to explode.
	Snap Tool Bar	
	Snap	Move a component by snapping to an exiting entity.
	Smart Move	Move a component by dragging and snapping. Notice there is not a tool labeled dumb move!

The **Scenes** Tool Bar

Tool Bar	Tool Name	Tool Definition
	Create Scene	Creates a new scene.

The **Constraints** Tool Bar

Assembly constraints are different than the constraints experience in the **Sketcher** and **Part Design Work Bench**.

Tool Bar	Tool Name	Tool Definition
	Coincidence Constraint	Creates a coincidence constraint.
	Contact Constraint	Creates a contact constraint.
	Offset Constraint	Creates an offset constraint.
	Angle Constraint	Creates an angular, parallelism or perpendicularity constraint.
	Fix Component	Fix the component position in the active component.
	Fix Together	Creates a fix together relationship between two components.
	Quick Constraint	Automatically creates constraints on the selected components.
	Flexible/Rigid Sub-Assembly	Allows overloading a position of child component of the product instance. Wow this definition helps a lot!
	Change Constraint	Change the type of selected constraint.
	Reuse Pattern	Reuse pattern for component installation.

The **Weld Planner** Tool Bar

Tool Bar	Tool Name	Tool Definition
	Weld Planner	Creates a weld planner.

The **Annotations** Tool Bar

This is the same tool bar found in the **Part Design Work Bench**.

Tool Bar	Tool Name	Tool Definition
	Text With Leader	Creates text with a leader. This allows the user to attach text to a feature in the **Part Design Work Bench**. The text can be placed in **Hide/Show** for future reference.
	Flag Note	Creates a flag note. The flag note can be linked to a file and/or a URL. The flag note can also be placed in **Hide/Show** for future reference.

The **Assembly** Tool Bar

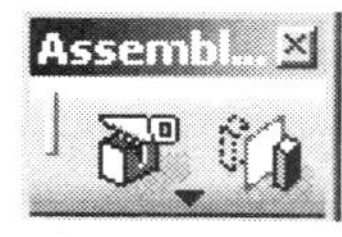

Tool Bar	Tool Name	Tool Definition
	Create Symmetry on Component	Creates symmetry on a child component.
Assembly Features Tool Bar		
	Split	Creates a split.
	Hole	Creates a hole.
	Pocket	Creates a pocket.
	Add	Adds a selected body.
	Remove	Removes a selected body.

The **Constraint Creation** Tool Bar

The default location for this tool bar is the bottom tool bar.

Tool Bar	Tool Name	Tool Definition
	Assembly Features Tool Bar	
	Default Mode	Create independent constraints in repeat mode.
	Chain Mode	Create independent constraints in chain mode.
	Stack Mode	Create independent constraints in stack mode in repeat mode.

The **Space Analysis** Tool Bar

The default location for this tool bar is the bottom tool bar.

Tool Bar	Tool Name	Tool Definition
	Clash	Analyzes interferences between different products.
	Sectioning	Manages sections and cuts of the products.
	Distance	Measures the distance between selected products.

The **Catalog Browser** Tool Bar

The default location for this tool bar is the bottom tool bar. This tool bar shows up on several other work benches.

Tool Bar	Tool Name	Tool Definition
	Catalog Browser	Imports selected components from the catalog.

The **Tools** Tool Bar

The default location for this tool bar is the bottom tool bar. This tool bar shows up on several other work benches.

Tool Bar	Tool Name	Tool Definition
	Update All	Updates All.

Steps To Creating An Assembly Using the Assembly Design Work Bench

In Lesson 5 you created the detail parts, "**Top U-Joint**," "**Bottom U-Joint**" and "**Swivel**." Now you get the opportunity to assemble them together. This is a simple assembly, but it will show you the basic process of bringing together parts and constraining them. Not all of the assembly tools will be used in this lesson. After completing the instructions in this lesson, you are encouraged to try some of the other tools. The instructions in this lesson will give you a good foundation to build on; it's up to you to expand on your basic understanding.

1 Entering the Assembly Design Work Bench

In Lesson 5 you created both the "**Top U-Joint**" and the "**Swivel**" in the same file. The file was saved as "**U-Joint Assembly.CATProduct**." For this step, **Open** the "**U-Joint Assembly.CATProduct**" created in Lesson 5, reference Figure 6.2. Check the top of the toolbar on the right side of the screen to verify which work bench is current. If the file opened in the **Assembly Design Work Bench**, you can move on to Step 2. If the file opened in the **Product Structure Work Bench**, you can select the base of the **Specification Tree** ("**U-Joint Assembly**") and then select the **Assembly Design Work Bench**. This will make the **Assembly Design Work Bench** the current work bench.

NOTE: In Lesson 1, you were taught two different methods of selecting a particular work bench. First, you can select the **Assembly Design Work Bench** by selecting **Start**, **Mechanical Design**, and then **Assembly Design Work Bench**. The second method depends on the default settings; you can only select a work bench if it shows up in the "**Welcome To CATIA V5**" window. Lesson 1 did not have you add the **Assembly Design Work Bench** to the "**Welcome To CATIA V5**" window, but it did give step-by-step instructions on how you can add a work bench to the "**Welcome To CATIA V5**" window.

Figure 6.2

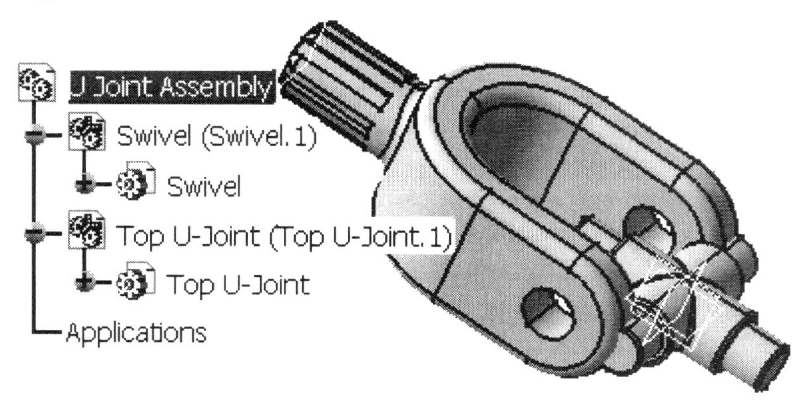

2 Inserting Components Into The Assembly Design Work Bench

So far you have only two of the assembly components in the **Assembly Design Work Bench**. To insert the third component, the "**Bottom U-Joint.CATPart**," complete the following steps.

2.1 The top of the **Specification Tree** must be selected before components can be inserted into the **Assembly Design Work Bench**. In this case, the top of the **Specification Tree** is labeled "**U-Joint Assembly**" as shown in Figure 6.2.

2.2 With "**U-Joint Assembly**" highlighted, select the **Existing Component** tool. The **File Selection** window will appear on the screen as shown in Figure 6.3.

NOTE: If you select the **Existing Component** tool and the **File Selection** window does not appear on the screen, it may mean that the base of the **Specification Tree**, the "**U-Joint Assembly**," is not highlighted.

Figure 6.3

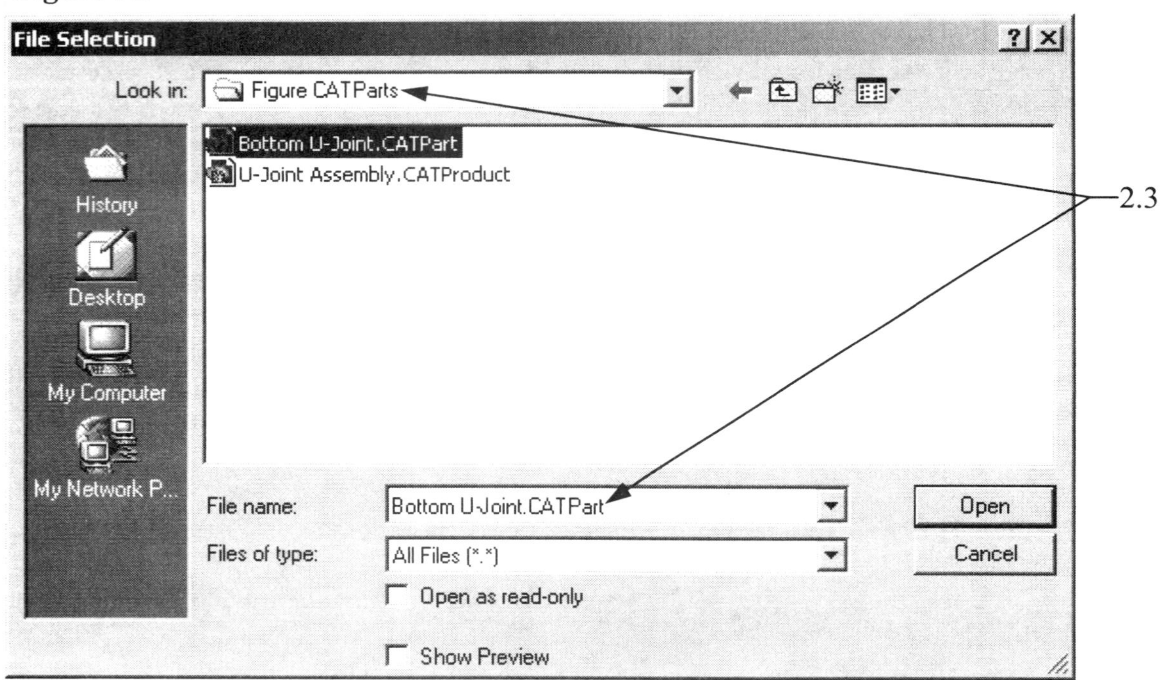

2.3 In the **File Selection** window, select the file that was saved earlier in Lesson 5 named "**Bottom U-Joint.CATPart**." Select the **Open** button. The part should appear on the screen. If it does not, verify you selected the correct file and then repeat Steps 2.1 thru 2.3. Your screen should show the "**Bottom U-Joint**" part similar to what is shown in Figure 6.4.

Figure 6.4

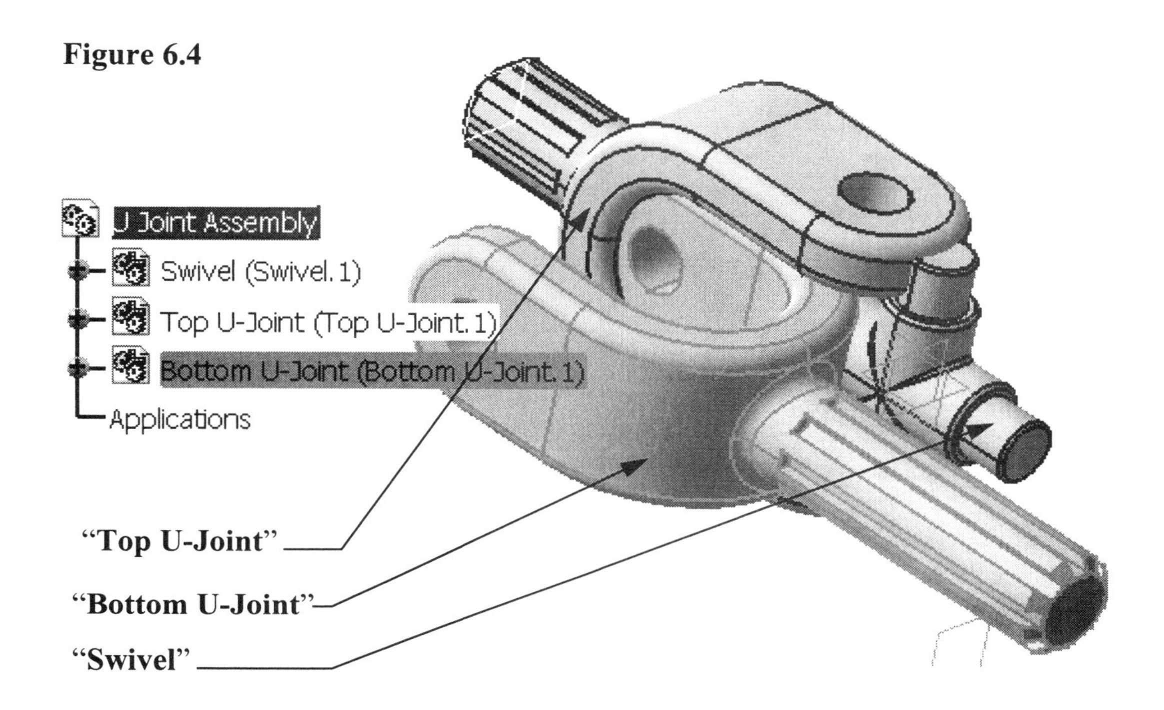

3 Moving Components In The Assembly Design Work Bench Using The Manipulation Tool.

When you insert an object into the **Assembly Design Work Bench**, the object does not always appear in the position and/or orientation that you expect. Usually, if you create the detail parts relative to the same axis, the parts will insert in the same orientation. There will be parts that you insert that have no relative orientation. This is not a problem. CATIA V5 has supplied tools that allow easy and accurate manipulation of part orientation. The following steps show you how to manipulate the orientation and location of a part that has been inserted into the **Assembly Design Work Bench**.

Figure 6.5

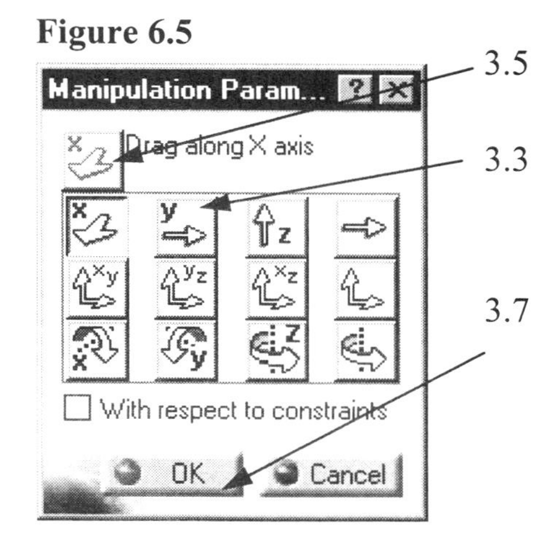

3.1 Select the **Manipulation** tool from the right side of the screen.

3.2 The **Manipulation Parameters** window will appear on the screen as shown in Figure 6.5. This window allows you to move or rotate the components along any of the three axes.

3.3 Select the "**Y**" direction box.

3.4 Move the pointer with the mouse onto the "**Top U-Joint**" part. The entire part should highlight. Drag it to the approximate location shown in Figure 6.6. Compare the location of the "**Top U-Joint**" to Figure 6.4.

Figure 6.6

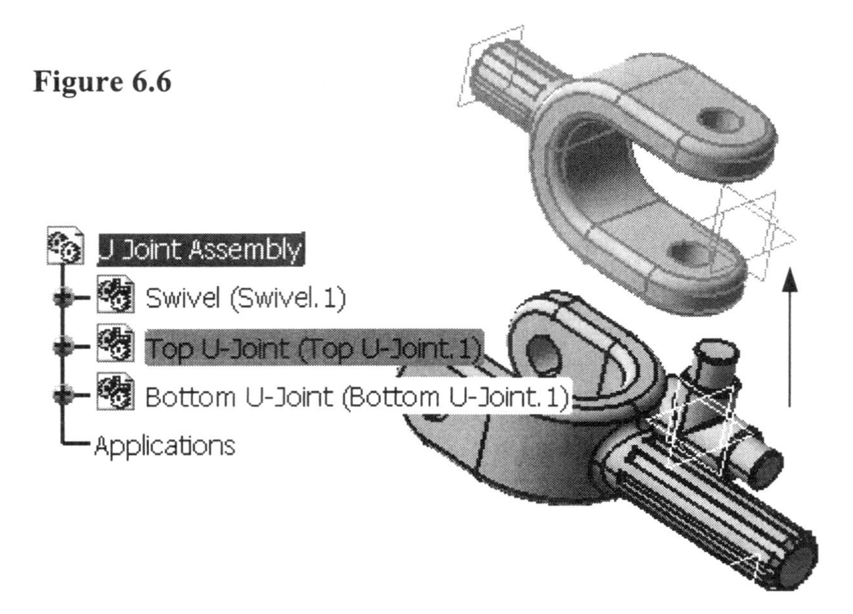

Figure 6.7

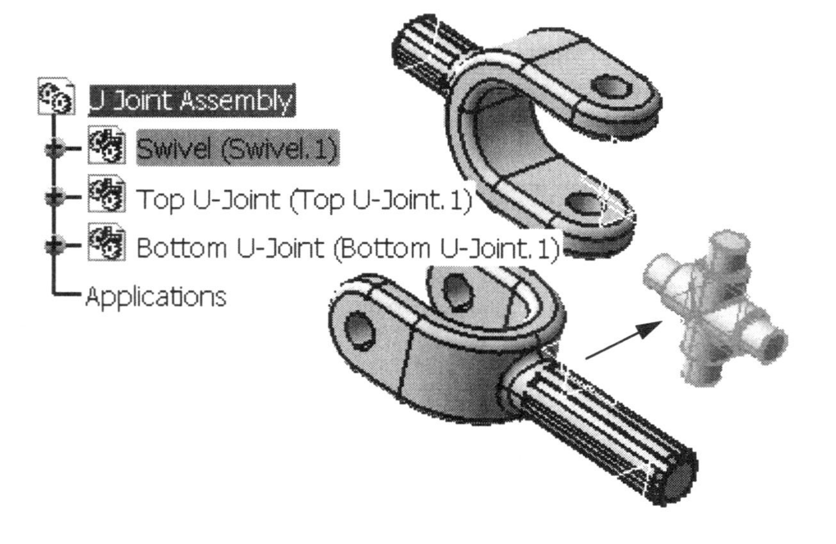

3.5 Select the "**X**" direction box from the **Manipulation Parameters** window as shown in Figure 6.6.

3.6 Move the pointer onto the "**Swivel**" part. Drag it to the approximate location shown in Figure 6.7.

3.7 Select the **OK** button in the **Manipulation Parameters** window to exit the **Manipulation** command. The three parts that were inserted are now separated so that each can clearly be seen.

4 Moving Components In The Assembly Design Work Bench Using The Compass

The location and orientation of the assembly components can also be manipulated using the **Compass** tool. The **Compass** tool was covered in detail in Lesson 2. The same rules apply in the **Assembly Design Work Bench** the only difference is you have several components to select from. The following steps will show you how to move the **Swivel** using the **Compass** tool.

4.1 From the **Specification Tree**, select the **Swivel** branch as shown in Figure 6.8.

4.2 Place the **Compass** tool on the **Swivel** surface as shown in Figure 6.8. The **Compass** tool will turn green indicating that the selected component can be manipulated. For detailed instructions on how to manipulate the **Compass** tool refer back to Lesson 2.

Figure 6.8

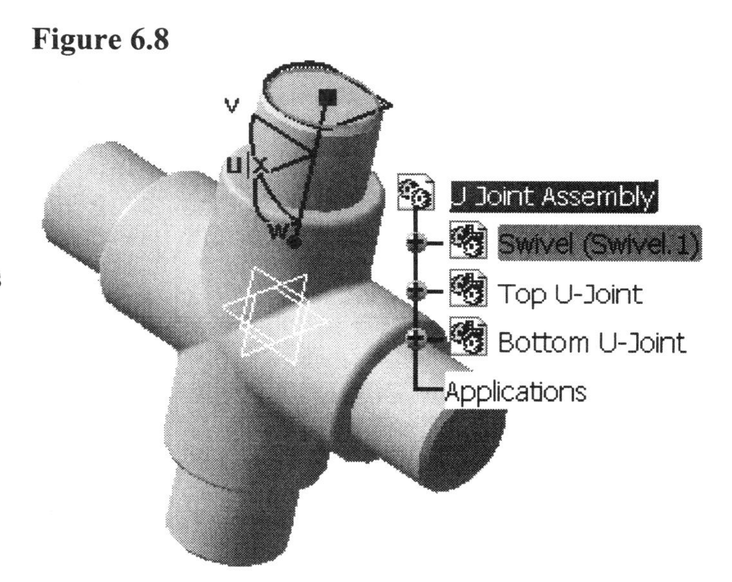

4.3 Select the **Z axis** on the **Compass** tool and drag it in the positive direction. The **Swivel** will move along with the **Compass**. Practice manipulating the orientation of the **Swivel** using the **Compass** tool.

4.4 Move the **Compass** to the surface of the "**Top U-Joint**" as shown in Figure 6.9. Select the **Z axis** on the **Compass** and drag it in the positive direction. Notice you are still manipulating the **Swivels** location, not the "**Top U-Joint.**" To move the "**Top U-Joint**" you must select the "**Top U-Joint**" branch from the **Specification Tree**.

The difference in using the **Compass** tool in the **Assembly Design Work Bench** and the **Part Design Work Bench** is that you must have the component you want to move highlighted in the **Specification Tree**. Step 5 shows you how to **Constrain** the assembly. You can use the **Compass** to move components even after they have been **Constrained**. To move the component back to the constrained position, you need to select the **Update** tool.

Figure 6.9

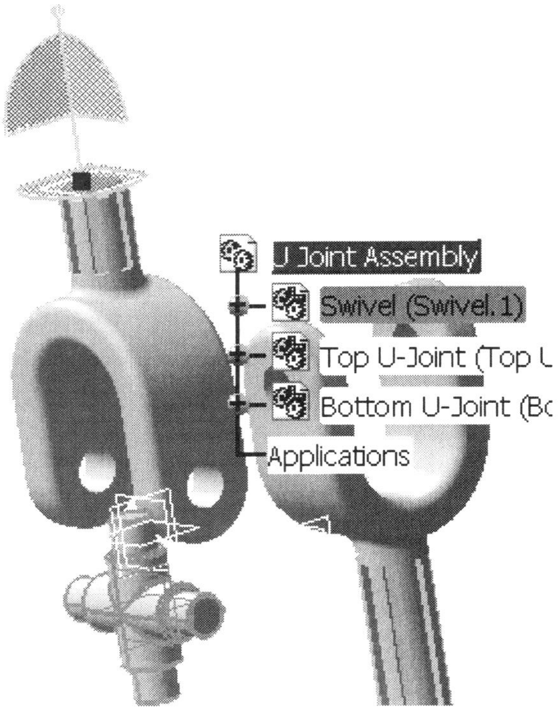

5 Assembling Existing Components

To assemble the components, **Constraints** will need to be created that define the relationship each object has with respect to the others. The steps below will show you how to create these **Constraints**.

5.1 Select the **Fix Component** tool from the right side of the screen and then select the "**Bottom U-Joint**" part. It can be selected from the **Specification Tree** or by selecting the object on the screen. An **Anchor** symbol will appear on the part to show that it has been fixed as shown in Figure 6.10. A new branch will be added to the **Specification Tree**, the **Constraints** branch.

Figure 6.10

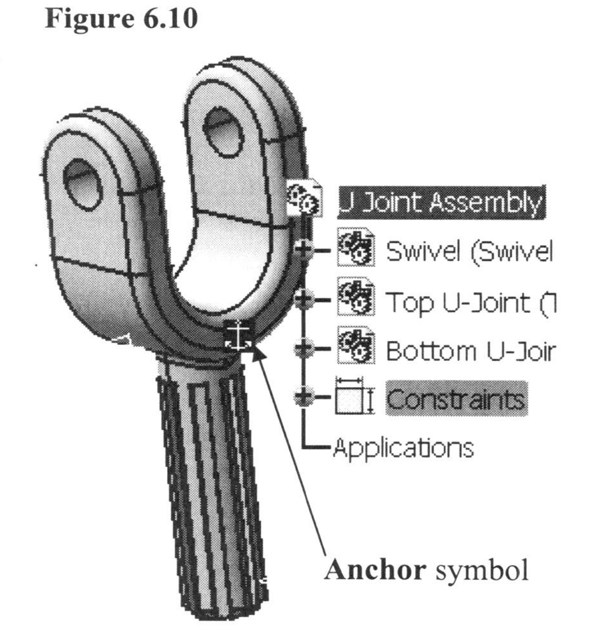

Figure 6.11

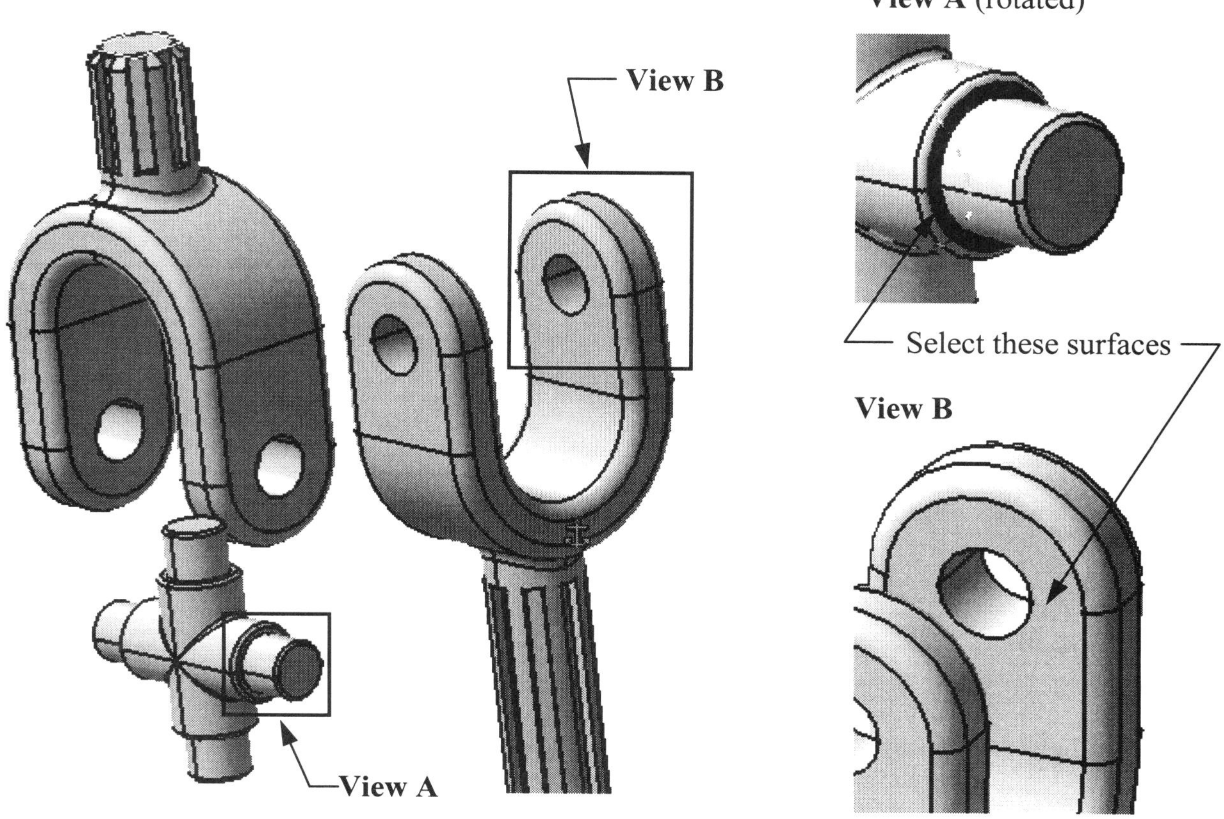

View A (rotated)

View B

Select these surfaces

View B

View A

5.2 Select the **Contact Constraint** tool. This tool will place the selected surface of one part, onto the selected surface of another part. Select the surfaces shown in Figure 6.11, **Views A** and **B**. You may have to **Zoom** in to select the surface shown in **View A**. The **Contact Constraint** symbol will appear on each surface selected. A line connecting the two surfaces will show that they share a common plane as shown in Figure 6.12. Figure 6.12 represents what the CATIA V5 screen will look like when you have made the correct selection. The selection is complete, but no parts have moved yet. To move the parts you have to select the **Update** tool. Figure 6.13 shows that the "**Swivel**" was moved or updated to the newest **Constraint** applied to it. What the **Contact Constraint** did was place both surfaces selected in Figure 6.11 on the same plane.

Figure 6.12 (Assembly with **Contact Constraint**)

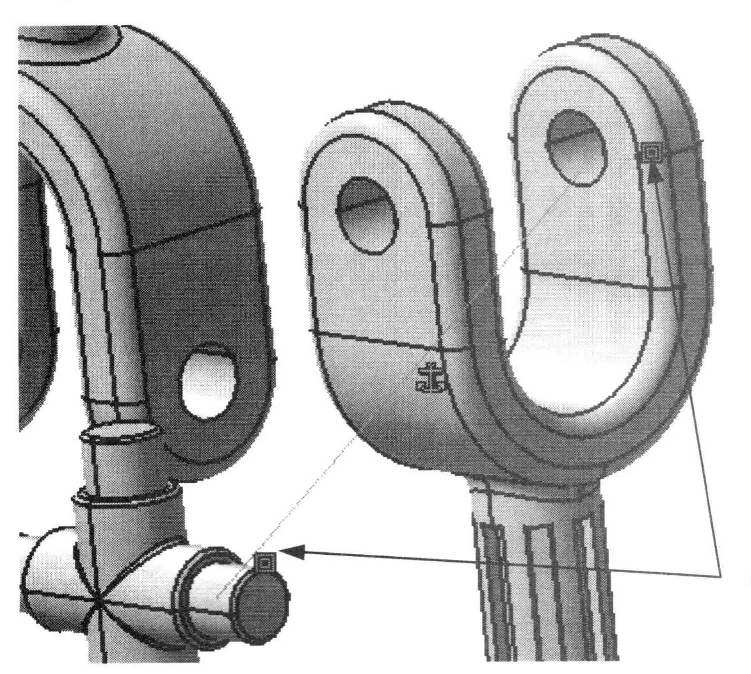

Figure 6.13

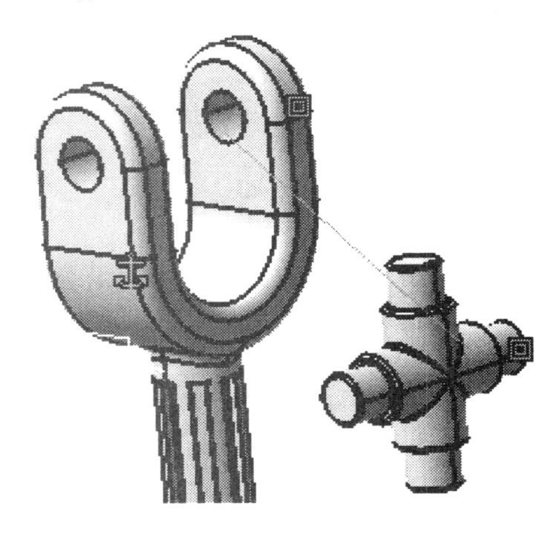

5.3 Select the **Coincidence Constraint** tool. This command will align the centers of two holes or cylinders. This command will be used to place the shafts of the "**Swivel**" part through the holes in the "**Bottom U-Joint**."

5.4 Place the curser on the hole of the "**Bottom U-Joint**" part and move it around until the axis that runs through the center of the holes appears as shown in Figure 6.14. This can sometimes be tricky; the axis is hard to activate to make it visible. The pointer needs to be in just the right place in order for the **Axis Line** to appear. Moving the mouse around in the area of the axis should allow you to pick it up. When it appears, press down the left mouse button to select it. Once the **Axis Line** is selected it will show up as highlighted.

Figure 6.14

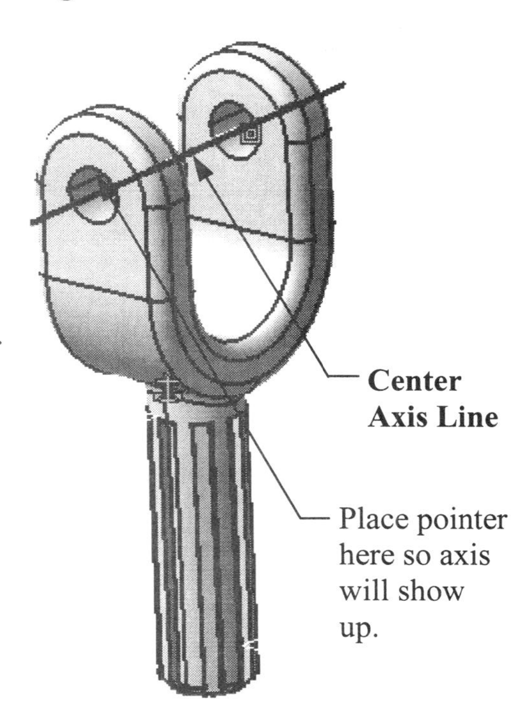

NOTE: If you have trouble getting the **Axis Line** to appear, **Zoom In** on the area and try again.

5.5 Repeat the same process to locate the **Axis Line** that passes through the center of the cylinder of the "**Swivel**" part as shown in Figure 6.15. Note that the Coincidence Constraint tool is still highlighted meaning you don't have to select it again. Make sure that you create the Axis Line through the same shaft as you constrained in Step 5.2; otherwise, you will get an error window.

5.6 The "**Swivel**" should now be lined up with the "**Bottom U-Joint**" so that the shafts of the "**Swivel**" are inside the holes in the "**Bottom U-Joint**" as shown in Figure 6.16. Remember, you may have to manually select the **Update** tool, depending on your default setting.

Figure 6.15

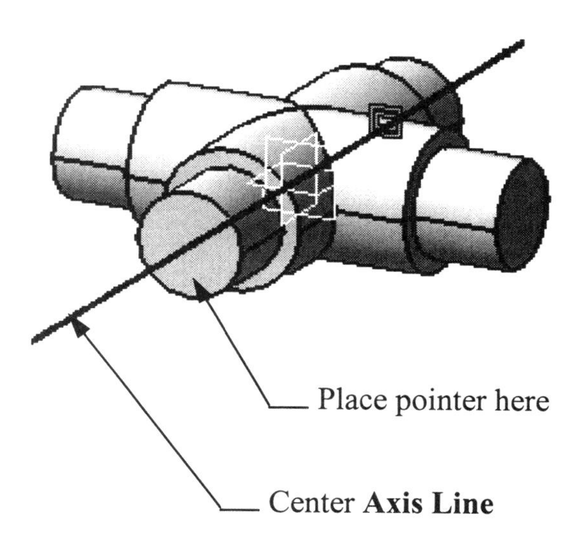

Figure 6.16

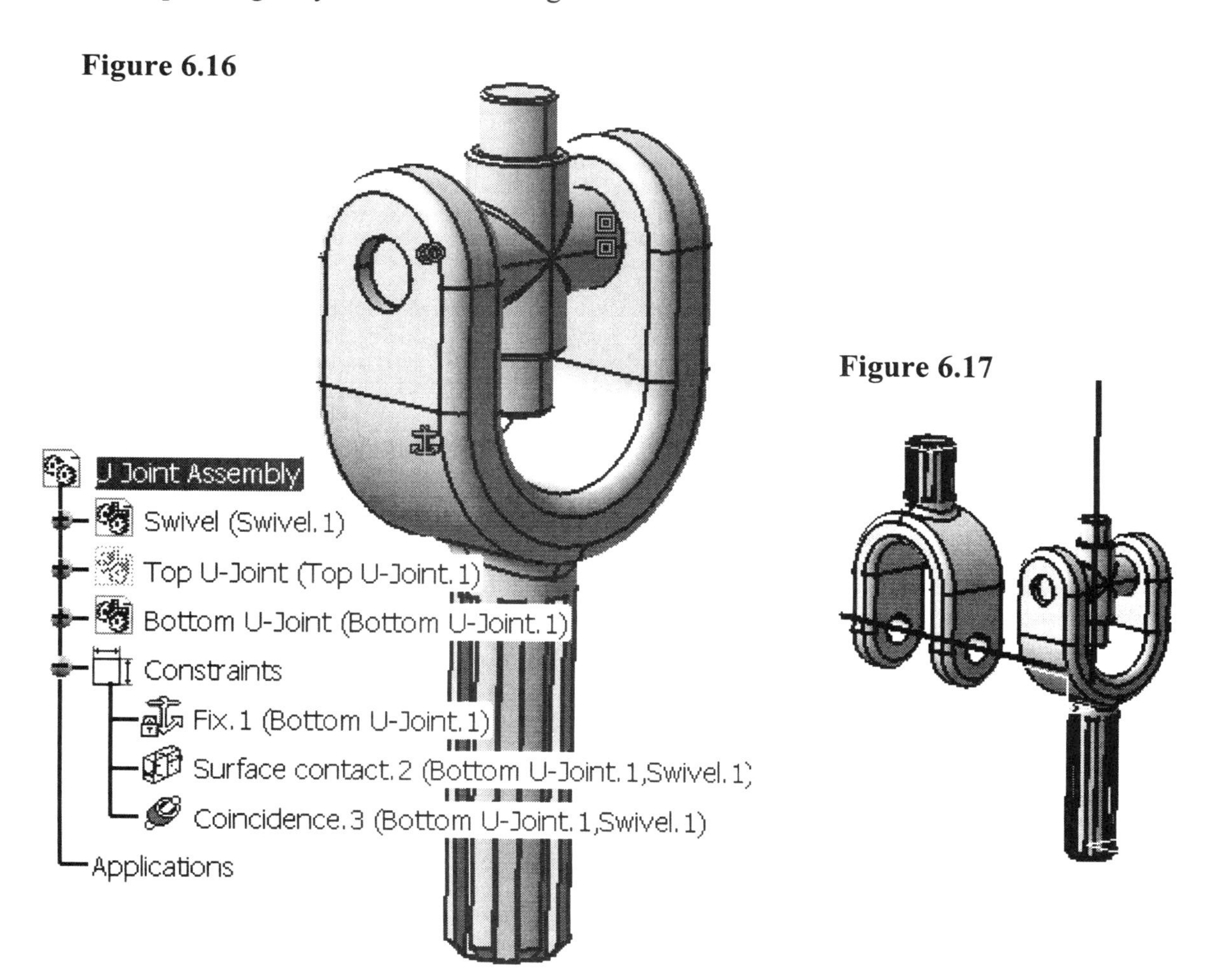

Figure 6.17

5.7 Repeat Steps 5.2 thru 5.4 to constrain the "**Top U-Joint**" to the "**Swivel**" as shown in Figure 6.17. When you are done your assembly should look similar to the one shown in Figure 6.18. Notice in the **Specification Tree** an additional **Surface Contact Constraint** between the "**Top U-Joint**" and the "**Swivel**" was required to properly place the correct contact surfaces together.

Figure 6.18

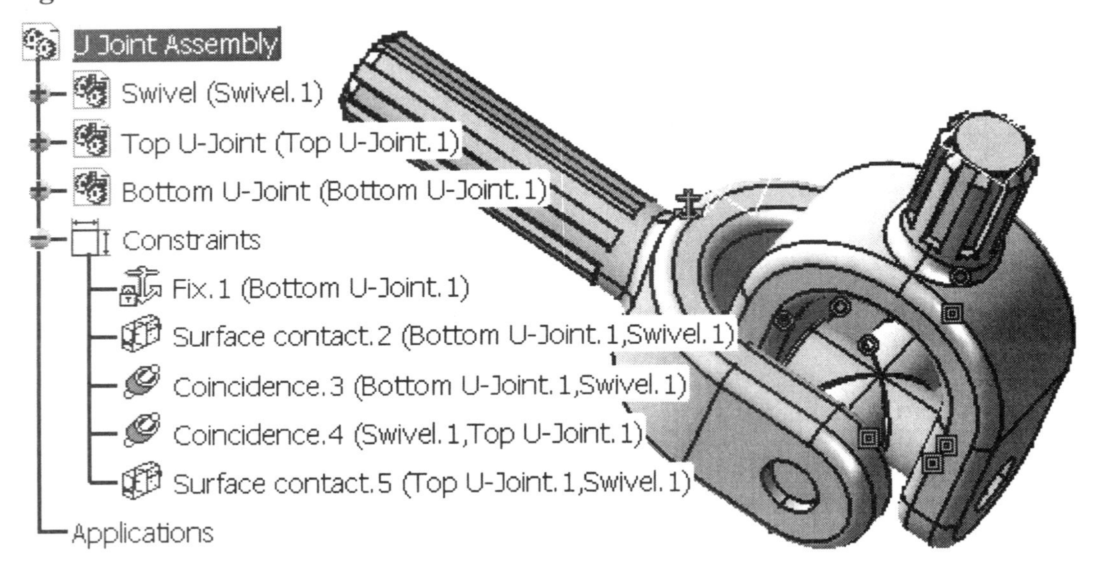

5.8 Figure 6.18 shows the **U-Joint Assembly** fully assembled. You can use the **Compass** to position the individual parts for a better presentation. To reposition the parts complete the following steps.

5.8.1 Place the **Compass** on the surface of the "**Swivel**" as shown in Figure 6.19.

5.8.2 Make sure the "**Swivel**" branch of the **Specification Tree** is highlighted.

5.8.3 Using the **Compass** rotate the "**Swivel**" 90 degrees, as shown in Figure 6.20. Remember you may have to manually select the **Update** tool. You can lock the position of the assembly by pushing down the right mouse button when it is on the **Compass**.

5.8.4 If your "**U-Joint Assembly**" looks similar to what is shown in Figure 6.20 you are ready to move on to the next step.

Figure 6.19

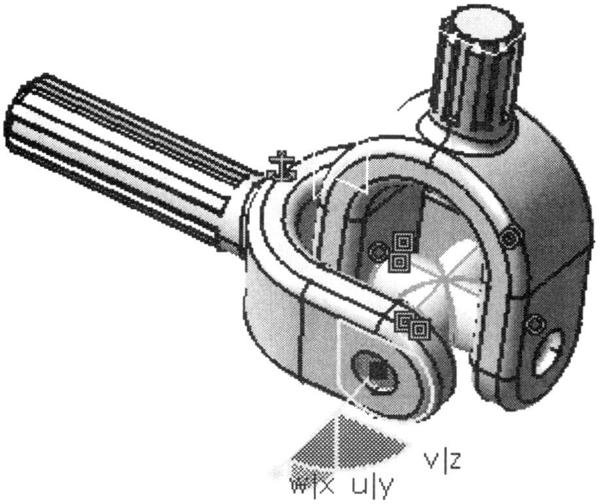

Figure 6.20

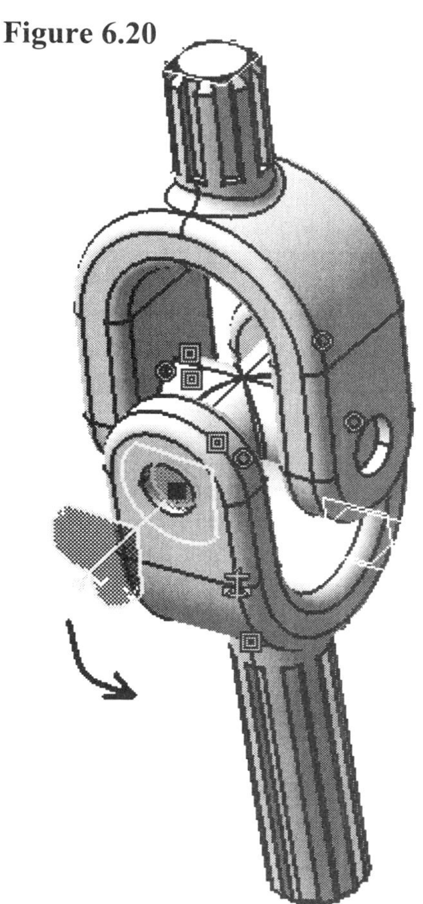

6 Modifying A Component In The Assembly Design Work Bench

The individual components were created in the **Part Design Work Bench**. Each individual component can still be modified from the **Assembly Design Work Bench**. The component is actually modified in the **Part Design Work Bench**, but is accessed from the **Assembly Design Work Bench**. This step will have you modify the "**Top U-Joint**" shaft radius and length.

6.1 To enter the **Part Design Work Bench** from the **Assembly Design Work Bench**, double click on the component you want to modify. For this step, double click on the "**Top U-Joint**" branch as shown in Figure 6.21.

6.2 The next step is to find where, in the **Specification Tree**, the parameters for the shaft radius and length were created. This is one reason renaming the branches of the **Specification Tree** can be helpful. Another method of finding the particular branch is to select the entity itself. For this step, expand the branch labeled "**Top U-Joint**" as shown in Figure 6.21. Figure 6.21 shows the "**Top U-Joint**" shaft highlighted. The shaft is **Pad.2** on the **Specification Tree**. The number assigned may differ depending on how and what order you created it. The important thing is that you select the correct **Pad** regardless of the number.

Figure 6.21

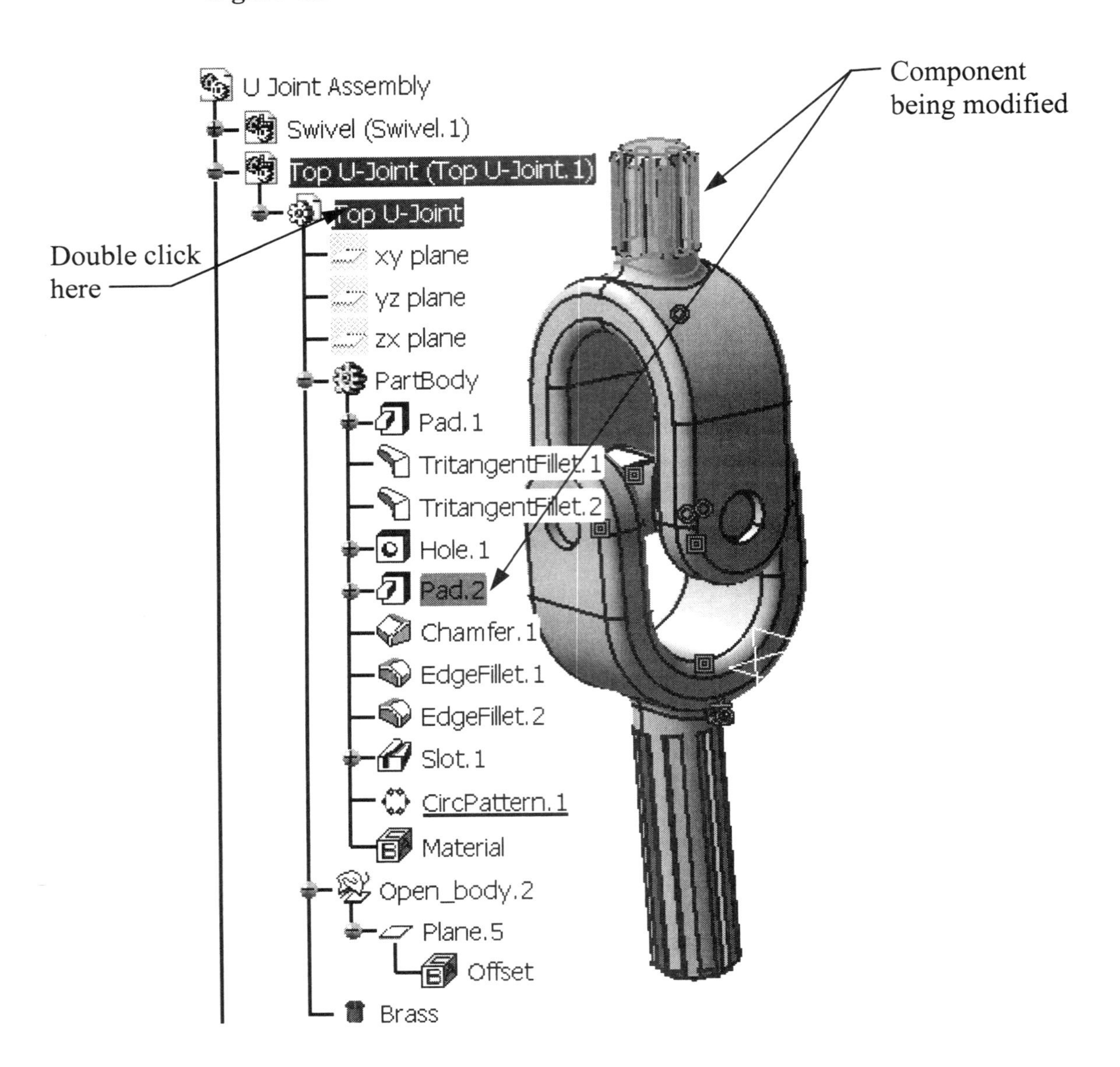

6.3 Expand the **Pad** branch so that the **Sketch** branch is visible. Double click on the **Sketch**. This will take you to the **Sketch** where the shaft radius was first created.

6.4 Expand the **Sketch** branch so the **Radius** branch is visible. Double click on the **Radius** branch. This will bring up the **Constraint Definition** window as shown in Figure 6.22.

Figure 6.22

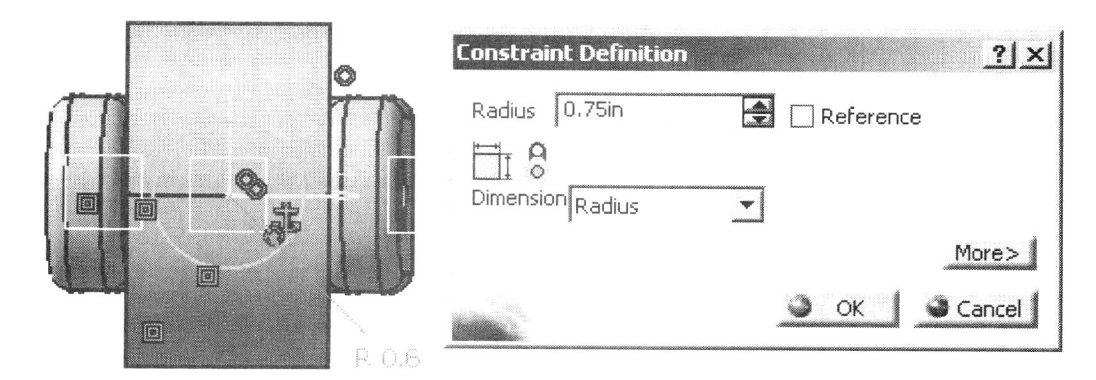

6.5 The **Radius** box currently is set at **.6in**. Replace the **.6in** with a **.75in**.

6.6 Select **OK** to modify the radius.

6.7 Exit the **Sketcher Work Bench**. When you exit the **Sketcher Work Bench**, the shaft you just modified will turn **Red**. Be patient! The **Red** color is CATIA V5 signaling you that it is updating the solid. When CATIA V5 is done updating, the solid will go back to its default color and the shaft will now have a radius of **.75″**. If the update causes any problems an **Update Diagnosis: Sketch.8** window will appear as shown in Figure 6.23. Modifying the sketch radius from **.6″** to **.75″** will cause an update error. The fillet that joins the two bodies together (the shaft and U shaped body) is too large. The buttons on the right side of the window give you several options. For this step select the **Edit** button. This brings up the **Edge Fillet Definition** window so the value can be edited, also shown in Figure 6.23.

Figure 6.23

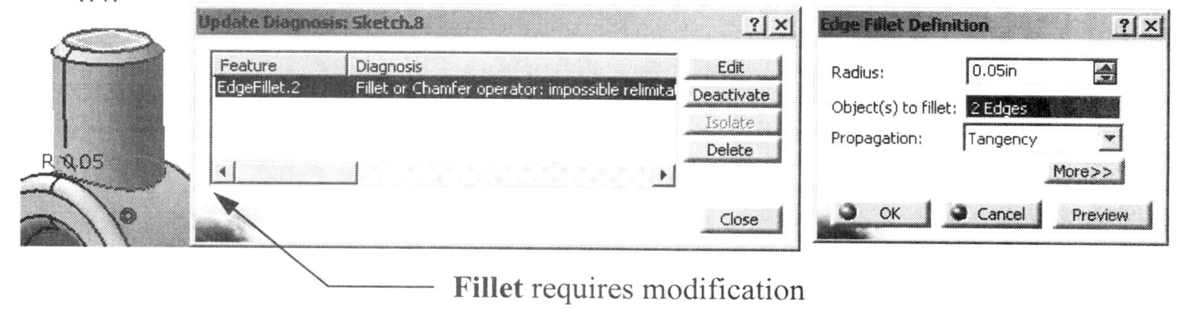

Fillet requires modification

6.8 Modify the **Fillet Radius** to .05″.

6.9 Select the **OK** button. The **Update Diagnosis** widow will disappear and the **Fillet** will be updated.

Figure 6.24

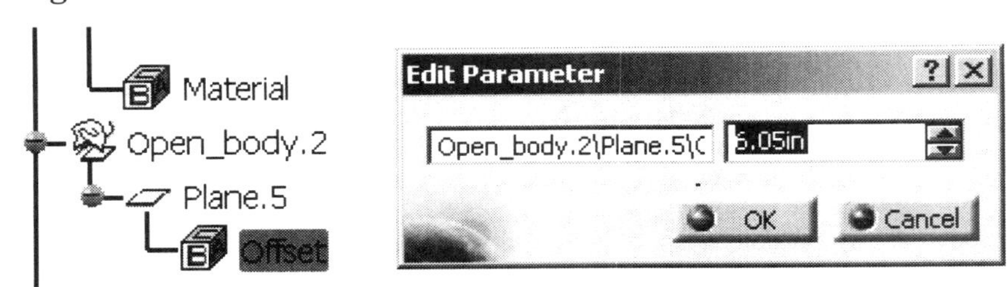

6.10 To modify the length of the shaft, expand the **Open_Body** branch of the **Specification Tree** (still in the "**Top U-Joint**" branch of the assembly). Double click on the **Offset** branch in the **Plane** branch as shown in Figure 6.24. This will bring up the **Edit Parameter** window also shown in Figure 6.24.

6.11 Modify the value of **6.05in** to **8.05in**. This is moving the **Plane** from 6.05 inches as it was originally created to 8.05 inches, 2 additional inches. The sketch that defines the shaft was created on this **Plane** and was extruded to the surface of the U-Shaped body. Modifying the location of the **Plane** also modifies the length of the **Pad**.

Figure 6.25

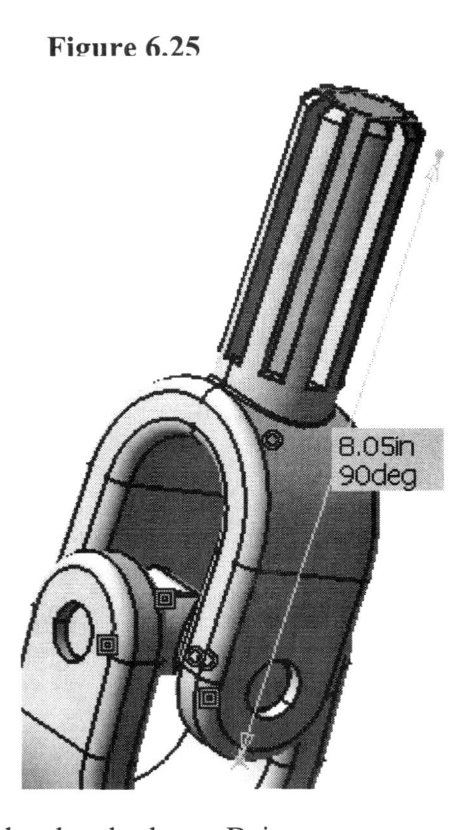

6.12 Select the **OK** button. The shaft will update, it will be 2 inches longer than before. Reference Figure 6.25.

NOTE: The **Circular Pattern** of the **Slot** was updated with the **Pad**. The automatic update of the **Slot** was determined by how the sketch making up the **Slot** was constrained. Figure 5.37 in Lesson 5 shows how the **Center Curve Sketch** was constrained. The **Center Curve** sketch was not constrained to an exact length but to entities that would update with the re-location of the sketch plane. Being more specific, the top endpoint of the line was constrained to 0 inches from the end of the shaft. The other end of the line was constrained 4.5 inches from the bottom of the part. This constraint method made the length of the line variable. The variable line defines the length of the **Slot**. Since that one **Slot** was used as a **Circular Pattern** all of the **Slots** were updated.

If your "**Top U-Joint**" looks similar to the one shown in Figure 6.26, you are ready to continue on to the next step. The changes were made in the **Part Design Work Bench** even though you started in the **Assembly Design Work Bench**. Getting back to where you started from is easy; double click on "**U-Joint Assembly**" in the **Specification Tree**. As you double click on "**U-Joint Assembly**," notice that the work bench changes back to the **Assembly Design Work Bench**.

Figure 6.26

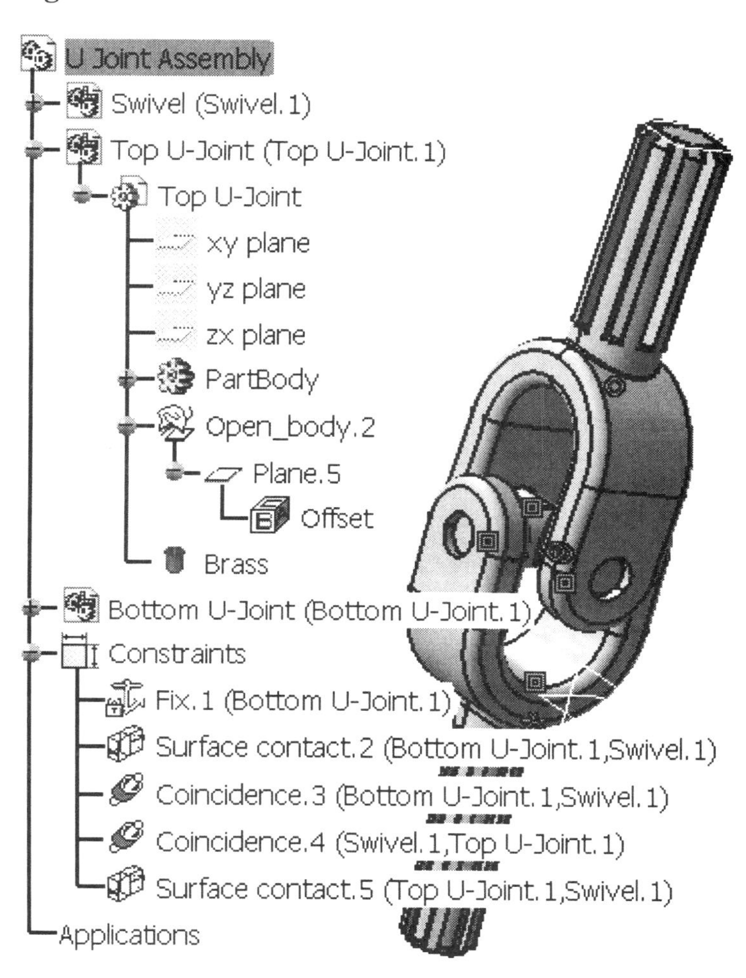

7 Creating A Bill Of Materials

CATIA V5 offers a multitude of assembly tools to help you create, control and document your assembly. This step will introduce you to a few of the more common and/or required tools. These tools are **Bill Of Material**, **Analyzing/Modifying Constrains**, **Compute Clash**, **Measuring Tools**, and **Product Management**.

7.1 Select the **Analyze** tool bar located in the top pull down menu. The **Analyze** tool menu is available only if you are in the **Assembly Design Work Bench**. This will bring up the menu shown in Figure 6.27.

7.2 Select **Bill Of Material**. This will bring up the **Bill Of Material: U-Joint Assembly** window as shown in Figure 6.28. As the **Bill Of Material** window shows, CATIA V5 has been keeping a record of everything you have inserted and or created in the **Assembly Design Work Bench**.

Figure 6.27

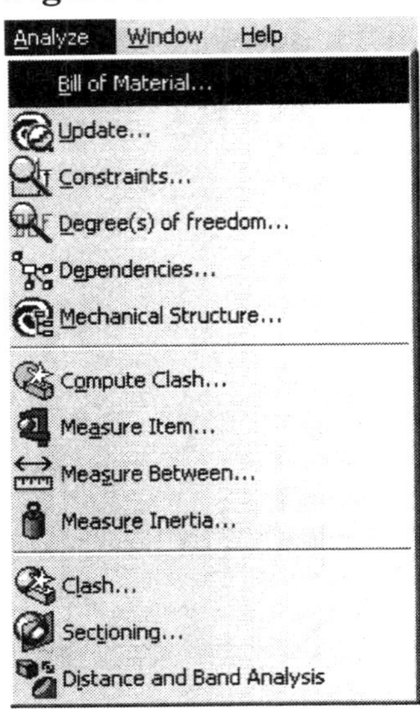

Figure 6.28

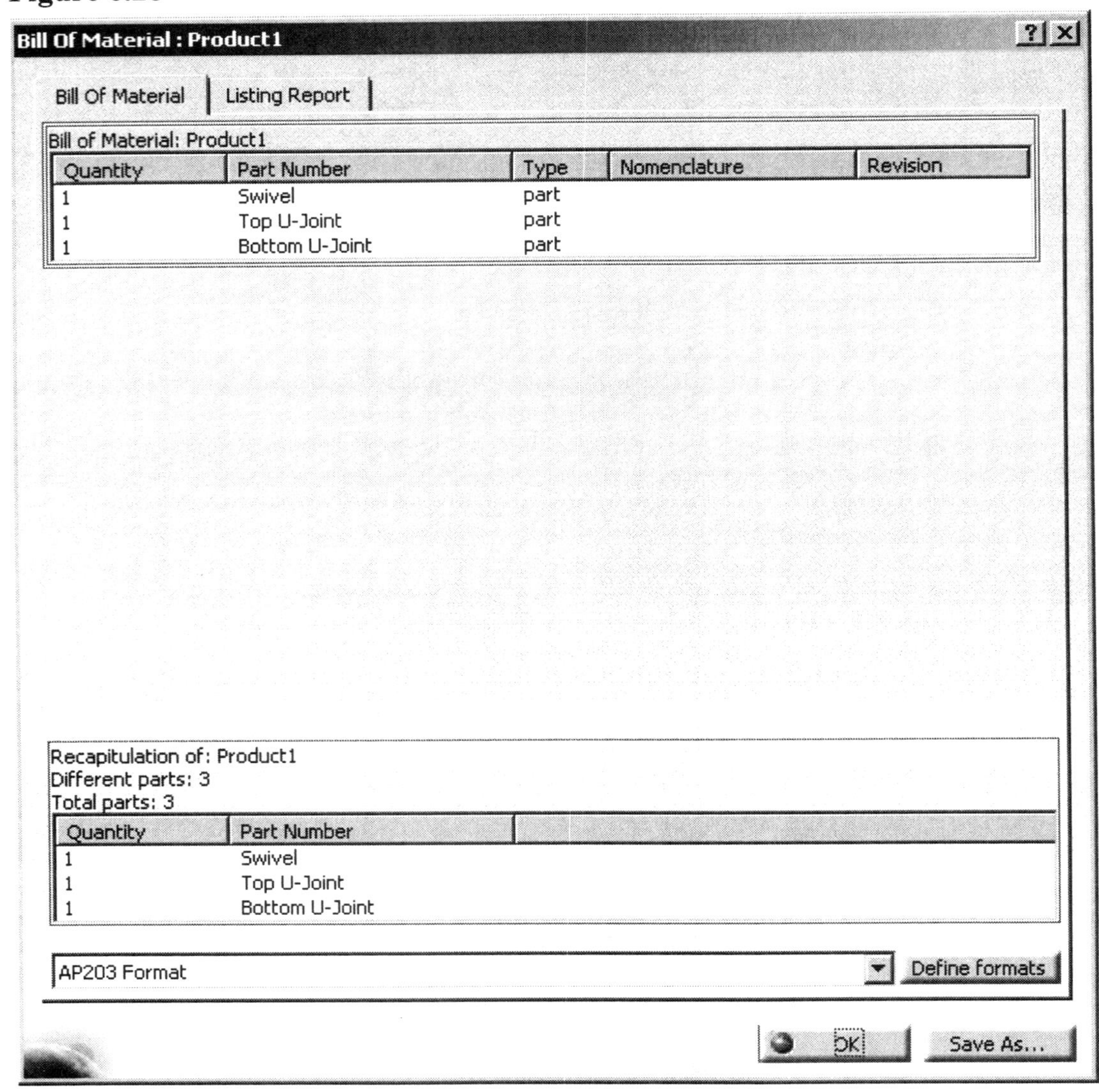

7.3 CATIA V5 allows you to export this information as a **.txt** file. Select the **Save As** button in the bottom right corner of the **Bill Of Material: U-Joint Assembly** window. Figure 6.29 shows an example of the **.txt** file from the **Bill Of Material: U-Joint Assembly** window. You are not required to do anything with the **Bill Of Materials**; this step is to show you that it is there and can be exported and/or printed.

Figure 6.29

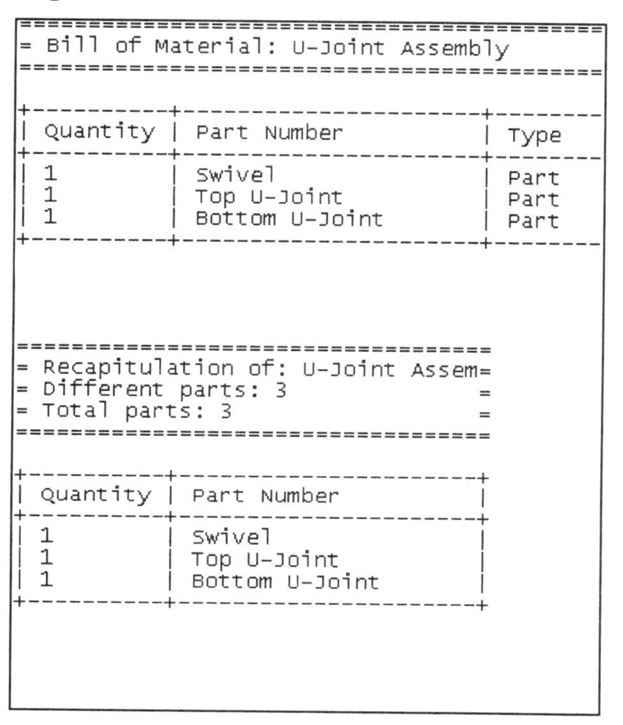

```
============================================
= Bill of Material: U-Joint Assembly
============================================

+----------+----------------------+--------
| Quantity | Part Number          | Type
+----------+----------------------+--------
| 1        | Swivel               | Part
| 1        | Top U-Joint          | Part
| 1        | Bottom U-Joint       | Part
+----------+----------------------+--------


===================================
= Recapitulation of: U-Joint Assem=
= Different parts: 3              =
= Total parts: 3                  =
===================================

+----------+---------------------+
| Quantity | Part Number         |
+----------+---------------------+
| 1        | Swivel              |
| 1        | Top U-Joint         |
| 1        | Bottom U-Joint      |
+----------+---------------------+
```

Figure 6.30

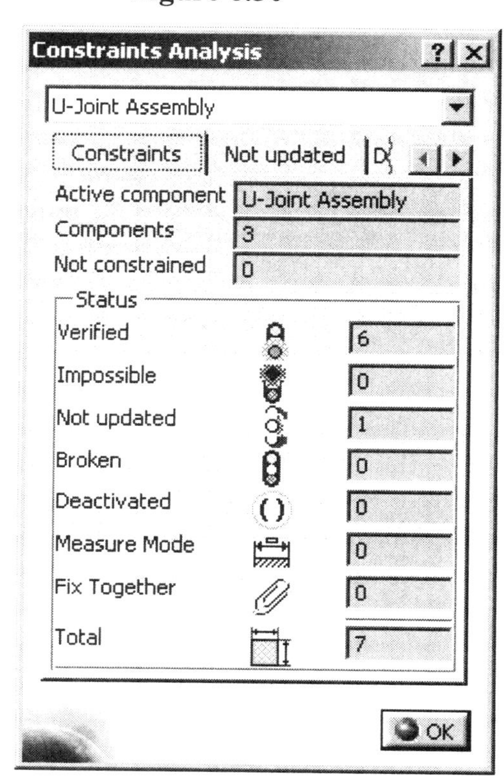

8 Analyzing/Modifying Assembly Constraints

8.1 The next analysis tool CATIA V5 offers is the **Constraints Analysis**. Select the **Constraints** option from the **Analyze** menu, reference Figure 6.27. This will bring up the **Constraints Analysis** window as shown in Figure 6.30. This window supplies information on how many and what kinds of Constraints were used in the assembly.

8.2 Take a minute to review the information contained in the **Constraints Analysis** window. Select the **OK** button to close out of the **Constraints Analysis** window. This method allows you to **Analyze** the **Constraints**. If you want to modify the **Constraints** complete the following steps.

8.2.1 Make sure you are in the **Assembly Design Work Bench**.

8.2.2 Select the **Constraint** and then double click on it. This will bring up the **Constraint Definition** window.

8.2.3 Select the **More** button. This will expand the **Constraint Definition** window as shown in Figure 6.31. The **Constraint Definition** window gives you the name, the type and what entities the constraint is using.

8.2.4 You can also get to this window by using the **Properties** window.

Figure 6.31

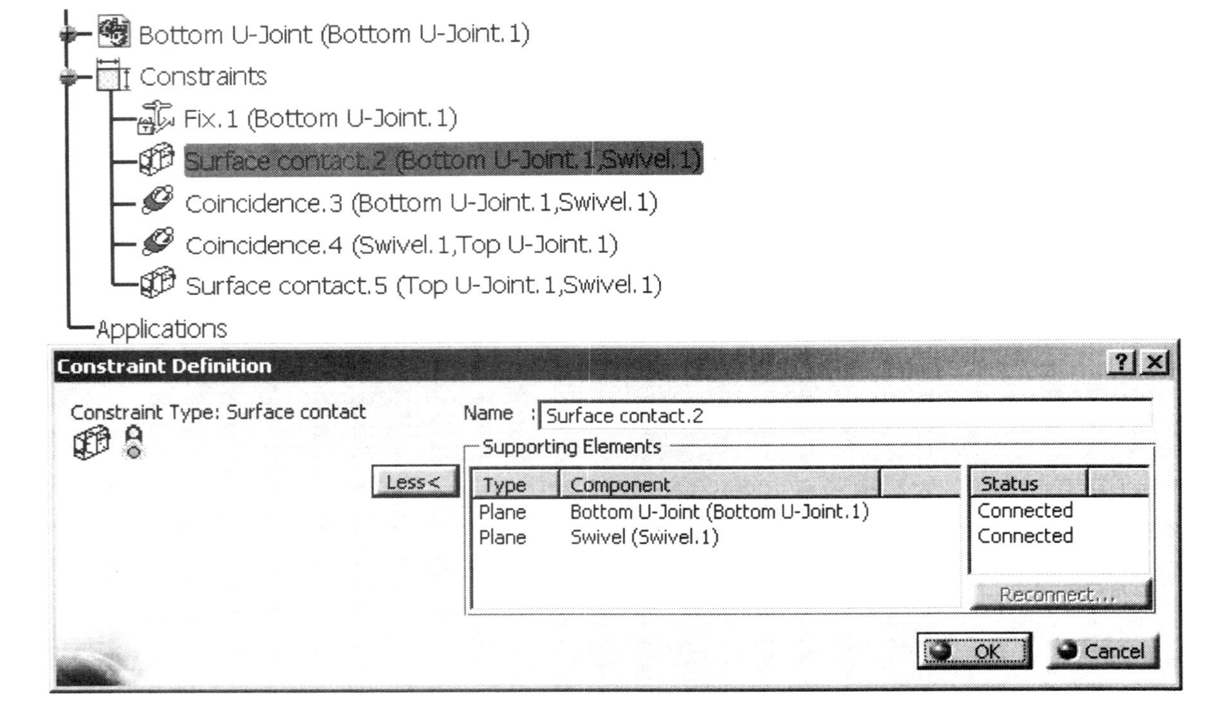

9 Clash Detection

9.1 Another handy tool located in the **Analyze** tool window is the **Compute Clash** option. Selecting the **Compute Clash** option will bring up the **Clash Detection** window as shown in Figure 6.32. The "**Bottom U-Joint**" has been placed in **Hide/Show** for clarification.

Figure 6.32

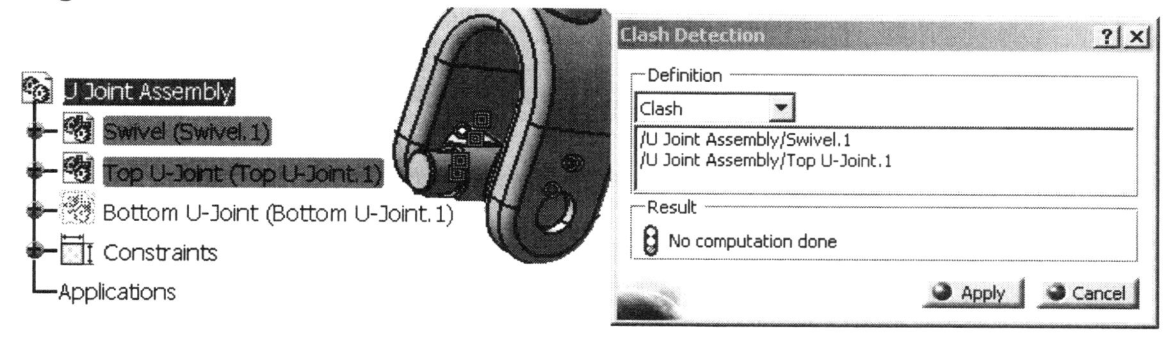

9.2 The **Prompt Zone** will prompt you to "**Select two components**." Select the "**Swivel**," hold down the Ctrl button and select the "**Top U-Joint**." Both selections will highlight in the **Specification Tree**. The two selections will also show up in the **Definition** box.

9.3 Select the **Apply** button. The **Result** box will display the kind of contact the two selected components share. This particular assembly has **Contact** only between the "**Swivel**" and "**Top U-Joint**" components. Notice that the **Contact** is also displayed on the solids. **Contact** is displayed in yellow by default. If the two solids clashed, the **Clash** would be represented in red and the **Results** box would also specify the **Clash**. Your assembly should only have **Contact** and no **Clash** between components.

9.4 To exit the **Clash Detection** window, select **Cancel**.

10 Measuring Tools

10.1 Two other options in the **Analyze** tool bar are the **Measure Between** and **Measure Item**. Both tools are used the same as described in the **Introduction**. The two tools are the same as the **Measure** tools found on the bottom tool bar. **Measure Item** measures a single length of an entity. **Measure Between** measures the distance between entities you select.

10.2 Another tool on the **Analyze** tool bar is **Measure Inertia**. Selecting **Measure Inertia** will bring up the **Measure Inertia** window as shown in Figure 6.33. You can measure the inertia of an individual component and/or the entire **Product**. You select the component you want the inertia information on and it will analyze it. Figure 6.33 shows the entire assembly inertia analysis. The centroid (center of mass) is displayed in the **Center Of Gravity** box as well as graphically on the assembly. You can customize the information displayed in the **Measure Inertia** window. You can also export the information as a **.txt** file.

Figure 6.33

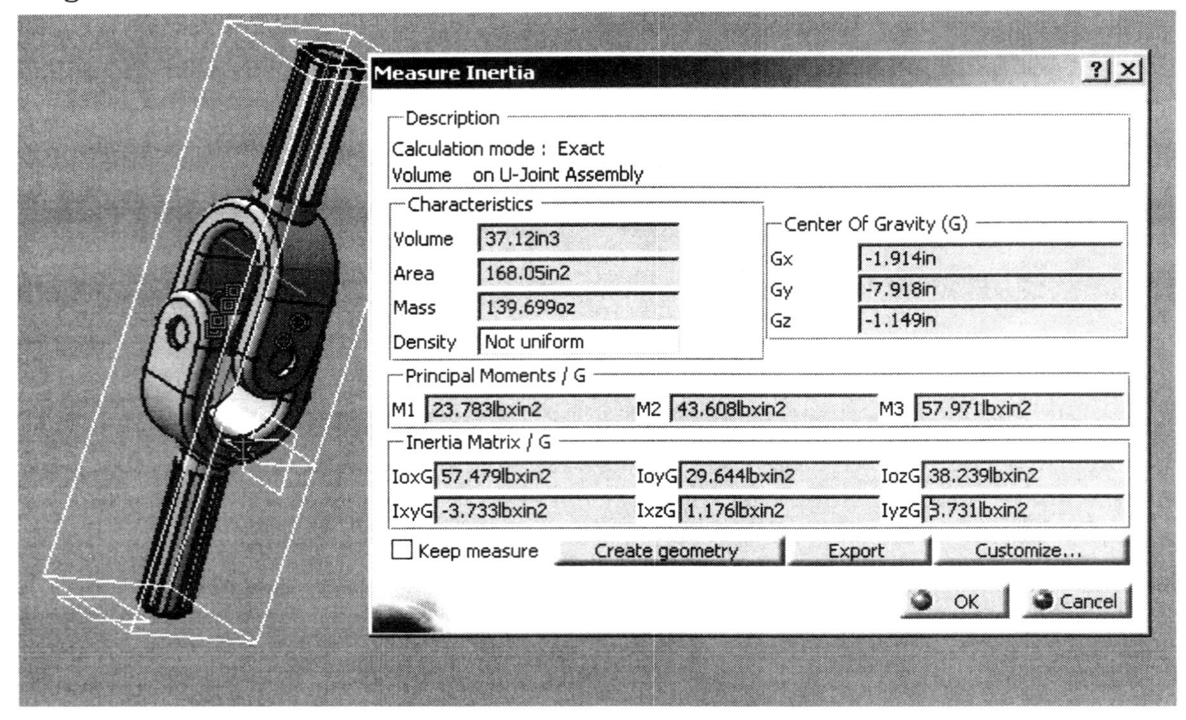

To get the most out of this step you need to work through it, explore it and test it.

Figure 6.34

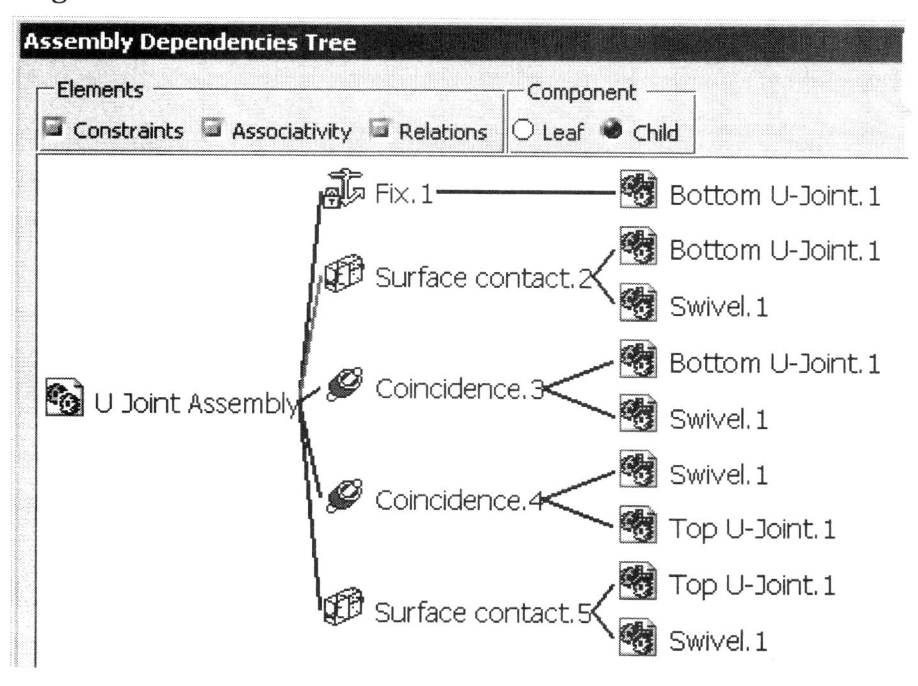

11 Assembly Dependencies

CATIA V5 gives you many different ways to view and analyze your assembly. The **Dependencies** tool is just another way of analyzing how your assembly is put together. This tool is found in the **Analyze** pull down window. Figure 6.34 shows how the "**U-Joint Assembly**" **Dependencies** are linked together. This tool is particularly helpful when you want to know which parts were used to create a particular **Constraint**. You will need to double click on the icons in the **Assembly Dependencies Tree** window to expand them.

12 Mechanical Structure

The **Mechanical Structure** tool is another method of analyzing the assembly. Figure 6.35 shows the **Mechanical Structure Tree** for the "**U-Joint Assembly**." This assembly is relatively simple. As you get to larger and more complex assemblies the value of this tool will become more important.

Figure 6.35

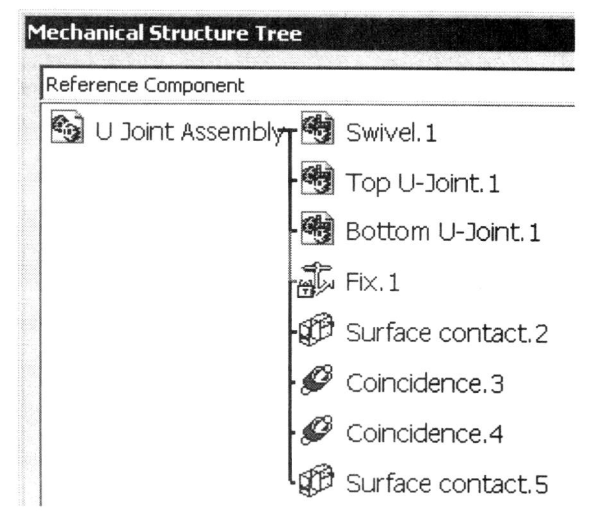

13 Adding Annotation

CATIA V5 has supplied you with a method to communicate with written text right in the assembly. You can do this by selecting the **Text With A Leader** tool from the right tool bar. Then all you have to do is select the entity that you want the text attached to. For this step select the .05 fillet radius as shown in Figure 6.36. This will bring up the **Text Editor** window. You type in the note you want attached and select the **OK** button. In this case make a note of the size of the radius fillet. Your note should look similar to what is shown in Figure 6.36.

Figure 6.36

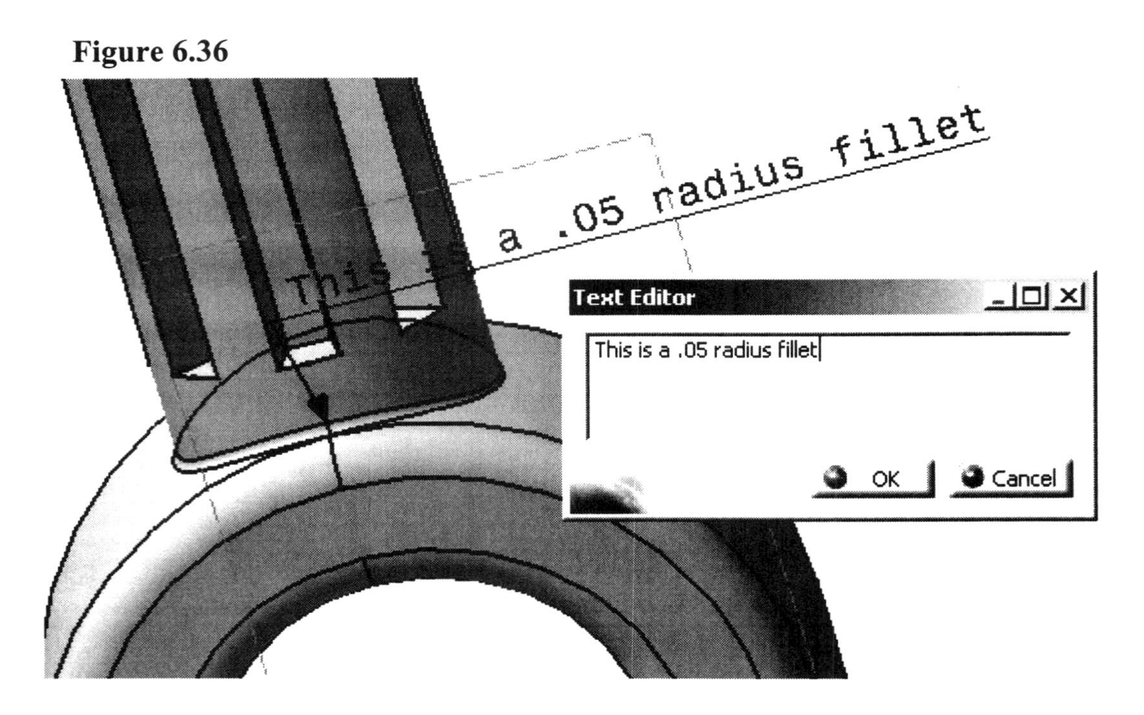

14 Saving A Specific View

The **Named Views** tool is a general tool available in most all of the CATIA V5 work benches. Even though this tool is not particular to the **Assembly Design** tool, now is a good time to put it to use. This tool is particularly useful when you have zoomed and oriented the part and/or assembly to exactly how you want it. If you could not save this exact zoom value and orientation, it might take a lot of extra time to get it close. Saving and naming the view will take you exactly to the way it was saved. Another application is if you were working on a large part and/or assembly and you needed to view one end and then the other end, you would not want to have to **Pan** from one end to the other every time! Saving a view will allow you to go directly from one end to the other, as you saved it. To create, save and name a view, complete the following steps.

Figure 6.37

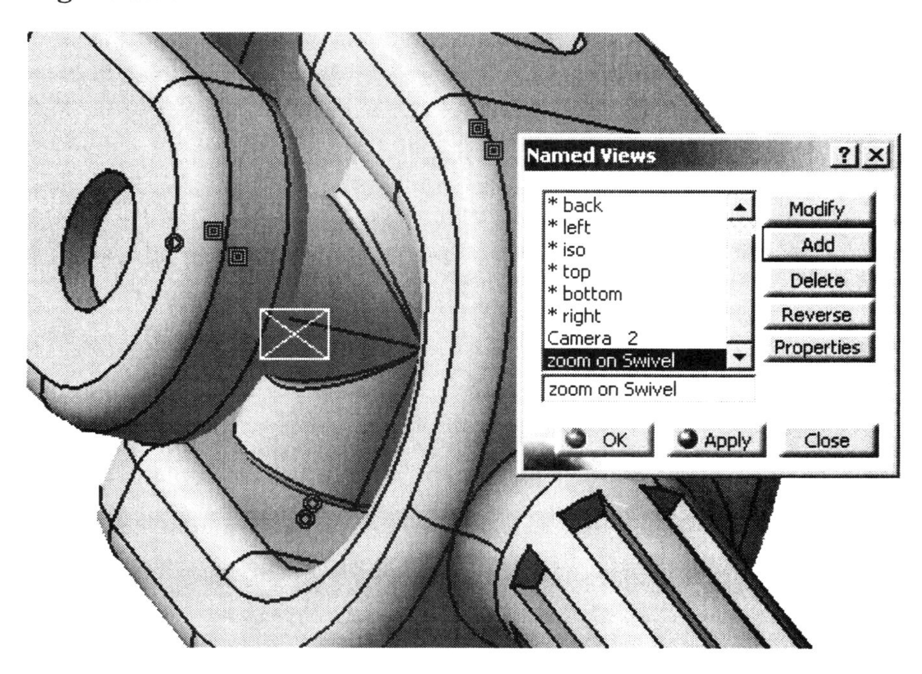

14.1 Zoom and orient the part/assembly or area of the part/assembly as you need it. For this step, zoom in on the "**Swivel**" as shown in Figure 6.37; it does not have to be exact.

14.2 Select the **View** pull down window (top menu).

14.3 Select **Named Views**. This will bring up the **Named Views** window, as shown in Figure 6.37.

14.4 Select the **Add** button. This will create the default view name "**Camera 1**." Replace "**Camera 1**" with the name "**Zoom on Swivel**" in the editing box above the **OK** button.

14.5 Select **Ok** to create the view.

14.6 Now you can rotate and zoom the assembly any way you want. If you want to get back to that exact view repeat the previous steps, except instead of creating a new view, select the view you just created "**Zoom on Swivel**" and select apply. The screen will update to that particular view, just the way you saved it.

15 Exploding The Assembly

You spent a lot of time and effort in creating and constraining the "**U-Joint Assembly**." CATIA V5 has created a tool that can explode the assembly automatically in a matter of several mouse clicks. To create an exploded view of your "**U-Joint Assembly**" complete the following steps.

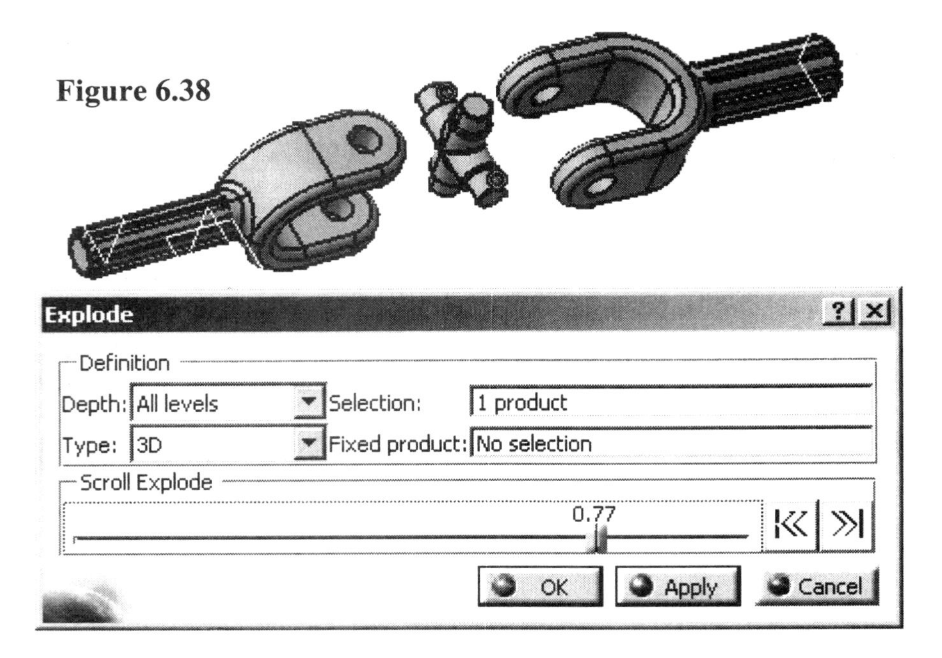

Figure 6.38

15.1 Select the **Explode** tool from the **Move** tool bar. This will bring up the **Explode** window as shown in Figure 6.38.

15.2 To view the default explode, select the **Apply** button. This will create the exploded assembly. Now you can select the **Play/Replay** button to re-assemble/explode the assembly. You can also use the mouse to manually control the explosion, by dragging the **Scroll Explode** button.

15.3 You can also customize the **Explode** tool by determining which entities will be fixed. For this step create a simple exploded assembly similar to the one shown in Figure 6.38.

16 Saving The Newly Created Assembly

Congratulations! You just completed your first assembly in CATIA V5. Save the file as "**U-Joint Assembly.CATProduct**." Notice CATIA V5 saves this as a separate file. The file has a "**CATProduct**" extension. This "**U-Joint Assembly.CATProduct**" file is directly linked to the "**CATPart**" file created in Lesson 5, the "**Bottom U-Joint.CATPart**."

To verify where all of the components are saved, select the **Tools**, **Product Management**. This will bring up the **Product Management** window. This window displays all of the assembly's components and where they are saved. If you are not in the **Assembly Design Work Bench** the **Product Management** option will not be accessible.

If you want to copy this assembly and all its linked components to another file, select **File**, **Send To**. This will allow you to copy it to another directory and/or mail.

Lesson 6 Summary

Though this was a simplified assembly, it did require you to go through the basic steps of inserting new and existing components (parts), re-orienting the components and bringing the components under specific restrictions using **Constraints**. Even complex assemblies require these basic steps. Now that you have a basic idea of how to create an assembly, you are strongly encouraged to go back and try different variations of accomplishing the same thing. Try some of the unused tools.

Lesson 6 Review

After completing this lesson, you should be able to answer the questions and explain the concepts listed below.

1. How many standard tool bars are there in the **Assembly Design Work Bench**?
2. What file extension does CATIA V5 give to assembly files?
3. Is the assembly file directly linked to the "**CATPart**" files that are used in the assembly file?
4. What option under the **Start** pull down menu gives you access to the **Assembly Design Work Bench**?
5. What must be selected before components can be inserted into the **Assembly Design Work Bench**?
6. What tool inserts parts into the **Assembly Design Work Bench**?
7. What tool was used to move the parts after they were inserted into the **Assembly Design Work Bench**?
8. What tool is used to anchor a part to a fixed location?
9. What does the **Contact Constraint** tool do?
10. What does the **Contact Constraint** symbol look like once it has been applied to the assembly?
11. What does the **Coincidence Constraint** tool do?
12. What does the **Coincidence Constraint** symbol look like once it has been applied to the assembly?
13. Why is it important to know how to move and orient parts after they have been inserted into the **Assembly Design Work Bench**?
14. **T** or **F** The **Axis Line** is always visible in the **Assembly Design Work Bench**.
15. What tool would you use if you want to find out how far apart two individual parts are?

16. **T** or **F** The **Specification Tree** in the **Assembly Design Work Bench** not only shows the history of the assembly, but also the individual part that was inserted into the assembly.

17. List the two different methods of moving the individual parts around in the **Assembly Design Work Bench**.

18. If you move a component using the compass after it has been located using a **Constraint**, how would you restore it to its original **Constraint** location?

19. **T** or **F** CATIA V5 allows you to modify a component without leaving the **Assembly Design Work Bench**.

20. **T** or **F** The **Measure Inertia** tool and the **Bill Of Materials** tool are found in the same tool bar.

Lesson 6 Practice Exercises

Now that you have created one assembly in CATIA V5, you can strengthen your newfound knowledge by completing the following practice exercises.

1 Insert the "**Lesson1 Exercise 1.CATPart**" and "**Lesson 1 Exercise 4.CATPart**." Assemble the two parts using the **Contact Constraint**. Your assembly should look similar to the one shown below. Save the assembly as "**Lesson 6 Exercise 1.CATProduct**."

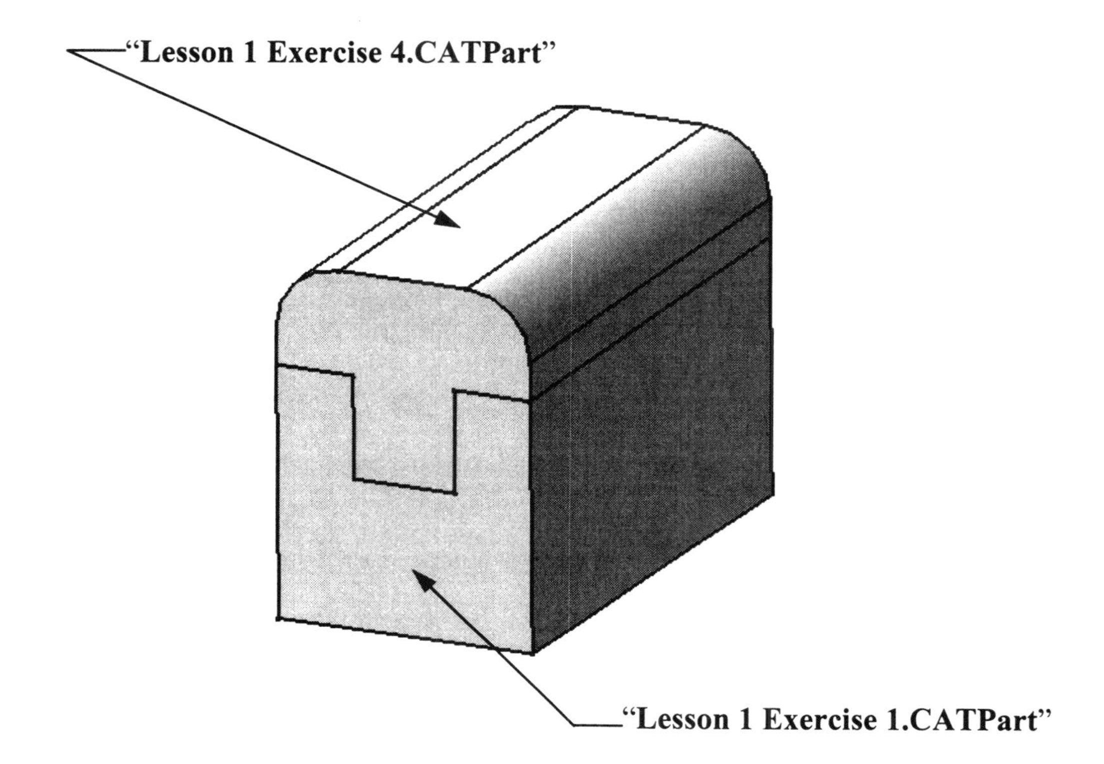

2 Insert the "**Lesson 1 Exercise 3.CATPart**" and "**Lesson 1 Exercise 5.CATPart**." Assemble the two parts using the **Contact Constraint** and the **Offset Constraint**. Your assembly should look similar to the one shown below. Save the assembly as "**Lesson 6 Exercise 2.CATProduct**."

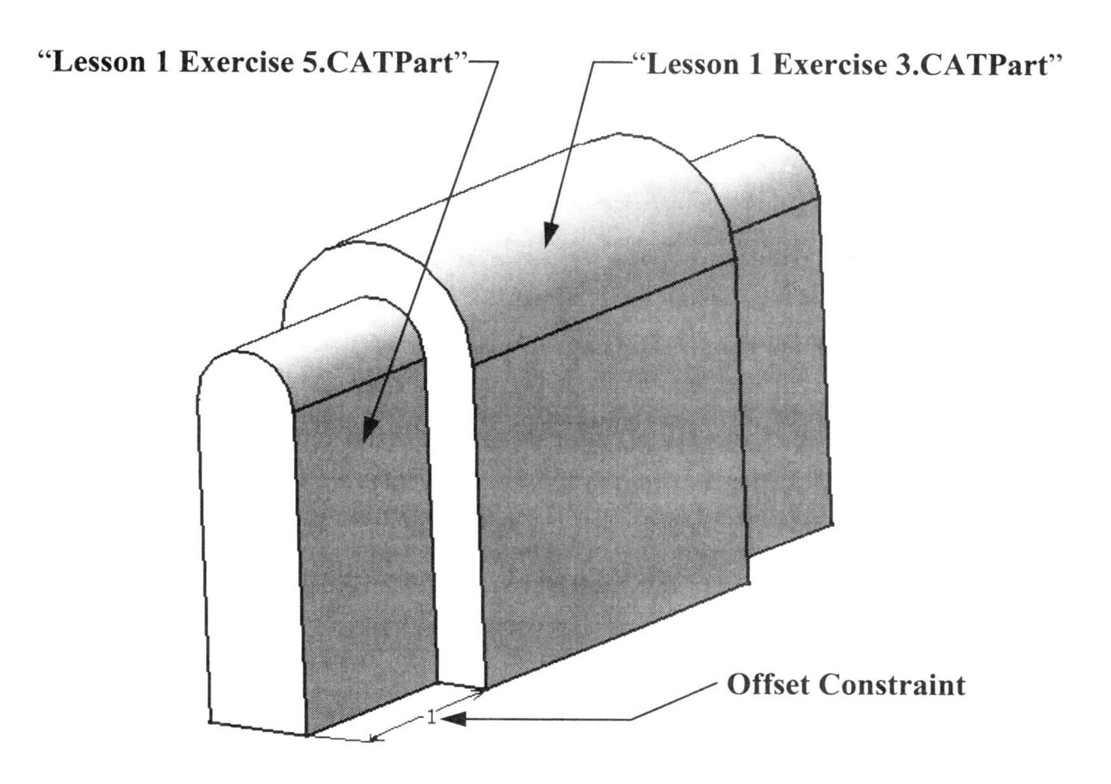

3 Insert the "**Lesson 5 Exercise 4.CATPart**" and "**Lesson 5 Extra Credit.CATPart**." Assemble the two parts using the **Coincidence Constraint** and the **Contact Constraint**. Your assembly should look similar to the one shown below. Save the assembly as "**Lesson 6 Exercise 3.CATProduct**."

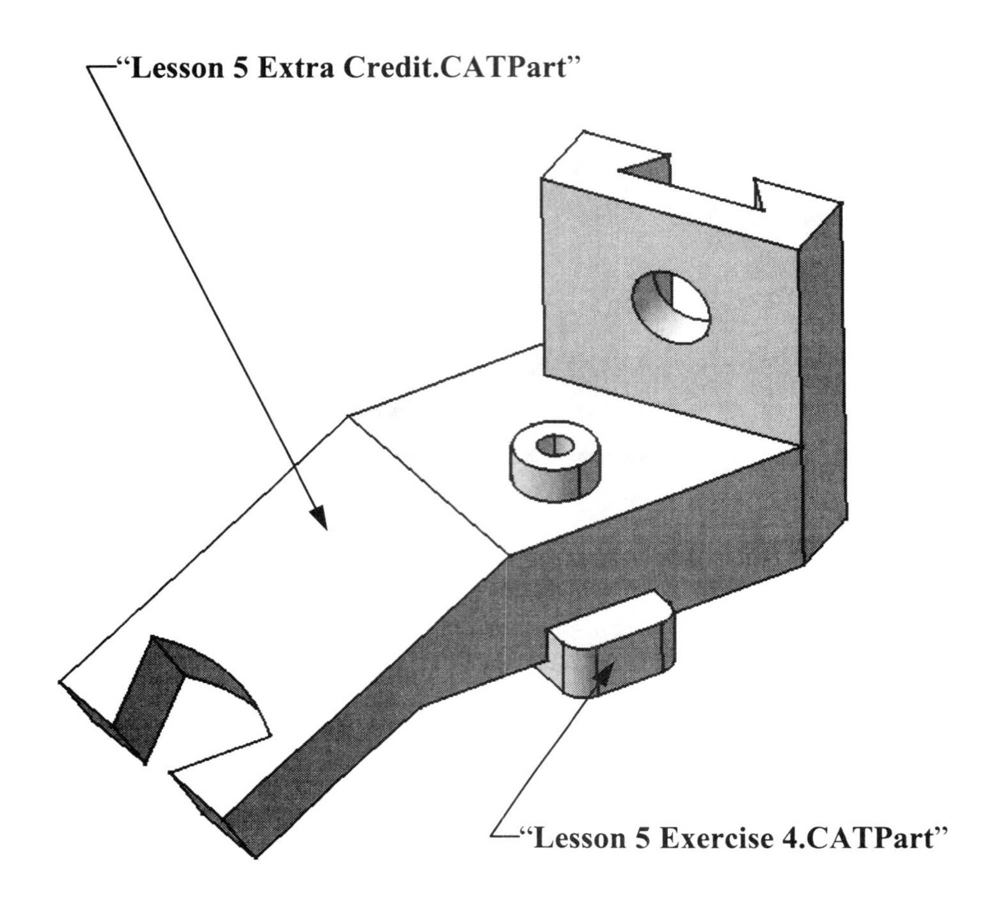

4 Insert the "**Lesson 5 Exercise 3.CATPart**" and "**Lesson 5 Exercise 5.CATPart**." Assemble the two parts using the **Coincidence Constraint** and the **Contact Constraint**. Your assembly should look similar to the one shown below. Save the assembly as "**Lesson 6 Exercise 4.CATProduct**"

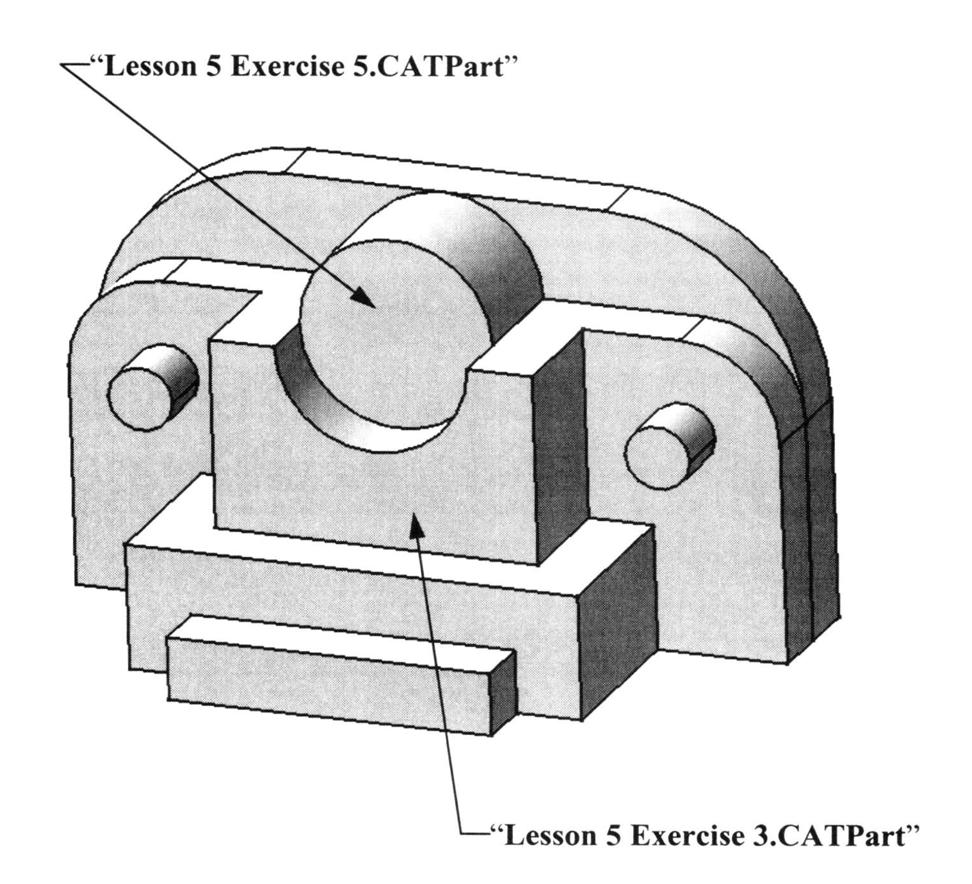

5 **Extra Credit**: If you are ready for a challenge, go to www.schroff1.com/catia and download all the documents (components) found under **Arbor Press Assembly**. The challenge is to correctly assemble all the components. The completed assembly should look similar to the one shown below. Good Luck.

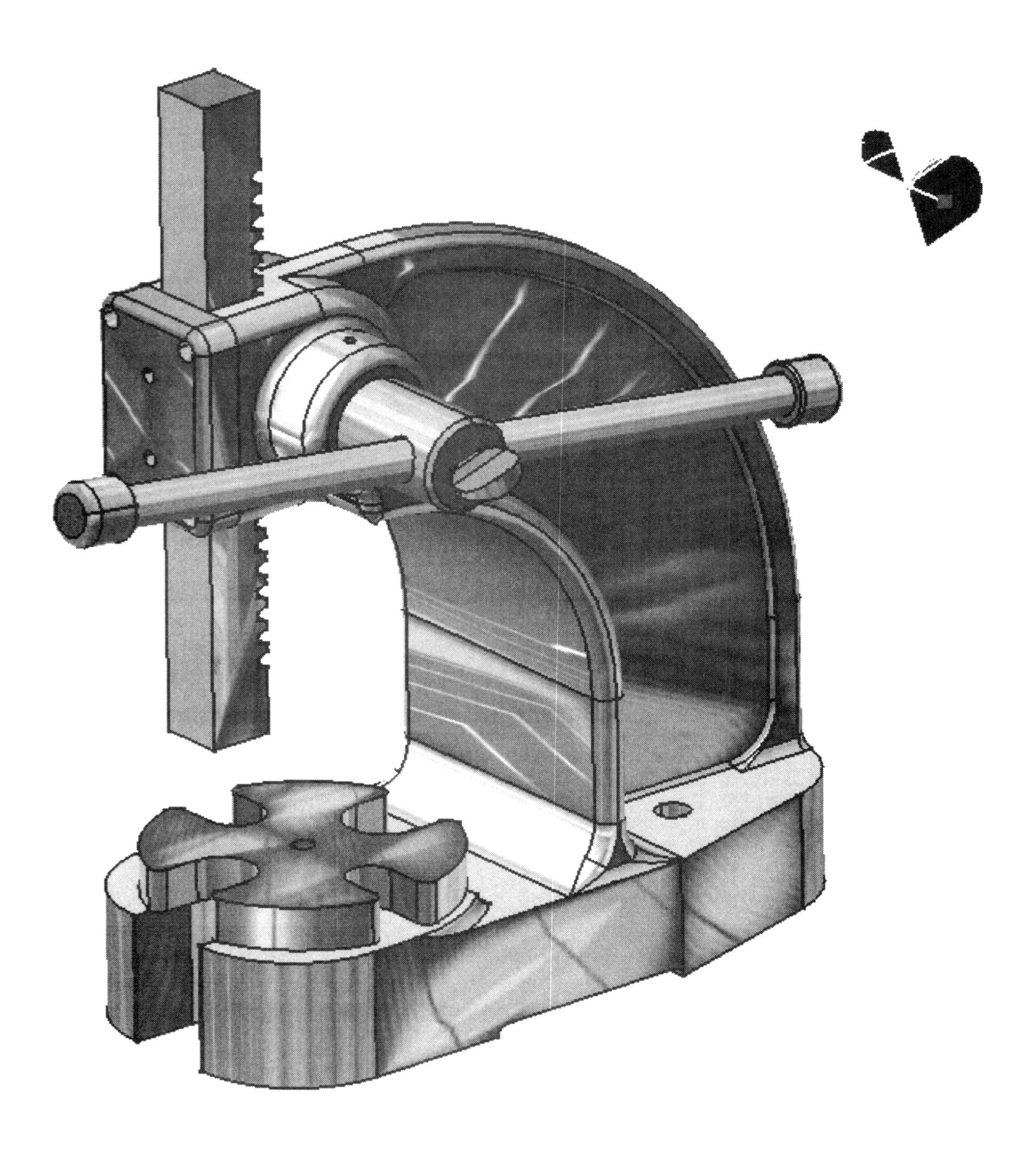

Lesson 7 Generative Shape Design Work Bench

Introduction To Creating Wireframe And Surface Geometry Using The Generative Shape Design Work Bench

This lesson describes to you, step-by-step how to create a local axis and how to create wireframe geometry using lines, points and corners. This lesson will then show you how to create surfaces from the existing wireframe. The end of the lesson shows you how to create a solid part from the wireframe and surfaces. This lesson does not cover all the tools found in the **Generative Shape Design Work Bench**. Lesson 8 is a continuation of the **Generative Shape Design Work Bench** and will cover additional tools.

You will see a significant difference in the **Specification Tree**. So far all of the objects have been created under the **PartBody** branch of the **Specification Tree**. As you create the wireframe all of the geometry will be created under the **Open_Body** branch, not the **PartBody** as in the previous lessons. The **PartBody** contains all of the solid geometry features such as **Pads, Fillets**, **Chamfers**, **Holes**, etc. The **Open_Body** contains everything that is not a solid such as **Lines**, **Curves**, **Points** and **Planes**. Since wireframe is not a solid object, it will be listed under the **Open_Body** branch in the **Specification Tree**.

The part that will be created in this lesson is shown in Figure 7.1. It will be called "**Sheetmetal Bracket"** and will be saved as "**Sheetmetal Bracket.CATPart**."

Lesson 7 Objectives

This Lesson will show you how to do the following:

- Select the **Generative Shape Design Work Bench**
- Create a **Local Axis System**
- Create 3D wireframe elements
- **Split** 3D elements
- Create simple **Surfaces**
- Create a **Fillet** between two surfaces
- **Split** surfaces
- Create a solid part using the **Thick Surface** Tool
- Create a part using the **Helix** tool

Generative Shape Design Work Bench Tool Bars

There are nine default tool bars found in the **Generative Shape Design Work Bench.** They are: **Select**, **Wireframe**, **Surfaces**, **Advanced Surfaces**, **Operations**, **Replication**, **Law**, **Analysis**, and **Tools**. The nine tool bars are shown on the following pages.

Just in case you forgot, the arrow at the bottom right of some of the tool icons is an indication of there being more than one variation of that particular type of tool.

The **Select** Tool Bar

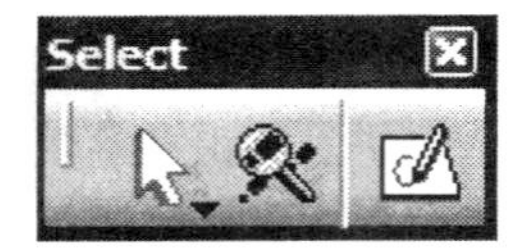

Note: All of the work benches have the **Select** tool bar; although many of the tools contained in the tool bar vary depending on the specific work bench.

Tool Bar	Tool Name	Tool Definition
	Select	Same tool as in all the previous work benches. To view the additional **Select** tools review the previous lessons.
	Quick Edit	This tool gives quick access to the generative specifications.
	Sketcher	This is the same **Sketcher** tool found in the previous lessons. The **Generative Shape Design Work Bench** includes it in the **Select** tool bar.

The **Advanced Surfaces** Tool Bar

Tool Bar	Tool Name	Tool Definition
	Develop	Creates a developed curve on a revolution surface.
	Junction	Creates a junction.
	Bump	Creates a bump.
	WrapCurve	Creates a wrap curve.

The **Wireframe** Tool Bar

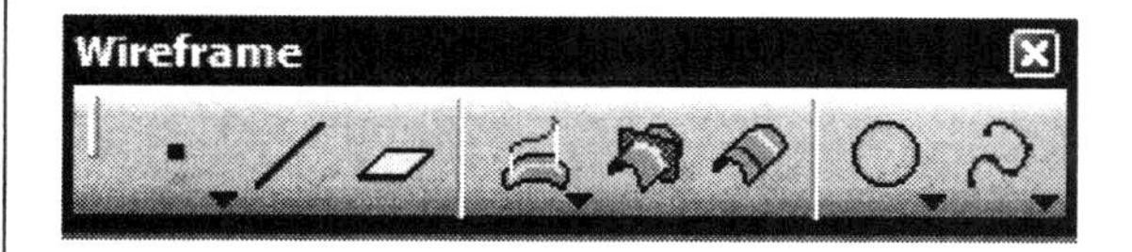

Tool Bar	Tool Name	Tool Definition
	Line	Creates a line in 3D space.
	Plane	Creates a plane.
	Intersection	Creates geometry by intersecting two geometric entities.
	Parallel Curve	Creates a curve that is offset from a reference curve.
	Points Tool Bar	
	Point	Creates one or more points.
	Points and Planes Repetition	Repeats several times the creation of points or planes.
	Extremum	Creates the extremum (extreme: max/min) point of a selected entity.
	ExtremumPolar	Gives the extremum of a selected element in polar coordinates.
	Project-Combine Tool Bar	
	Projection	Projects a point or curve onto a selected support element.
	Combine	Combines two curves along two directions.
	ReflectLine	Creates a reflect line.
	Circle-Conic Tool Bar	
	Circle	Creates a circle or circular arc.
	Corner	Creates a corner between two curves.
	Connect Curve	Creates a connect between two curves.
	Conic	Creates a conic.
	Curves Tool Bar	
	Spline	Creates a spline curve.
	Helix	Creates a helix.
	Spiral	Creates a spiral.
	Spine	Creates a spine.
	Polyline	Creates a polyline.

The **Surfaces** Tool Bar

Tool Bar	Tool Name	Tool Definition
	Offset	Creates a surface that is offset from a reference surface.
	Fill	Creates a fill surface inside a closed boundary.
	Loft	Creates a lofted surface.
	Blend	Creates a blended surface.
	Extrude-Revolution Tool Bar	
	Extrude	Creates an extruded surface.
	Revolve	Creates a surface by revolving a profile around an axis.
	Sphere	Creates a spherical surface.
	Sweeps Tool Bar	
	Sweep	Creates a swept surface.
	Adaptive Sweep	Creates a swept adaptive surface.

The **Analysis** Tool Bar

Note: This tool bar appears in the bottom tool bar but is specific to the **Generative Shape Design Work Bench**.

Tool Bar	Tool Name	Tool Definition
	Connect Checker	Checks surface connections.
	Curve Connect Checker	Checks curve connections.
	Draft Analysis	Analyses the draft direction on selected surfaces.
	Surfacic Curvature Analysis	Analyses the Gaussian curvature on a shape.
	Porcupine Curvature Analysis	Performs a porcupine curvature analysis on any curve.
	Geometric Information	Provides geometric information a specific face/edge.

The **Operations** Tool Bar

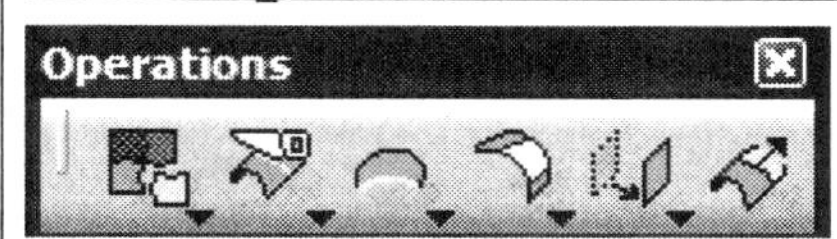

Tool Bar	Tool Name	Tool Definition
	Extrapolate	Creates a surface or curve by extrapolation.
	Join–Healing Tool Bar	
	Join	Joins curves or surfaces.
	Healing	Heals surfaces.
	Curve Smooth	Smoothes curves.
	Untrim Surface Or Curve	Untrims previously trimmed surfaces or curves.
	Disassemble	Disassembles multi-cell bodies into mono-cell bodies.
	Trim–Split Tool Bar	
	Split	Cuts and trims an element using a cutting element.
	Trim	Cuts and assembles two elements.
	Extracts Tool Bar	
	Boundary	Creates a boundary from a surface.
	Extract	Extracts a face or surface edge.
	Multiple Edge Extract	Extracts a group of faces or surfaces edges.
	Fillets Tool Bar	
	Shape Fillet	Creates a fillet between two surfaces.
	Edge Fillet	Creates an edge fillet.
	Variable Radius Fillet	Creates a variable radius fillet.
	Face–Face Fillet	Creates a face–face fillet by selecting two faces.
	Tritangent Fillet	Creates a fillet by removing a surface you specify.
	Transformations Tool Bar	
	Translate	Translates an element by a specified direction.
	Rotate	Rotates a specified element around a specified axis.
	Symmetry	Transfers an element by symmetry.

	Scaling	Transfers an element by scaling.
	Affinity	Transfers an element by affinity.
	Axis to Axis	Transfers an element from one axis system to another.

The **Replication** Tool Bar

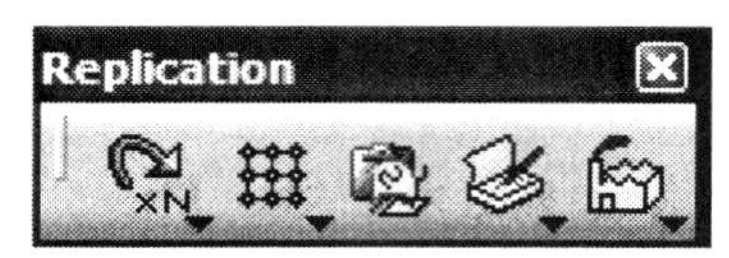

Tool Bar	Tool Name	Tool Definition
	Duplicate Open Body	Creates a duplicate open body.
	Repetitions Tool Bar	
	Object Repetition	Repeats the creation of an object a specified number of times.
	Points And Planes Repetition	Repeats the creation of points and planes a specified number of times.
	Planes Repetition	Creates a specified number of planes between two exiting planes.
	Patterns Tool Bar	
	Rectangular Pattern	Creates a rectangular pattern to repeat the creation of a specified object.
	Circular Pattern	Creates a circular pattern to repeat the creation of a specified object.
	Power Copy Tool Bar	
	PowerCopy Creation	Creates a power copy.
	Save In Catalog	Saves an object in a catalog.
	UserFeature Tool Bar	
	UserFeature Creation	Creates a user feature.
	Save In Catalog	Saves an object in a catalog.

The **Law** Tool Bar

Tool Bar	Tool Name	Tool Definition
	Law	Creates a law.

The **Tools** Tool Bar

Note: This tool bar appears in the bottom tool bar but is specific to the **Generative Shape Design Work Bench**.

Tool Bar	Tool Name	Tool Definition
	Update All	Same as previous work benches. Updates all entities that require updating.
	Axis System	Same as previous work benches. Creates a new axis system.
	Show Historical Graph	Displays a historical graph on selected features.
	Create Datum	Creates a datum feature. One click creates one feature, two clicks creates all features.
	Insert Mode	Created features are inserted into tree.
PartBody	**Select Current Tool**	Selects current tool.
Grid Tool Bar		
	Work On Support	Allows working on a support.
	Snap To Point	Allows snapping to a point.
	Working Supports Activity	Toggles the working support on and off. Allows the working on support systems.
	Create A New Set	Allows the creation of new sets.
Instantiation Tool Bar		
	Open Catalog	Opens a specific catalog.
	Instantiate From Document	Instantiates an element stored in a part.

Steps To Creating A Simple Wireframe Part Using The Generative Shape Design Work Bench

1 Select The Generative Shape Design Work Bench 

This step will use the **Generative Shape Design Work Bench** to create a "**Sheetmetal Bracket**" shown in Figure 7.1. Selecting the **Generative Shape Design Work Bench** is similar to selecting the **Sketcher Work Bench** and **Part Design Work Bench** like you did in Lessons 1 and 2. If the **Generative Shape Design Work Bench** does not appear in the "**Welcome To CATIA V5**" window you will have to select the **Generative Shape Design Work Bench** from the **Start** pull down menu and **Shape** function. For reviewing this process reference Lesson 1 Step 2. Select the **Generative Shape Design Work Bench**. Your screen should look similar to the one shown in Figure 7.2 with the exception of the tool bars being two tool icons wide. By default the tool bars will be one tool icon wide. The screen in Figure 7.2 has been modified to two tool icons wide so all of the standard **Generative Shape Design Work Bench** tools can be represented. The bottom tool bar has also been modified to show all of the tools available. The process of moving the tool bars around on the screen is covered in detail in Lesson 2 Step 10.4.

Figure 7.1

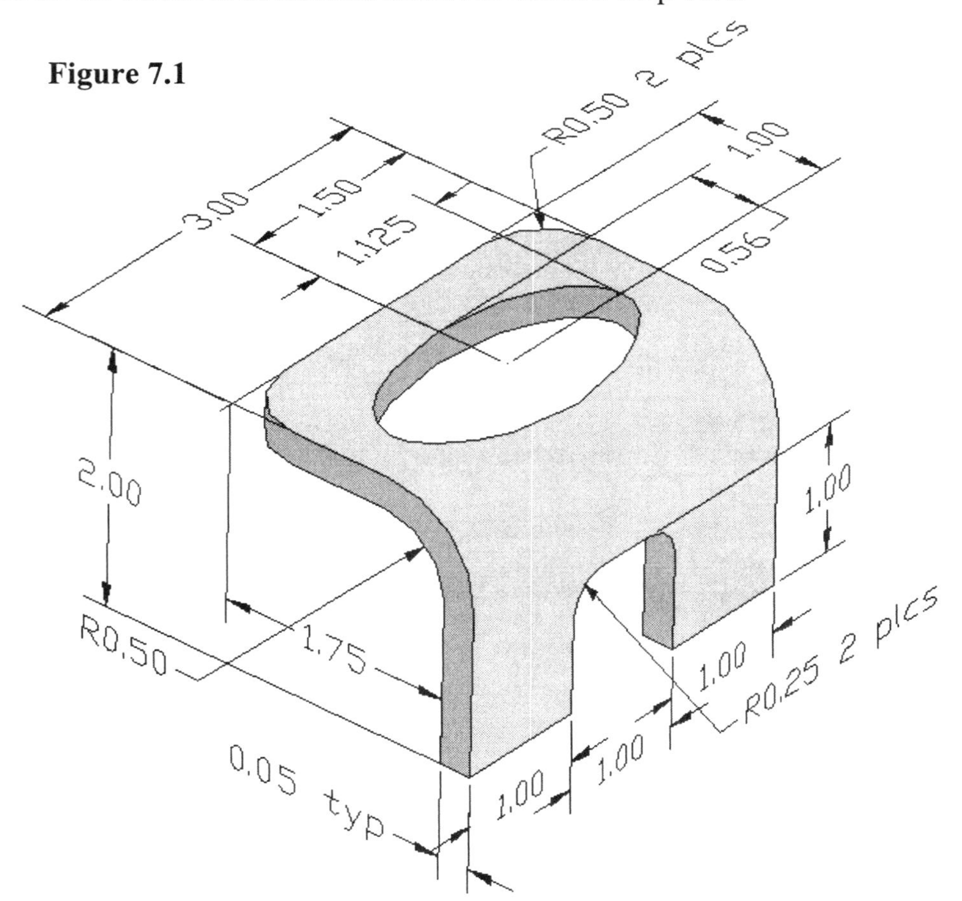

Notice the **Specification Tree**. The top of the tree is **Part1** and under the **PartBody** branch you have an **Open_Body.1** branch. It is under this branch that all of the wireframe geometry will be placed.

Figure 7.2

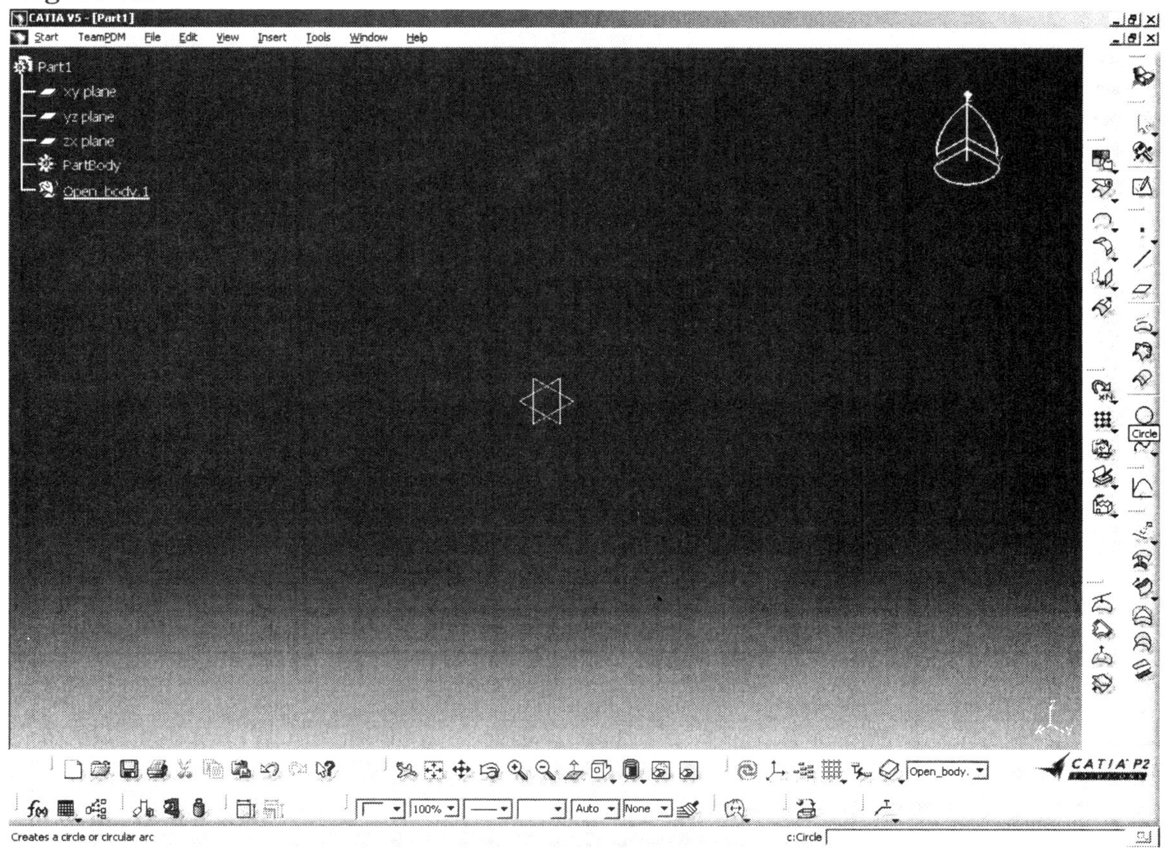

NOTE: The **Generative Shape Design Work Bench** screen looks somewhat similar to the **Part Design Work Bench** screen. Notice that the tool bars on the right side of the screen are different and the **Specification Tree** has an **Open_Body** branch added to it.

2 Creating A Local Axis System 

This step is not absolutely necessary for creating wireframe objects, but it is a useful tool and will help you in your wireframe and surface creation. Creating a local axis system can be applied in any of the CATIA V5's workbenches. The axis systems will be used later on in the lesson. As you create and use the axis systems in this lesson, be thinking how local axis systems might be used in past and future lessons and/or projects. This step will have you create two local axis systems before creating the wireframe for the "**Sheetmetal Bracket**."

The default 0,0,0 location for CATIA V5 is the intersection of the **XY**, **YZ**, and **ZX Planes**. Since there is no coordinate system visible other than the representative planes, create a point at the 0,0,0 location.

2.1 Create a point by completing the following steps.

Figure 7.3

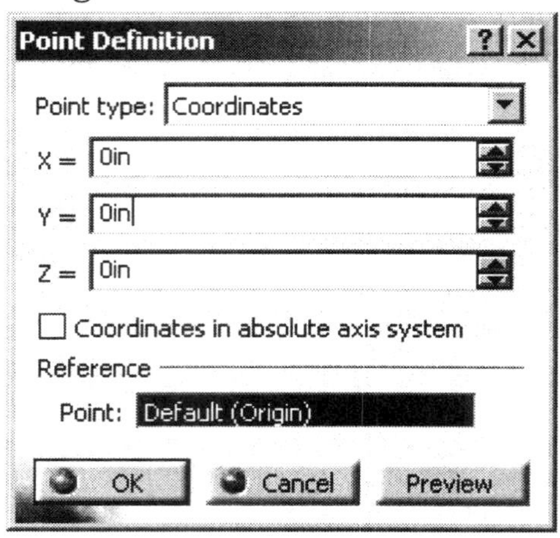

2.1.1 Select the **Point** tool. This brings up the **Point Definition** window as shown in Figure 7.3.

2.1.2 Select the **Point Type** box. This displays all the different methods available for creating points.

2.1.3 Select the **Coordinates** option.

2.1.4 For the **X** value leave it at the default value of 0.

2.1.5 For the **Y** value leave it at the default value of 0.

2.1.6 For the **Z** value leave it at the default value of 0.

2.1.7 Select **OK**. This will create a point at the 0,0,0 location as shown in Figure 7.4.

2.1.8 Notice where the newly created point appears, the intersection of the three planes. Now that you have proven the location of the 0,0,0 location, lets move on.

Figure 7.4

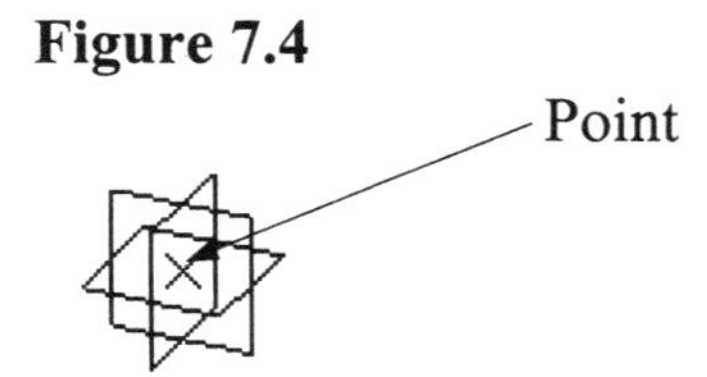

2.2 The first axis system you will create is at the original 0,0,0 location, which is the intersection of the three planes and the location of the point you just created and will be the absolute axis system. To create an axis at the original 0,0,0 location, complete the following steps.

2.2.1 Select the **Axis** tool at the bottom of the screen. This will bring up the **Axis System Definition** window as shown in Figure 7.5.

2.2.2 Leave **Standard** as the default in the **Axis System Type** box.

2.2.3 Select the **Origin** box. With the box highlighted, select the point you just created at the 0,0,0 location. **Point.1** should now appear in the **Origin** box as shown in Figure 7.5. This locates the origin of the axis.

2.2.4 Leave **No Selection** as the default in the **X Axis** box.

2.2.5 Leave **No Selection** as the default in the **Y Axis** box.

2.2.6 Leave **No Selection** as the default in the **Z Axis** box.

2.2.7 Select the **OK** button to create the axis.

Figure 7.5

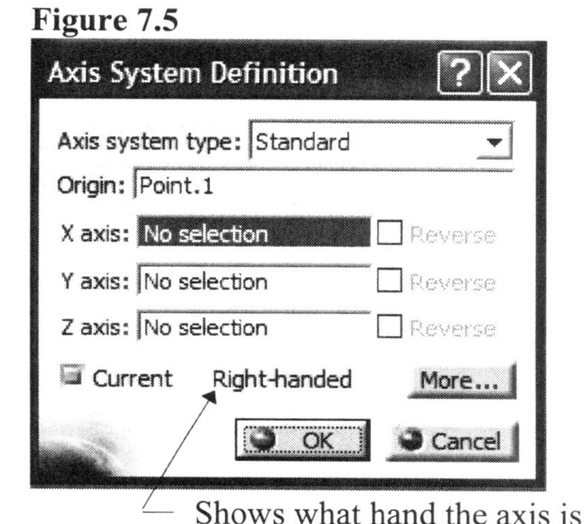

NOTE: You can reverse the direction of any of the axis by selecting that particular axis or by selecting the **Reverse** box to the right of each axis value box in the **Axis System Definition** window. It is important for you to know that changing the direction of a particular axis in relationship to one of the other two axis systems could change the axis system from a right hand axis system to a left hand axis system, reference Figure 7.6. The **Axis System Definition** window displays the type of axis being created.

Figure 7.6

If you want the X axis pointing in the direction of the Y axis, you must select the **X Axis** box and then select a geometrical entity that will indicate the direction of the axis. Since there are no other entities to select, you can select the plane that is normal to the direction you want the axis to point. If you define the direction of just one axis, CATIA V5 will rotate the other axis systems to create a right hand axis. To create a left hand axis, you must define the direction of the other two axis systems. Reference the **Axis System Definition** window. The bottom middle of the window will indicate **Right Hand** or **Left Hand** (reference Figure 7.5).

2.2.8 When you get the axis system oriented as shown in Figure 7.7, you are ready to go on to the next step.

Figure 7.7

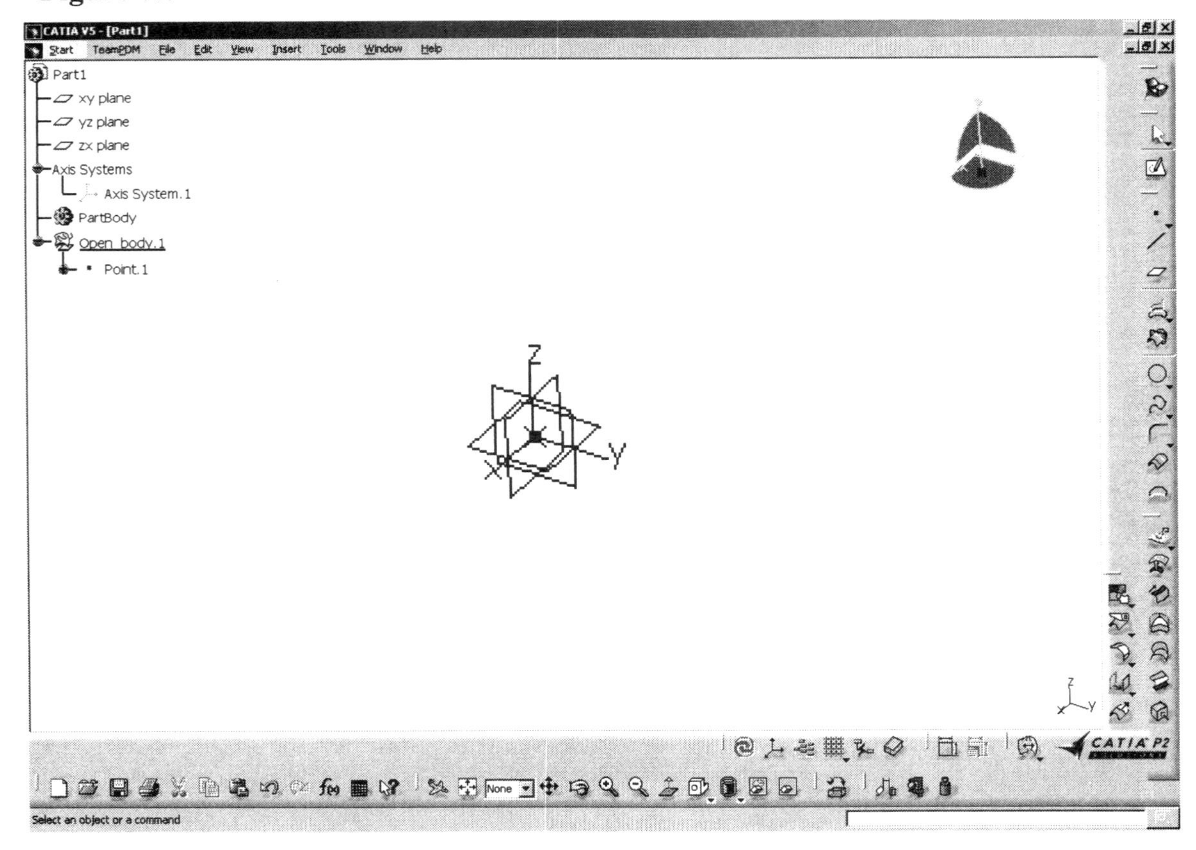

2.3 The "**Sheetmetal Bracket**" you are going to create in this section of the lesson will start at the 10,0,10 location. Some of the points will be created using X,Y and Z coordinates. To make the point creation easier, you will need to create a local axis system at the 10,0,10 location. Having a local axis system will save you from having to calculate the points from the original 0,0,0 point in the absolute axis system. Create the local axis system by completing the following steps.

2.3.1 Create a point in the 10,0,10 location using the **Point** tool and the same process used in Step 2.1. The point will be used to locate the **Origin** of the local axis. (You may need to select the **Fit All In** tool to see the newly created axis.)

Figure 7.8

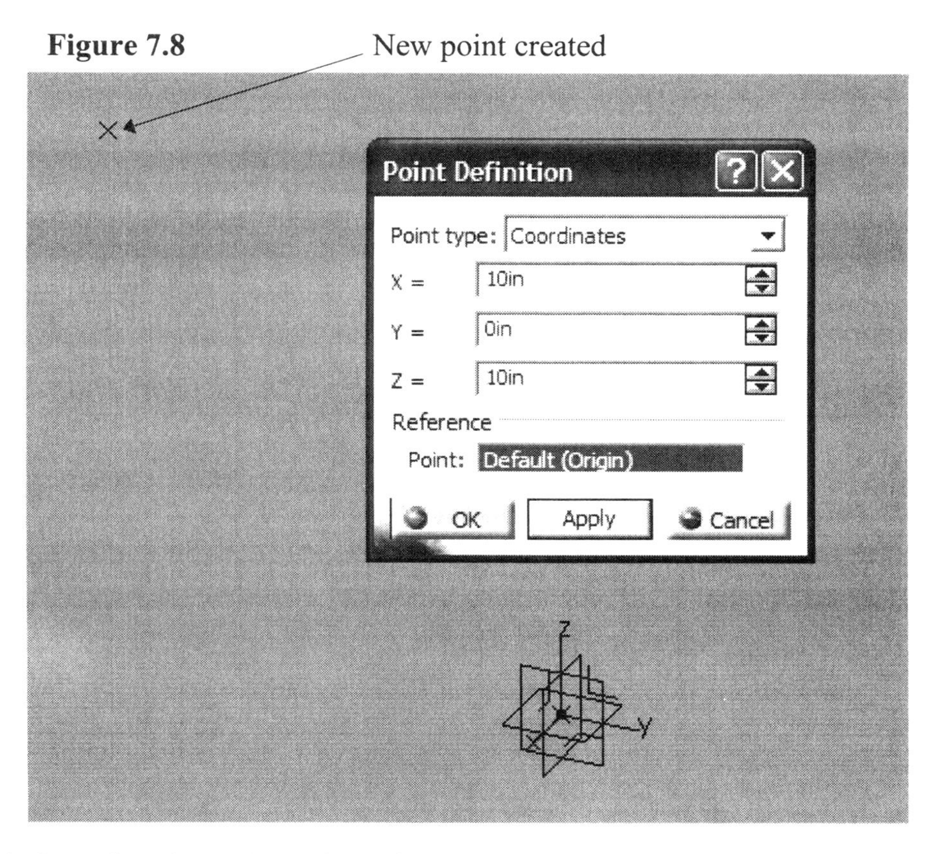

2.3.2 Complete Steps 2.2.1 through 2.2.7 to create a local axis system. The only difference in creating this axis system will be the **Origin**. Select the point created in Step 2.3.1 for the **Origin**.

2.3.3 Default values can be used for all of the other variables in the **Axis System Definition** window. Basically you are creating the same axis; the only difference is the **Origin**.

2.3.4 If your newly created axis system looks similar to Figure 7.9, you are ready to move on to the next step

Figure 7.9

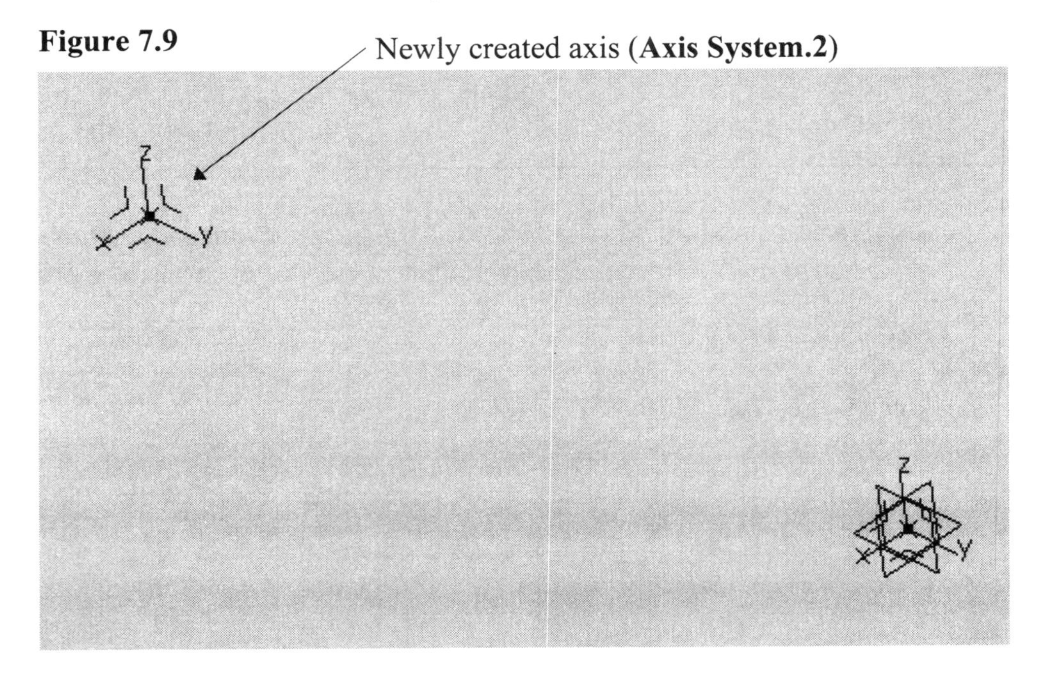

2.4 You now have two axis systems created. If you created a new point, the coordinates you assign to the point would be relative to the axis system that is current. In this context current means active. How do you tell what axis system is current? How do you make an axis system current? Double click on the axis you want to make current. This will bring up the **Axis System Definition** window. The bottom left of the **Axis System Definition** window has a toggle box labeled **Current**. By selecting this box, you make that particular axis system current. This means, if you create a point at the 0,0,0 location, it will be created in relation to that axis system. If you unselect it, it will be an inactive axis system. If you leave this box blank, it will default to the previous axis. For this section of the lesson, make the local axis system current (**Axis System.2**).

Figure 7.10

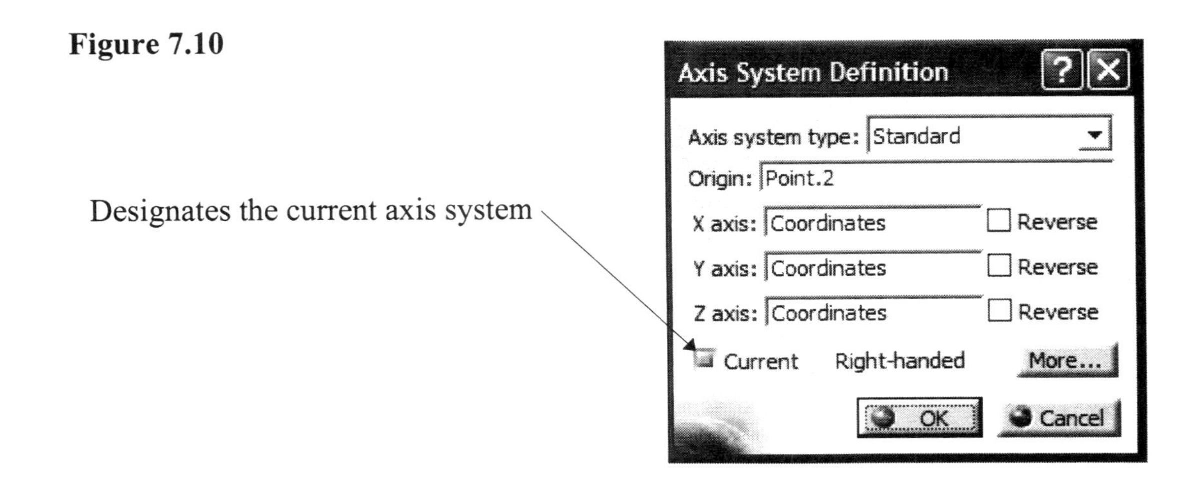

You can verify which axis system is current by using the **Specification Tree**. Figure 7.11 shows **Axis System.1** and **Axis System.2** in the **Specification Tree**. Notice the difference between the two axis system icons. One is thin and dark while the other is thick and orange in color. The thick orange axis system icon is the current axis system. Double clicking on the axis system in the **Specification Tree** or on the **Axis** its self can change this. This will bring up the **Axis System Definition** window where you can toggle on or off the **Current** box as described in Step 2.4.

Figure 7.11

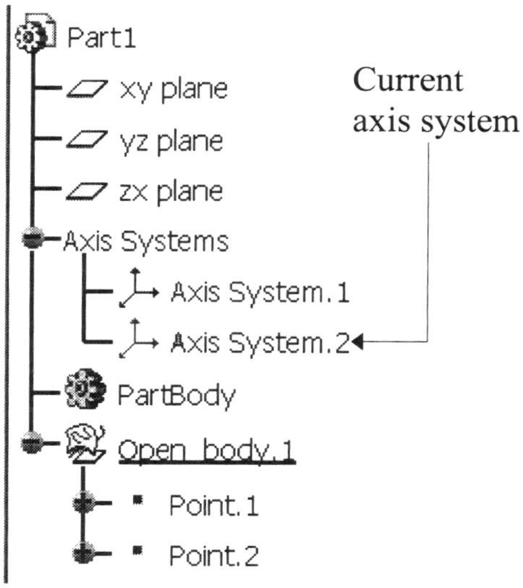

Part1
xy plane
yz plane
zx plane
Axis Systems
Axis System.1
Axis System.2
PartBody
Open_body.1
Point.1
Point.2
Current axis system

2.5 Your **Specification Tree** now has a branch labeled **Axis System.1** and a branch labeled **Axis System.2**. It will not take long to forget which axis is which. It would be a good idea at this point to rename the two axis systems so that they retain some logical meaning. This is done the same as renaming any other entity in the **Specification Tree**. Complete the following steps to rename **Axis System.1** and **Axis System.2**.

2.5.1 Highlight **Axis System.1** in the **Specification Tree**.

2.5.2 Push the right mouse button. This brings up the **Edit** window (Figure 7.12).

2.5.3 Select the **Properties** option.

2.5.4 In the **Properties** window select the **Feature Properties** tab (Figure 7.13).

2.5.5 Edit the **Feature Name** box. Change the name from **Axis System.1** to **Main Axis**.

2.5.6 Select the **OK** button to complete the task.

2.5.7 Complete the same steps as described above to change **Axis System.2** to **Sheetmetal Bracket Axis**.

Figure 7.12

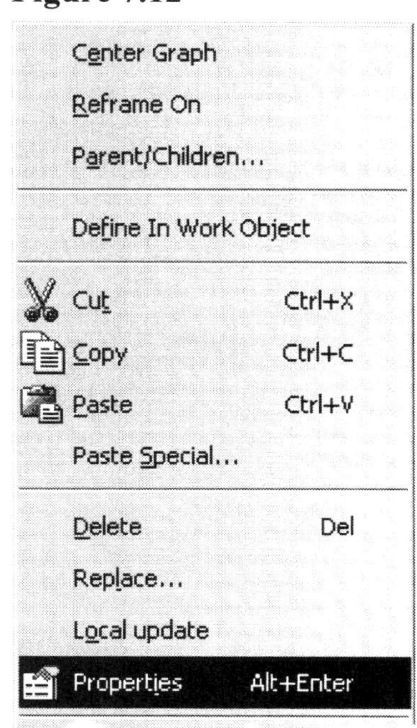

With the two axis systems created and renamed and the **Sheetmetal Bracket Axis** selected as the current axis system, you are ready to start creating wireframe geometry.

Figure 7.13

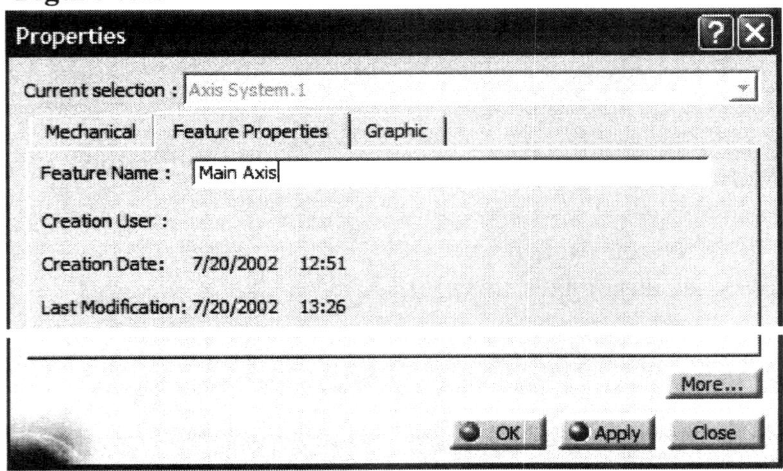

3 Creating Points

You must first determine where the part will be created in relationship to the 0,0,0 point and how it will be oriented. After the location and orientation has been decided you can start creating wireframe geometry. For this section of the lesson you will need to create three points. Create the points by completing following steps.

3.1 First, verify that the **Sheetmetal Bracket Axis** is the current axis, reference Step 2.

3.2 Using the **Point** tool, create the three points using the coordinates below. Instructions to creating points are covered in detail in Step 2.1.

Point 1 (0,0,0) already created in Step 2.1
Point 2 (1,0,0)
Point 3 (0,0,2)

NOTE: The **Point Definition** window has changed slightly. A toggle box has been added (**Coordinates In Absolute Axis System**). This option is not used in this lesson.

NOTE: The newly created points are added to the **Specification Tree**. They will be labeled as **Point.3** and **Point.4**. You have already created **Point.1** and **Point.2** in the previous steps. If the points are not visible in the **Specification Tree**, you may need to select the plus sign to the left of the branch. This should expand the **Specification Tree** so all of the points are visible. Rotate the screen to get a better view of the points. Your screen should look similar to the one shown in Figure 7.14.

Figure 7.14

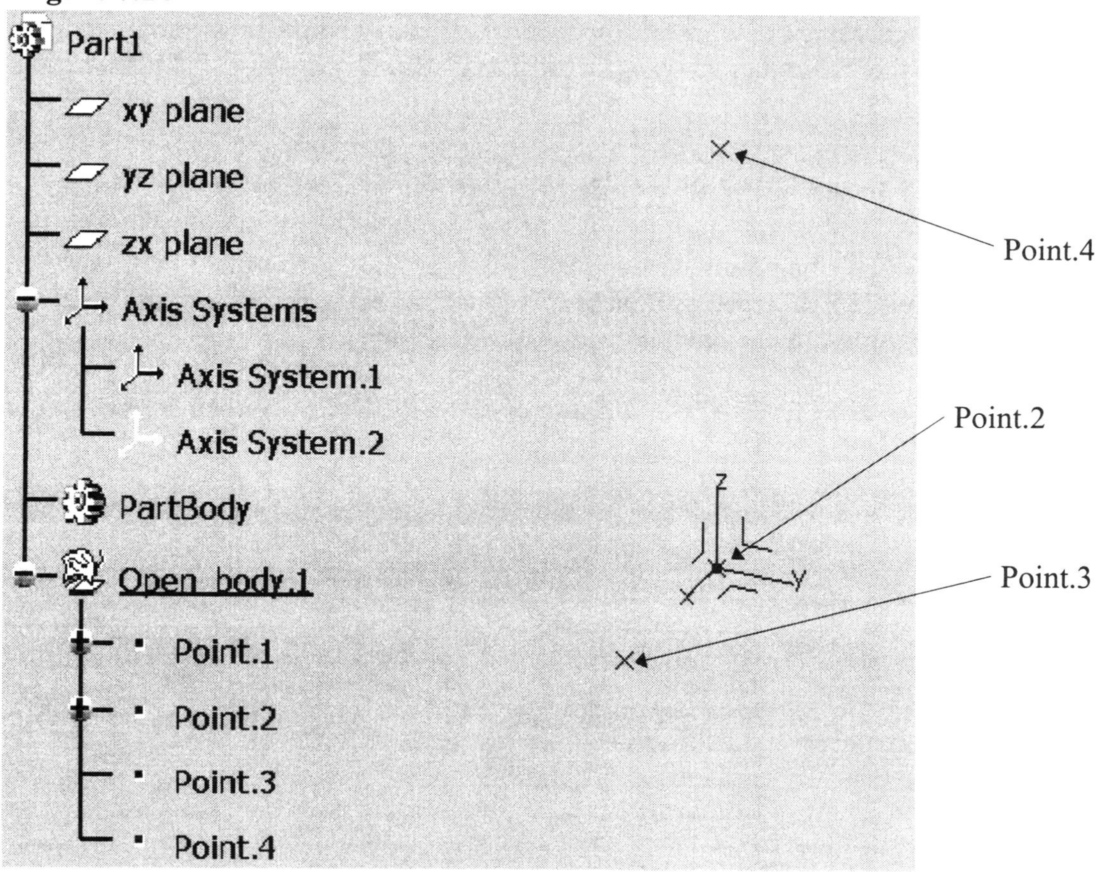

3.3 Rename the three points using the same process described in Step 2.6. Name the points accordingly.

Point at 0,0,0 = **First Point**
Point at 1,0,0 = **Second Point**
Point at 0,0,2 = **Third Point**

After renaming the points, the **Specification Tree** should look similar to Figure 7.15.

Figure 7.15

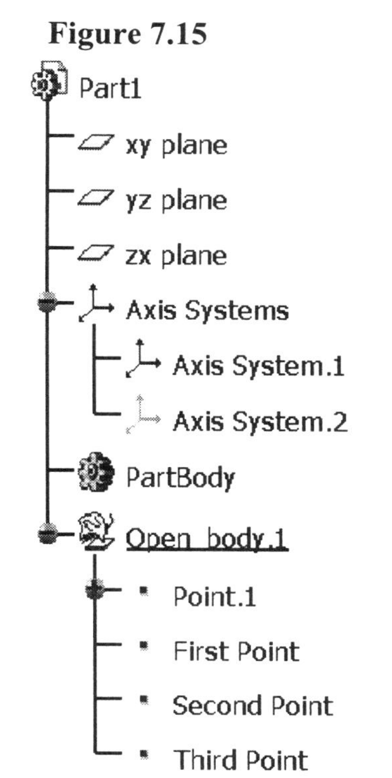

3.4 The point location can be modified by double clicking on any one of the points in the **Specification Tree**. This will bring up the **Point Definition** window for that particular point. The **Point Definition** window shows the point X, Y and Z values; it is in this window that you can edit the X, Y and Z values. You can also expand the **Point** branch out to the individual X, Y and Z values. Double clicking on the individual value branches will bring up the **Edit Parameter** window. You can change the individual value in this window also.

4 Creating Lines Using The Point–Point Method 

Now you will use the points created in Step 3 to define the lines that make up the wireframe of the "**Sheetmetal Bracket**." Remember, the points and lines you are creating are three-dimensional, they are not **Sketcher** entities.

4.1 Select the **Line** tool. This will bring up the **Line Definition** window as shown in Figure 7.16.

Figure 7.16

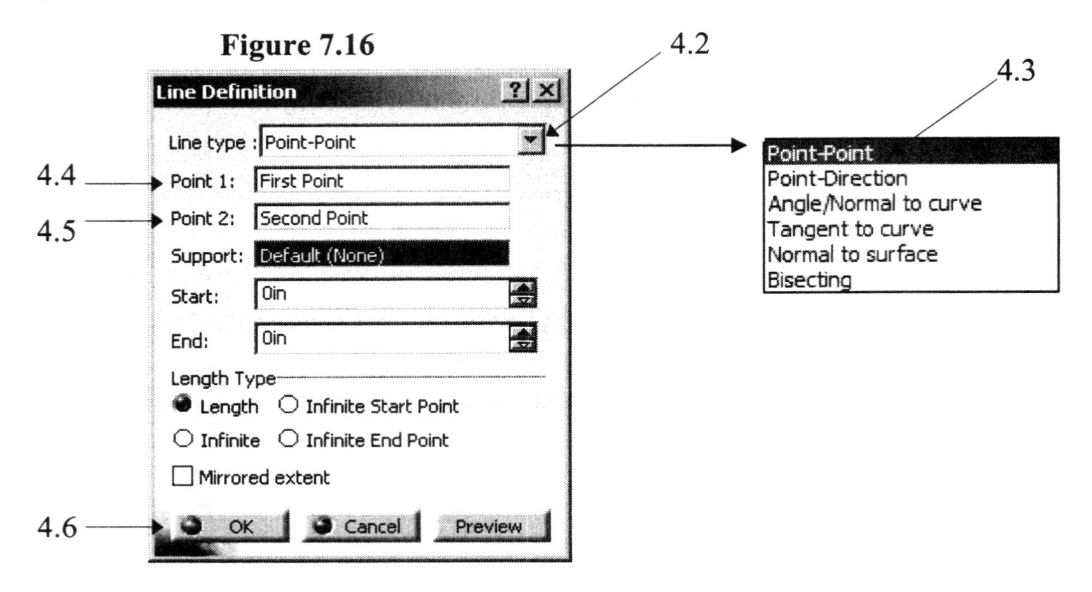

4.2 The first box in the **Line Definition** window is the **Line Type** box. Select the down arrow to the far right of the **Line Type** box. This will display the six different options (methods) for creating a line.

4.3 Select the **Point-Point** method. This selection requires only two other inputs by you the user, **Point 1** and **Point 2**. The starting point of the line and the ending point of the line.

4.4 For **Point.1** select the point you created and renamed as **First Point** in Step 3. The point may be selected graphically or by selecting it from the **Specification Tree**.

4.5 For **Point 2** select the point you created and renamed as **Second Point** in Step 3.

4.6 Select the **OK** button to create **Line.1**. The **Support**, **Start** and **End** boxes are optional and not required to create a **Point-Point** type line.

4.7 Create **Line.2** using the **Point-Point** method. Use the **First Point** and **Third Point** for the **Point 1** and **Point 2** input boxes.

If your wireframe looks similar to the one shown in Figure 7.17 you are ready to move on to the next step.

Figure 7.17

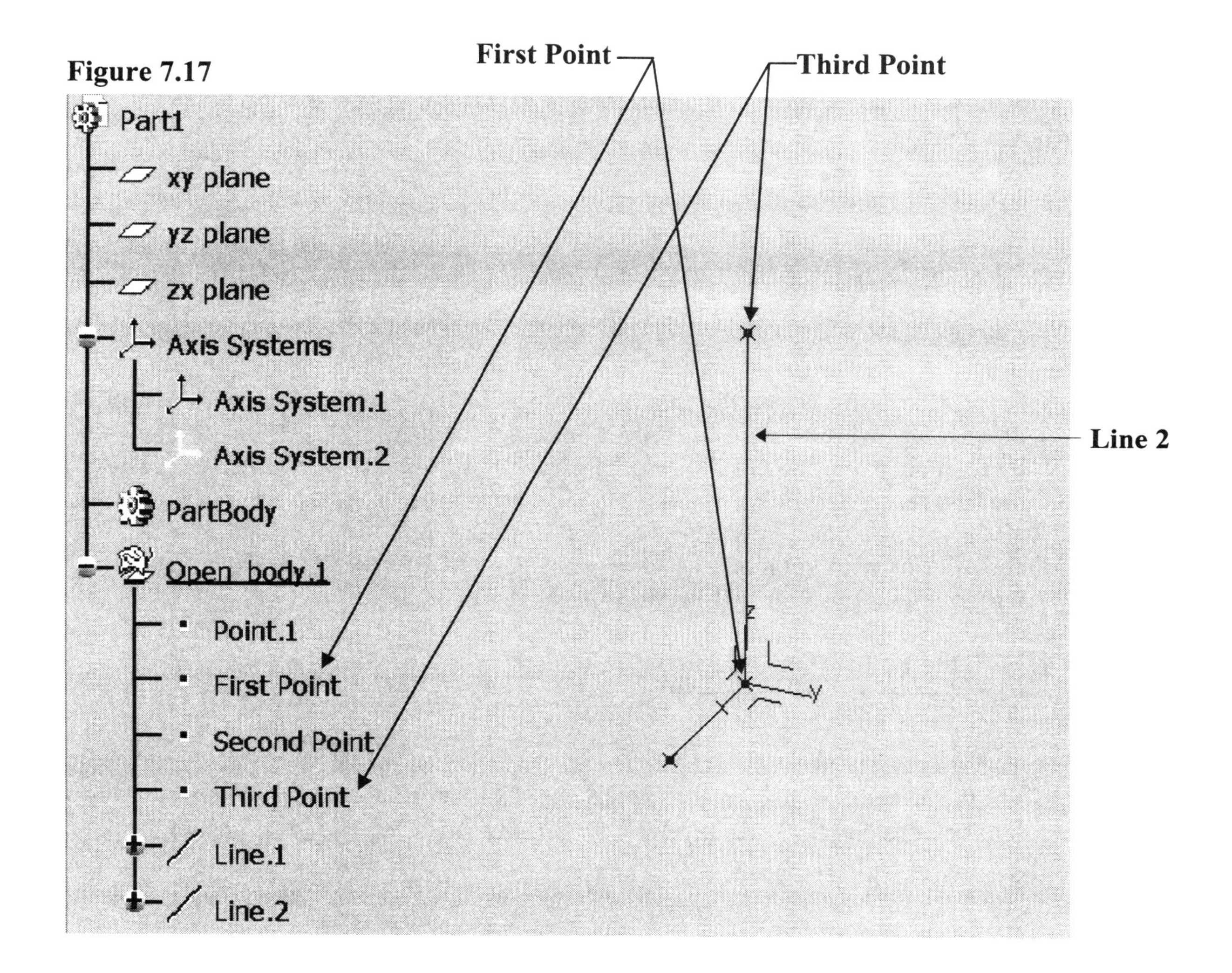

5 Creating Line 3 Using The Point–Direction Method

The **Point-Point** method in Step 4 required predefined points; points created using the **Point** tool. Points created using the **Point** tool are represented by an **X**. These points have there own branch in the **Specification Tree** and are labeled as **Points** sequentially. As you created lines using these points, you noticed that CATIA V5 added dots to both ends of the lines you created. These dots can also be used as points. You could select them as points when using the **Point-Point** line creation method. The points (dots) are found under the **Line** branch of the **Specification Tree** and labeled **Start** and **End**. Figure 7.14 shows points (X's) created using the **Point** tool, Figure 7.17 shows points (dots) created at the start and end of each line. In this step you will create **Line.3** using the **Point–Direction** method. Complete the following steps.

5.1 Select the **Line** tool.

5.2 From the **Line Definition** window, select the down arrow to the right side of the **Line Type** box. This will expand the box so it will list all six methods of creating lines.

5.3 Select the **Point–Direction** method. Notice, as you select a different method, the other boxes within the **Line Definition** window change. The **Point–Point** method only required you to select **Point 1** and **Point 2**. The **Point–Direction** method requires a **Point** (anchor point), a **Direction**, a **Start** and an **End**. It also gives you an option to **Reverse Direction** of the line. Another option is to specify **Support** geometry. Reference Figure 7.18 for the **Line Definition** window with the **Point–Direction** line type selected.

5.4 With the **Point–Direction** line type selected make sure the **Point** box is highlighted. Select **Second Point**. **Second Point** will show up in the **Point** box.

5.5 The **Direction** box will highlight as soon as the **Point** box has an entity selected. For the **Direction** box, select **Line.2**. All **Line.2** does is give a direction for the line you are creating. As Figure 7.19 shows **Line.3** is parallel to **Line.2**.

Figure 7.18

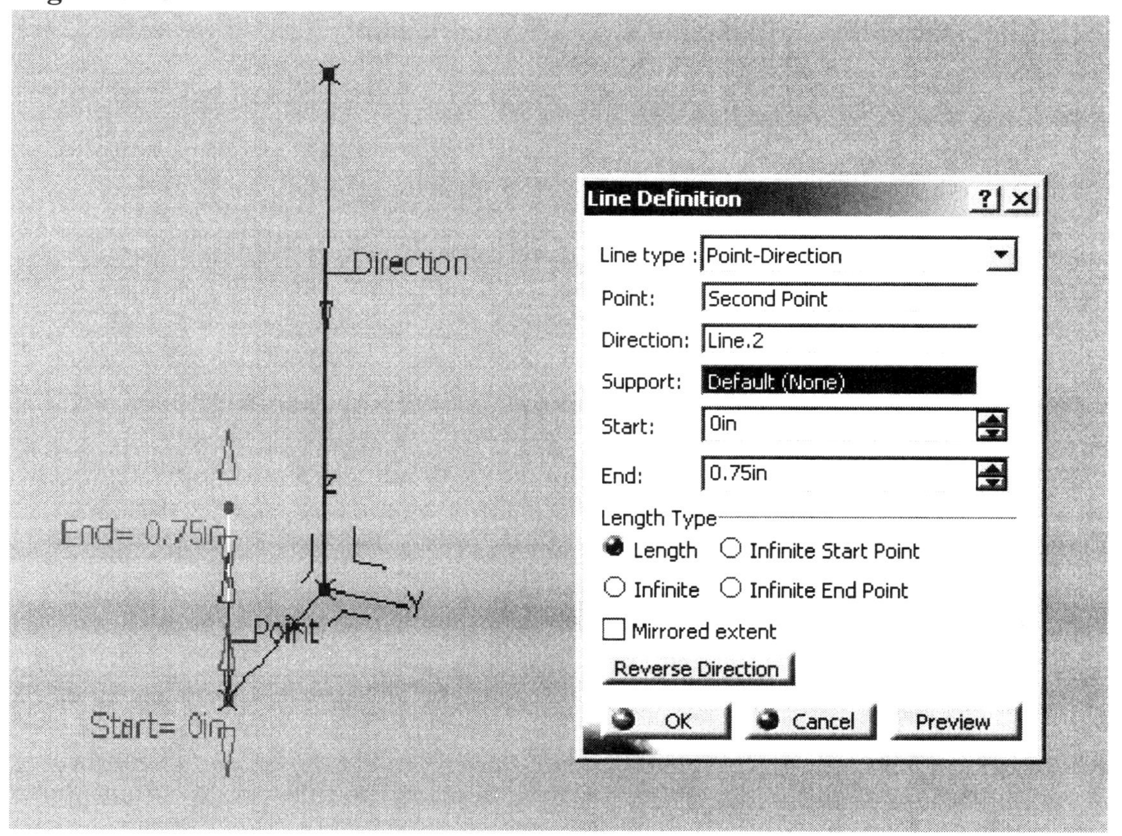

5.6 When you select the direction, CATIA V5 provides a preview of the line you are creating. Figure 7.18 shows the default **Start** point and the default **End** point. The default **Start** point is the same point you select for the **Point** box. For this section of the lesson the default **Start** point will suffice. Tab or use the mouse to move the cursor over the **End** box.

5.7 For the **End** box type in **.75 in**. This gives the length of the line.

5.8 Hit **Enter** or select **OK** to create the line. **Line.3** is now complete. Your wireframe so far should look similar to the one shown in Figure 7.19.

5.9 Use the same process to create **Lines 4**, **5**, **6** and **7**. Apply the following information to the **Line Definition** window using the **Point-Direction** method. For the **Point** box select the point at the end of the line just created. For example to create **Line.4** you will select the point at the end of **Line.3** as shown in Figure 7.19. Reference Figure 7.20 for all of the lines called out in the following table.

Figure 7.19

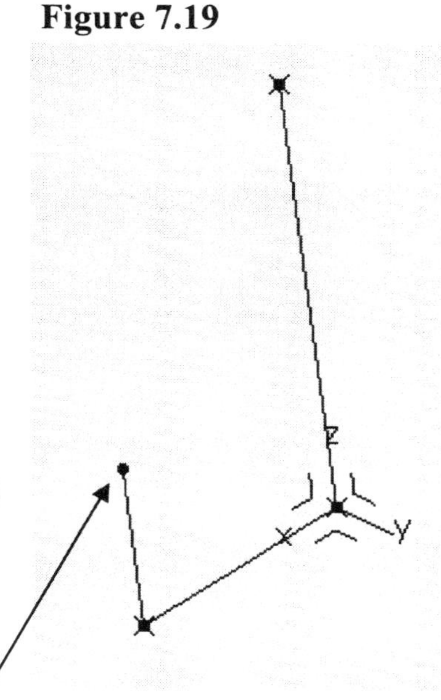

NOTE: Check the direction the line is oriented, you may need to **Reverse Direction.**

Line	Point	Direction	End (length)
Line.4	Point at the end of **Line.3**	**Line.1**	1″
Line.5	Point at the end of **Line.4**	**Line.3**	.75″
Line.6	Point at the end of **Line.5**	**Line.4**	1″
Line.7	Point at the end of **Line.6**	**Line.2**	2″

6 Creating Line 8 Using The Same Point–Direction Method 

The process to create **Line.8** using the **Point–Direction** method is almost identical to Step 5, except that the **Direction** box will use a third-dimension. Step 5 **Directions** were all on the same two-dimensional plane. There is no reference line pointing in the same direction as **Line.8**. This step will show you that other reference elements can be used in the place of a line. You will be expected to create the remaining lines for the "**Sheetmetal Bracket**" using the methods described so far. Create **Line.8** by completing the following steps. (Note**:** If you think you can create the geometry on your own skip the next couple of pages and duplicate the dimensional lines in Figure 7.21).

6.1 Select the **Line** tool.

6.2 From **Line Type** box (in the **Line Definition** window), select the **Point–Direction** method as shown in Figure 7.18.

6.3 For the **Point** box, select the point at the top of **Line.7**. If it is unclear which line is **Line.7** reference Figure 7.21.

6.4 For the **Direction** box, select the **Y Axis**. In Step 5 you had lines (existing geometry) to select from. The **Y Axis** is the only geometry that defines the direction of **Line.8**.

NOTE: This is the difference between Step 5 and Step 6. You can use almost any CATIA V5 geometry to define the direction of a line.

6.5 Reference Figure 7.20 to make sure **Line.8** is being created in the correct direction. If the default direction is opposite of what is shown in Figure 7.20, select the **Reverse Direction** button in the **Line Definition** window.

6.6 For the **End** box value type in **2in**.

6.7 If the wireframe looks similar to what is shown in Figure 7.20 select **OK** to create **Line.8**.

Figure 7.20

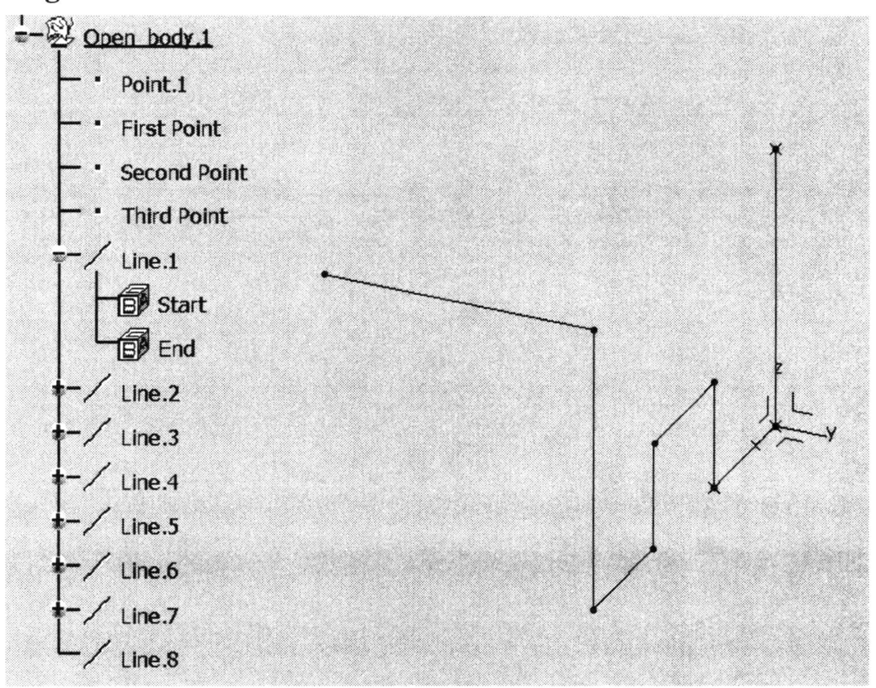

6.8 If there are any questions on the lengths of the lines created thus far use the **Measure** tool located at the bottom of the screen to verify line length.

7 Completing The Wireframe By Creating Lines 9, 10 And 11 Using The Point-Direction Method

Now that you have some experience using several variations of the **Line** tool, complete the wireframe by creating **Lines 9**, **10** and **11**. There are other methods available to create the remaining lines that weren't covered and you are encouraged to explore them. This step supplies the basic information to complete the "**Sheetmetal Bracket**" wireframe using **Point-Direction** and **Point-Point** methods.

Line	Point	Direction	End (length)
Line.9	Point at the end of **Line.8**	**Line.1**	3″
Line.10	Point at the end of **Line.9**	**Line.8**	2″
Line.11	Point at the end of **Line.10**	**Line.9**	3″

If your "**Sheetmetal Bracket**" wireframe looks similar to the one shown in Figure 7.21 you are ready to continue to the next step.

NOTE: You could have created both **Line 10** and **11** using the **Point-Point** method.

NOTE: Be sure that your completed "**Sheetmetal Bracket**" wireframe looks similar to the one shown in Figure 7.21. If your completed "**Sheetmetal Bracket**" wireframe does not look similar to the one shown in Figure 7.21 the remaining steps may not work as described.

Figure 7.21

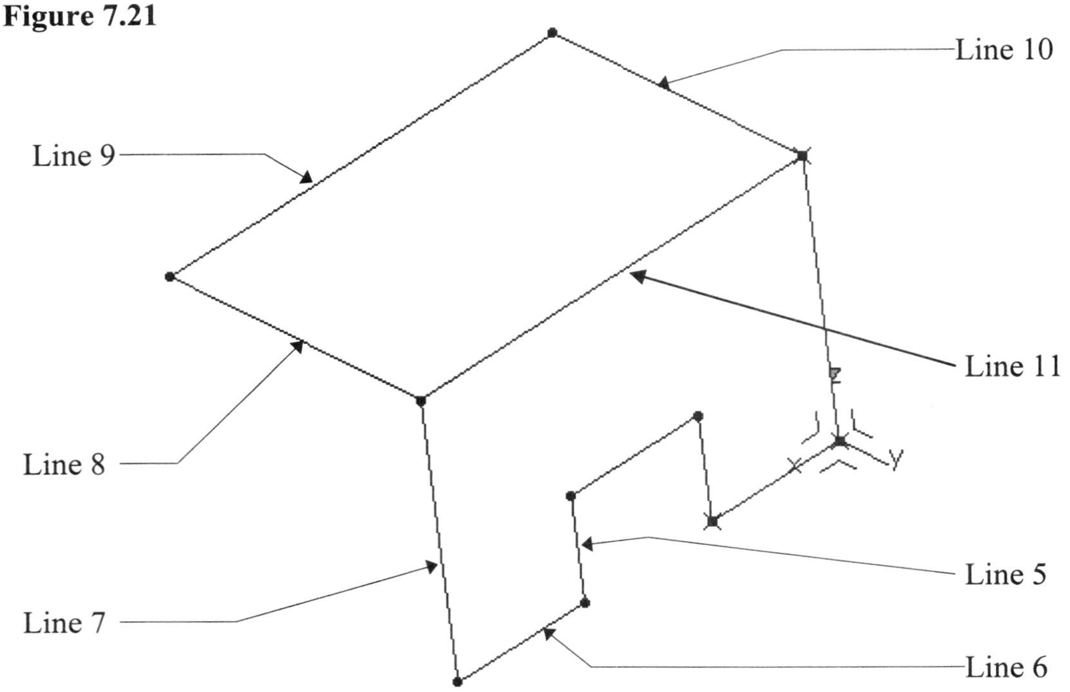

Line 10
Line 9
Line 11
Line 8
Line 5
Line 7
Line 6

8 Creating Planes

The basic wireframe is complete, but before you apply a surface to the wireframe you will need to touch it up. Planes are sometimes required for the **Corner** tool. It is more efficient to create the support entity prior to getting half way through the **Corner** process and realizing you do not have a support entity. The support entity that is best suited for the **Corner** process is the **Plane**. The **Corner** tool will be covered in Step 9. To create the **Planes** complete the following steps.

8.1 Select the **Plane** tool.

8.2 The **Plane Definition** window will appear. If you select the down arrow to the far right of the **Plane Type** box, CATIA V5 will display all of the different methods to create a plane. The method you will use in this step is the **Through Two Lines** method. Figure 7.22 displays the **Plane Definition** window along with the many different methods of creating planes.

Figure 7.22

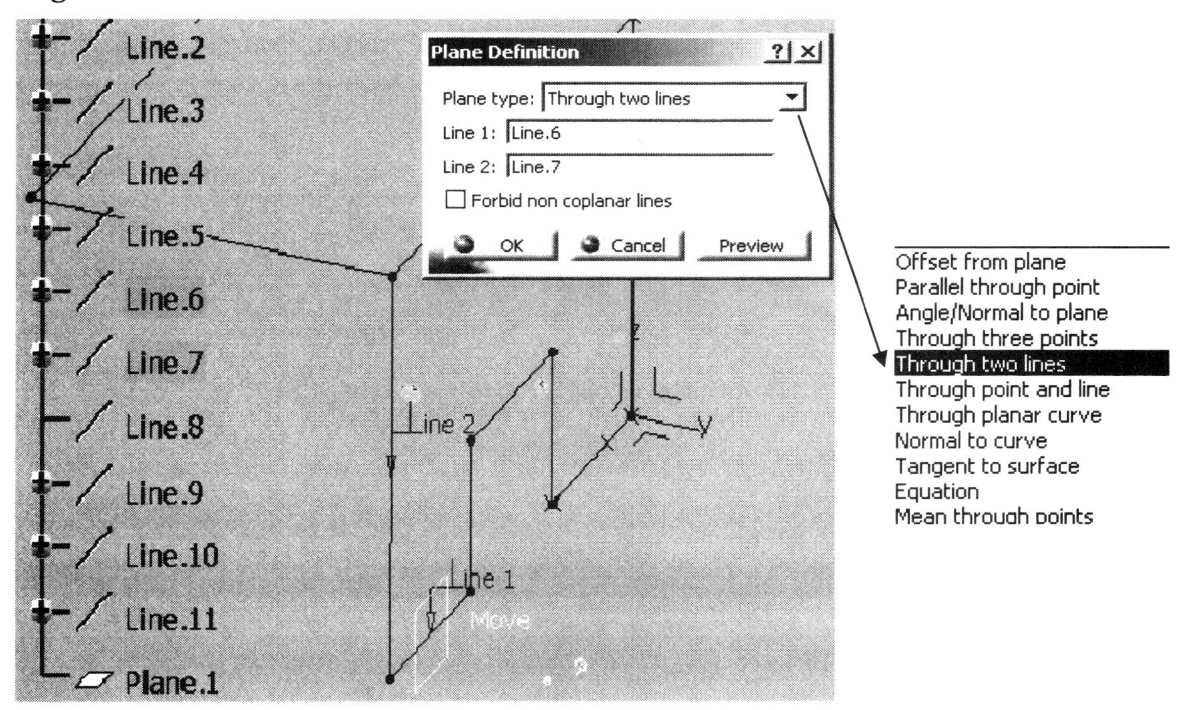

8.3 Select the **Through Two Lines** method. CATIA V5 does not require you to select the option. If you selected the **Plane** tool and then just selected the first line, CATIA V5 will automatically select the **Through Two Lines** method.

8.4 To create the first plane select **Line.6** and **Line.7** as shown in Figure 7.22. CATIA V5 will then place a plane in the window. Attached to the plane will be the word **Move**. This is prompting you to move the plane to another location. To move the plane, move mouse over the **Plane** until it turns to an orange color. Select the plane and drag it to the desired screen location. Once you have the plane placed where you want it, select the **OK** button to create the plane. The **OK** button has to be selected to complete the **Plane** creation.

8.5 CATIA V5 numerically names the planes you create i.e. **Plane.1**, **Plane.2** …and so on. When you have multiple planes on the screen, it can be difficult to know what each plane represents. It is a good idea to rename the planes so that you can readily select the correct plane. This would be a good time to rename the plane you created in Step 8.4. The name should help you understand what the plane represents. Rename the plane to **Cutout Flange**. If you do not remember how to rename an element in the **Specification Tree,** refer back to Step 2.6.

8.6 The next required plane represents the **Top Flange** of the wireframe. Create a plane using the same process as the steps described above, except select **Line.8** and **Line.9**. Rename this plane as **Top Flange**. Your "**Sheetmetal Bracket**" wireframe and **Specification Tree** should look similar to Figure 7.23.

Figure 7.23

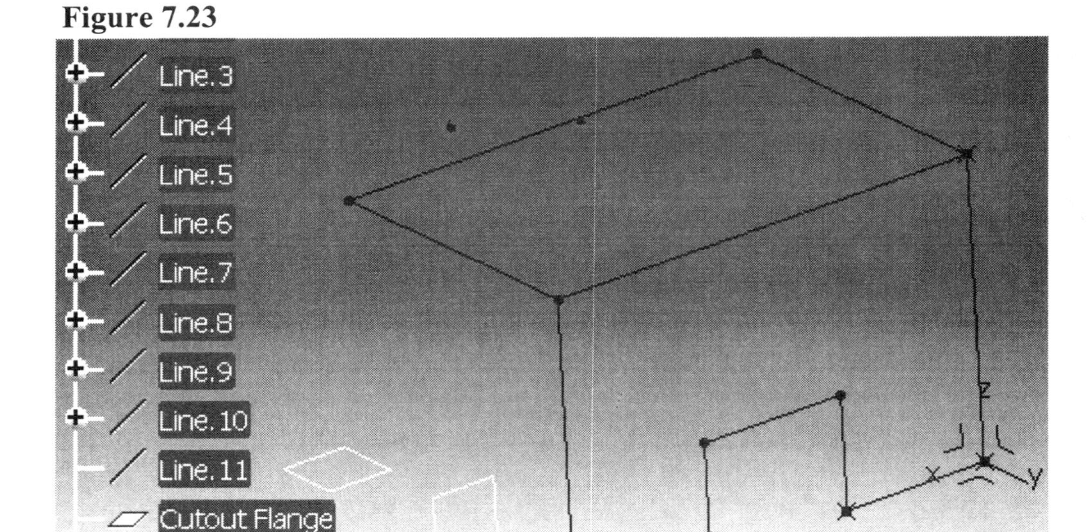

9 Creating Corners

Now that the **Support** entities (the **Planes**) have been created, you can add the **Corner** to the "**Sheetmetal Bracket**" wireframe. To create the **Corners** complete the following steps.

9.1 Select the **Corner** tool.

9.2 This will bring up the **Corner Definition** window as shown in Figure 7.24. The **Corner Definition** window has four boxes that will need to be filled in before the **Corner** can be created.

Figure 7.24

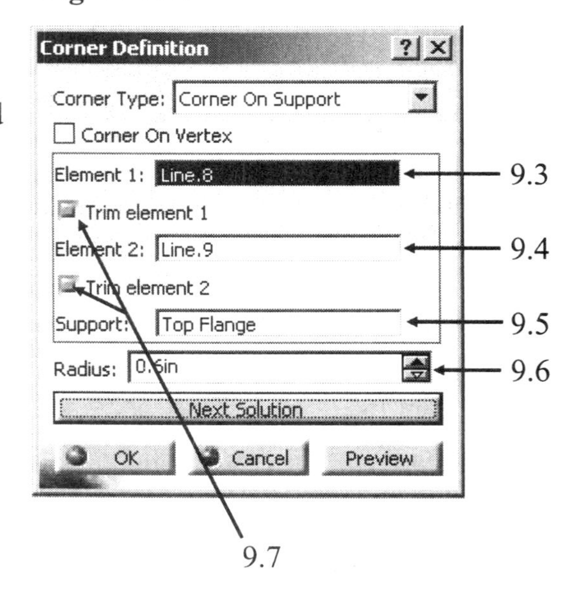

9.3 The first **Corner** you are going to create is between **Line.8** and **Line.9**. Highlight the **Element 1** box and select **Line.8**.

9.4 The **Element 2** box will automatically highlight. Verify that it is highlighted and select **Line.9**.

9.5 Make sure the **Support** box is highlighted and then select the **Plane** you renamed as **Top Flange**.

NOTE: The lines selected for **Element 1** and **Element 2** must be on the same **Plane.**

9.6 For the **Radius** box type **.6in**.

9.7 For this particular operation you will want to trim the excess lines. Select the **Trim Element** boxes. Select **OK** to create the **Corner**. This will complete your first **Corner** operation. If your "**Sheetmetal Bracket**" wireframe looks similar to the one in Figure 7.25, you are ready to create the other three **Corners**. Notice that a **Corner** branch has been added to the **Specification Tree** and that **Line.8** and **Line.9** are now part of that **Corner** branch. **Line.8** and **Line.9** are the parents of the **Corner**. **Note:** When the corner is created the parent elements (in this case, lines 8 & 9) are automatically placed in "Hide" mode. They are part of the specification tree and cannot be deleted.

Figure 7.25

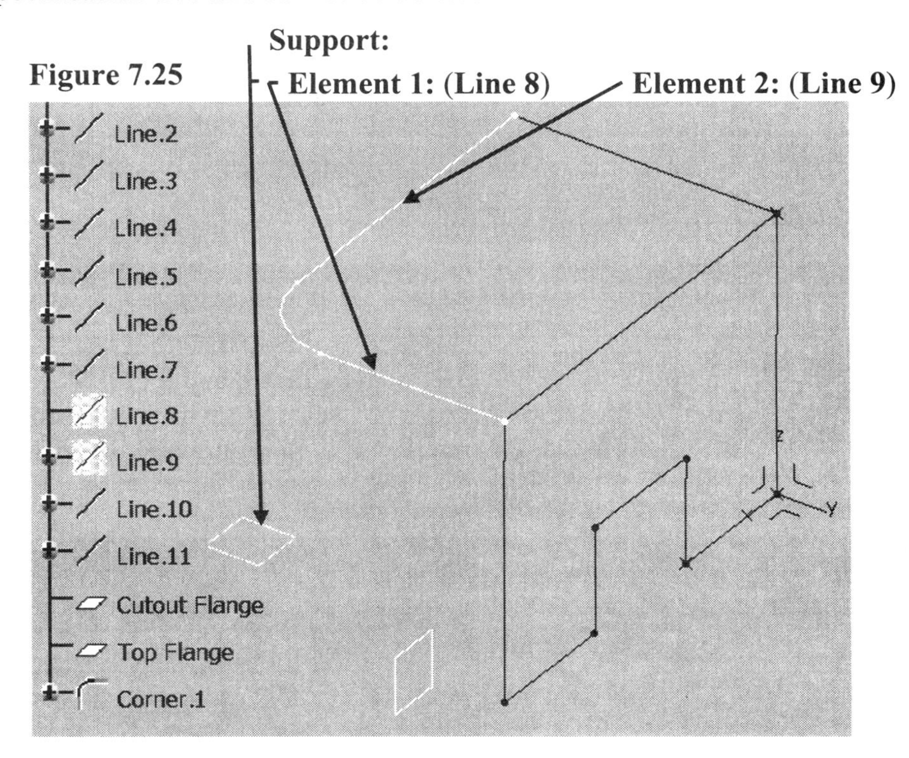

9.8 In this step you will create a **Corner** using **Line 3** and **Line 4**. Complete the same steps used in Steps 9.1 – 9.7 except select the entities listed below.

9.8.1 **Element 1** = **Line.3**
9.8.2 **Element 2** = **Line.4**
9.8.3 **Support** = **Plane** labeled **Cutout Flange**

9.8.4 **Radius** = **.25in**
9.8.5 **Trim Element** = Check both boxes
9.8.6 Select **OK** to create the **Corner**

9.9 In this step you will create a **Corner** using **Corner.2** and **Line.5**. Complete the **Corner** using the entities listed below.

9.9.1 **Element 1** = **Corner.2**
9.9.2 **Element 2** = **Line.5**
9.9.3 **Support** = **Plane** labeled **Cutout Flange**
9.9.4 **Radius** = **.25in**
9.9.5 **Trim Element** = Check both boxes
9.9.6 Select **OK** to create the **Corner**

9.10 In this step you will create a **Corner** using **Corner.1** and **Line.10**. Complete the **Corner** using the entities listed below.

9.10.1 **Element 1** = **Corner.1**
9.10.2 **Element 2** = **Line.10**
9.10.3 **Support** = **Plane** labeled **Top Flange**
9.10.4 **Radius** = **.6in**
9.10.5 **Trim Element** = This time **Do Not** check the boxes
9.10.6 Select **OK** to create the **Corner.**

9.11 If your "**Sheetmetal Bracket**" wireframe looks similar to Figure 7.26 you are ready for Step 10.

NOTE: When you created a **Corner** using **Lines 8** and **9** both lines became **Corner.1**. You can prove this by selecting what was formally known as **Line.8** and see that **Corner.1** will highlight on the **Specification Tree**.

Figure 7.26

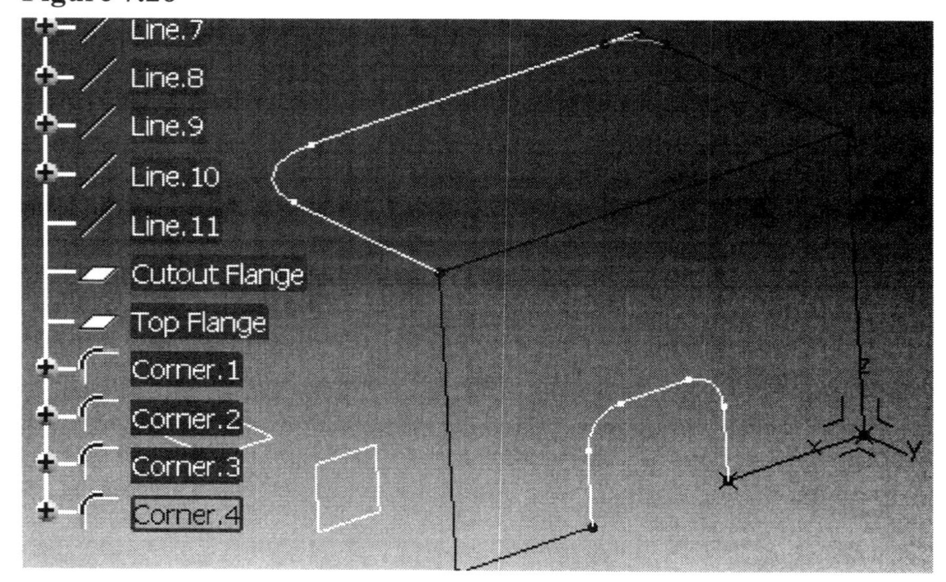

10 Using The Split Tool 

In the previous step you did not trim **Lines 9 & 10**. In this step you will manually trim the lines using the **Split** tool. To **Split** the **Corner** and **Lines 9 & 10,** complete the following steps.

10.1 Select the **Split** tool.

10.2 This will bring up the **Split Definition** window. Select **Corner.1** for the **Element To Cut** box. Reference Figure 7.27 for **Corner** selection and **Split Definition** window.

Figure 7.27

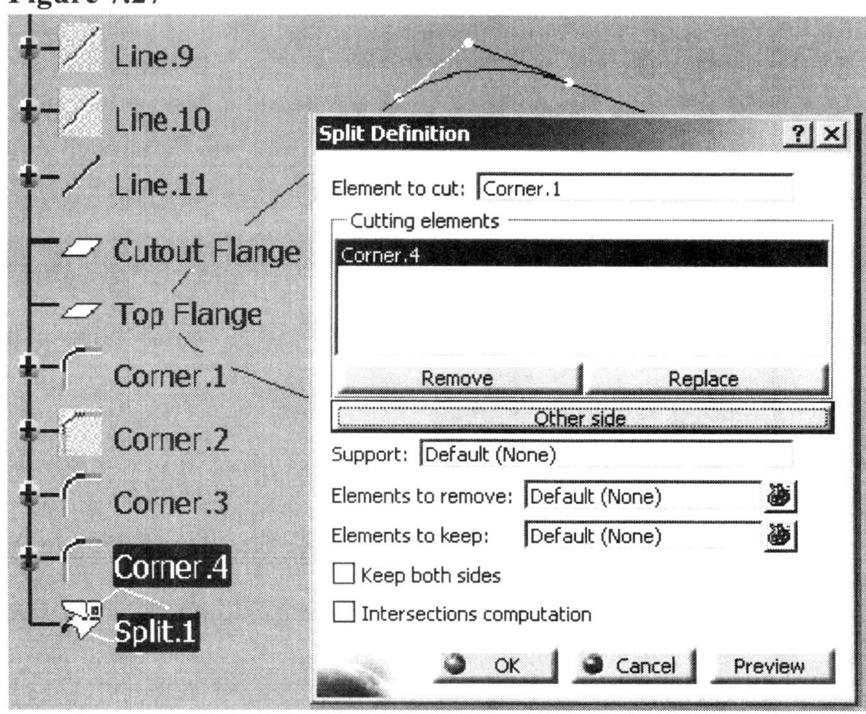

10.3 Select **Corner.4** for the **Cutting Elements**.

NOTE: The line that is highlighted during the **Split** process is the line that will remain. The line on the other side of the **Cutting Element** will be trimmed unless you select the **Other Side** button that will swap the selection of line to keep and line to trim. Remember the parent elements are not deleted, but merely placed into "**Hide**" mode.

10.4 Select **OK** to complete the **Split**.

NOTE: The **Split** tool trimmed of the excess line.

10.5 Now you need to **Split Line.10**. To **Split Line.10,** repeat the previous step except change the **Element To Cut** to **Line.10** and the **Cutting Elements** to **Corner.4**.

10.6 Select **OK** to complete the **Split**. If your "**Sheetmetal Bracket**" wireframe looks similar to the one shown in Figure 7.28 you are ready to move on to Step 11.

Figure 7.28

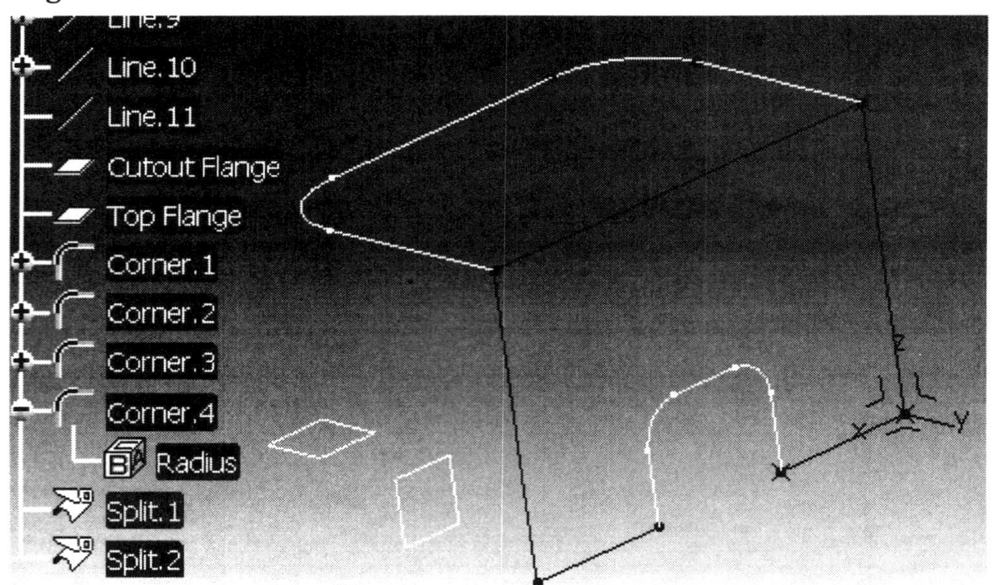

11 Adding An Elliptical Wireframe Using The Sketcher Work Bench

You may not have noticed, but the **Tool Bar** section of this lesson listed the **Sketcher Work Bench** as one of the tools available in the **Generative Shape Design Work Bench**. This step will demonstrate how useful it can be in creating wireframe geometry. The part in Figure 7.1 shows the "**Sheetmetal Bracket**" has an elliptical cut out on the top flange. This is the step that will define the ellipse. Since the **Sketcher Work Bench** has been covered in **Lesson 1**, instructions related to the **Sketcher Work Bench** will be less detailed. To create the ellipse, complete the following steps. Use Figure 7.30 as a reference.

11.1 Select the **Sketcher Work Bench** icon.

11.2 Select the **Plane** labeled **Top Flange**.

11.3 Select the **Ellipse** tool.

11.4 The **Prompt Zone** will prompt you to "**Click To Define The Ellipse Center**." For this step, select near the center of the flange. An exact location is not required at this time. The exact location along with the exact ellipse size can be modified and constrained later.

11.5 The next prompt is "**Click To Define The Major Axis And Ellipse Orientation**." The major axis is parallel with **Line.9**, so select an approximate location near **Line.9**.

11.6 The next prompt is "**Click To Define A Point on the Ellipse**." Select a point somewhere between **Line.10** and the center of the ellipse. Your ellipse is now created.

11.7 To modify the ellipse, double click on it. This will bring up the **Ellipse Definition** window. Modify the **Major Radius** to **1.125in**.

11.8 Modify the **Minor Radius** to **.563in**.

11.9 Leave the **Angle** at **0.**

11.10 If you know or want to calculate the exact center of the top flange, you can modify the center point of the ellipse using **Cartesian** or **Polar** coordinates. Figure 7.29 shows the center point of the ellipse as located using the **Constraint** tool. You know the dimensions of the top flange, so constrain them accordingly. If the **Constraint Definition** window shows the constraint value as a reference, toggle the **Reference** box off. If your sketch looks similar to the one shown in Figure 7.29 you are ready to continue on to the next step.

NOTE: Remember, when entering values, you can use math operators. For example, if you know the diameter is 1.125in, but you were prompted for a radius value, you could use a calculator to figure the radius value or you could enter "1.125/2" and CATIA V5 will calculate and enter it for you.

Figure 7.29

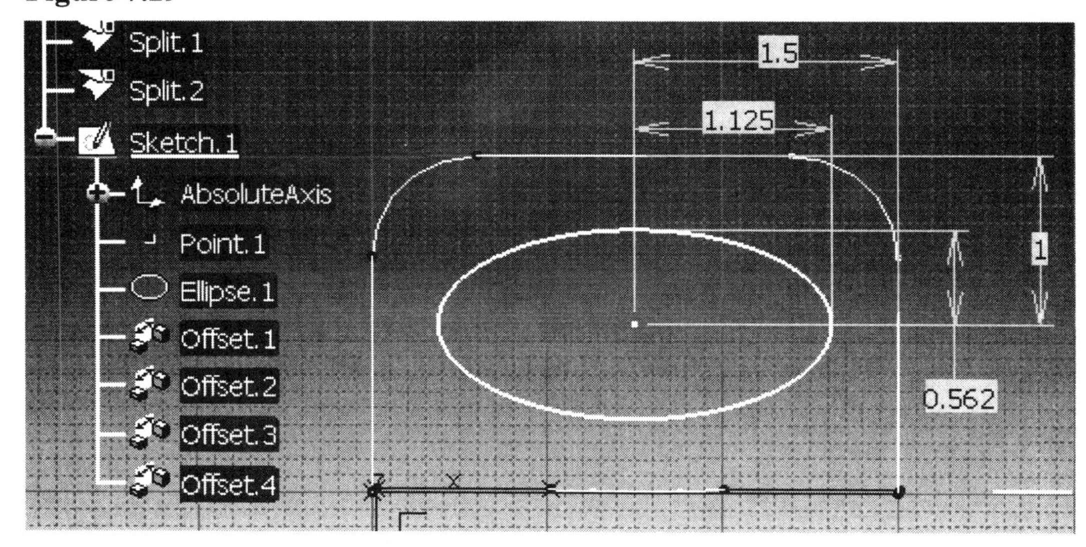

12 Appling A Surface To The Wireframe Using The Fill 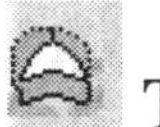 Tool

Creating the wireframe geometry was the difficult part of this lesson. If the wireframe was created correctly, applying the surface will be simple and fun. Complete the following steps to apply a surface to the "**Sheetmetal Bracket**" wireframe.

12.1 Select the **Fill** tool in the **Generative Shape Design Work Bench**.

12.2 This will bring up the **Fill Surface Definition** window as shown in Figure 7.30.

12.3 Select all of the wireframe entities that define the boundary of the top part of the "**Sheetmetal Bracket**" as shown in Figure 7.30. If you followed this lesson closely, the entities should be identified the same as the selected entities in the **Fill Surface Definition** window in Figure 7.30. If you have explored beyond what this lesson covers, your geometry may be identified a little bit differently. For example, if you created additional lines, your wireframe may be using a **Line.21** or so. The important thing in this step is to select the geometry that defines the boundary of the top flange of the "**Sheetmetal Bracket**" wireframe similar to what is shown in Figure 7.30.

NOTE: The geometry listed in **Boundary** section of the **Fill Surface Definition** window is also highlighted in the **Specification Tree** and the wireframe geometry itself.

NOTE: The geometry defining the boundary cannot have any gaps. If the boundary has gaps you will get a **Boundary Definition Error**. All boundary geometry must be connected with the previous selection. You cannot select a line, skip a line, and then select another line or you will get a **Boundary Definition Error**.

Figure 7.30

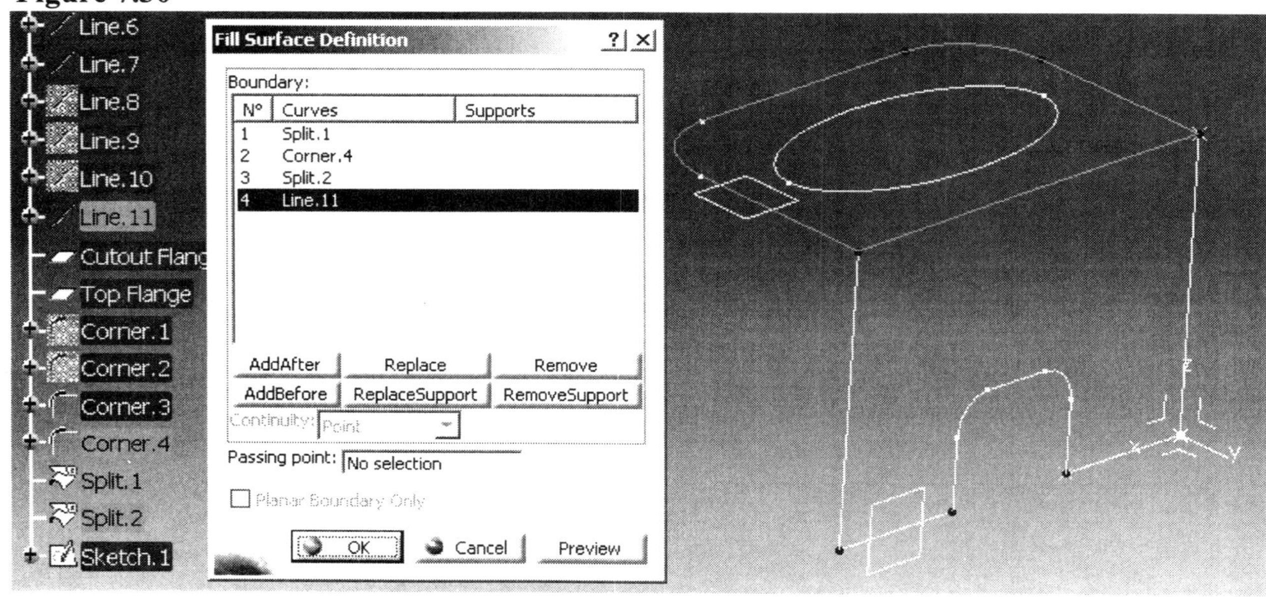

12.4 For this particular step no **Support** elements are required.

12.5 Leave the **Continuity** box at **Point**.

12.6 When you have defined the boundary as shown in Figure 7.30 select **OK** to create a surface. Notice that a **Fill.1** branch has been added to the **Specification Tree**. If your newly created surface looks similar to the one shown in Figure 7.31 you are ready to apply a surface to the **Cutout Flange**.

Figure 7.31

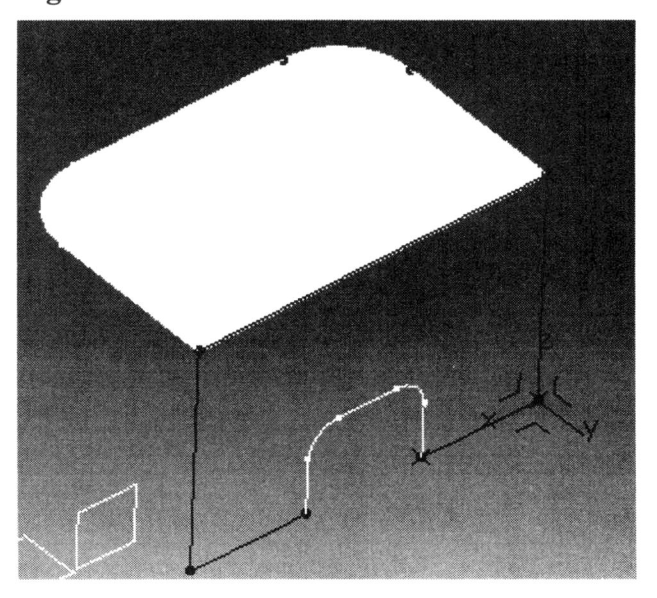

12.7 Use the **Fill** tool to apply a surface to the **Cutout Flange**. The **Boundary** should consist of **Line.1**, **2**, **11**, **7**, **6** and **Corner.3**. **Line.11** is also a shared boundary for **Fill.1**, the surface on the **Top Flange**.

12.8 If your "**Sheetmetal Bracket**" looks similar to the one shown in Figure 7.32, you are ready to move on to Step 13.

Figure 7.32

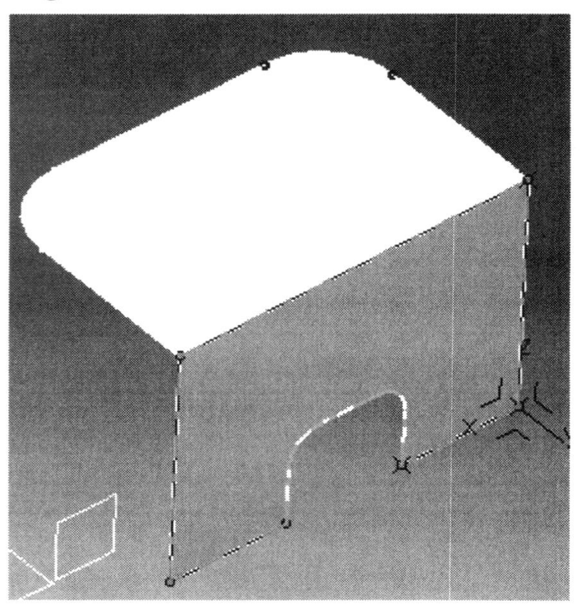

13 Creating A Fillet Between Two Surfaces 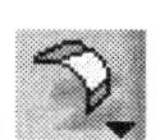

The two **Fill** surfaces you created in Step 12 are at a right angle to each other. CATIA V5 represents this with no problem, but in the manufacturing world this is not a true representation of an actual part. By adding a **Fillet** (radius) between the **Fill.1** surface and the **Fill.2** surface, your wireframe will be accurately represented. To create the **Fillet**, complete the following steps.

13.1 Select the **Shape Fillet** tool.

13.2 This will bring up the **Fillet Definition** window as shown in Figure 7.33.

13.3 Select **Fill.1** for the **Support 1** box.

13.4 Select **Fill.2** for the **Support 2** box.

13.5 Type in **.25** for the **Radius** box.

13.6 Keep the default **Smooth** for the **Extremities** box.

13.7 Verify the **Trim Support Element** boxes are selected.

13.8 Red directional arrows will appear out of the two **Support** surfaces you selected. Make sure they are pointed in the direction shown in Figure 7.33. If the arrows are pointed in the opposite direction, select the arrow to reverse the direction.

13.9 When the **Fillet Definition** window is set up as shown in Figure 7.33, select **OK** to create the **Fillet**.

13.10 The excess surfaces are trimmed, but the lines that were used to create the initial surface are not. **Line.11**, the line that shared the boundary of both **Fill.1** and **Fill.2**, and others are still visible. You **cannot** delete these elements because they help to define **Fill.1** and **Fill.2**. Since you cannot delete the supporting elements, place them in "**Hide**" mode. You will need to select the **Hide/Show** tool between each element selected or double click on it for multiple selections.

Figure 7.33

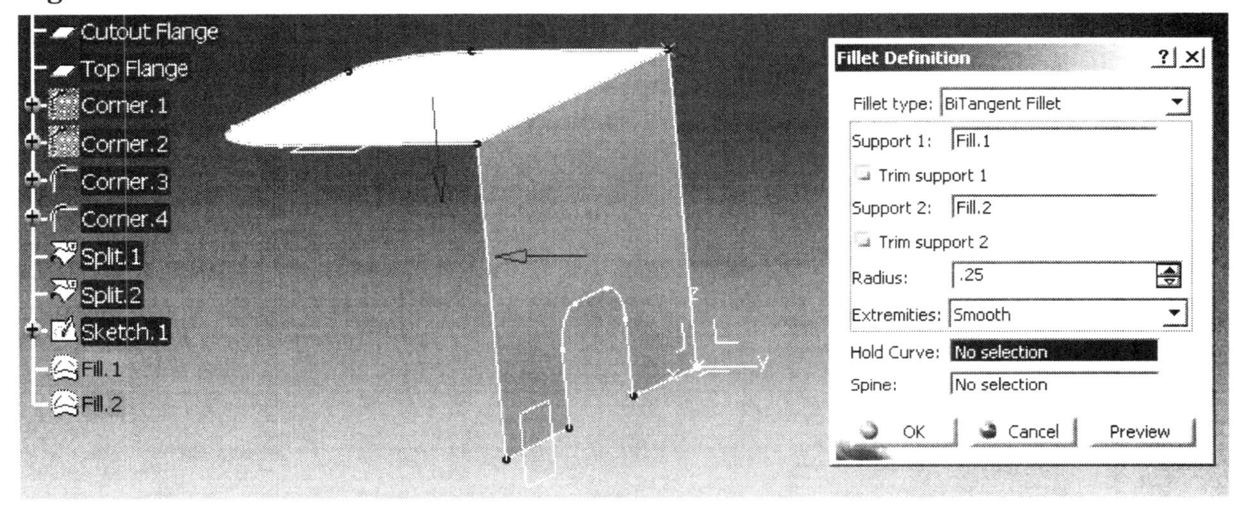

13.11 Figure 7.34 shows the "**Sheetmetal Bracket**" right after the **Fillet** operation. Figure 7.35 shows the "**Sheetmetal Bracket**" after **Hiding** the supporting elements. If your "**Sheetmetal Bracket**" looks similar to the one shown in Figure 7.35, you are ready to move on to Step 14.

Figure 7.34

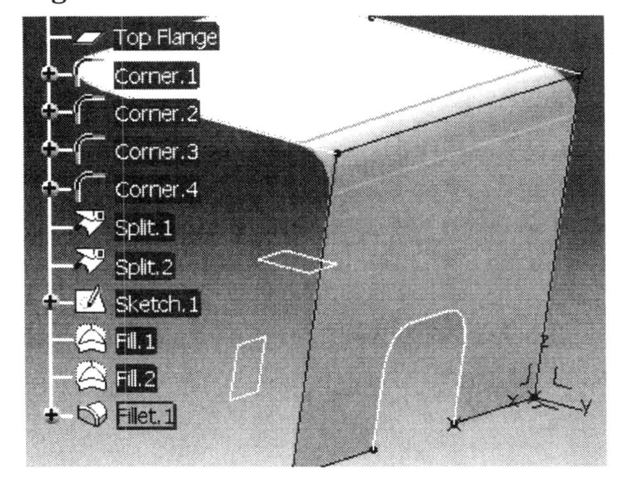

Figure 7.35

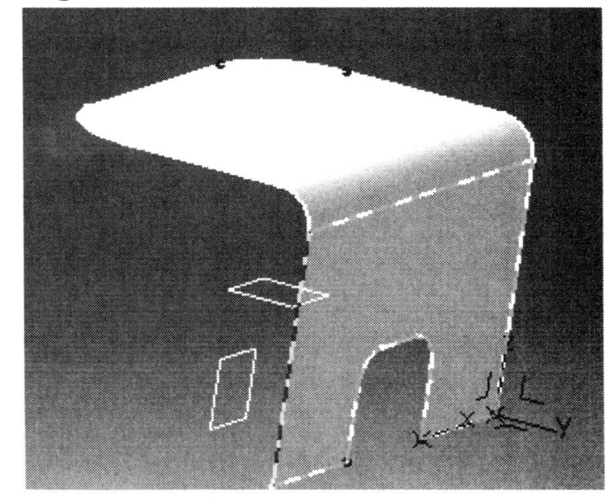

14 Creating The Elliptical Cutout Using the Split Tool

In this step you will use the sketch you created in Step 11 to remove the surface inside of the **Ellipse**. The tool required to complete this task is the **Split** tool. To remove the surface from inside of the ellipse, complete the following steps.

14.1 Select the **Split** tool. Notice this tool is a variation of the **Trim** tool. Do not confuse them: they are similar but produce different results.

14.2 This will bring up the **Split Definition** window. Select **Fillet.1** for the **Element To Cut** box.

14.3 Select **Sketch.1** (ellipse) for the **Cutting Element** box.

14.4 In this case you want to remove the surface inside of the ellipse, so do not select the **Keep Both Sides** box.

14.5 If the dimmed surface is the same as shown in Figure 7.36, you will not have to select the **Other Side** button. If the dimmed surface is on the outside of the ellipse then you will have to select the **Other Side** button.

14.6 If you have everything selected as shown in Figure 7.36, you can select the **OK** button to create the **Split**.

Figure 7.36

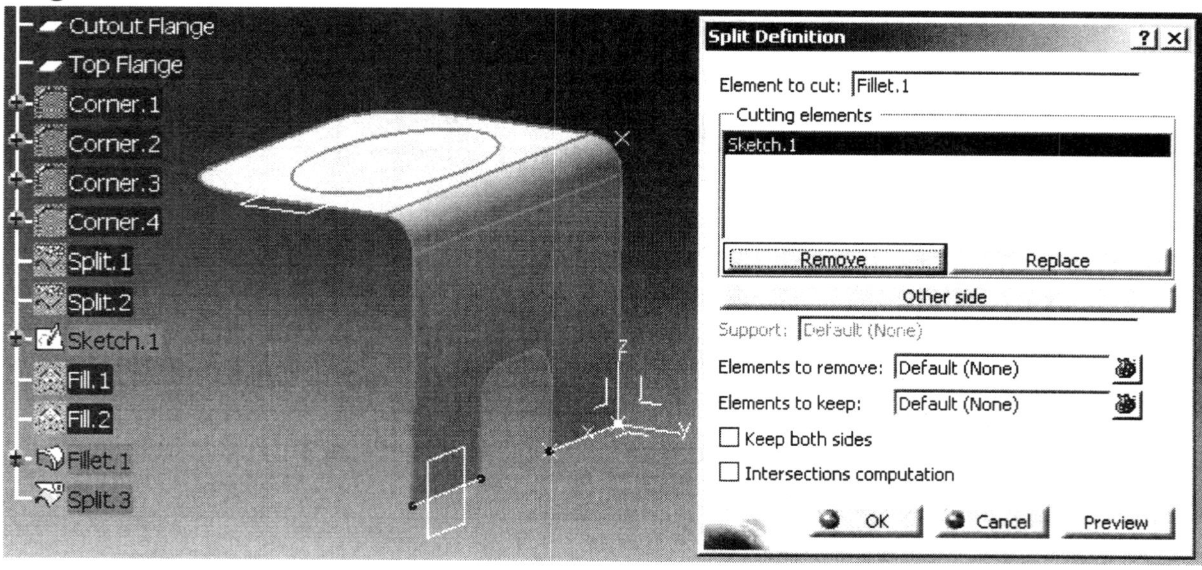

14.7 If your "**Sheetmetal Bracket**" looks similar to the one shown in Figure 7.37, you are ready to move on to the next step.

Figure 7.37

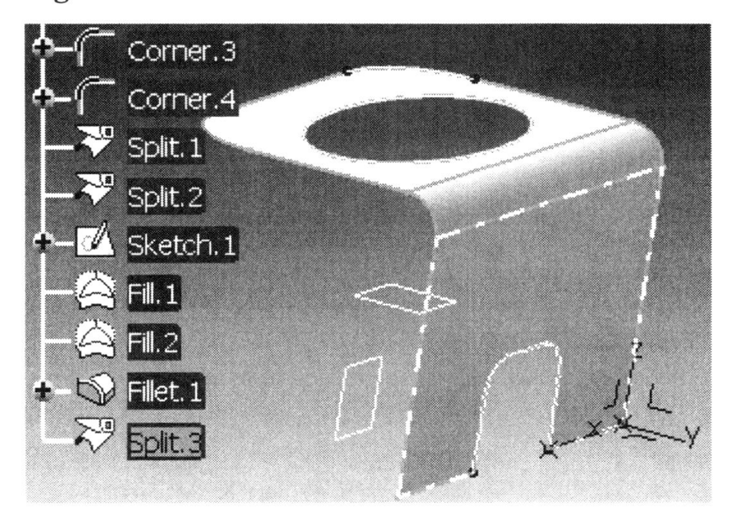

15 Creating Surface Thickness Using The Thick Surface Tool

CATIA V5 has made this powerful tool easy to use. CATIA V5 takes the selected surface and gives it thickness. You as the user define the **Surface**, the **Direction** and the **Thickness**. To complete the "**Sheetmetal Bracket**" part, complete the following steps.

15.1 In this step you will be switching from the **Generative Shape Design Work Bench** to the **Part Design Work Bench**. The **Thick Surface** tool is found in the **Part Design Work Bench** because it actually creates a solid from the parent surface. Highlight **Part.1** on the **Specification Tree**.

15.2 With **Part.1** (top of the **Specification Tree**) highlighted select the **Part Design Work Bench**. Your "**Sheetmetal Bracket**" in the work area will not change, but the workbench and workbench tools will.

15.3 Select the object to thicken: **Split.3**.

15.4 Select the **Thick Surface** tool. There will only be a few tools available for you to choose from. The tools that don't apply to surfaces will be dimmed meaning they are not selectable. Do not confuse the **Thick Surface** tool with the **Thickness** tool.

15.5 This will bring up the **ThickSurface Definition** window as shown in Figure 7.38.

15.6 Type in **.05in** for the **First Offset** box. This offset value is the part thickness.

15.7 You can leave the **Second Offset** box set at **0**. This part requires you to offset the thickness in only one direction.

15.8 For the **Object To Offset**, select the "**Sheetmetal Bracket**" currently represented in the **Specification Tree** as **Split.3**.

15.9 Verify the arrows are pointing inward of the part as shown in Figure 7.38. If the arrows are pointing in the opposite direction, select the **Reverse Direction** button.

15.10 When all of the variables are defined as they are shown in Figure 7.38, select the **OK** button to create the **Surface Thickness** (solid).

Figure 7.38

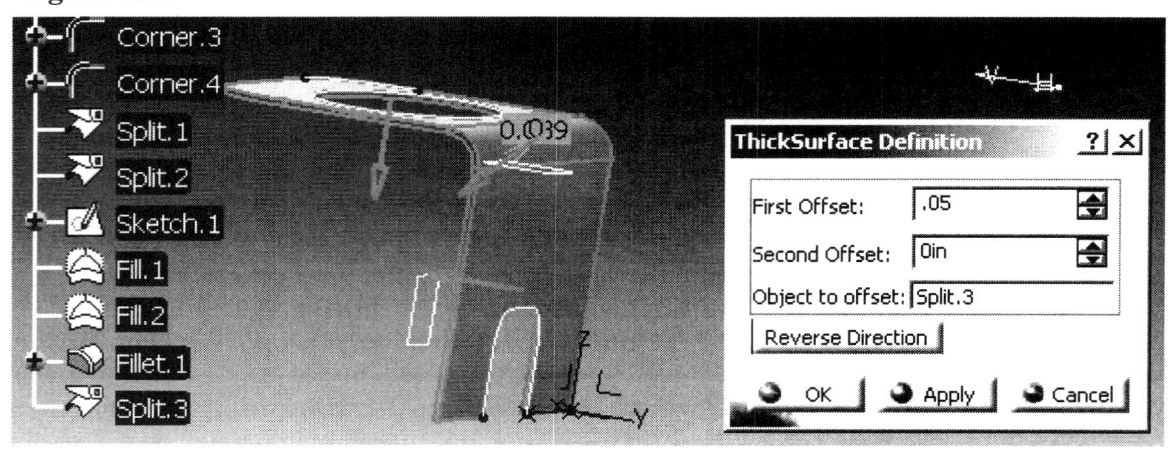

15.11 At this point, both the solid and the surface share the same space. It may be difficult to differentiate between the two entities. To simplify the workspace, select **Split.3** (the surface) from the **Specification Tree** and put it in "**Hide**" mode. The solid will be the only thing left on the screen.

NOTE: A quick way to hide everything other than the solid is to select the **Open_body.1** branch from the specification tree and use the contextual menu to hide all non-solid elements.

NOTE: The **Specification Tree** will now have a new branch on it. It will not be found under the **Open_Body** branch, but the **PartBody** branch. The solid will exist under the **PartBody** branch, as did the other solids created in the **Part Design Work Bench**. The only branch under the **PartBody** branch will be the **Thickness** branch.

15.12 Apply **Aluminum** material to the "**Sheetmetal Bracket.**" Reference Lesson 2 Step 18 for instructions.

15.13 Change the **Custom View Modes** window to show the **Material.** If your "**Sheetmetal Bracket**" looks similar to what is shown in Figure 7.39 you have successfully completed the "**Sheetmetal Bracket**."

15.14 Save your part as "**Sheetmetal Bracket.CATPart**."

NOTE: Geometry created in the **Generative Shape Design Work Bench** is saved with the same extension as geometry created in the **Part Design Work Bench**. The extension is "**CATPart.**"

Figure 7.39

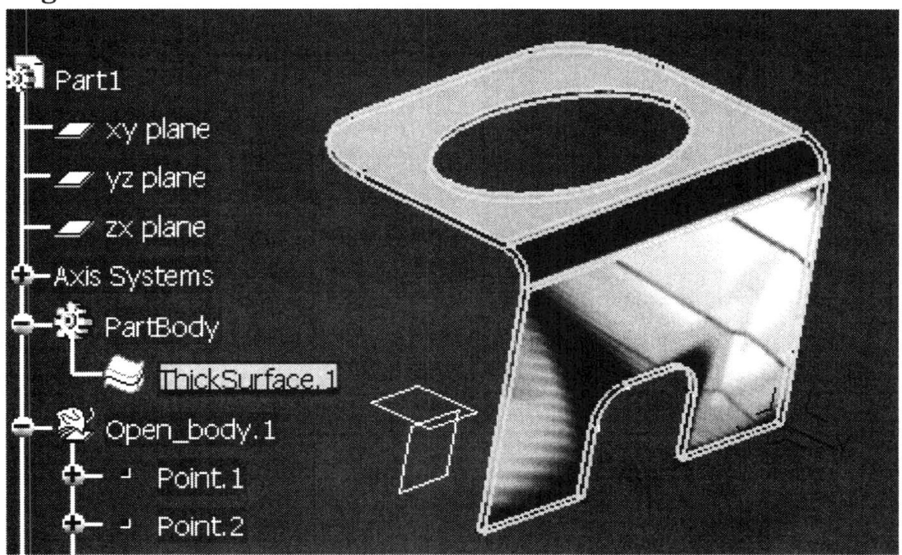

16 Creating A Helix 

In this section you will create a **Slinky** toy as a way of introducing you briefly to the **Helix** tool found in the **Curves** tool bar. The **Slinky** unlike the previous part (Sheetmetal Bracket) could be created entirely within the **Part Design** and **Sketcher Work Benches**, this part will require you to use wireframe. This part will also require you to use some new tools, not used in the previous steps. To create the **Slinky** complete the following steps.

16.1 Open a new window in the **Generative Shape Design Work Bench**.

16.2 Create an **Axis** at the 0,0,0 location.

16.3 Create a starting **Point** for the helix at 0,1.5,0.

16.4 Select the **Helix** tool. This will bring up the **Helix Curve Definition** window as shown in Figure 7.40.

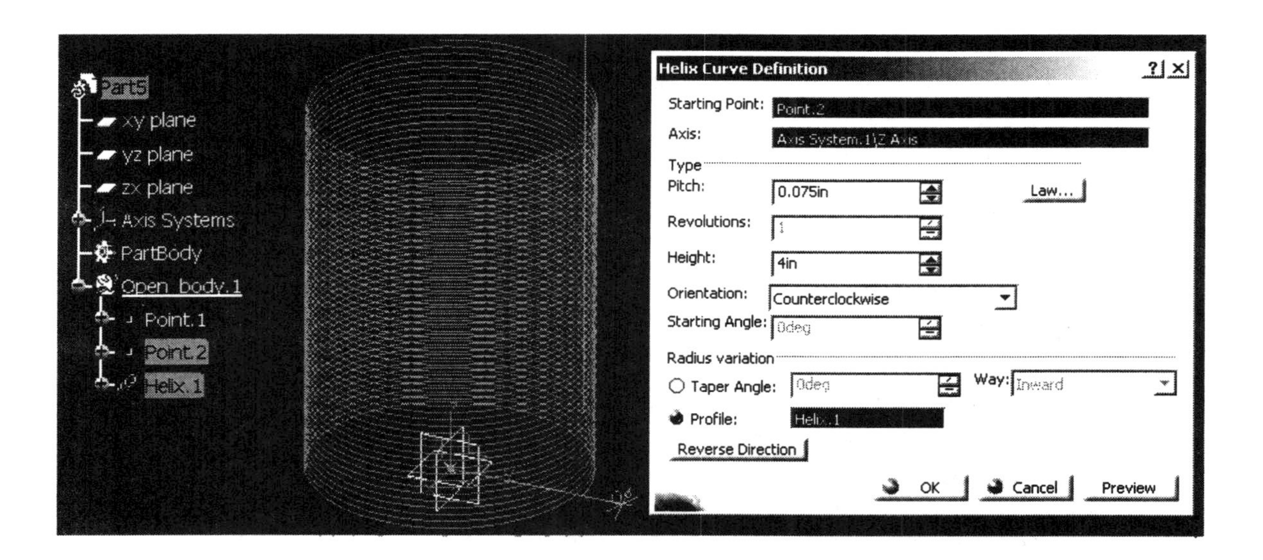

16.5 Click in the **Starting Point** box and then select **Point.2**.

16.6 For the **Axis** box select the Z axis.

16.7 For the **Pitch** type in .075″.

16.8 For the **Height** type in 4″.

16.9 Select the **Preview** button to view the helix.

16.10 Select **OK** if your Slinky looks similar to the one shown in Figure 7.40.

16.11 Now create three more points for the wireframe profile. Create points at the following locations: (0, 1.75, 0), (0, 1.5, .025), (0, 1.75, .025). Zoom in on the newly created points so they are visible.

16.12 Create lines connecting the four points as shown in Figure 7.41

16.13 Use the **Join** tool to create one entity using the four lines, reference Figure 7.42. Notice you now have a **Join.1** added to the **Specification Tree**.

16.14 Go to the **Part Design Work Bench.**

16.15 Select the **Rib** tool. This will bring up the **Rib Definition** window as shown in Figure 7.43.

16.16 For the **Profile** box select **Join.1**

16.17 For the **Center Curve** select **Helix.1**. If you receive an error window stating "It is impossible to compute the geometry" select **OK**.

16.18 For the **Profile Control** box select the **Pulling Direction** option.

16.19 Select the **Z Axis**.

Figure 7.41

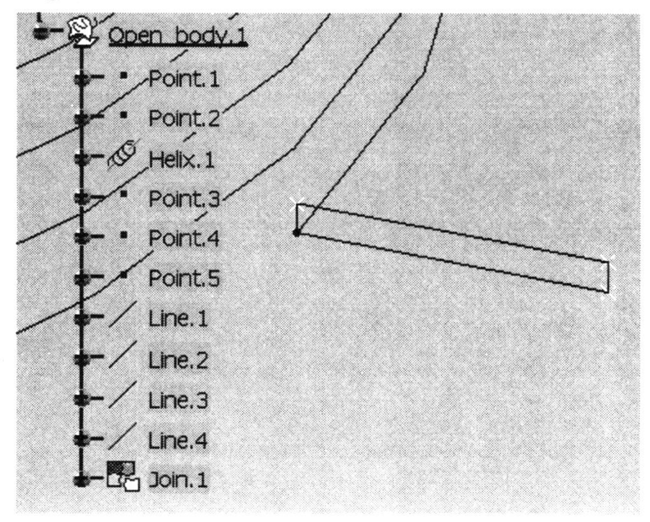

Figure 7.42

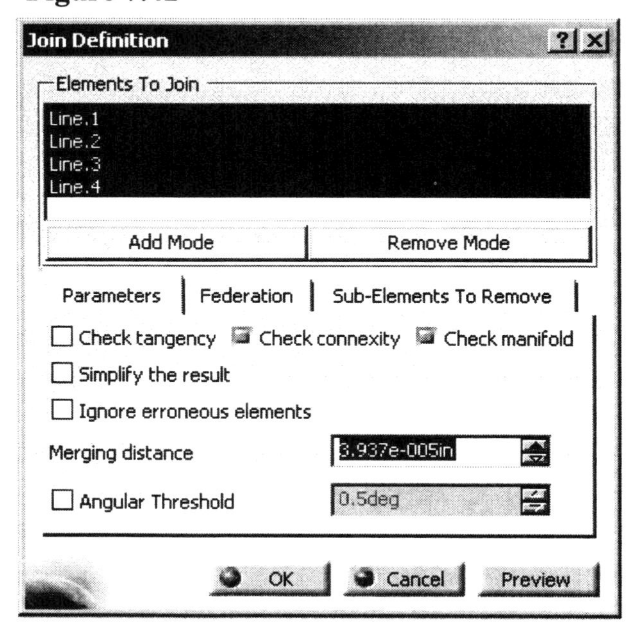

16.20 Select the **Preview** button. The part should look similar to the on shown in Figure 7.43.

16.21 Select the **OK** button to create the **Rib**.

16.22 Clean up the part by selecting **Helix.1** and moving it to **Hide/Show**.

16.23 Add material of your choice to the newly created Slinky

16.24 Save your new part as "**Slinky.CATPart**."

Figure 7.43

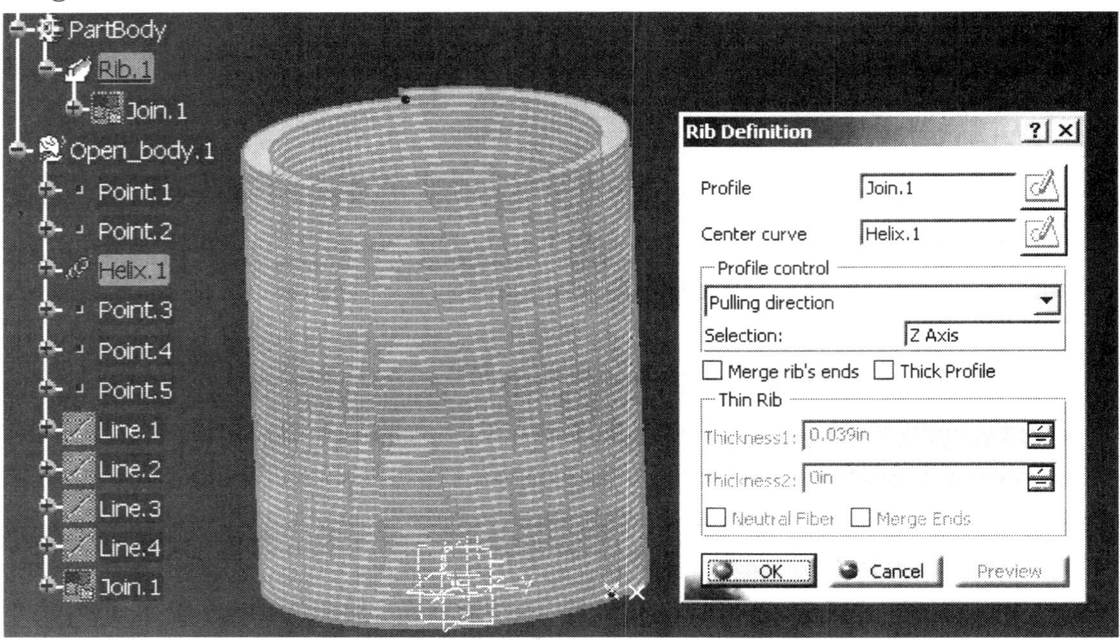

Lesson 7 Summary

There are other possible methods of creating this part; this lesson only covers a few of the possibilities. This lesson is designed to give you enough basic information so that you can explore the other possibilities on your own. You are strongly encouraged to explore the other methods. Lesson 8 will cover additional **Generative Shape Design** tools and applications. Congratulations and good luck on the **Lesson Review** and **Practice Exercises**.

Lesson 7 Review

After completing this lesson, you should be able to answer the questions and explain the concepts listed below:

1. What two **Work Benches** are used in this lesson?
2. How many standard toolbars are there in the **Generative Shape Design Work Bench**?
3. What are the 5 options available in CATIA V5 for creating a line?
4. What is the difference between a **PartBody** and an **Open_Body**?
5. Which tool may be selected to trim a line that is too long?
6. When trimming a line, what color is the part of the line that will be deleted?
7. When trimming a line, what color is the part of the line that will be kept?
8. Which tool should be selected to see the objects that have been hidden?
9. What are the six options available in CATIA V5 for creating a plane?
10. Which plane creation methods were used in this lesson?
11. Why did planes have to be created before the **Corner** tool could be used in this lesson?
12. How many different ways can you create a point in CATIA V5?
13. What is the difference between the **Spit** tool and the **Trim** tool?
14. **T** or **F** The **Fill** tool is used to create a solid from an existing surface.
15. **T** or **F** You can create a **Left Hand Axis** in CATIA V5.
16. When creating an axis, how do you define the direction of the **X Axis**?
17. **T** or **F** In CATIA V5 the trimmed elements are deleted.
18. Explain the relationship of parent and children geometry.
19. What tool is used to create a solid (**PartBody**) from a surface (**Open_Body**)?
20. What workbench is the tool in question 19 found in?

Lesson 7 Practice Exercises

Now that your CATIA V5 toolbox has more tools in it, put these tools to use on the following practice exercises. Each of the practice exercises will need to be done in the **Generative Shape Design Work Bench** and created as a wireframe.

1. You may recognize the part below; it was a practice exercise in Lesson 1. This part is much easier to create in the **Sketcher Work Bench**. Completing this part as a wireframe will give you the opportunity to compare methods (sketcher vs. wireframe).

   1.1 Create a wireframe of the front surface using the **Point** and **Line** tools.
   1.2 Create a surface over the wireframe using the **Fill** tool.
   1.3 Create a solid from the surface using the **Thick Surface** tool.
   1.4 Apply a material of your choice to the solid.
   1.5 Set the **Visualization** to display the material applied.
   1.6 Save the completed part as "**Lesson 7 Exercise 1.CATPart**."

   NOTE: You could copy the exercise from Lesson 1, but the **Specification Tree** shows the history of how the part was created.

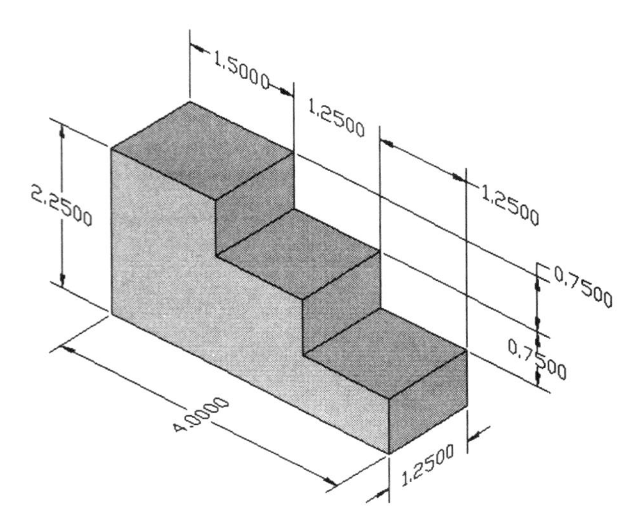

2. This exercise is also from Lesson 1. The geometry is getting a little more complex. This part has the same requirements as stated in Exercise 1. When you are done, save the part as "**Lesson 7 Exercise 2.CATPart**."

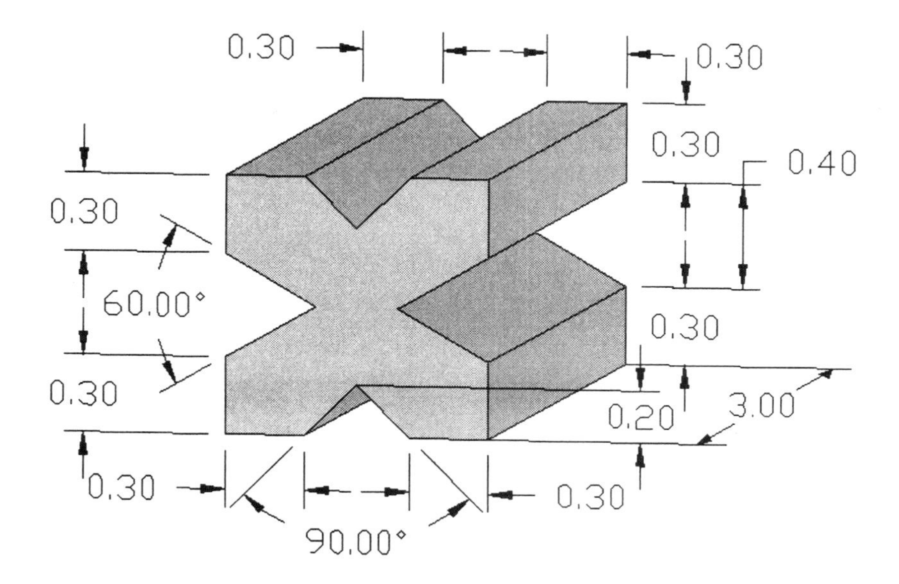

3. This part is more of a challenge than the first three. This part requires drawing lines, offsetting them and then rotating them. There are several possible methods of creating this part. You can use any method that works for you. When you are done, save the part as "**Lesson 7 Exercise 3.CATPart**."

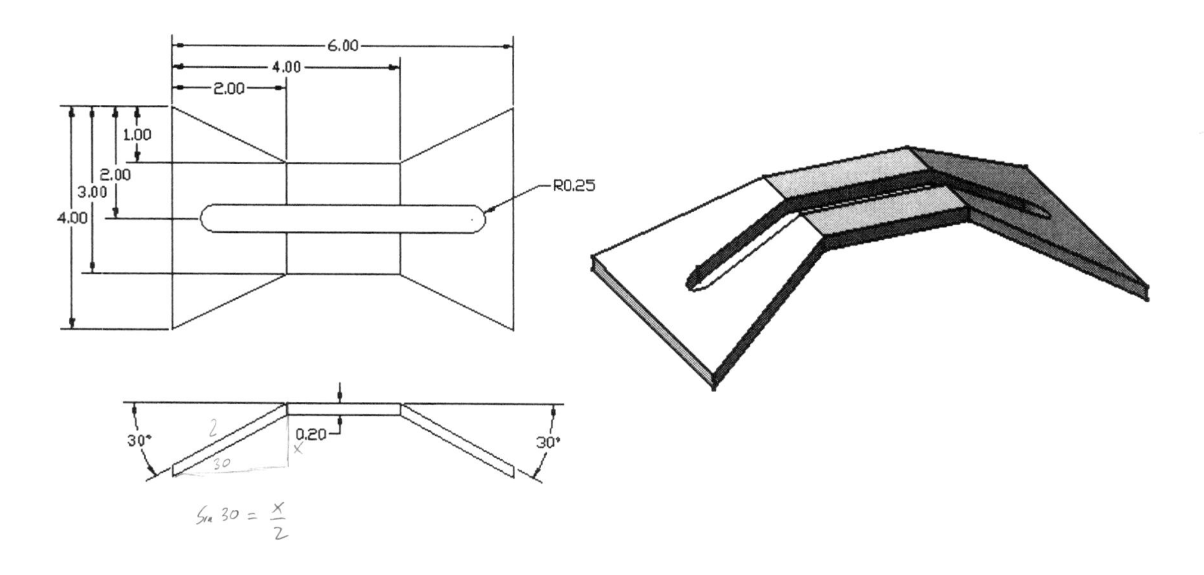

4. This exercise is for those who thrive on exploring and trying new things. When you are done save the part as "**Lesson 7 Exercise 4.CATPart**."

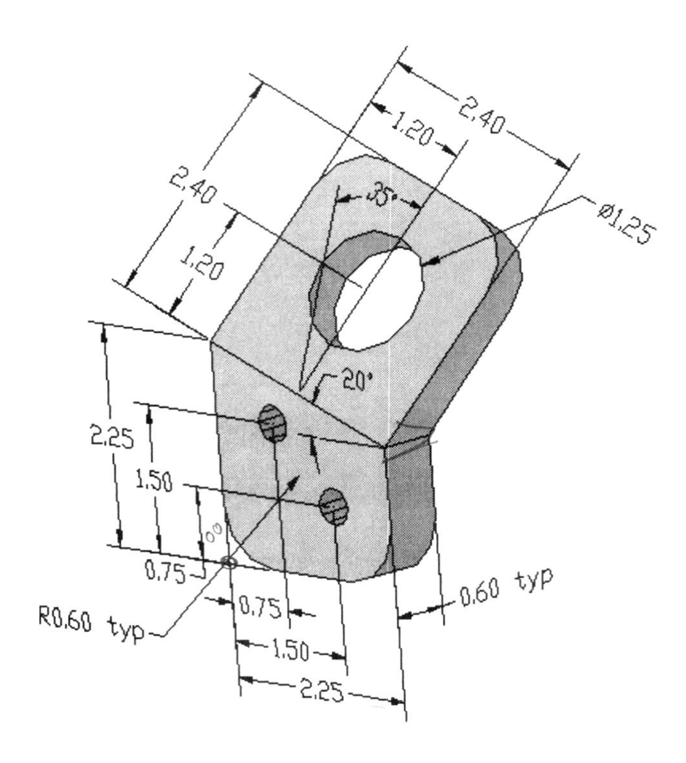

HINT: To create the 20 deg. line, use the **Angle/Normal to curve** option in the **Line Type** box of the **Line Definition** window. Another hint is that you may have to use tools in the **Part Design Work Bench** prior to using the **Thick Surface** tool.

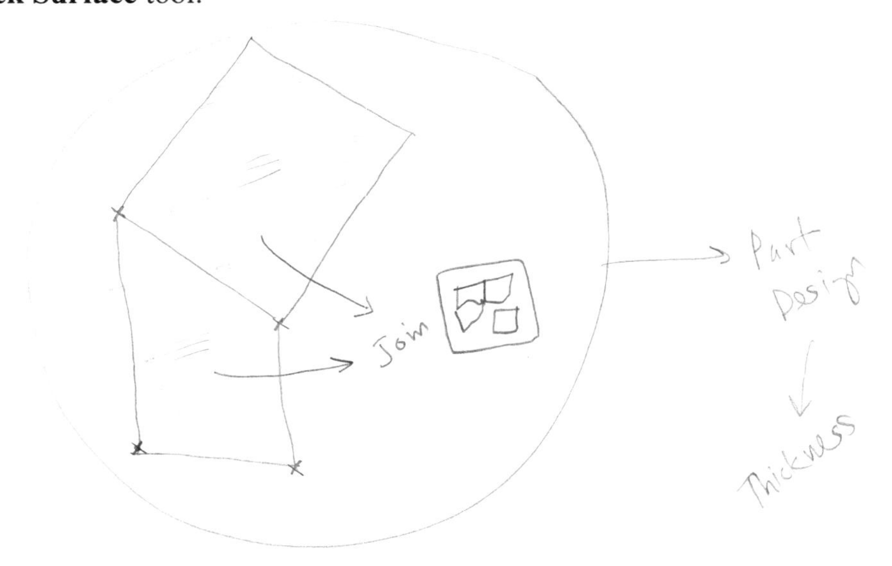

5. The dimensions give the information you need to create the point locations for this part. You could also create the lines using the coordinates, skipping the need for points. Create the wireframe for the part shown below. You can use the methods covered in this lesson and/or you can explore and try your own method. The thickness of the part shown below is .125 inches. When you are done, save the part as "**Lesson 7 Exercise 5.CATPart**."

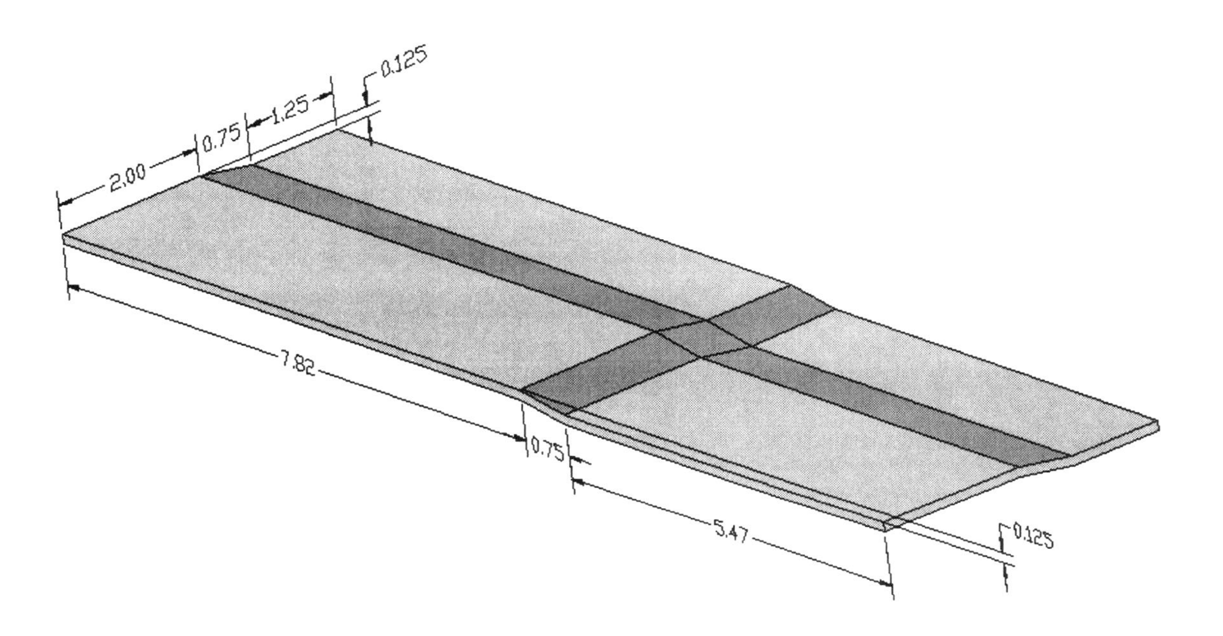

6. Create another **Slinky** using the following criteria.

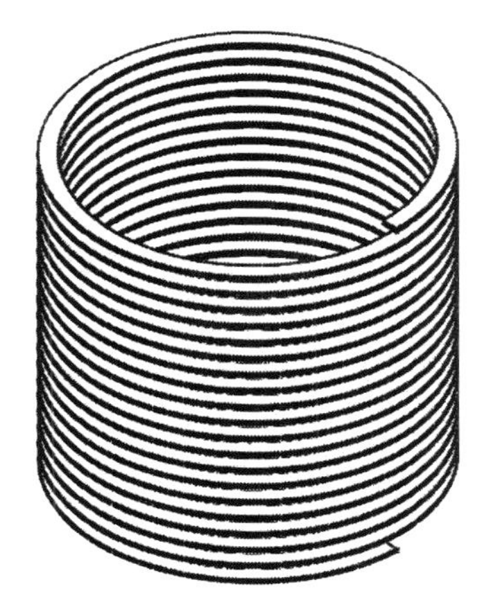

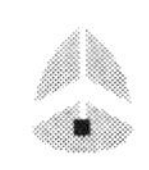

6.1 Axis point at 0,0,0.
6.2 Starting point for the helix at 0,2,0.
6.3 Create your helix with a Pitch of .400 in.
6.4 Height of 4.0 in.
6.5 Counterclockwise Orientation.
6.6 Create 4 points for the wireframe profile for the Rib creation. Create points that are .5 inches across and .25 inches in height for the contour.
6.7 Connect the points with four lines join them using the **Join** tool.
6.8 Go the **Part Design Work Bench**.
6.9 Create a **Rib** from the **Helix** and the **Join** contour

When you are done, save the part as "**Lesson 7 Exercise 4.CATPart.**"

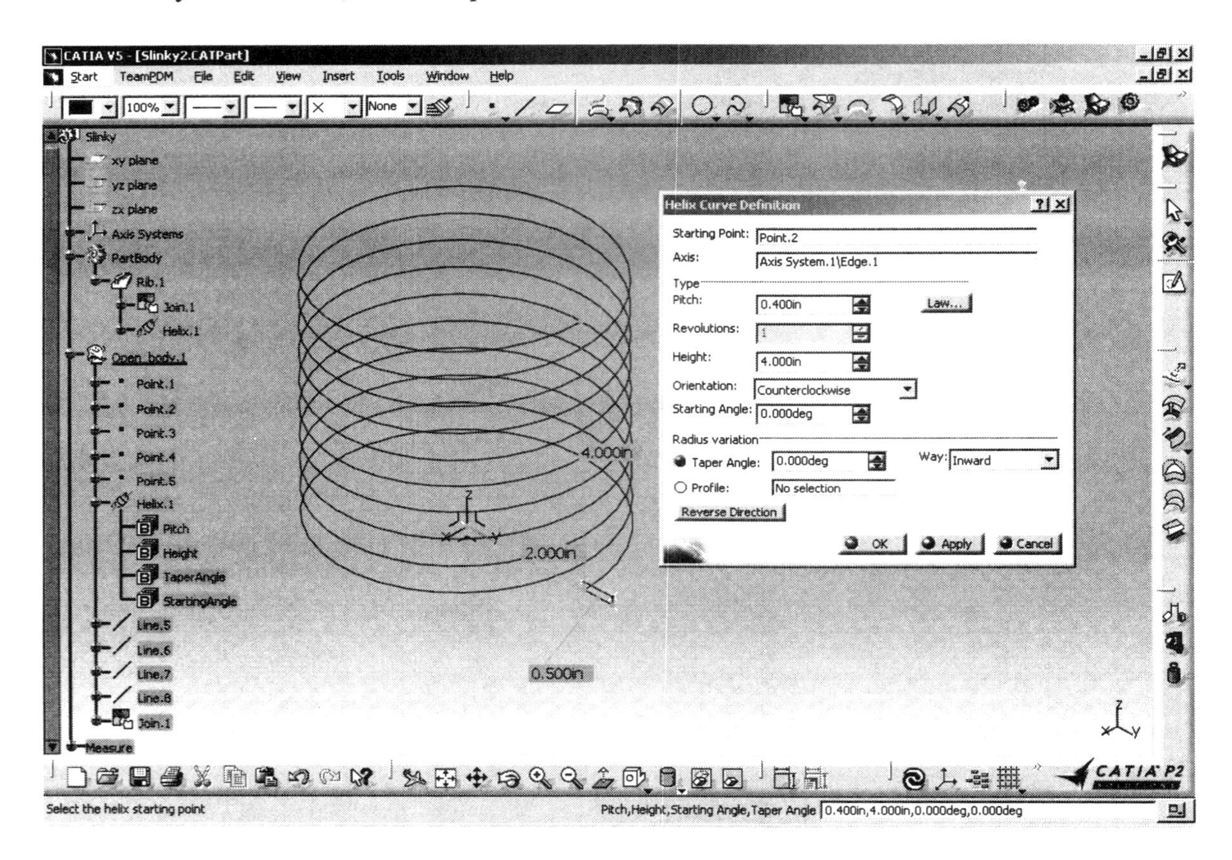

Lesson 8 Generative Shape Design Work Bench

Creating Wireframe And Surface Geometry Using The Sweep Tool

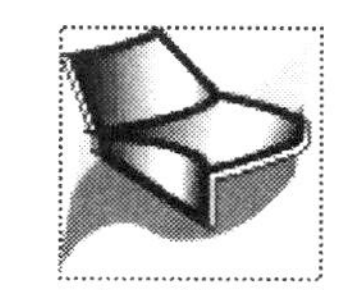
Generative Shape Design

This lesson gives step-by-step instructions on how to create the wireframe geometry required for creating a **Swept** surface, using the **Sweep** tool. There are other **Generative Shape Design** tools used in this process, but the main focus will be the **Sweep** tool. Some of the processes and tools used in Lesson 7 could be applied in this lesson. There are many different methods that could be used to create the part used in this lesson. The purpose of this lesson is to show you how to create a part using the **Sweep** tool. An additional purpose of this lesson is to show you several different methods of accomplishing the same and/or similar tasks. Once you have completed the different methods, you can choose which method is best for you.

As in Lesson 7, you will convert the wireframe/surface into a solid part. When you save your **CATPart,** save it as the "**Joggled Extrusion.CATPart**." The "**Joggled Extrusion"** is shown in Figure 8.1.

Figure 8.1 (Dimensions for the "**Joggled Extrusion**" that will be created.)

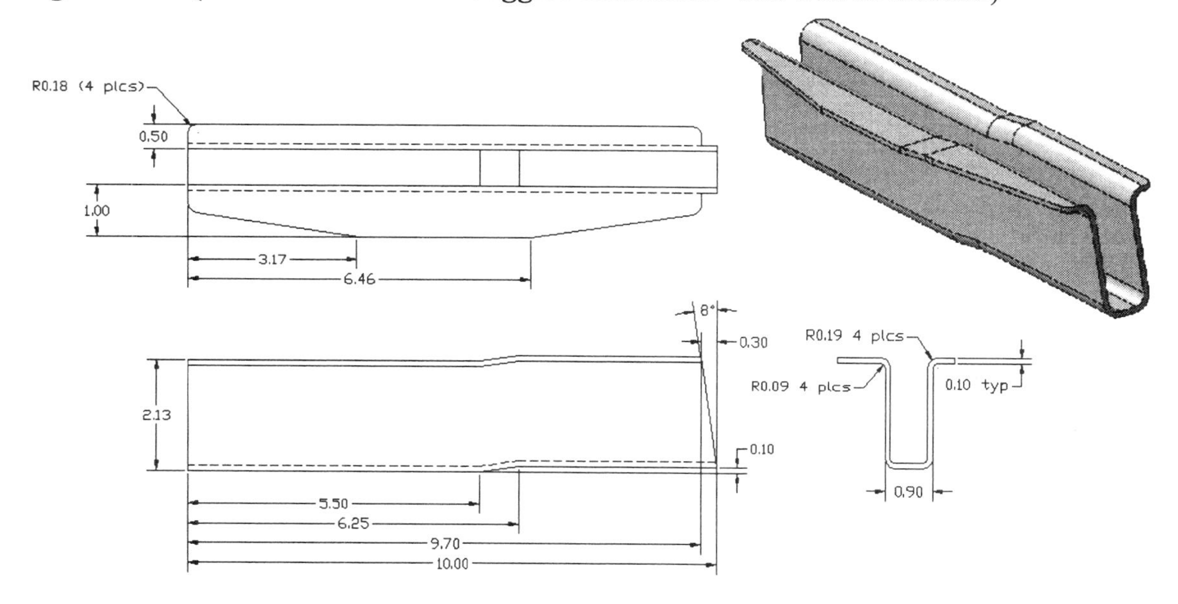

Lesson 8 Objectives

This Lesson will show you how to do the following:

- Create a **Profile** Sketch
- Create a **Guided Curve** Sketch
- **Join** curve entities
- Create surfaces using the **Sweep** tool
- Create a **Trim** sketch
- **Extrude** a sketch into a surface
- **Split** surfaces
- Create surface **Offset**
- Create surface with the **Fill** tool
- **Join** surfaces
- **Close** Surfaces to create a solid
- Modify entities using the **Pocket** tool (Part Design Work Bench)
- Create surface **Fillet**
- **Apply Material**

Generative Shape Design Work Bench Tools Bars

There are five standard tool bars found in the **Generative Shape Design Work Bench**. They are **Surfaces**, **Wireframe**, **Operations**, **Replication** and **Law**. The five tool bars are shown in Lesson 7.

Steps To Creating A Part Surface Using The Sweep Tool

This lesson uses the **Generative Shape Design Work Bench** as well as the **Sketcher** and **Part Design Work Benches**. The main focus will be in the **Generative Shape Design Work Bench**. It is in the **Generative Shape Design Work Bench** that the **Sweep** tool is found. In this lesson you will be creating the "**Joggled Extrusion**" which is shown in Figure 8.1.

It is important for you to understand what you are going to create prior to starting. What is required will determine the method you use to create the part. Requirements for this lesson are shown in Figure 8.1. A basic outline of the processes used in this lesson is listed below.

a. Create an open profile of the "**Joggled Extrusion**" using the inside surface dimensions (inside mold line). Create the open profile in the **Sketcher Work Bench**.

b. Create a path to extrude the open profile along (**Sweep**). This path is not a straight line; it will include the joggle as shown in Figure 8.1.

c. Trim the top flange as defined in Figure 8.1 using the **Plane** tool, **Sketcher Work Bench**, **Extrude** tool and the **Split** tool.

d. Create the surface thickness (the outside mold line dimensions) using the **Offset** tool.

e. Complete the surface using the **Fill** and **Join** tools.

f. Switch to the **Part Design Work Bench,** close the surface and create a solid using the **Close** tool.

g. Clean up the **CATPart**.

There are similarities and significant differences between the "**Joggled Extrusion**" and the part you created in Lesson 1 and Lesson 2: the "**L Shaped Extrusion**." Figure 8.2 shows the comparison between the two parts.

Figure 8.2

L Shaped Extrusion	Joggled Extrusion
required a closed profile in the **Sketcher Work Bench**	required a open profile in the **Sketcher Work Bench** or wireframe in the **Generative Design Work Bench**
extruded with the **Pad** tool in **the Part Design Work Bench**	requires geometry to define the path the profile will be extruded along (**Sweep**)
extrusion direction limited, normal to the sketch plane only	extrusion direction limited only by the path selected
result of the extrusion is a solid	the result of the **Sweep** is a surface, additional steps are required to create a complete solid

There are advantages and disadvantages to both. The **Pad** tool is faster for creating a complete solid, but it is also limited. The "**Joggled Extrusion**" could not be created in the **Part Design Work Bench** because of the joggle.

1 Select The Generative Shape Design Work Bench 

Make sure the **Units** are set to inches and the grid is set to 1″ **Primary Spacing** and 10 **Graduation**.

Select the **Generative Shape Design Work Bench.**

2 Creating The Profile In The Sketcher Work Bench 

In this step you will create the profile required by the **Sweep** tool. The profile is similar to the profile you created in the **Part Design Work Bench**. Reference Figure 8.2 for the similarities and the differences. The profile could be created in the **Generative Shape Design** workspace using the tools in the **Wireframe Tool Bar**. This lesson will have you create the profile in the **Sketcher Work Bench** because a planner sketch/wireframe is quicker and easier in the **Sketcher Work Bench**. Use the dimensions given in Figure 8.1 and the sketch in Figure 8.3. When using the dimensions, remember that the part is .10 of an inch thick. This means that when you create the overall height of the part (2.13″), you will need to subtract .10″. The .10″ is the thickness of the "**Joggled Extrusion**." Creating the part thickness will be one of the last steps in creating the "**Joggled Extrusion**." CATIA V5 makes it easy to modify your sketch, but it is still easier to do it right the first time. Create the profile by completing the following steps.

2.1 Create the sketch on the **ZX Plane**.

2.2 Create the basic outline of the cross section as shown in Figure 8.3.

2.3 **Constrain** the cross section as shown in Figure 8.3. Remember, when adding constraint values you can enter algebraic formulas. A simple example is typing in the part height dimension 2.13″ minus the part thickness .10″ and CATIA V5 will calculate the resulting value, 2.03″.

Figure 8.3

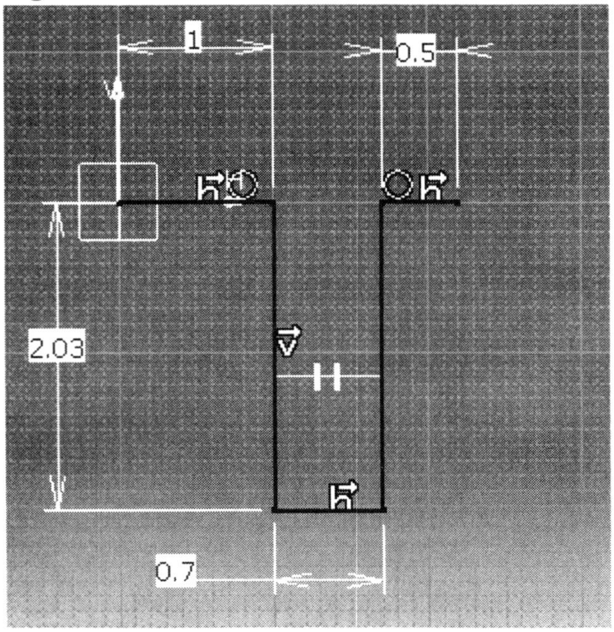

2.4 After you have your sketch constrained similar to what is shown in Figure 8.3, add the radii using the **Corner** tool. When you create the radii, if you will get a message as shown in Figure 8.4, select **Yes**. All this does is convert your constraint to a distance constraint.

Figure 8.4

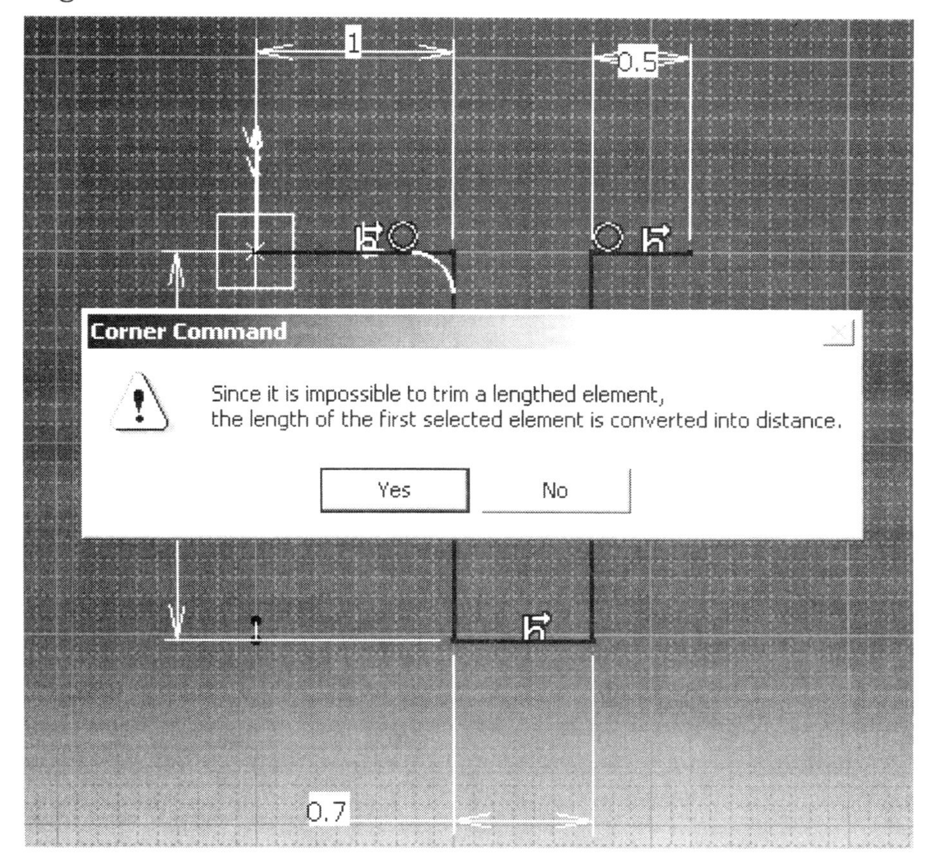

2.5 Add the radii as shown in Figure 8.5 using the **Corner** tool. The .19″ radii will be the outside mold line of the surface, while the .09″ radii will be the inside mold line of the surface. When you get your sketch to look similar to the one shown in Figure 8.5, exit the **Sketcher Work Bench**.

2.6 CATIA V5 will automatically name this sketch **Sketch.1**. **Sketch.1** does not really give you any information. Rename **Sketch.1** to **Profile Sketch**. **Profile Sketch** contains more information. At the very minimum, it gives you an idea of what is contained in the sketch.

Figure 8.5

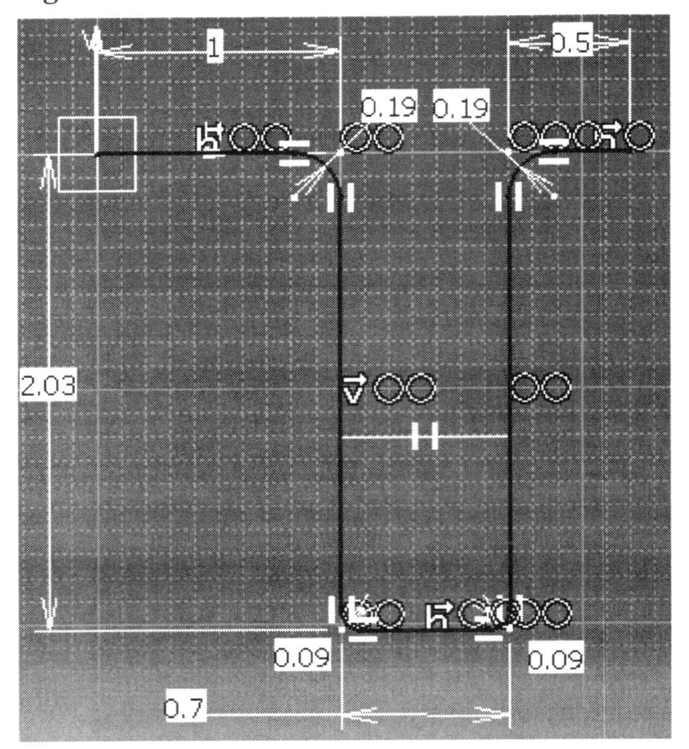

3 Creating The Guided Curve In The Sketcher Work Bench 

In this step you will create the **Guided Curve** required for the **Sweep** tool. The **Guided Curve** will be created using the dimensions shown in Figure 8.1. To create the **Guided Curve,** complete the following steps.

3.1 Create a sketch on the **YZ Plane**.

3.2 Create the basic outline of the **Guided Curve** as shown in Figure 8.6. Creating the **Guided Curve** and constraining it as shown can be tricky. The following steps might be helpful, refer to Figure 8.6.

3.2.1 Create a rough sketch of the three lines similar to Figure 8.6.

3.2.2 Constrain **Line.1** to 2.03″ from the **H Axis**.

3.2.3 Constrain the starting point of **Line.1** (use **Point.1**) from the **V Axis** to 0.0″. Anchoring this point will over constrain the sketch. Setting the Constraint to 0.0″ will accomplish the same thing without over constraining the sketch.

3.2.4 Constrain the length of **Line.1** to 5.5″.

3.2.5 Constrain the starting point of **Line.3** to 6.25″ from the **V Axis**.

3.2.6 Constrain **Line.3** to 1.93″ from the **H Axis**.

3.2.7 Constrain the ending point of **Line.3** to 10.0″ from the **V Axis**.

3.3 When you get the sketch to look similar to what is shown in Figure 8.6, exit the **Sketcher Work Bench**.

3.4 Rename the sketch "**Guided Curve Sketch**."

Figure 8.6

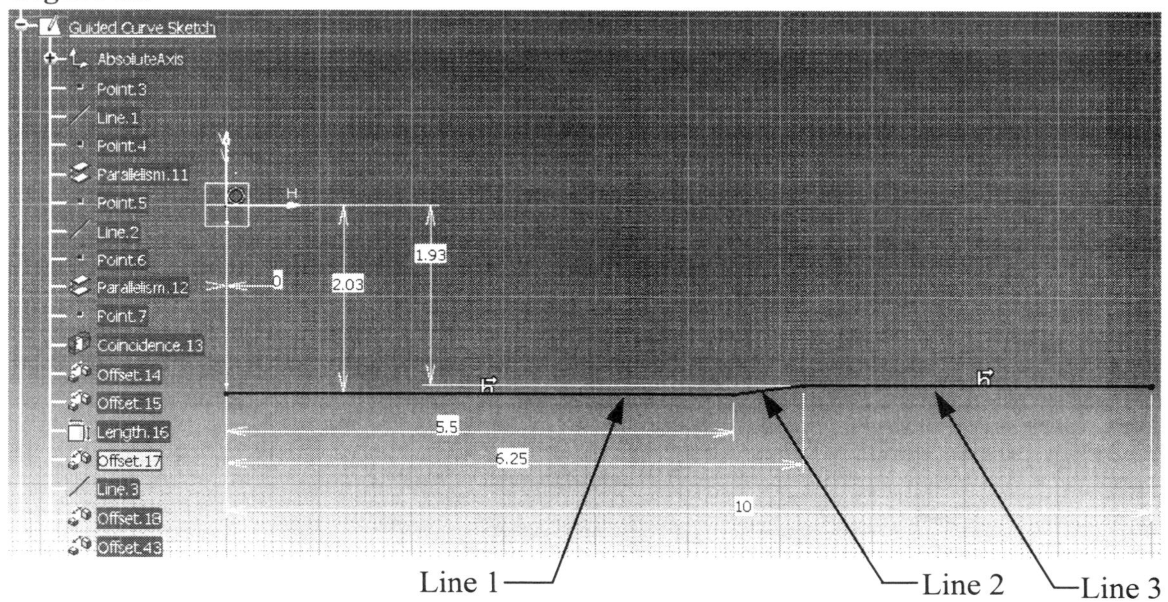

4 Joining The Guided Curve Entities Using The Join Tool

The path you are going to sweep (extrude) the **Profile Sketch** along is now complete. If you attempted to sweep the profile now you would only sweep it along **Line.1**. The **Guided Curve Sketch** must be joined into one entity so the profile can be swept along the entire length of the **Guided Curve Sketch**. To **Join** the **Guided Curve Sketch** entities, complete the following steps.

4.1 Select the **Join** tool icon.

4.2 This will bring up the **Join Definition** window as shown in Figure 8.7.

4.3 Select all three sketch lines that make up the **Guided Curve Sketch**. (Depending on how the lines were created, the selectable elements may change. You may only be able to select the **Guided Curve Sketch**; therefore, having only one element in the **Elements To Join** box. The three lines selected will be listed in the **Elements To Join** box as shown in Figure 8.7.

4.4 Leave the default value in the **Merging Distance** box.

4.5 Select **OK** to join the sketch entities.

4.6 A **Join** branch will appear on the **Specification Tree**.

Figure 8.7

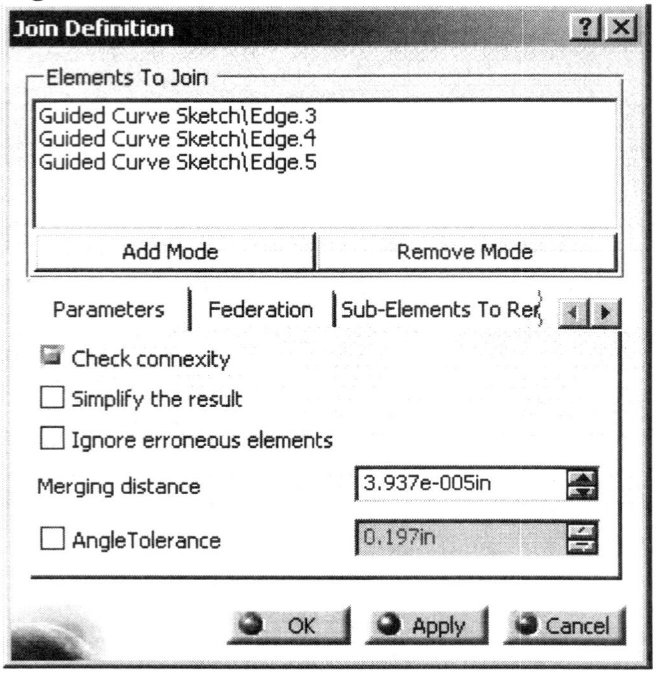

Join Definition
Elements To Join
Guided Curve Sketch\Edge.3
Guided Curve Sketch\Edge.4
Guided Curve Sketch\Edge.5
Add Mode
Remove Mode
Parameters
Federation
Sub-Elements To Re
Check connexity
Simplify the result
Ignore erroneous elements
Merging distance
3.937e-005in
AngleTolerance
0.197in
OK
Apply
Cancel

5 Creating A Surface Using The Sweep  Tool

You are now ready to **Sweep** the **Profile Sketch** along the **Guided Curve Sketch**. To complete this task, perform the following steps.

5.1 Select the **Sweep** tool icon.

5.2 This will bring up the **Swept Surface Definition** window as shown in Figure 8.8.

5.3 Select the **Explicit** icon for the **Profile Type**. It is the first of the three icons.

5.4 Select the **Profile Sketch** for the **Profile** box.

5.5 Select **Join.1** for the **Guided Curve** box. Remember, the **Guided Curve Sketch** is the parent entity of **Join.1**.

5.6 The remaining options can be left at their default values. This step does not require any **Optional Elements**.

5.7 When your selection looks similar to Figure 8.8, select the **OK** button to create the **Swept Surface**.

Figure 8.8

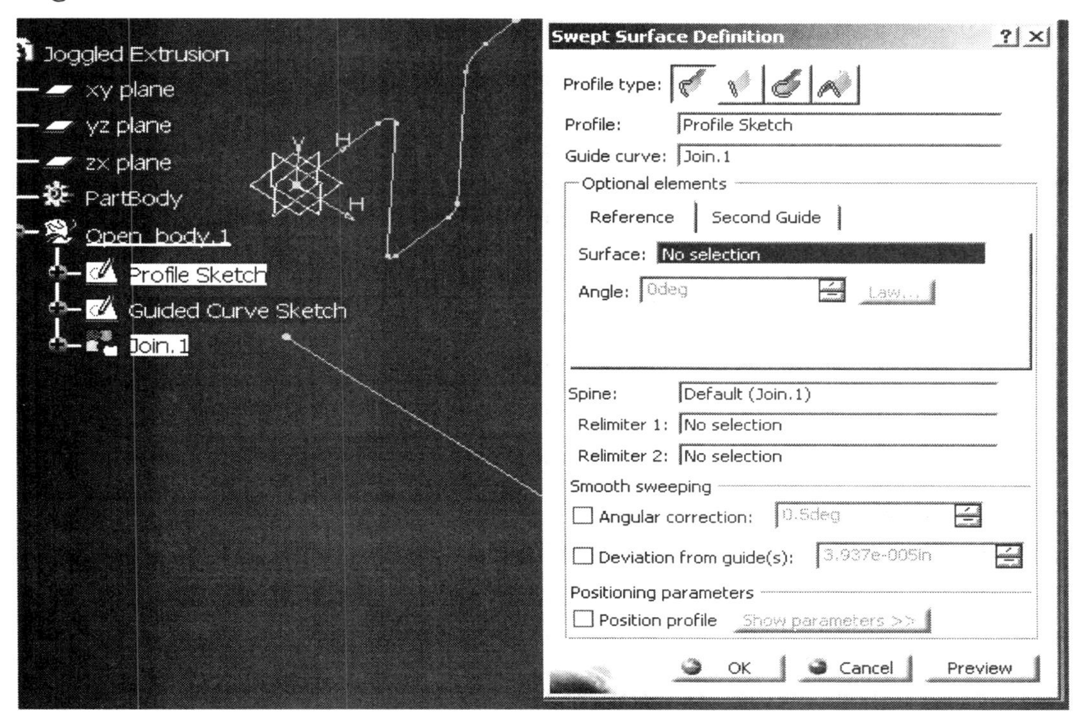

5.8 The resulting **Swept Surface** should look similar to the one shown in Figure 8.9. Notice the **Sweep.1** branch was added to the **Specification Tree**.

5.9 Before you continue on to the next section, **No Show** the **Profile Sketch** and **Guided Curve Sketch**. This lesson will require you to create additional sketches. **No Showing** the existing sketches will make the working space less confusing.

Figure 8.9

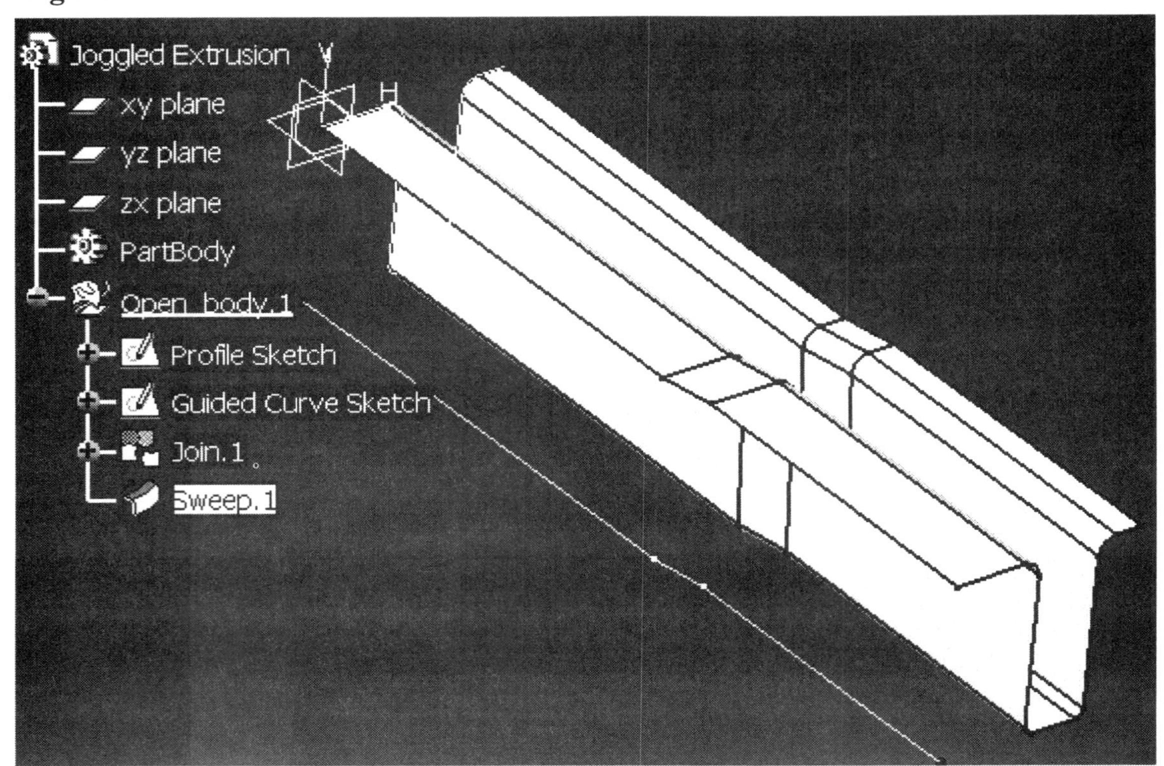

6 Creating The Trimmed Profile For The Top Flange Using The Sketcher Work Bench

The part surface for the "**Joggled Extrusion**" has now been created, but the part is not complete. If you reference Figure 8.1 you will notice that the top 1″ wide flange has two different angles on it. In this step you will create a sketch that defines the trimmed flange, extrude the sketch and trim the extruded surface. The following steps are to help guide you through this process.

6.1 Create a **Plane** that represents the top flange surface as shown in Figure 8.10.

Figure 8.10

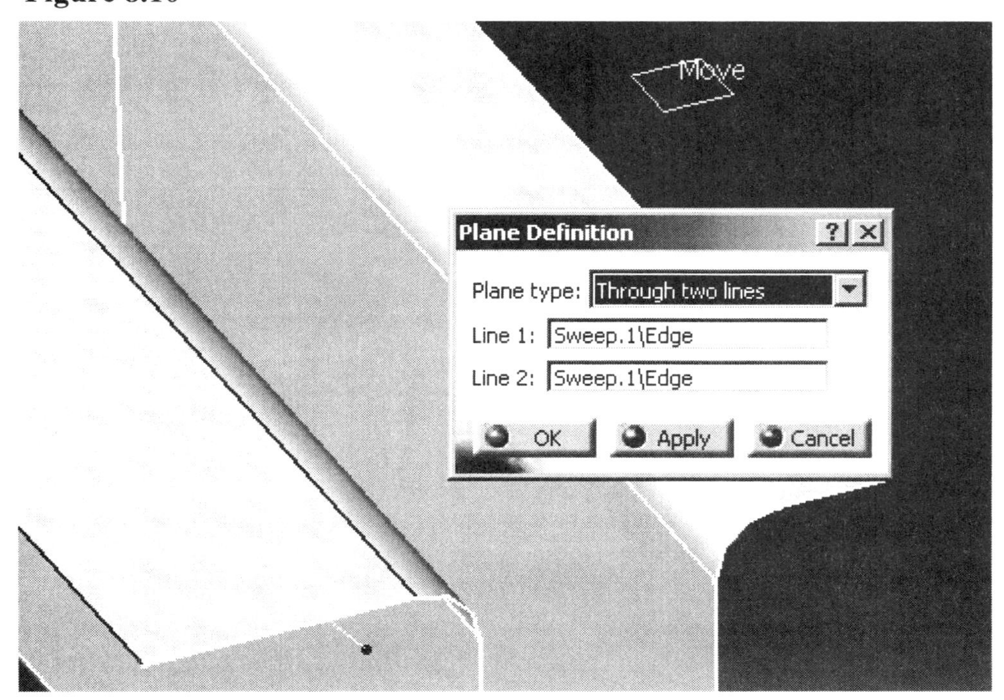

6.2 Create a sketch using the plane you created in Step 6.1. Use the dimensions in Figure 8.1 and the constraints in Figure 8.11 to create the flange profile. Constrain the sketch as shown in Figure 8.11. The .5″ constraint located at the bottom right of Figure 8.11 is not located at the bottom of the part, but .3″ up from the bottom of the part. The reason for this is, that end of the part has an angle cut into it a total length of .3″ at the top of the part, reference the front view of Figure 8.1. The angle will be trimmed off in the **Part Design Work Bench**.

Figure 8.11

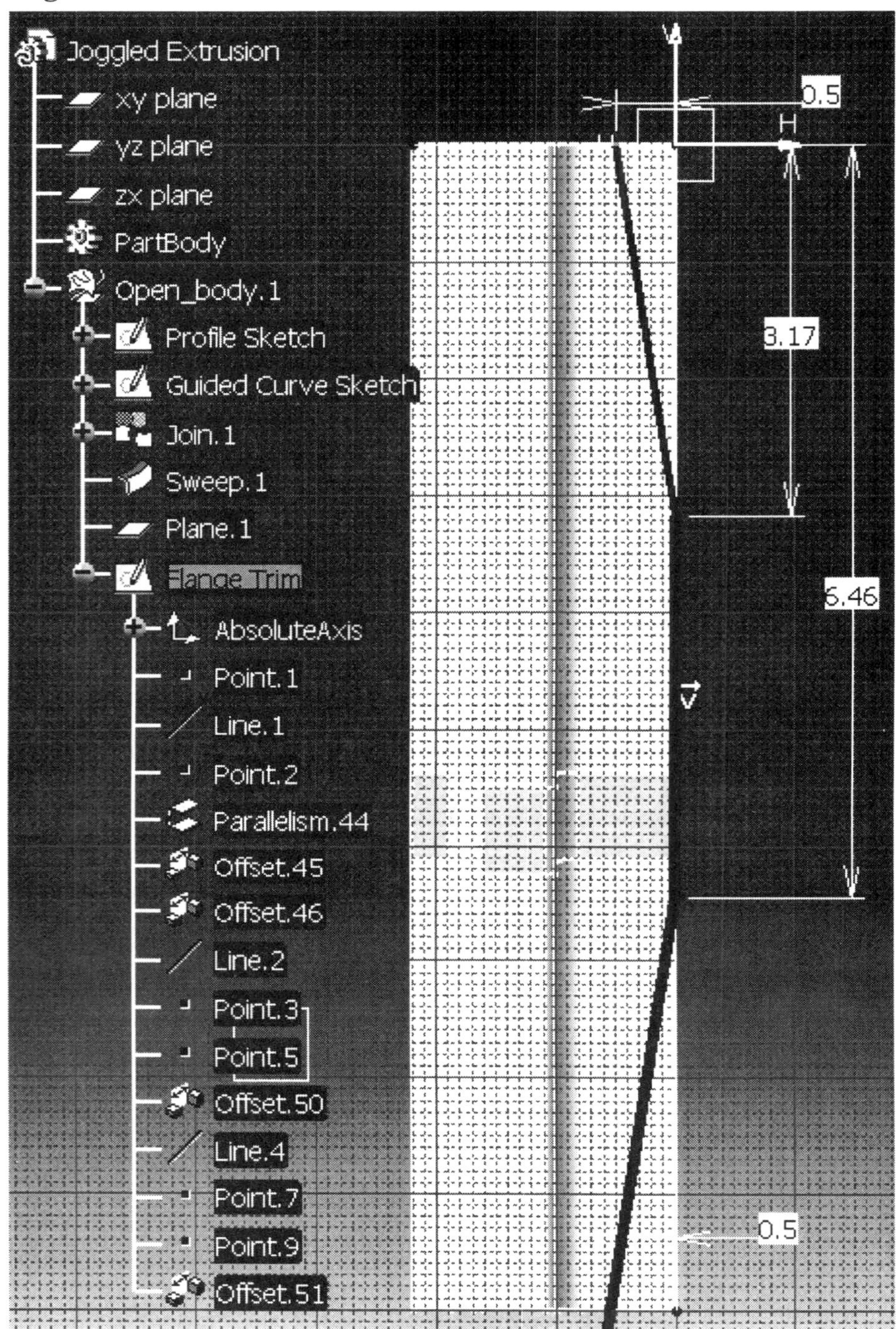

6.3 Exit the **Sketcher Work Bench**.

6.4 **No Show Line.1**.

6.5 Rename the newly created sketch to "**Flange Trim**."

6.6 Change the color of the "**Joggled Extrusion**" surface (**Sweep.1**). You can determine the color. When all of the surfaces have the same properties (color), they can be confusing and/or hard to differentiate. Remember, this is done through **Properties**, **Graphic** and then **Color**.

7 Extruding The "Flange Trim" Sketch Using The Extrude  Tool

In this step you will extrude the sketch created in Step 6. To extrude the "**Flange Trim**" sketch, complete the following steps.

7.1 Select the **Extrude** tool.

7.2 This will bring up the **Extruded Surface Definition** window as shown in Figure 8.12.

7.3 For the **Profile** box, select the "**Flange Trim**" sketch.

7.4 The **Direction** box will default to **Plane.1**, the plane the sketch was created on.

7.5 The **Extrusion Limits** box allows you to type in values for both directions, **Limit 1** and **Limit 2**. Another method of defining **Limit 1** and **Limit 2** is moving the mouse over the green arrow, selecting the arrow and then dragging the arrow in the direction and distance you want the sketch extruded. For this step you can extrude the sketch using either method. Extrude the sketch to the approximate distances shown in Figure 8.12. An exact distance is not necessary for this step. The only requirement is that the extruded surface passes through the "**Joggled Extrusion**" surface.

7.6 If your extruded surface looks similar to what is shown in Figure 8.12, you are ready for the next step, trimming the surfaces.

Figure 8.12

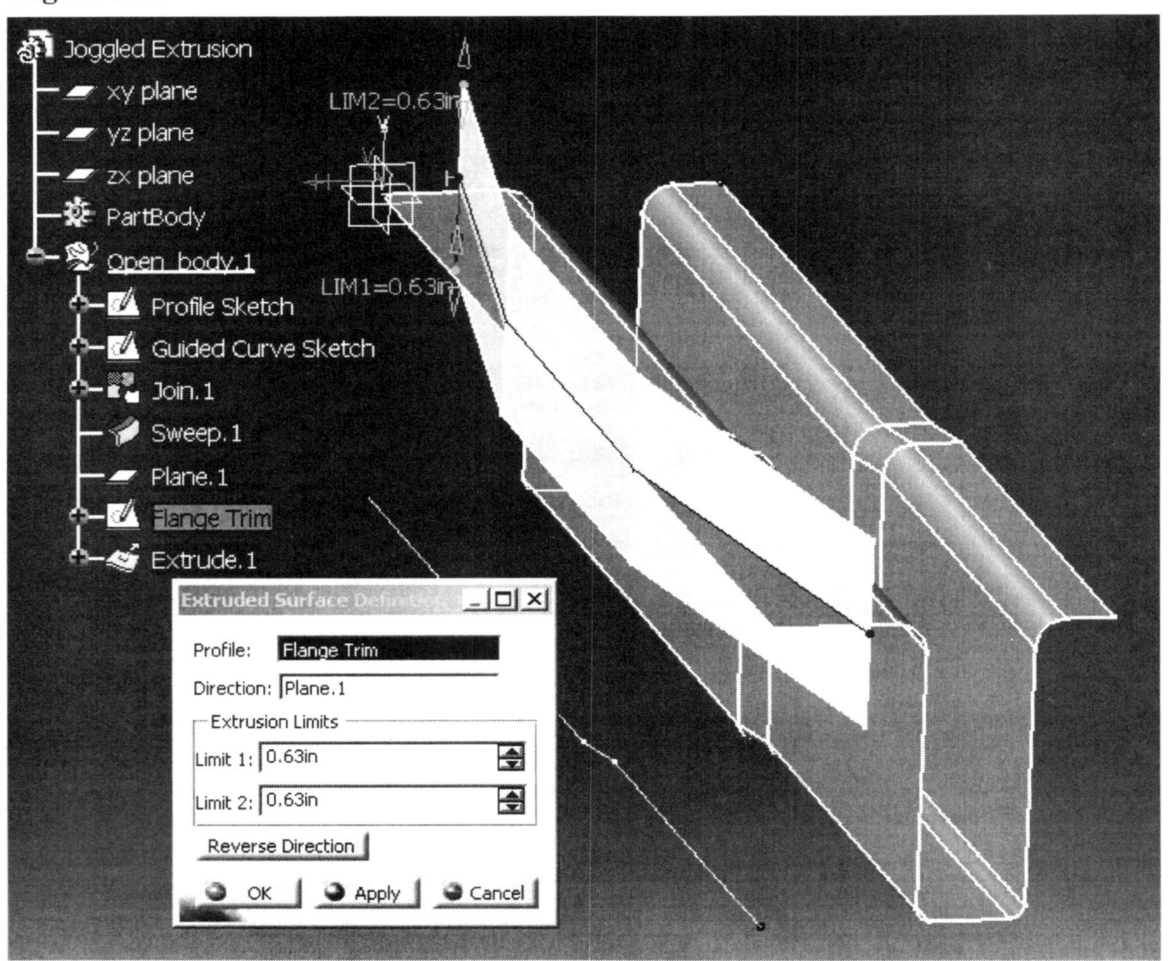

8 Trimming The Surfaces Using The Split  Tool

In this step you will trim the flange of the "**Joggled Extrusion**" by splitting the "**Joggled Extrusion**" surface with the surface you extruded in Step 7. This will be done using the **Split** tool. A similar tool is the **Trim** tool. You could complete this process using either tool. The **Split** tool in this situation is the quicker tool. To split the two surfaces, complete the following steps.

8.1 Select the **Split** tool.

Figure 8.13

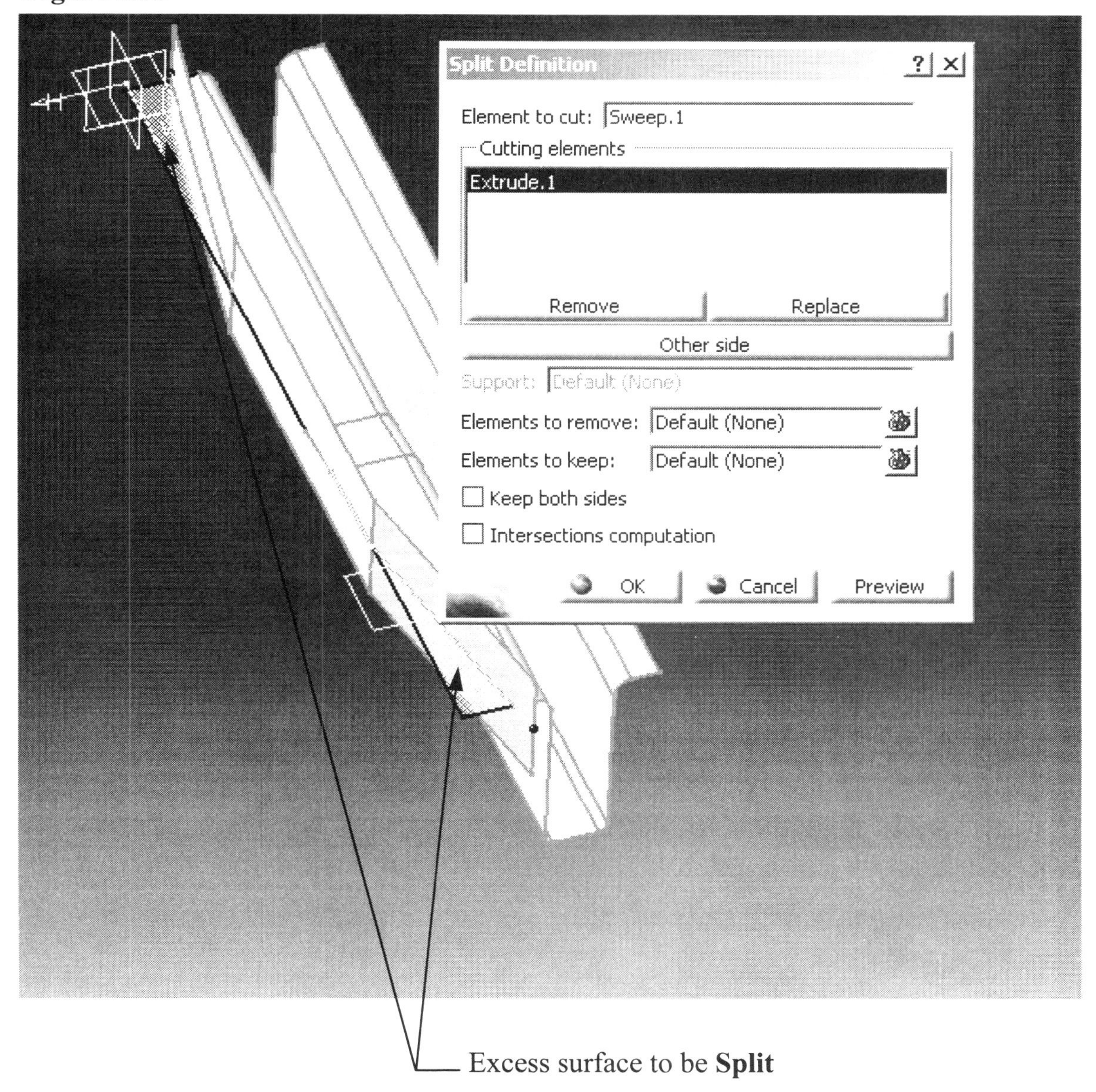

8.2 This will bring up the **Split Definition** window as shown in Figure 8.13.

8.3 Select the **Sweep.1** branch of the **Specification Tree** for the **Element To Cut** box. Remember, **Sweep.1** is the "**Joggled Extrusion**" surface.

8.4 Select the **Extrude.1** branch of the **Specification Tree** for the **Cutting Element** box. The excess surface on the "**Joggled Extrusion**" will dim. The dimmed surface is the surface that will be trimmed away. Make sure the correct surface is dimmed prior to completing the **Spit** process. If the wrong surface is dimmed, select the **Other Side** button in the **Split Definition** window. For the correct surface, reference Figure 8.13.

8.5 After verifying you have the correct surface being **Split**, select **OK** to complete the process.

NOTE: The **Extrude.1** surface is still there and the remaining "**Joggled Extrusion**" has turned back to its default color. The flange should have been trimmed as shown in Figure 8.14. The surface that you changed colors is now in **No Show**. The current "**Joggled Extrusion**" surface is represented in the **Specification Tree** as the **Split.1** branch. If you select the **Split.1** branch, the "**Joggled Extrusion**" part will also highlight.

8.6 You cannot delete the **Extrude.1** surface. **Extrude.1** is a parent of your current "**Joggled Extrusion**" surface. Although you cannot delete it, you can get it out of the way by placing it in **No Show**. Also, **No Show** your "**Flange Trim**" sketch.

8.7 If your "**Joggled Extrusion**" looks similar to the one shown in Figure 8.14, you are ready to continue to the next step.

Figure 8.14

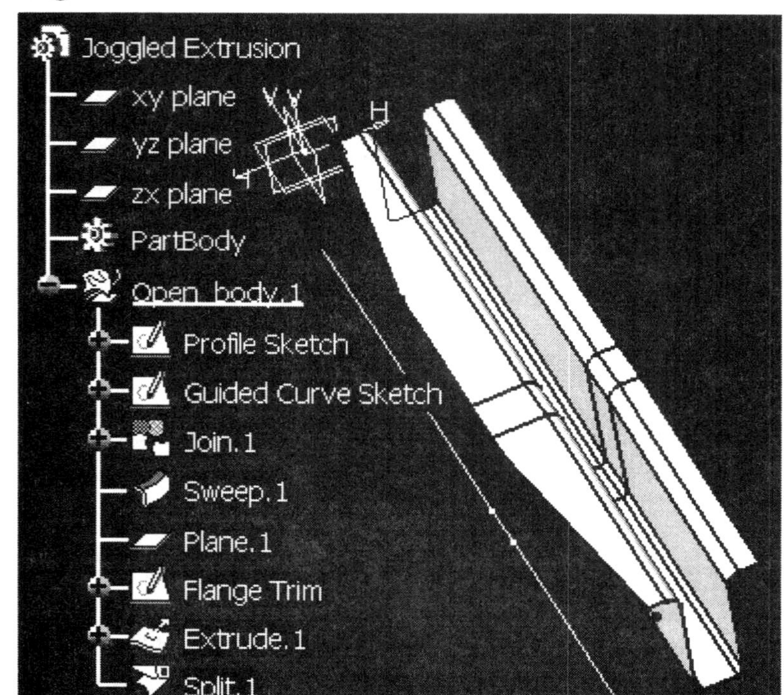

9 Creating Surface Thickness Using The Offset  Tool

Using the **Offset** tool to create an additional surface is the first step in giving the "**Joggled Extrusion**" part actual part thickness. Remember, the original surface you created was the **Inside Mold Line** of the part. This determination made a significant difference in applying dimensions to the **Profile Sketch**. The surface you create in this step will be the **Outside Mold Line** of the "**Joggled Extrusion**." When you are done with Step 9 you will have a surface that represents both inside and outside surfaces of the "**Joggled Extrusion**." To create an offset surface, complete the following steps.

9.1 To help you visualize the different surfaces, change the color of the **Split.1** surface ("**Joggled Extrusion**").

9.2 Select the **Offset** tool.

9.3 This will bring up the **Offset Surface Definition** window as shown in Figure 8.15.

9.4 For the **Surface** box, select the **Split.1** surface.

9.5 CATIA V5 gives you the two methods of defining the offset distance. One method is moving your mouse over the green arrow labeled **Dist** and dragging the arrow to the desired value and in the desired direction. The other method, a more precise method, is to type the exact value in the **Offset** box in the **Offset Surface Definition** window. For this step, type **.1in.** for the **Offset** value.

9.6 If your offset surface looks similar to what is shown in Figure 8.15, select **OK** to create the offset surface. You are ready to continue on.

NOTE: The new offset surface is created in the default surface color. The **Specification Tree** has added an **Offset.1** branch.

Figure 8.15

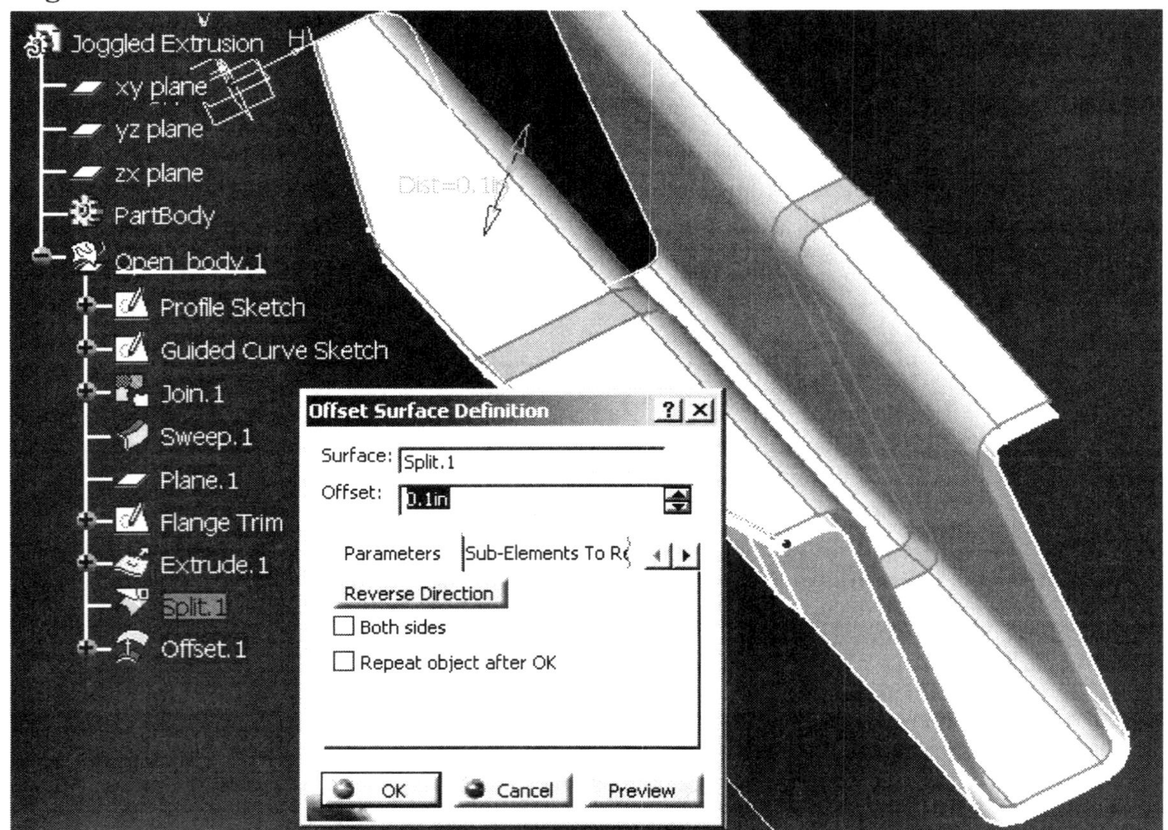

10 Closing The Surfaces Using The Fill  Tool

Your "**Joggled Extrusion**" now has a surface representing both the inside and outside of the part. To create a solid part (**PartBody**), the surface must be completely closed by surfaces. For detailed instructions on how to use the **Fill** tool, refer to Step 12 in Lesson 7. The following instructions are general steps to close the part with surfaces.

10.1 This step requires you to create six small surfaces. Change the color of the **Offset.1** surface. This will help you differentiate between the existing surfaces and the new small surfaces. The color is not important, as long it contrasts the existing and default surface colors.

10.2 Before you use the **Fill** tool, you will have to create the boundaries for the surfaces. The **Line** tool can be used to create the boundaries. Create lines between the inner surface (**Split.1**) and the outer surface (**Offset.1**) as shown in Figure 8.16. Use the **Point-to-Point** method of creating lines.

Figure 8.16

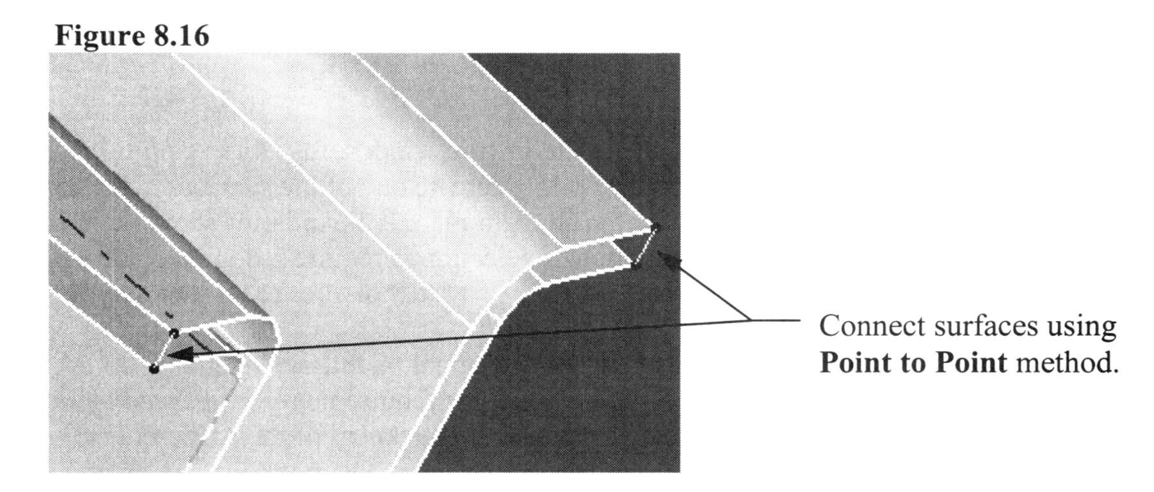

10.3 Create lines 1 through 10 in the locations shown in Figure 8.17.

Figure 8.17

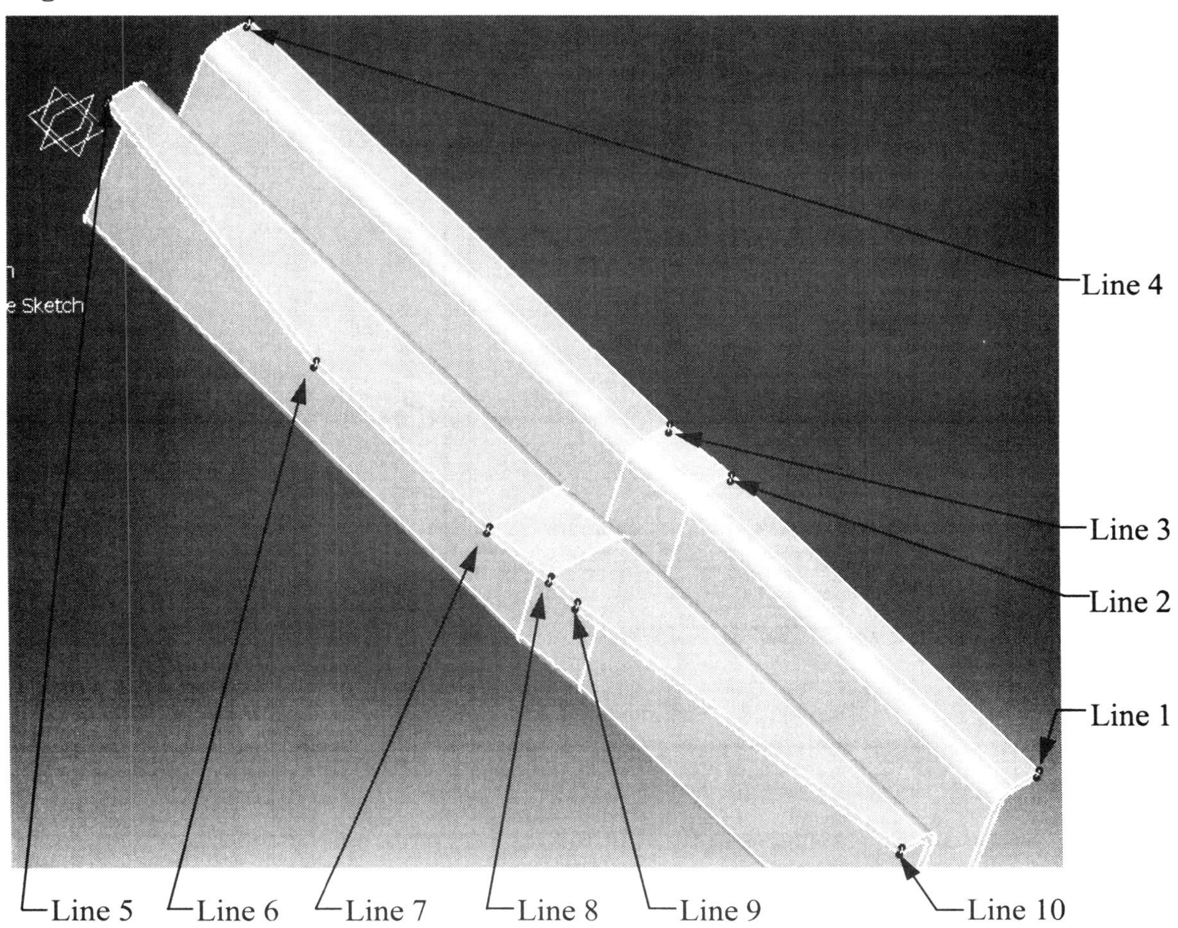

10.4 When all ten lines are created, you are ready to create surfaces. Select the **Fill** tool.

10.5 This will bring up the **Fill Surface Definition** window. As an example, select the boundaries as shown in Figure 8.18 to create one of the surfaces. Use the same process to create surfaces for all eight planner boundaries shown in Figure 8.19. Remember, when repetitive tool selection is required, double clicking on the particular tool will automatically reselect the tool after each completion.

NOTE: You may be required to add an additional boundary depending on how you created your **Sweep.1** sketch. The requirement for this step is to create surfaces that totally enclose the thickness of the part. You are creating an enclosed volume. The volume must not have any open surfaces. Another way of looking at it is the surfaces must hold air, you must patch all air leaks using the **Fill** tool.

10.6 Do not create surfaces on either end of the "**Joggled Extrusion**" part, only the side that represents the thickness of the part.

Figure 8.18

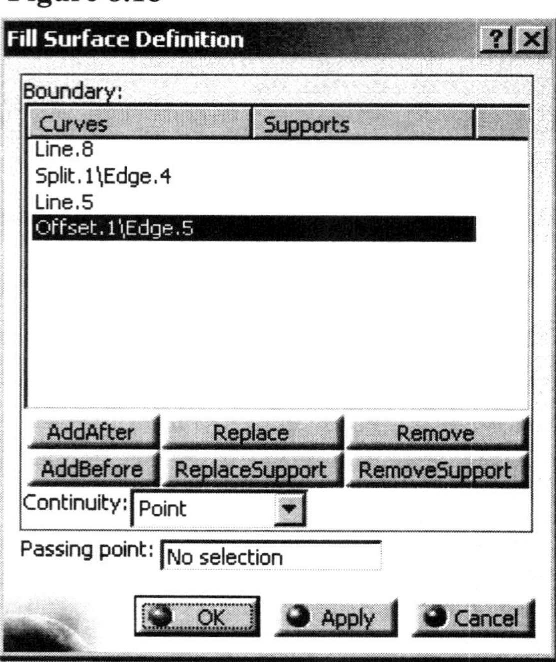

10.7 When all eight surfaces defined in Figure 8.19 have been created, you can go on to Step 11. It is critical that all eight surfaces have been created correctly or the process used in Step 11 will not work. **No Show** the 10 lines that were created. They could cause problems in later steps.

Figure 8.19

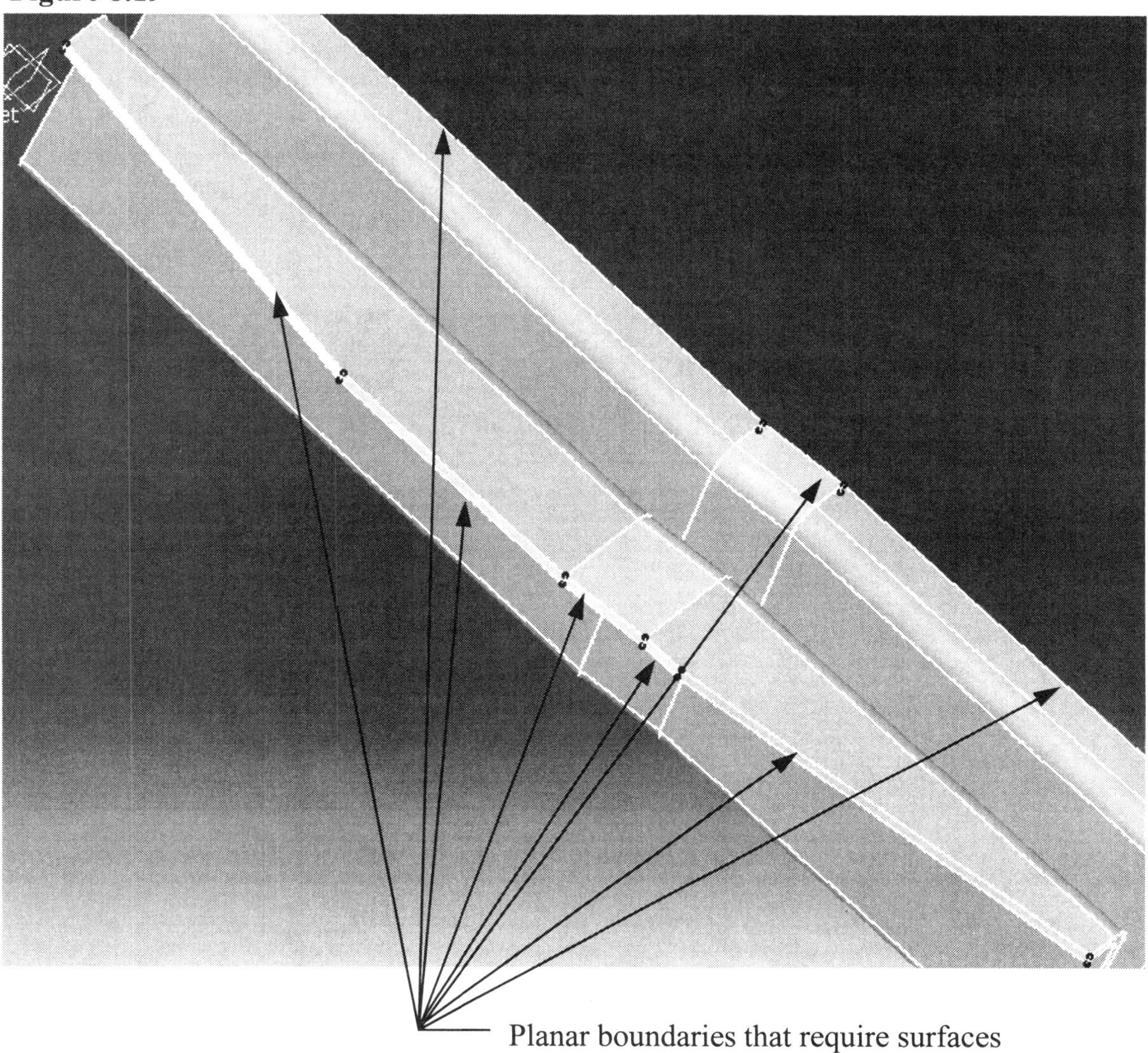

11 Joining The Surfaces Using The Join Tool

Before the surfaces can be converted to a solid, the surfaces must be joined into one complete surface. This step will have you join all the surfaces into a **Join.2** surface. The following instructions are a basic guideline to completing this step. If you need detailed instruction on using this tool, refer back to Step 4.

11.1 Select the **Join** tool.

11.2 This will bring up the **Join Definition** window.

11.3 Select the **Split.1** surface, **Offset.1** surface and **Fill.1** through **Fill.8** surfaces. Make sure the selected surfaces show up in the **Elements To Join** box. Selecting each individual surface could be difficult. In this case it is easier to select the surfaces from the **Specification Tree** as shown in Figure 8.20.

11.4 If you have all ten surfaces selected, hit the **OK** button to join the surfaces.

Figure 8.20

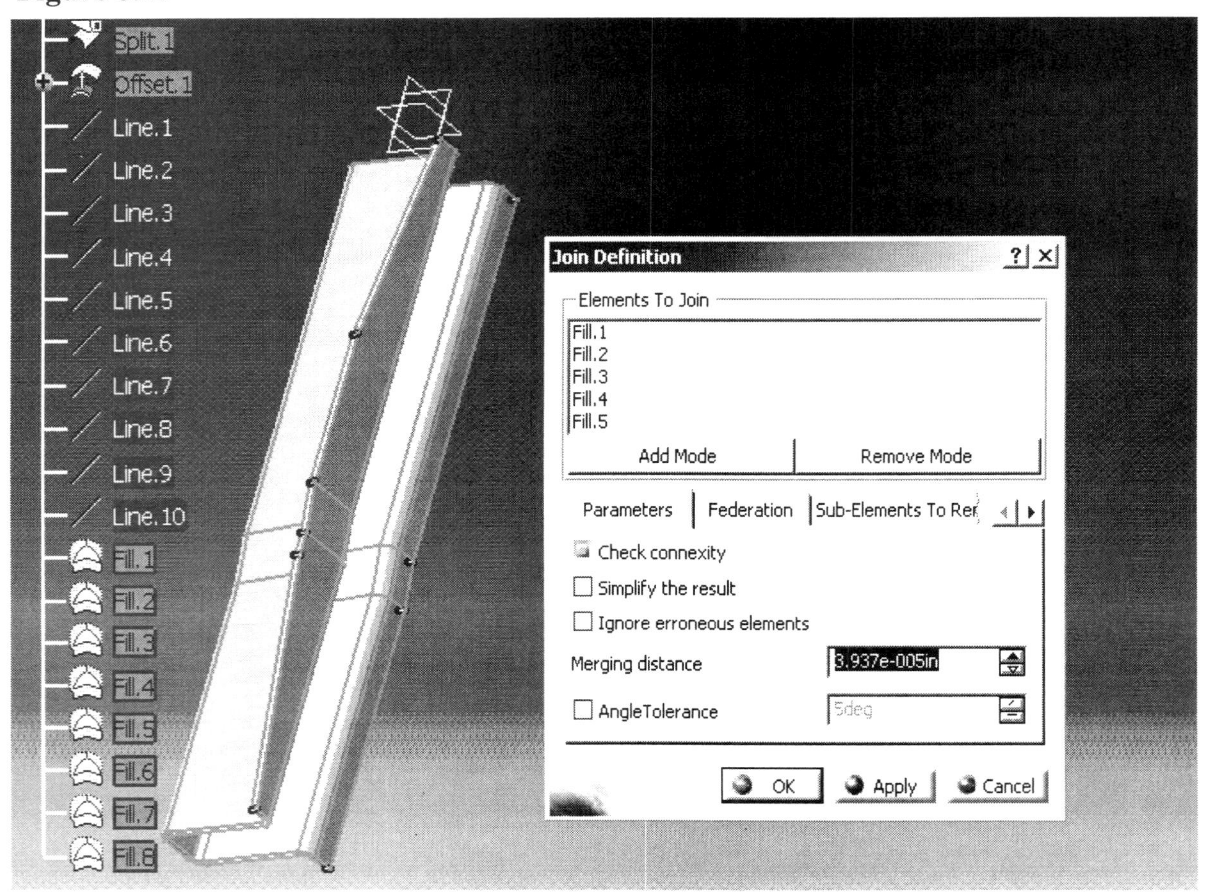

NOTE: You modified the colors of the **Split.1** and **Offset.1** surfaces. Those surfaces are no longer visible, but not deleted. CATIA V5 moved the **Split.1** and **Offset.1** surfaces to the **No Show** workspace. If they were deleted, all of the surfaces that were built using them would also be deleted. In this case, it would be the **Join.2** surface. The result of the **Join** tool is the new resulting surface. The **Join.2** surface shows up in the default colors.

11.5 If your **Join.2** surface looks similar to what is shown in Figure 8.21, you are ready for Step 12.

Figure 8.21

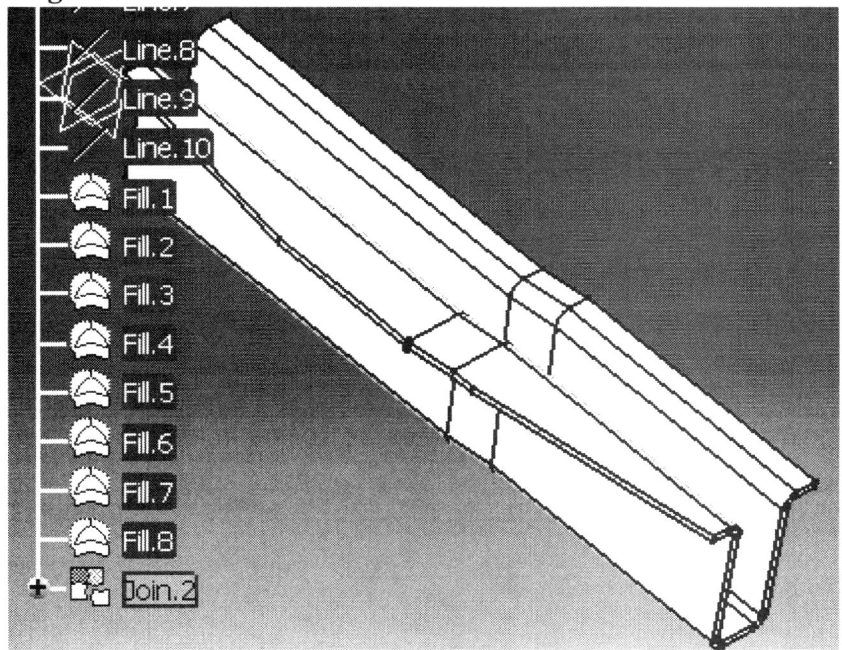

12 Closing The Surfaces Using The Close Surface  Tool

At this point your "**Joggled Extrusion**" represented by the **Join.2** surface is completely enclosed except for the two ends of the part. You could have completely enclosed the surface using the process defined in Step 11, but that would have required more detailed and time consuming work. It would have required 6 additional lines for boundaries and 18 additional surfaces. The **Close Surface** tool combines all that work into several easy steps. To use this tool you must bring up the **Part Design Work Bench**. Close the "**Joggled Extrusion**" surface by completing the following steps.

12.1 The first step is to bring up the **Part Design Work Bench**. Select the "**Joggled Extrusion**" at the top of the **Specification Tree**. With the "**Joggled Extrusion**" highlighted, select the **Part Design Work Bench**.

12.2 The tool bars will change from the **Generative Shape Design Work Bench** to the **Part Design Work Bench**.

12.3 From the **Part Design Work Bench** select the **Close Surface** tool. This tool is in the **Surface Based Features** tool bar. The default tool that shows up under that tool bar is the **Split** tool. Select the arrow at the bottom right of the icon to access the **Close Surface** tool.

12.4 CATIA V5 will prompt you in the prompt zone to "**Select the surface to close.**" Select the **Join.2** surface.

12.5 This will bring up the **CloseSurface Definition** window as shown in Figure 8.22. Make sure that the **Join.2** surface is entered in the **Object To Close** box and the geometry is highlighted as shown in Figure 8.22.

Figure 8.22

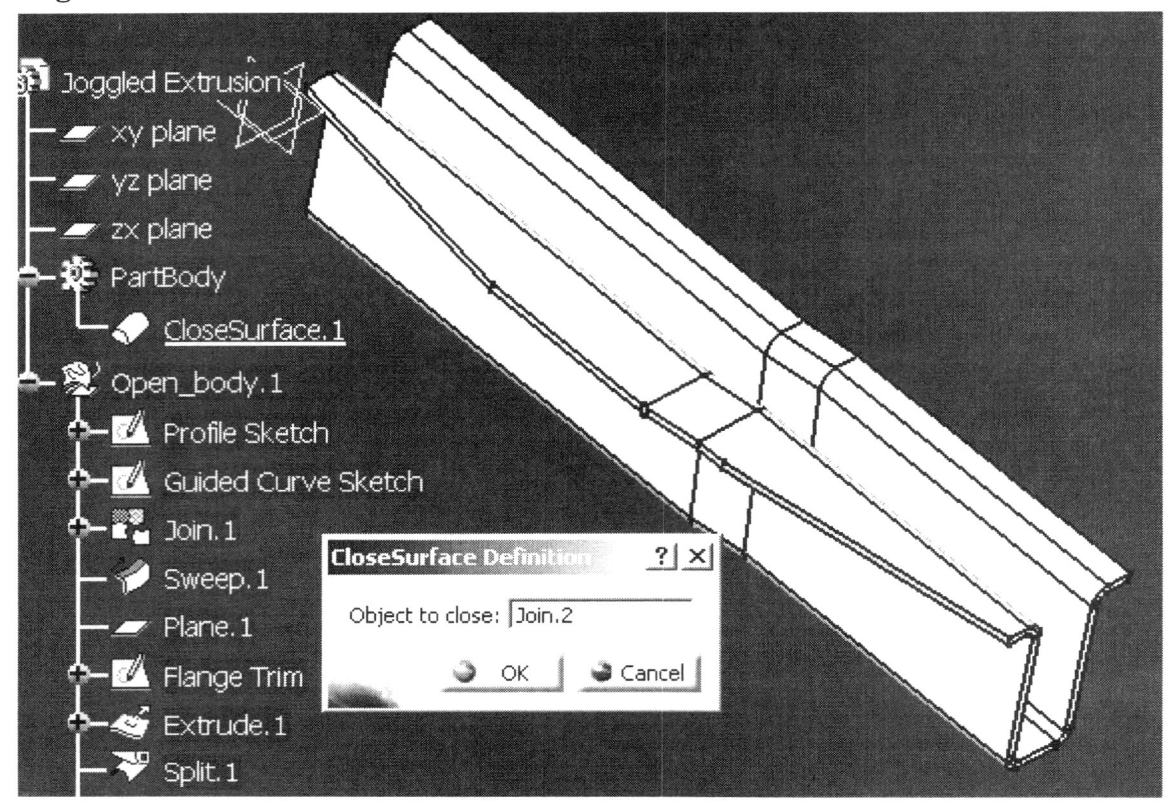

12.6 Select the **OK** button to close the surface and create a solid.

NOTE: The **CloseSurface.1** branch was added to the **PartBody** branch of the **Specification Tree** not the **Open_Body.1** branch. The reason is, that it was created in the **Part Design Work Bench**. Your "**Joggled Extrusion**" now has a surface and a solid visible on the screen.

12.7 At this point you are done working with the surface, so place the **Join.2** surface in the **No Show**.

12.8 At this point the "**Joggled Extrusion**" solid should be the only geometry visible. If your "**Joggled Extrusion**" solid looks similar to what is shown in Figure 8.23, you are ready to proceed to the next step. Just a few more steps and you will be done with the "**Joggled Extrusion**." Figure 8.23 shows the **CloseSurface.1** surfaces highlighted.

Figure 8.23

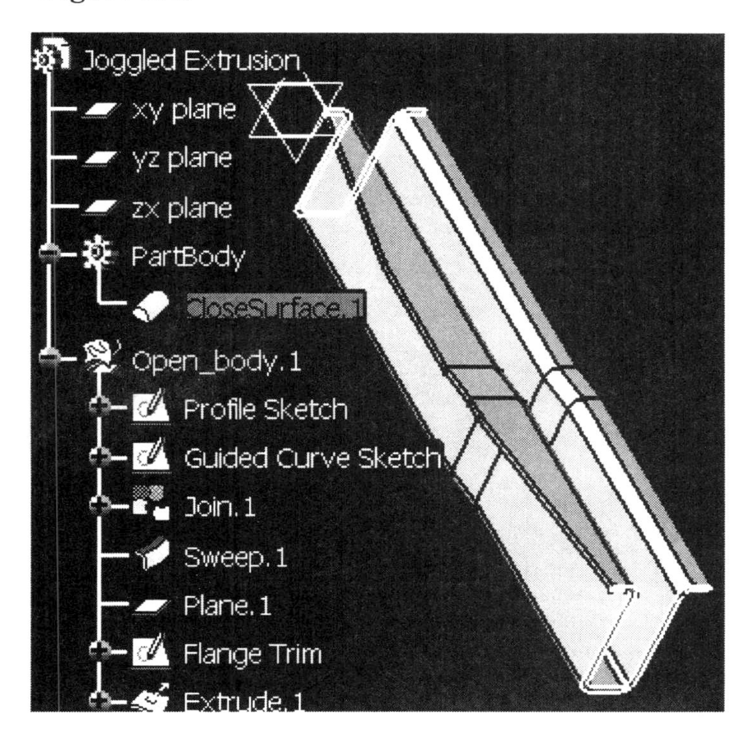

13 Adding The 8-Degree Angle Using The Plane And Pocket Tools

If you reference Figure 8.1, you will notice the right side of the **Front View** shows that the "**Joggled Extrusion**" has an 8-degree angle on it. This angle could have been created in the **Generative Shape Design Work Bench,** but would have added additional steps. This step uses the **Generative Shape Design Work Bench** to define the **Plane** that will create the 8-degree angle. This step uses the **Pocket** tool found in the **Part Design Work Bench**. To create the required angle, complete the following steps.

13.1 The **Plane** tool is found in the **Generative Shape Design Work Bench,** so bring up the **Generative Shape Design Work Bench** up.

13.2 Select the **Plane** tool.

13.3 This will bring up the **Plane Definition** window. For the **Plane Type** box, select the **Angle/Normal To Plane** option.

13.4 For the **Rotation Axis** box, select the bottom horizontal edge of the "**Joggled Extrusion.**" Reference Figure 8.24.

13.5 For the **Reference** box, select the edge surface as shown in Figure 8.24.

13.6 For the **Angle** box, type in **-8deg**. Verify that the **Plane** is rotated 8 deg in the correct direction. If it is not, type in **8deg** and select the **Plane**. The represented **Plane** will be updated to show the revised angle.

13.7 When you get the **Plane** to look similar to what is shown in Figure 8.24, select the **OK** button to create the **Plane**. If you want to move the **Plane** away from the part, you need to select it prior to selecting the **OK** button.

Figure 8.24

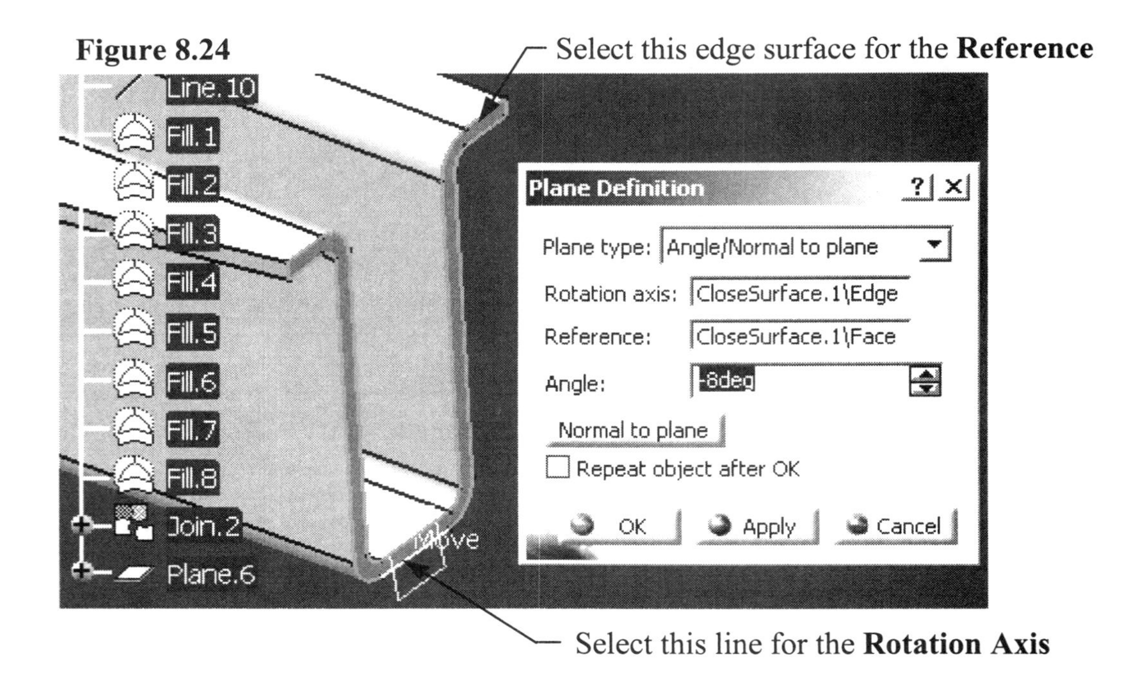

13.8 Now that you created the geometry that will be used to define the 8-degree angle, you can bring up the **Part Design Work Bench**.

13.9 In the **Part Design Work Bench,** select the **Sketcher Work Bench**.

13.10 Select the **Plane** you created in the previous steps.

13.11 All that is required in this step is to create a rectangular sketch that encompasses the end of the "**Joggled Extrusion**." Create a sketch similar to the one shown in Figure 8.25.

13.12 Exit the **Sketcher Work Bench**.

Figure 8.25

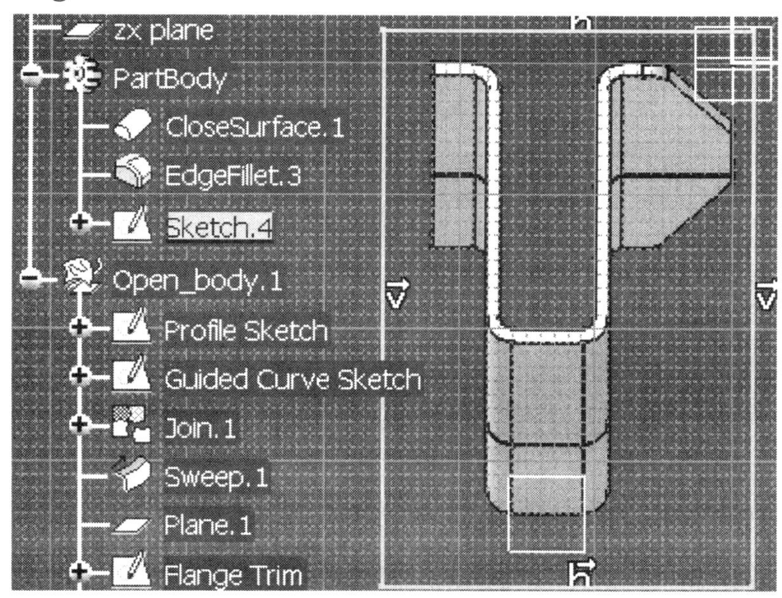

13.13 With the **Sketch** highlighted, select the **Pocket** tool.

13.14 This will bring up the **Pocket Definition** window as shown in Figure 8.26. The exact values are not critical as long as the direction is correct and the **Depth** is enough to encompass the part of the angle that is being trimmed. The values shown in Figure 8.26 are sufficient.

13.15 If your "**Joggled Extrusion**" looks similar to the one shown in Figure 8.26, select the **OK** button to create the **Pocket**.

NOTE: No pocket was created! The 8-degree angle was trimmed off. You could have created a **Pad** and then used a **Boolean Operation** to subtract it out, but that would have required additional steps. As this demonstrates the **Pocket** tool can be a handy tool.

Figure 8.26

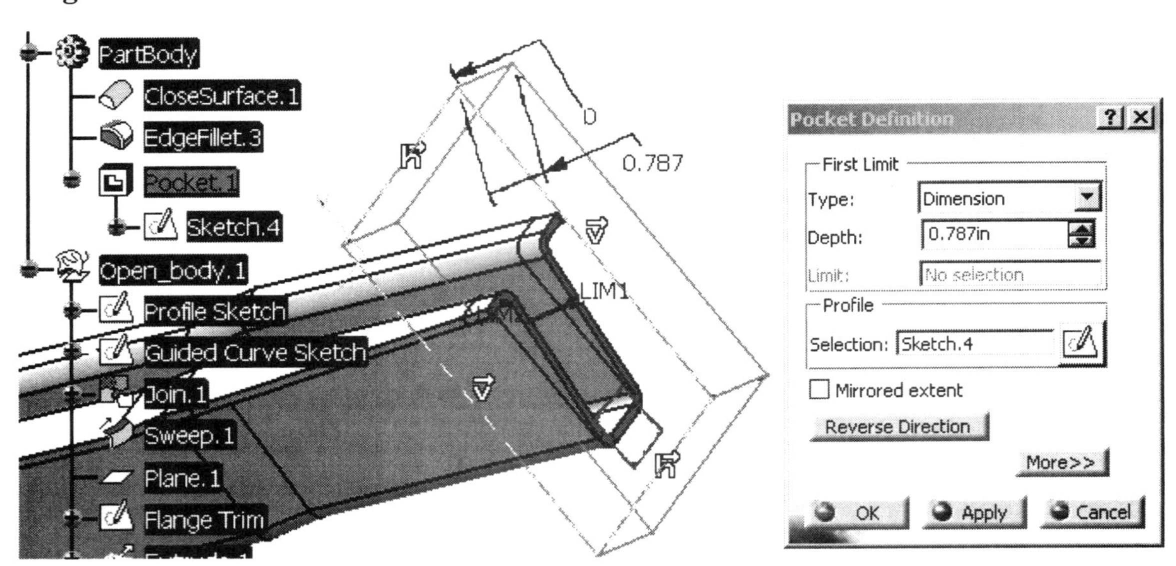

13.16 If your "**Joggled Extrusion**" looks similar to what is shown Figure 8.27, you are ready to create the **Fillets**, Step 14.

NOTE: It is important for you to know that there is a quicker way to trim off the 8-degree angle. The quicker method is to use the **Split** tool in the **Part Design Work Bench**. You are encouraged to try this method. It is a quicker method; no **Sketch** or **Pocket** selection is needed. Even though it is a quicker method in some situations, the trim operation will extend the edge curve representation. For the "**Joggled Extrusion**," the **Pocket** method was a cleaner method. You are encouraged to try trimming the 8-degree angle using the **Split** tool or try one of the many other possible methods. Trying the different variations will give you a chance to compare the different options.

Figure 8.27

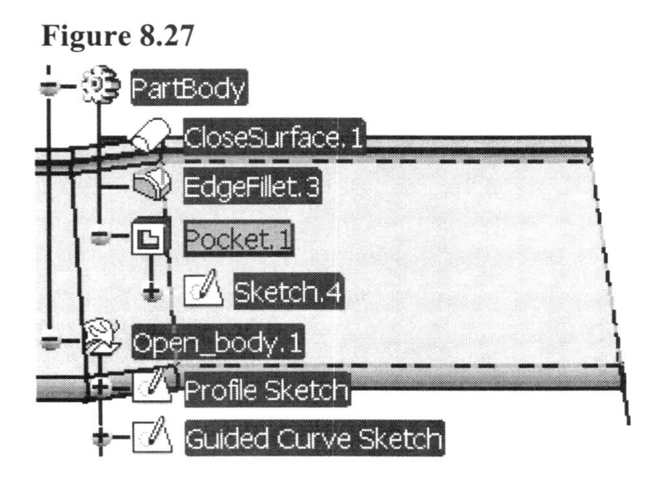

14 Dressing Up The "Joggled Extrusion" Solid Using The Fillet Tool 

The **Fillet** tool was covered in detail in Lesson 2. This process should be a review. The four fillets you will be creating in this step could have been created as surfaces in the **Generative Shape Design Work Bench**, but would have required additional sketches, extruded surfaces, split surfaces, lines defining boundaries, filling surfaces and adding them to the join step. Creating the fillets in the **Part Design Work Bench** requires only a few simple steps. The steps are listed below.

14.1 Select the line defining the corner of the flange as shown in Figure 8.28.

14.2 Select the **EdgeFillet** tool.

14.3 This will bring up the **Edge Fillet Definition** window. For the **Radius:** box, type in **.18in**. Reference Figure 8.1 for part dimensions.

14.4 The edge you selected in Step 13.1 will show up in the **Object(s) To Fillet** box.

14.5 Leave the **Propagation** box on **Tangency**.

14.6 Select **OK** to create the **Edge Fillet**.

Figure 8.28

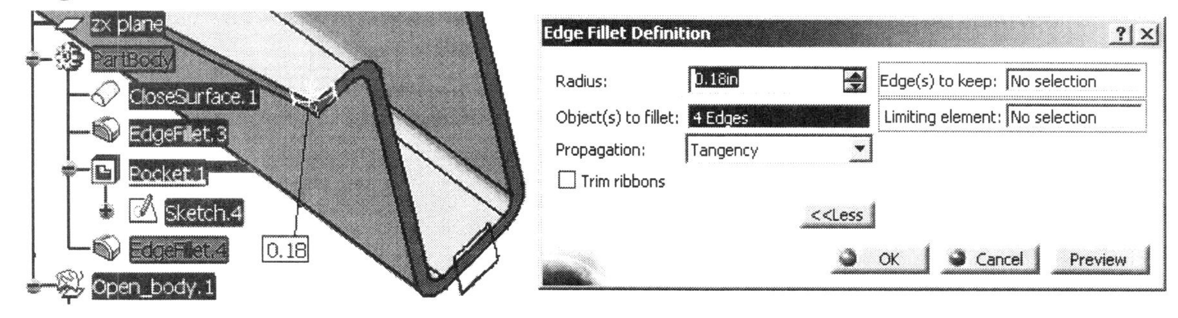

14.7 Select the **EdgeFillet.1** you just created.

14.8 Press the **Ctrl C** keys simultaneously. This copies the **EdgeFillet.1** to the **Windows Clip Board**.

14.9 Select one of the other edges that require an .18″ edge fillet. The dimensioned **Top View** in Figure 8.1 displays the corners that require an .18″ radius edge fillet.

14.10 With one of the other edges highlighted, press the **Ctrl V** keys simultaneously. This pastes the **EdgeFillet.1** properties copied in Step 13.8 to the selected edge. You should now have two .18″ radius edge fillets.

14.11 Repeat the process described in Step 14.10 to the remaining two edges. When you have completed creating the four edge fillets, your "**Joggled Extrusion**" should look similar to the one shown in Figure 8.29.

Figure 8.29

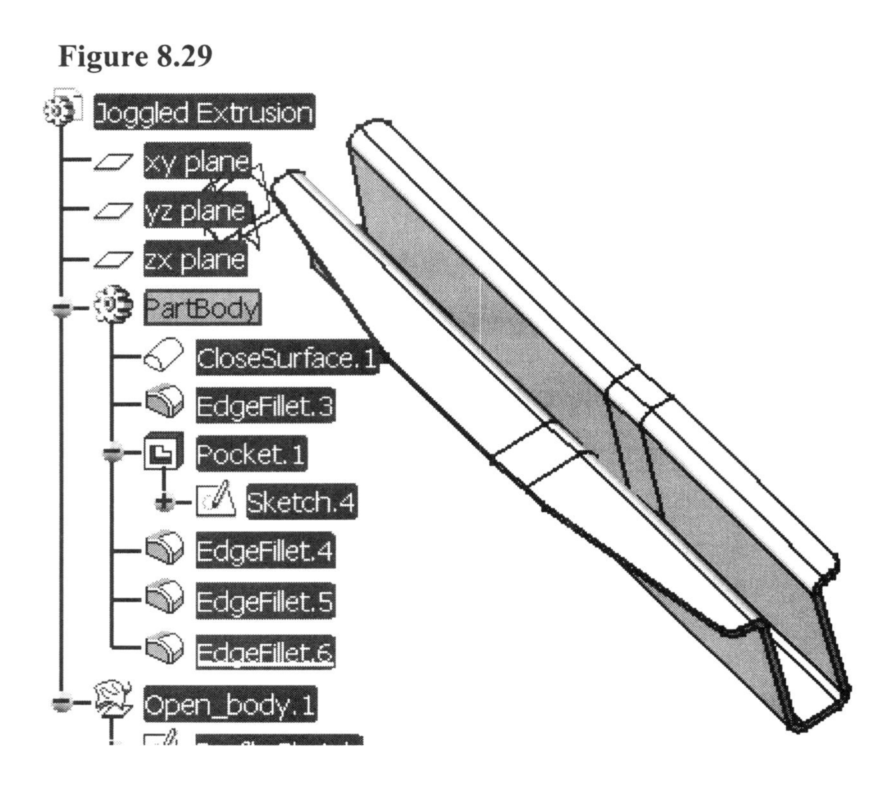

15 The Finishing Touch: Appling Material

As you complete the lessons, you can see how important it is to have a good understanding of the **Sketcher** and **Part Design Work Benches**. Even though the majority of this lesson discussed **Generative Shape Design Work Bench** tools, a good portion of the tools was covered in previous chapters. Lessons 7 and 8 should have given you ideas on new and different ways to create and design three-dimensional geometry. The last step in completing the "**Joggled Extrusion**" is to apply **Aluminum** material and change the **View Parameters** to show the material. If your "**Joggled Extrusion**" looks similar to the one shown in Figure 8.30, you are officially done with the Lesson 8 instruction exercise. You are now ready to tackle the **Lesson 8 Review** questions and **Lesson 8 Practice Exercises.** Congratulations! Don't forget to save your "**Joggled Extrusion.CATPart**."

Figure 8.30

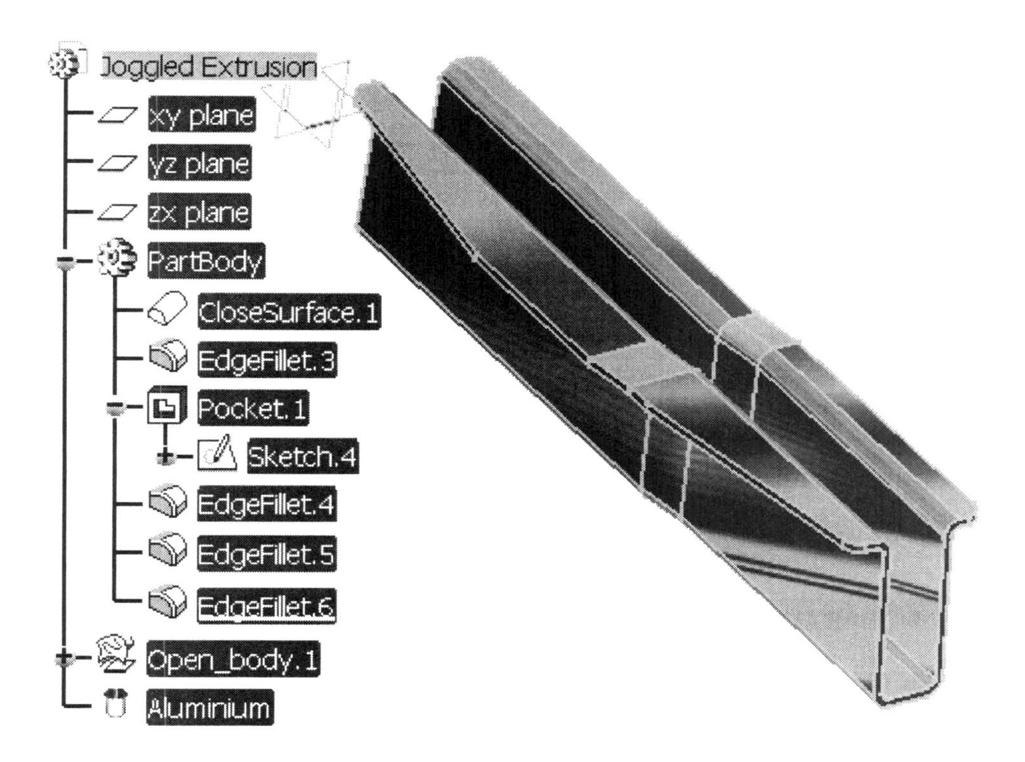

Lesson 8 Summary

As in Lesson 7 there are other possible methods of creating this part. This lesson only covers a few of the possibilities. The **Generative Shape Design Work Bench** is a work bench with a wide variety of applications and possibilities, the last two lessons only scratched the surface. This lesson is designed to give you enough basic information so that you can explore the other possibilities on your own. You are strongly encouraged to explore the other methods.

Lesson 8 Review

After completing this lesson you should be able to answer the questions and explain the concepts listed below.

1. **T** or **F** In the **Generative Shape Design Work Bench,** a sketch can be extruded into a surface.
2. **T** or **F** Geometry created in the **Generative Shape Design Work Bench** is placed under the **Open_Body** branch of the **Specification Tree.**
3. In relationship to the **Sweep** tool, what is the **Profile**?
4. In relationship to the **Sweep** tool, what is the **Guided Curve**?
5. Explain the relationship between the **Profile** and the **Guided Curve**.
6. Why wasn't the "**Joggled Extrusion**" created in the **Part Design Work Bench** like the "**L Shaped Extrusion**" was?
7. **T** or **F** The **Join** tool can join surfaces only.
8. **T** or **F** There is a **Split** tool in the **Generative Shape Design Work Bench**.
9. **T** or **F** There is a **Split** tool in the **Part Design Work Bench**.
10. In this lesson, what tool was used to create a solid (**PartBody**) from a surface (**Open_Body**)?
11. What work bench is the tool in question 19 found in?
12. Most radii and/or fillets can be created at what stage of the design process?

    a. In the **Sketcher Work Bench**.
    b. During the surface creation; in the **Generative Design Work Bench**.
    c. During the solid creation; in the **Part Design Work Bench**.
    d. Both a and b.
    e. All the above.

13. When a surface is trimmed away, is it deleted?
14. Explain your answer to question 13.

15. When extruding a surface, CATIA V5 allows you which methods of determining the length (limit) of the extrusion.

    a. Typing in the **Limit 1** value in the **Extruded Surface Definition** window.
    b. Typing in the **Limit 2** value in the **Extruded Surface Definition** window.
    c. Selecting the green arrow and dragging it to the desired extrusion length value.
    d. Both a and b.
    e. All of the above.

16. **T** or **F** The result of using the **Split** tool in the **Generative Surface Design Work Bench,** is one combined surface.

17. **T** or **F** In this lesson, when you used the **Sweep** tool, a **Plane** was required to complete the process.

18. What **Generative Shape Design Work Bench** tool was used to create and define the part thickness?

19. When creating several sketches or other similar entities, why is it a good idea to rename them instead of leaving them at the default name? For example Sketch.1, Sketch.2 and so on.

20. What workbench is the **Close Surface** tool located in?

Lesson 8 Practice Exercises

Now that your CATIA V5 tool box has more tools in it, put them to use on the following practice exercises. Each of the practice exercises will require the **Generative Shape Design Work Bench** and the **Part Design Work Bench**.

1. Create a sketch on the **ZX Plane** as shown in Figure 8.31. Name the sketch, "**Guided Path.**"

2. Create a sketch on the **YZ Plane** as shown in Figure 8.32. Name the sketch, "**Profile.**" Notice where the 0,0 point is on both sketches. It is important that the sketches are in correct relationship to each other, reference Figure 8.33.

Figure 8.31

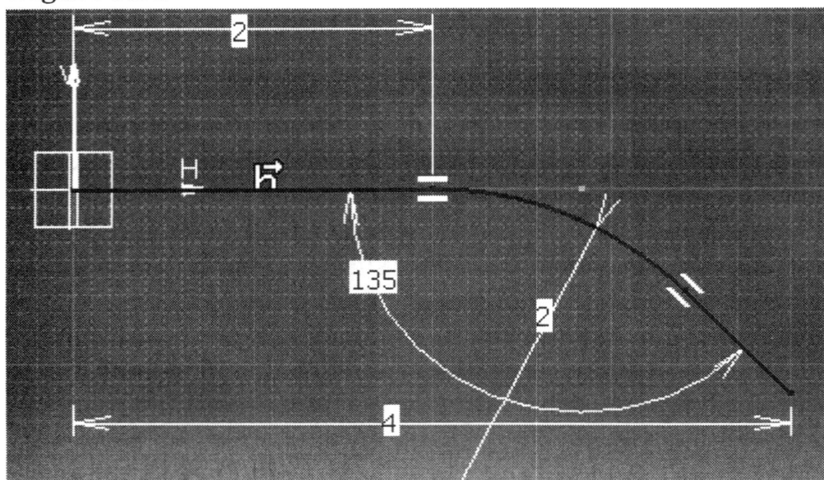

Figure 8.32

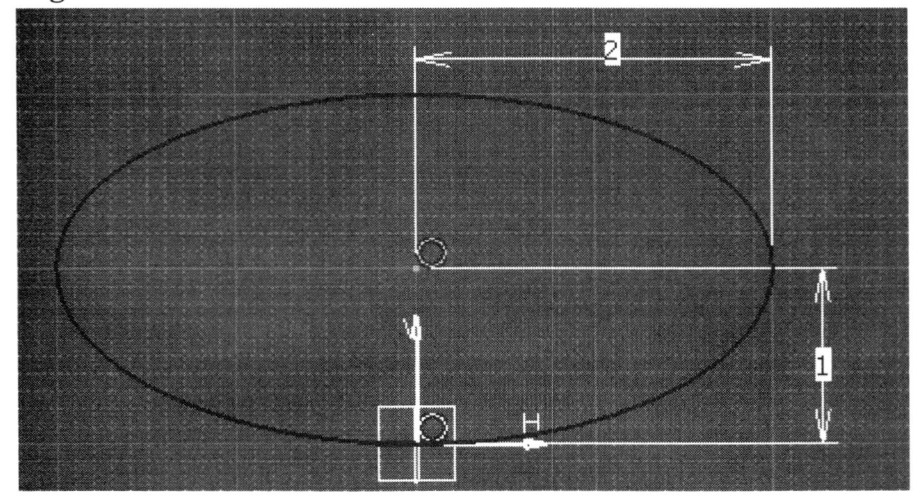

Figure 8.33

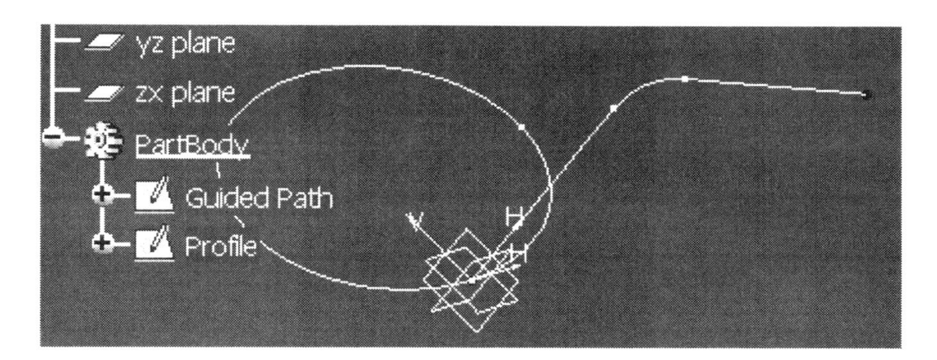

3. Use the **Sweep** tool to create a swept surface. Select the **Profile** sketch for the **Profile** box and the **Guided Path** sketch for the **Guided Path** box. Rename the **Sweep** branch to "**Pipe Surface**."

4. **No Show** the "**Pipe Surface**" surface created in Exercise 3. Create another surface using the **Sweep** tool. This time select the **Guided Path** sketch for the **Profile** box and the **Profile** sketch for the **Guided Path** box. Rename this surface to "**Flange Surface**."

5. Select the **Part Design Work Bench**. Give the **Flange Surface** a **.075in** thickness. The **Flange Surface** is the inside surface. **Hint:** Use the **Thick Surface** tool. After **No Showing** the **Open_Body** entities, your part should look similar to the one shown in Figure 8.33.

6. **No Show** the **ThickSurf** branch created in the **PartBody** in Exercise 5. Use the **Offset** tool and **Close** tool in the **Part Design Work Bench** to accomplish the same thing. Save your completed part as "**Lesson 8 Exercise 6.CATPart**."

7. Go back into the **Guided Path** sketch and modify the **2″** radius to **1″**. Then modify the **ThickSurf** value from **.075″** to **.125″**. Notice how CATIA V5 automatically updates the surfaces and solids to the new values. Save your completed part as "**Lesson 8 Exercise 7.CATPart**."

8. Observe the differences between the two processes. What extra steps were required to complete the exercise using the process required in Exercise 7?

Figure 8.34

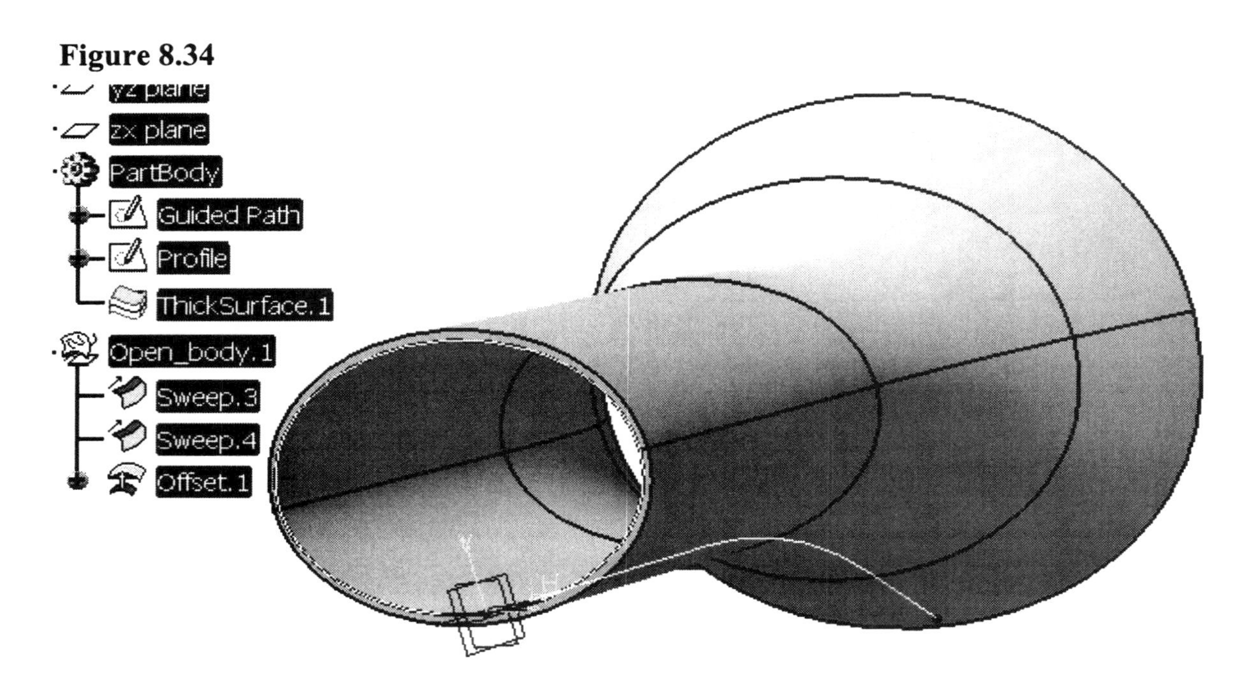
yz plane
zx plane
PartBody
Guided Path
Profile
ThickSurface.1
Open_body.1
Sweep.3
Sweep.4
Offset.1

Lesson 9 DMU Navigator

Introduction To The DMU Navigator Work Bench

The previous lessons should be a solid testament to the power of parametric modeling. The **DMU Navigator Work Bench** supplies the tools to take the power of parametric modeling to even a higher level. The tools covered in this lesson allow the user to check and verify the design concept virtually taking the place of a physical mock up.

In order for you to complete this lesson, you must download a zipped file (**SuperMileage.zip**) from the SDC Publications website. This lesson uses twelve CATPart files and one CATProduct file that make up a concept car known as the **SuperMileage.CATProduct** as shown in Figure 9.1. Using this larger assembly within the DMU Navigator will allow you to better understand how the DMU fits into the product review process.

Figure 9.1 SuperMileage.CATProduct

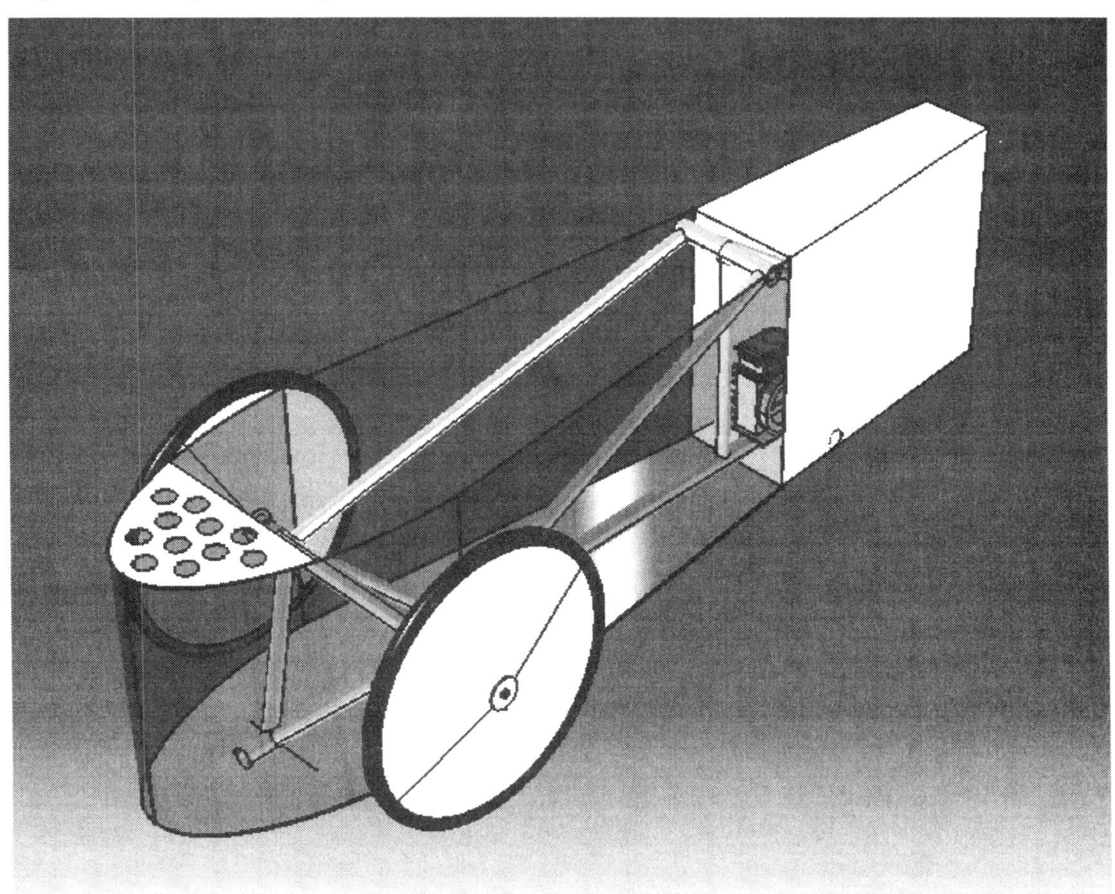

Lesson 9 Objectives

This Lesson will show you how to do the following:

- Download the **SuperMileage.zip** file
- Unzip the **SuperMileage.zip** file
- Enter the **DMU Navigator Work Bench**
- Introduce you to all The **DMU Navigator** tool bars & tools.
- How to insert components into the **DMU Navigator Work Bench**
- How to use the **Fly Mode** and **walk Mode**
- How to create and modify **2D** and **3D Annotation Views**
- How to create, use and modify **Scenes**
- How to create a simple **Animation**

DMU Navigator Work Bench Tool Bars 

There are seven standard tool bars found in the **DMU Navigator Work Bench**. Five tool bars are located on side bar and the other two on the bottom horizontal bar. The **View** tool bar is briefly covered in the **Introduction**. This lesson covers the **View** tool bar in a little more depth.

The **DMU Data Navigation** Tool Bar

Tool Bar	Tool Name	Tool Definition
	Search	Searches and selects objects according to specified criteria (name, type, color).
	Proximity Query	Retrieves the neighbor's components in session from some reference components.
	Current Selection Panel	Shows the objects in the current selection.
	Go To Hyperlinks	Jumps to hyperlinks.
	Start Publish	Starts a publish session.

The **DMU Navigator** Tools Tool Bar

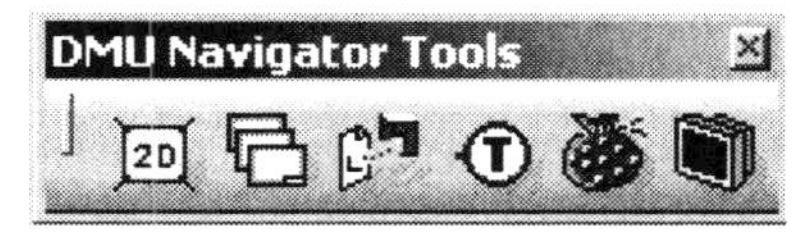

Tool Bar	Tool Name	Tool Definition
	Create An Annotated View	Creates an annotated view. This tool will also give you access to the **DMU 2D Marker** tool bar. The tools in the **DMU 2D Marker** tool bar are listed below (the indented tool icons).
	Draw Line	Draws a line similar to line created in a paint program.
	Draw Free Hand Line	Draws a free hand line similar to line created in a paint program.
	Draw Circle	Draws a circle similar to line created in a paint program.
	Draw Arrow	Draws an arrow similar to line created in a paint program.
	Draw Rectangle	Draws a rectangle similar to line created in a paint program.
	Annotation Text	Adds text similar to line created in a paint program.
	Insert A Picture Marker	Allows you to insert a picture file in the view.
	Create A Audio Marker	Allows you to insert a audio file in the view
	Delete Marker	Deletes all view markers created by the tools in the DMU 2D Marker tool bar.
	Manages Annotated Views	Manages all the annotated views. Selecting this tool will bring up a window listing all the created views. You can select the annotated view to make it active.
	Add Hyperlinks	Adds hyperlinks.
	3D Annotation	Associates 3D annotation text. This text will attach to a selected entity so as you rotate the model the text stays normal to your view.
	Group	Edits a group of objects.
	Create Scene	Creates a new scene. You can create multiple scenes and move from scene to scene.
	Sectioning	Creates a sectioning plane through the CATPart/CATProduct.

The **DMU Generic Animation** Tool Bar

Tool Bar	Tool Name	Tool Definition
	Play Simulation	Plays a simulation.
	Edit Sequence	Edits a specified sequence.
	Record ViewPoint Animation	Records viewpoint animation.
	Simulation Actions Icon Box Tool Bar	
	Track	Edits the track of a moving object.
	Color Action	Creates color action.
	Visibility Action	Create visibility action.
	Automatic Clash Detection Tool Bar	
	Clash Detection (Off)	This is a toggle tool, it turns the clash detection off.
	Clash Detection (On)	This is a toggle tool, it turns the clash detection on.
	Clash Detection (STOP)	Activates automatic clash detection in stop mode.

The **Filter** Tool Bar

This tool bar has been covered in previous lessons.

Tool Bar	Tool Name	Tool Definition
	Products Selection	Selects only products.

The **DMU Move** Tool Bar

Tool Bar	Tool Name	Tool Definition
	Translation Or Rotation	Moves a component by translation or rotation.
	Snap	Moves a component by snapping.
	Creates A Symmetry On A Component	Creates a symmetry of a child component.
	Reset The Position Of Products	Resets the position of all the products.

The **DMU Viewing** Tool Bar

This tool bar is located in the bottom horizontal tool bar.

Tool Bar	Tool Name	Tool Definition
	Look At	Allows you to view the document from a specific direction. It is a view port.
	Previous View	Allows you to return to the previous view.
	Next View	Switches to the next view.
	Magnifier	Allows you to magnify the current view.
	Depth Effects	Creates depth effects in the current view.
	Horizontal Ground	Shows/Hides the ground.
	Lighting	Creates lighting effects using lighting sources.

The **View** Tool Bar

Selecting the **Fly Mode** tool will change the **View** tool bar as shown. Only the tools related to the fly thru mode are discussed in this lesson. The remaining tools have been covered in the previous lessons.

Tool Bar	Tool Name	Tool Definition
	Fly Mode or Walk Mode	Sets the fly mode. This tool is a toggle between the **Examine Mode** and the **Fly Mode**. To access **Walk Mode** select **View**, **Navigation Mode** and then **Walk Mode**. Then the **Walk** tool will appear in the place of the **Fly** tool
	Examine Mode	Sets the examine mode. This tool is a toggle between the **Examine Mode** and the **Fly Mode**.
	Fit all In	Same as discussed in previous lessons it brings all existing geometry into view of the screen. CATIA V5 may zoom in or out automatically to accomplish this.
	Turn Head	Allows you to view the entire model from a fixed position. You can rotate your view from that one position.
	Fly/Walk	This tool allows you to fly through the model (left, right, up and down). **Walk Mode** locks you down to a defined plane.
	Accelerate	This tool allows you to increase your fly speed. The **PageUp** key accomplishes the same thing.
	Decelerate	This tool allows you to decrease your fly speed. The **PageDown** key accomplishes the same thing.

Steps To Navigating The Assembly Using The DMU Navigator Work Bench

This lesson requires that you download a file form the www.schroff1.com/catia website. The file that you download will be the model (CATProduct) you use throughout the lesson.

1 Downloading The File From The Schroff Development Company Website

Before you can begin using the **DMU Navigator**, you must first download the zipped file (**SuperMileage.zip**) from the SDC Publications website to your hard drive.

1.1 Create a directory called "download" in your **Temp** directory on your hard drive **(i.e. - C:\Temp\Download)**. Use this directory for your download location.

1.2 You can access the file by going to the website: **http://www.schroff1.com/catia.**

You can also access the file through: **http://www.schroff.com.** This website takes a bit of navigation to get to the file. The Suggested path is as follows:

1.2.1 Select **SDC Educational Books**.
1.2.2 Select **CATIA**.
1.2.3 Select **CATIA V5 Workbook Release 10 & 11**.
1.2.4 Select **Resources**.

1.3 Locate the CATIA V5 workbook within the site and find the "**Resources**" section with the link labeled **SuperMileage.zip.**

1.4 Start the download process to your download directory that you created (C:\Temp\Download**)**. Note: The zipped file is 2 megabytes in size and therefore will take time to download.

2 Unzip The SuperMileage.zip File To Your Working Directory

Once you have successfully downloaded the **SuperMileage.zip** file to your download directory, you must then extract the CATProduct and CATPart files in order to use the **DMU Navigator Work Bench**. Extract the files to a directory to the **Temp** directory and at the conclusion of the process a directory called SuperMileage will be created in the Temp directory **(i.e. - C:\Temp\SuperMileage).**

3 Entering The DMU Navigator Work Bench

Previous lessons taught two different methods of selecting a particular work bench. In this lesson you will be inserting an entire assembly as an existing component into the **DMU Navigator Work Bench**. The following steps explain how to activate the **"DMU Navigator Work Bench."**

3.1 Prior to inserting an existing component (**SuperMileage.CATProduct**) into the **DMU Navigator** the first step is enter the **DMU Navigator Work Bench**. To do this, select the **Start** pull down menu from the top of the screen as shown in Figure 9.2

3.2 From the **Start** pull down menu select the **Digital Mockup** option shown in Figure 9.2.

3.3 From the **Digital Mockup** menu select the **DMU Navigator** option. This will bring you into the **DMU Navigator Work Bench**.

NOTE: The name at the top of the **Specification Tree** is "**Product 1.**" You might have noticed that this applied to the **Assembly Design Work Bench** in Lesson 6. Remember, you can use the "**contextual menu**" to change the properties and rename **Product 1** to **Super Mileage**.

Figure 9.2

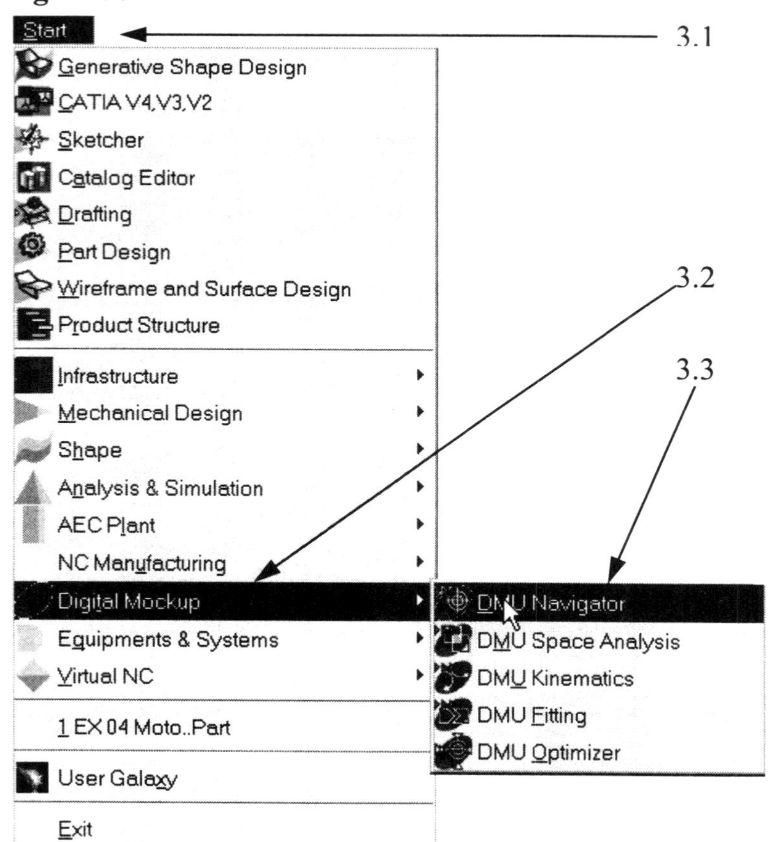

4 Inserting Objects Into The DMU Navigator Work Bench

In Lesson 6 you learned how to insert detail parts to make an assembly. Although you can insert detail parts into a **DMU Navigator** session, during this lesson you will insert an entire assembly (**CATProduct**) as an existing component into the DMU Navigator. Follow the steps below to bring the **"SuperMileage.CATProduct"** assembly file into the **DMU Navigator Work Bench**.

4.1 The top of the **Specification Tree** must be selected before objects can be inserted into the **DMU Navigator Work Bench**. The top of the **Specification Tree** is labeled as **Product1** (or **Super Mileage** if you renamed it using the **Contextual Menu** with **Properties**).

4.2 Use your right mouse button to activate the menu then scroll down and select the **Existing Component** icon as shown in Figure 9.3.

Figure 9.3

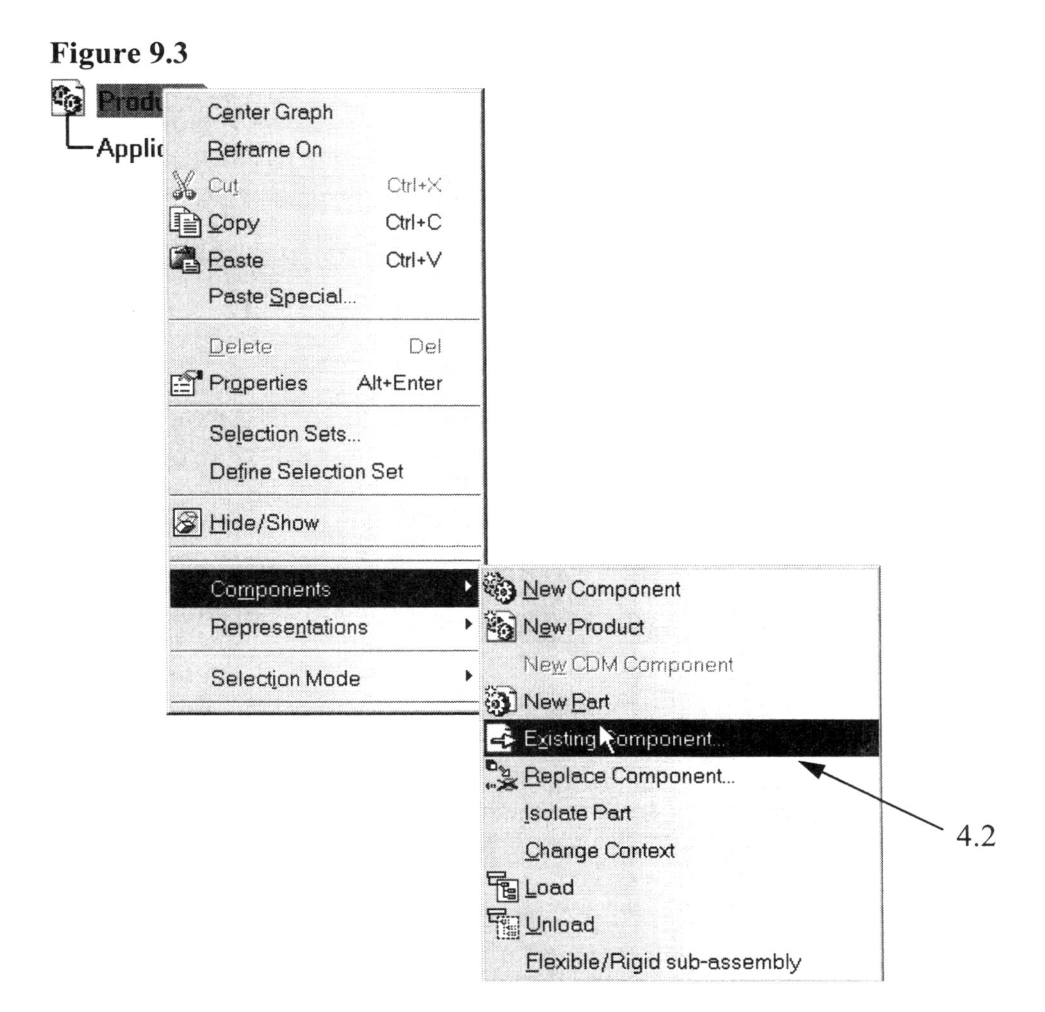

4.3 The **Insert an Existing Component** window will appear as shown in Figure 9.4.

Figure 9.4

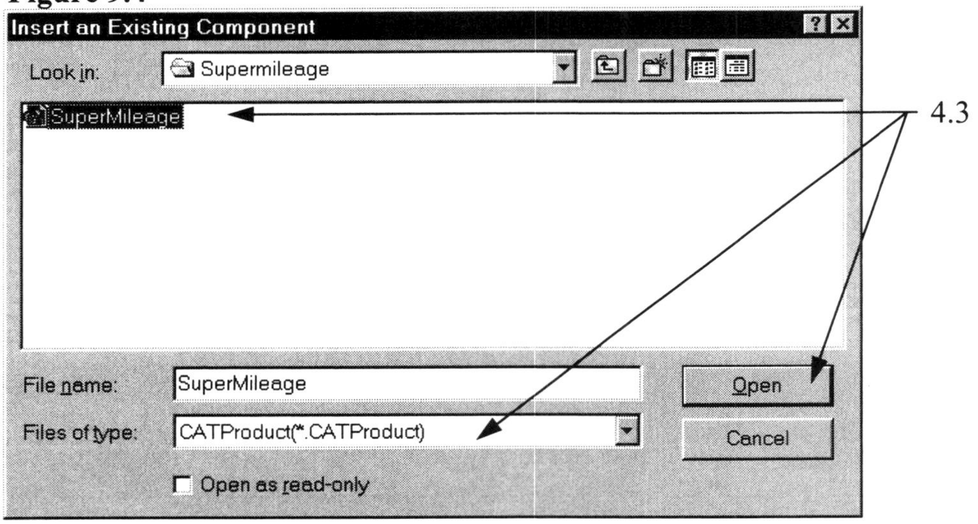

4.4 In the **Insert An Existing Component** (Figure 9.4) window find the file named "**SuperMileage.CATProduct.**" Select the **Open** button. The assembly should appear on the screen. If it does not, then select the **Fit All In** icon and then select the **ISO** icon from the **View Toolbar**. Your screen should show the SuperMileage assembly similar to what is shown in Figure 9.1.

5 Navigating Through The Assembly Using The Fly Mode

Up until this point you have been able to manipulate your part and/or assembly in the **Examine Mode**. You have done so by using your mouse buttons to rotate, pan and zoom parts and assemblies. Now you will learn to fly freely through a digital mockup of the SuperMileage assembly. You will use your mouse buttons to start the **Fly Mode**; learn to use the mouse like a joystick while controlling the speed with the **Page Up** and **Page Down** keys on your keyboard (to the left of your numeric keypad). The steps below will show you how to navigate using the **Fly Mode** from the **Front View** of the assembly. Once inside the assembly, we will stop to use several tools on the **Viewing Tool Bar** to give you a better perspective on how the **DMU Navigator** can be used in the design review process. You will then repeat the steps starting from the **Back View** to give you more experience in the **Fly Mode**.

NOTE: The response you get from the computer will be directly related to the following; speed of the processor, RAM and the size of the file and if mater.

5.1 First, turn off the **Specification Tree** to free up screen space by pressing the **F3** key. (Note: Also use **View** – **Specifications** popdown.)

5.2 Look at the assembly to verify that it is not one gray shaded mass, but has the materials files activated. If not, then select the **Applied Customized View Parameters** icon located in the **View** tool bar. Refer to the **CATIA Standard Menu** and **Tools** section in the **Introduction** section of this book.

5.3 Select the **Sets The Fly Mode** icon which is located on the far left of the **View Tool Bar**. You will receive a prompt indicating that it will only work if in perspective projection. Select **Yes**. Look at the same location of the tool bar and you will see a new icon appear called the **Sets The Examine Mode** . Select this icon to take you out of the **Fly** Mode so we can make a small change to help in the visualization of the assembly. Notice when you are in fly mode the rotation tool does not work the same.

5.4 Select the **Horizontal Ground** from the **DMU Viewing Menu** which will then display a black grid underneath the SuperMileage assembly similar to what is shown in Figure 9.5.

Figure 9.5

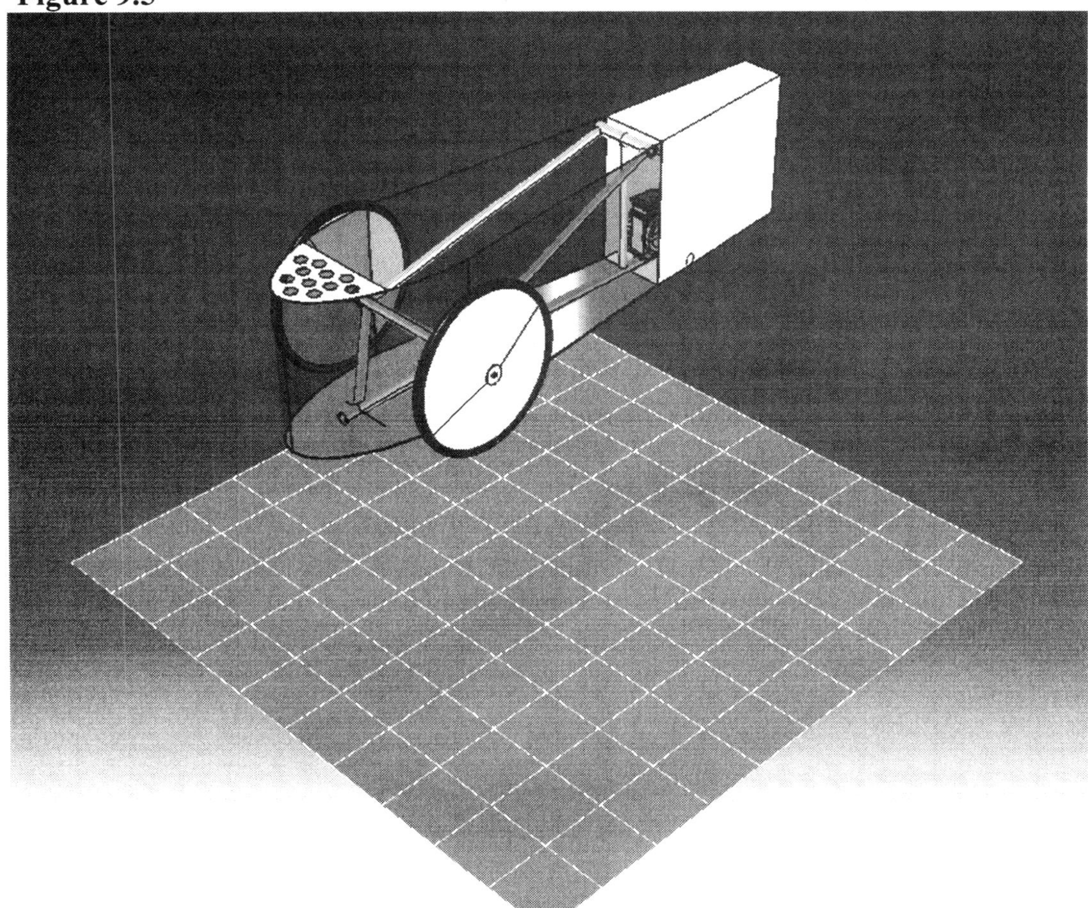

5.5 Select the **Front View** icon in the **View Tool Bar**. This will reorient the assembly so you can move the **Horizontal Ground** closer to the bottom of the vehicle.

5.6 Place your cursor over the **Horizontal Ground** and move the ground plane closer to the bottom of the vehicle. Move the ground plane by holding down the Left Mouse button and moving your mouse reference Figure 9.6.

Figure 9.6

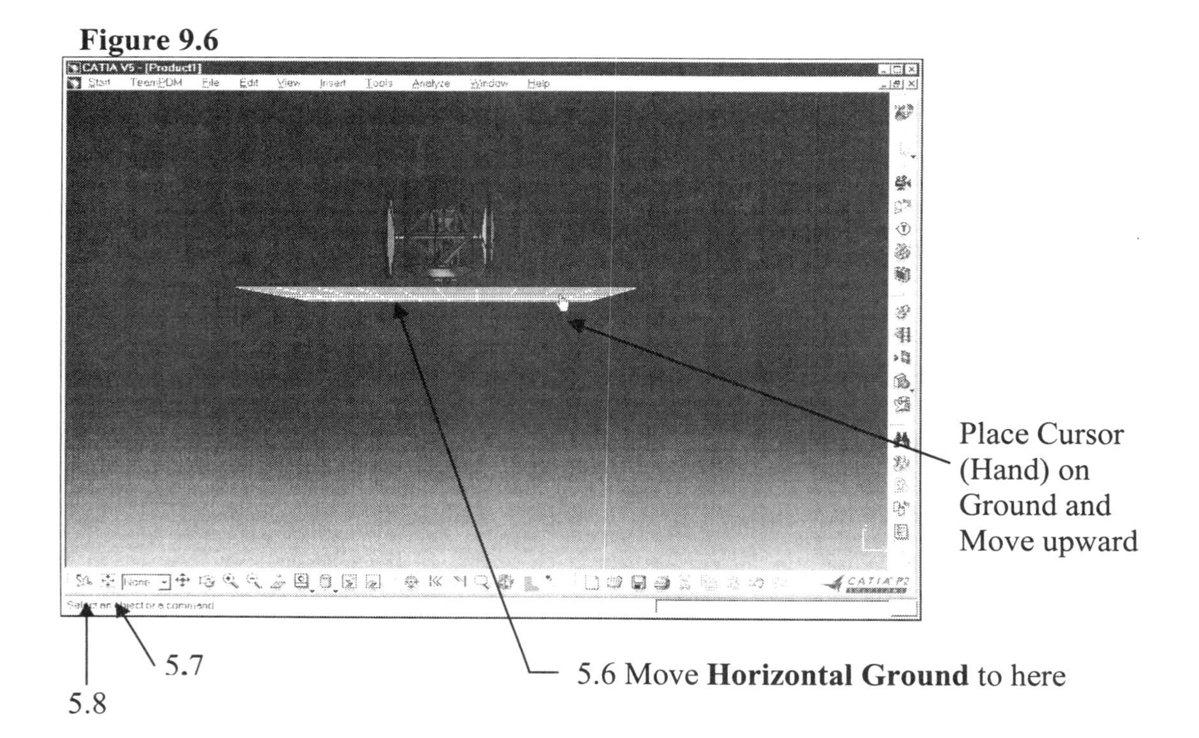

5.7 Select the **Fit All In** icon to position the vehicle and the **Horizontal Ground** in the middle of your window.

5.8 Select the **Sets The Fly Mode** icon and place your cursor on the **Horizontal Ground** and centered between the two wheels.

5.9 For the first **Fly Mode**, DO NOT move your mouse. While holding down the **Middle** mouse button, quickly tap the **Left** mouse button and continue to hold down the Middle button. Watch the vehicle fly over your head on the screen similar to Figure 9.7.

NOTE: A small white direction arrow and a circular target site appear on the **Horizontal Ground**. The large **Green Arrow** indicates direction controlled by the mouse and current speed. Speed can be controlled by quickly tapping the **Page Up** and **Page Down** keys.

Figure 9.7

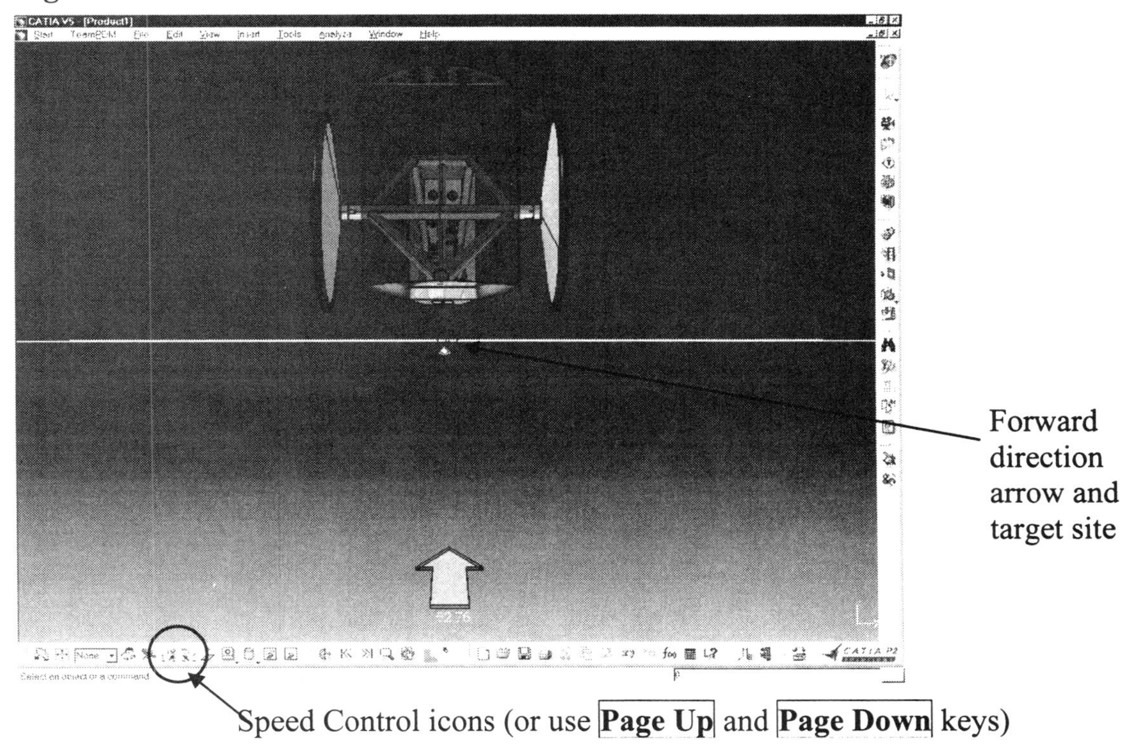

5.10 Select the **Fit All In** icon and then select the **Front** icon to position the vehicle and the **Horizontal Ground** in the middle of your window once again.

5.11 Repeat the exact same process in step 5.9 but when one half of the vehicle passes overhead, quickly tap the **Left** mouse button to reverse the direction. Watch the vehicle move back over your head. Try changing directions several times until you get the feel of the mouse.

5.12 Select the **Fit All In** icon and then select the **Front** icon to position the vehicle and the **Horizontal Ground** in the middle of your window once again.

5.13 Repeat the process as in step 5.9, but this time you will move your mouse like a joystick to simulate left and right and up and down. Watch how the large **Arrow** changes shape in relation to the movement of the mouse.

5.14 Select the **Fit All In** icon and then select the **Front** icon to position the vehicle and the **Horizontal Ground** in the middle of your window once again.

5.15 This time, place your cursor in the exactly in the middle of the front axle and then start the **Fly Mode**. Fly through the front of the vehicle and stop when your screen looks similar to **Figure 9.8**.

Figure 9.8

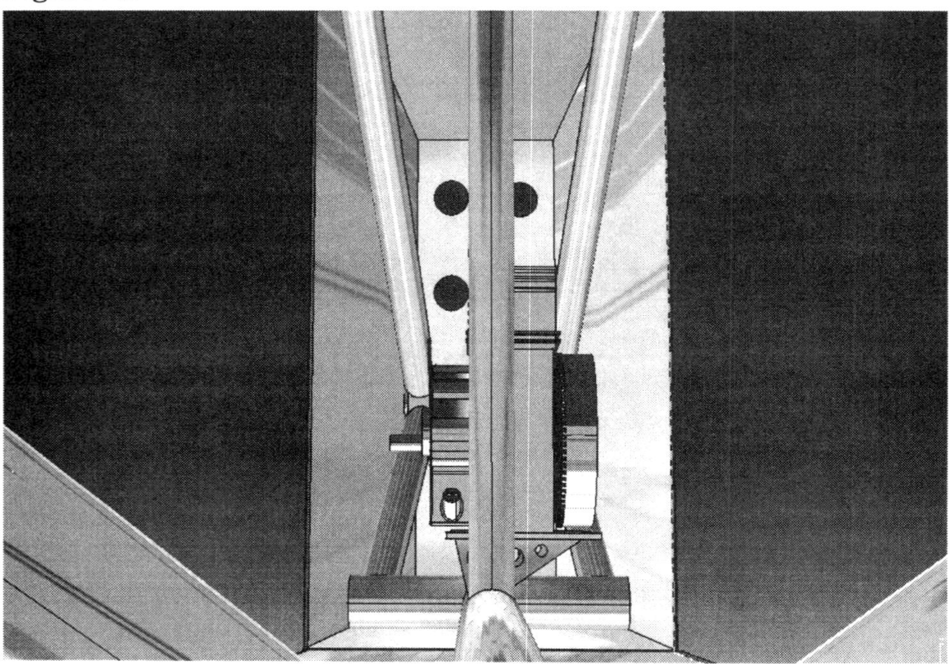

5.16 Select the **Lighting** icon from the **Viewing** toolbar. This will open the **Light Sources** window.

5.17 The **Light Sources** window will appear on the screen (Figure 9.9). Select the **Two Lights** icon. Did you notice a lighting change? You can modify the direction of the light source by placing your mouse over the directional light source and dragging it to a new location around the sphere. Again, notice the lighting change in the Super Mileage model as you modify the light source.

Figure 9.9

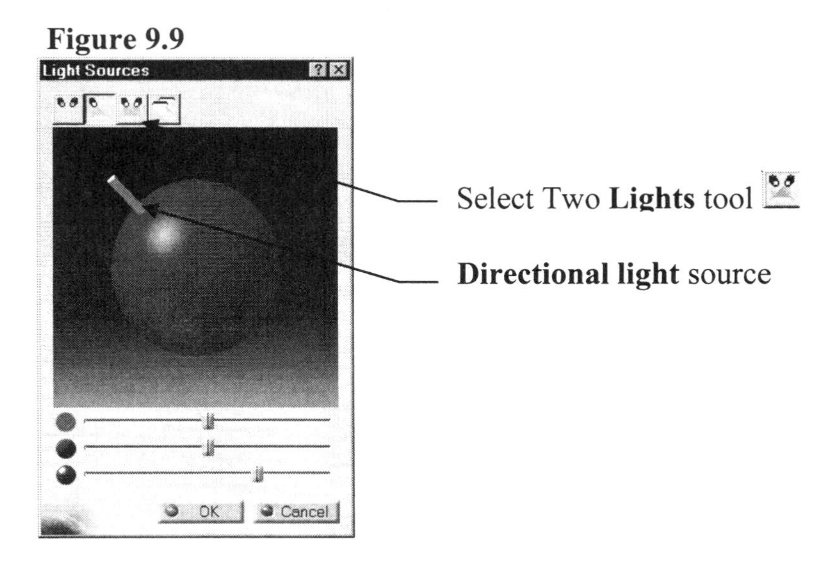

5.18 Close the **Light Sources** window and select the **Magnifier** icon from the **DMU Viewing Tool Bar**.

5.19 The **Magnifier** window will appear on the screen. Position the window in the upper left hand corner as shown in Figure 9.9.1. Drag the **Square** over the shaft of the engine. This will put the shaft in the **Magnifier** window. You can grab one of the corners of square and drag it to enlarge its size.

Figure 9.9.1

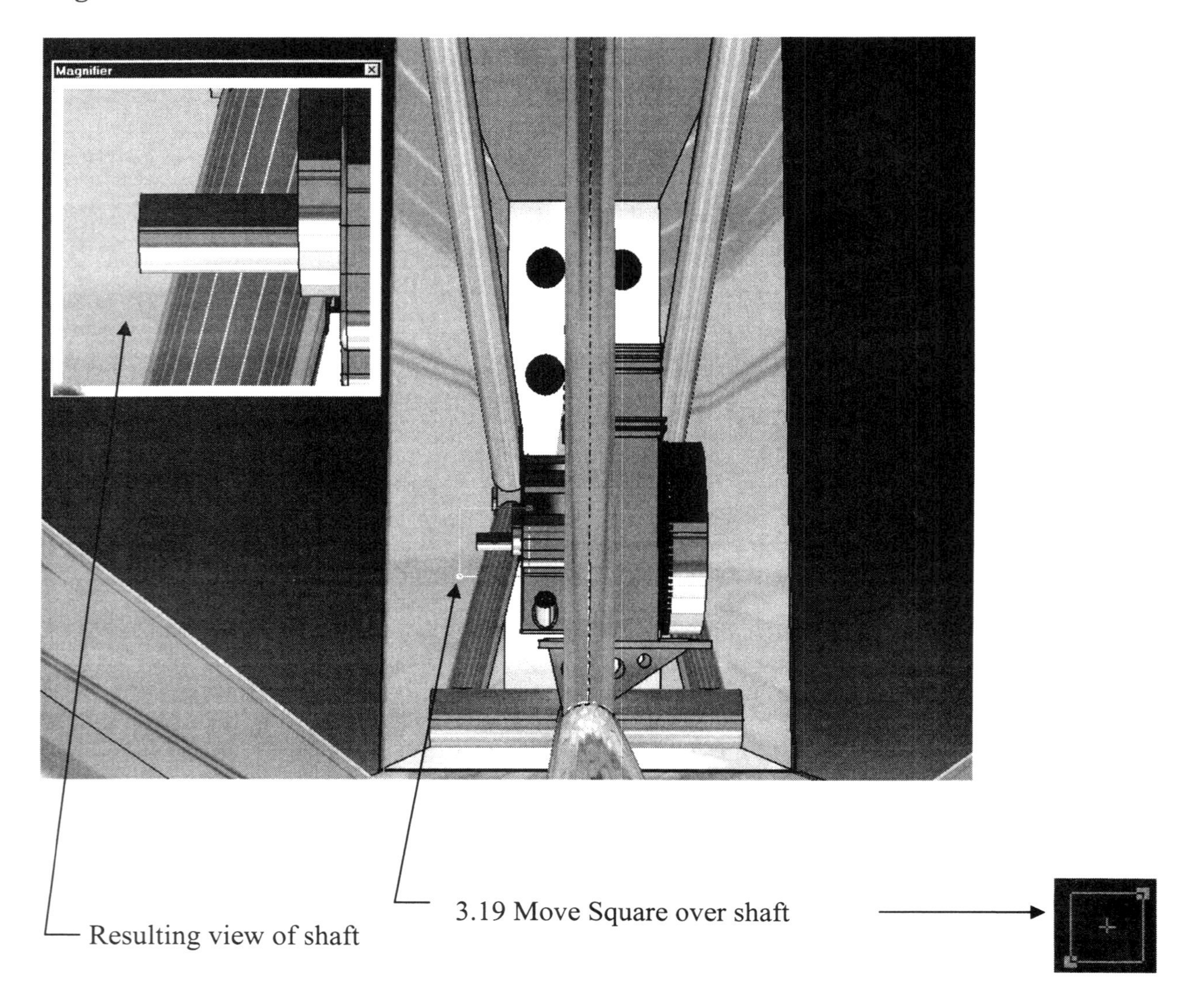

5.20 Select the **Measure Between** icon from the **DMU Measure** tool bar.

5.21 The **Measure Between** window will appear on the screen (Figure 9.9.2).

Figure 9.9.2

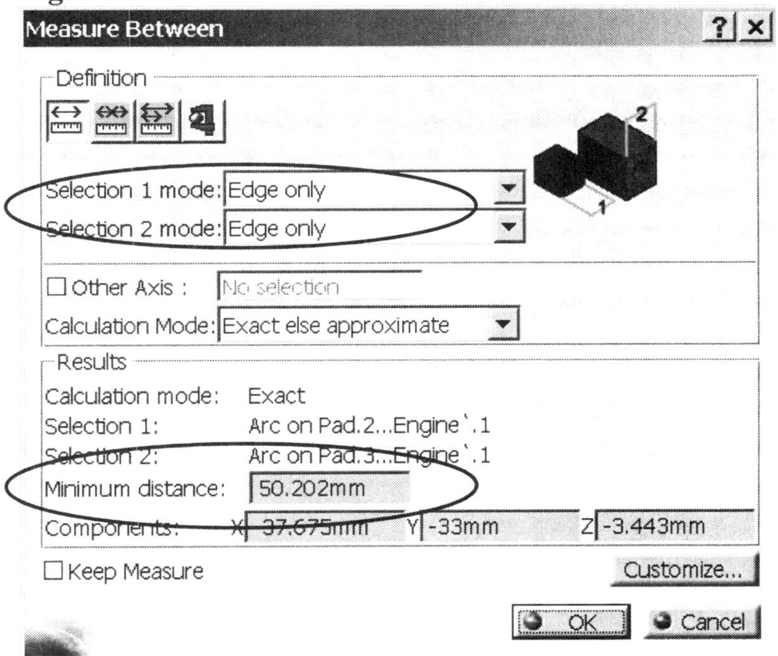

5.22 You will now measure the length of the shaft that sticks out from the engine. Select "**Edge Only**" for **Selection 1 & 2 Modes**. Take your cursor and select the circular geometry as show in **Figure 9.9.3** and notice the results that appear in the **Magnifier** window. This same distance information appears in the **Minimum distance** area on the **Measure Between** window (Figure 9.9.2). Take a minute and try some of the additional options in the Measure Between window. Notice that you can keep the measurements taken even after closing the window. Remember the units of measurement will depend on the default units unless you changed them as explained in Lesson 1.

Figure 9.9.3

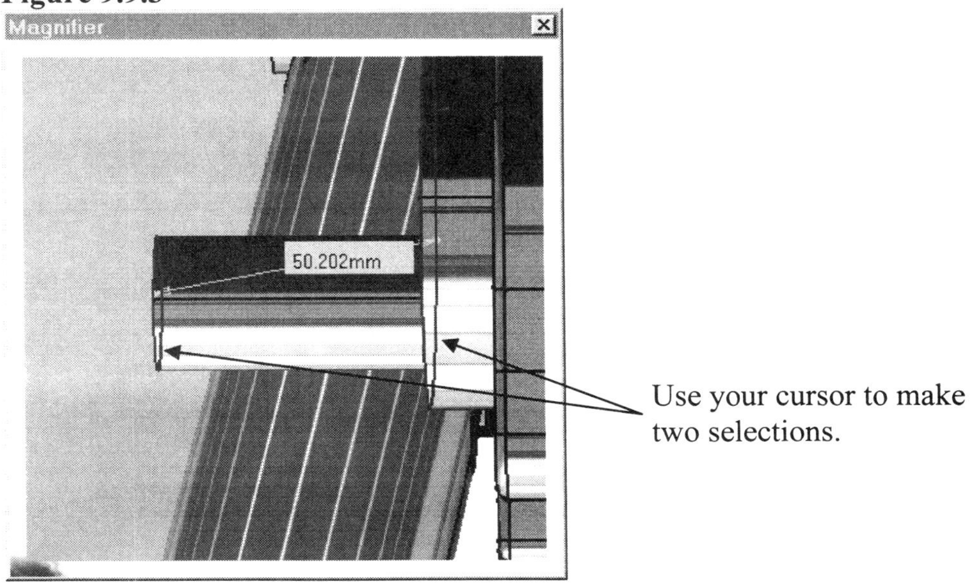

5.23 Close the **Magnifier** and **Measure Between** windows.

5.24 Select the **Turning Head** icon from the **View Tool Bar**.
Place your cursor in the middle of the vertical tubing (refer back to Figure 9.9.1) and hold down the **Left** mouse button. See if you can rotate to your left using your mouse. Rotate until you can see the front right wheel as shown in **Figure 9.9.4**. Note: if you release the mouse button during the rotation, you need to reselect the **Turning Head** icon in order to continue.

Figure 9.9.4

5.25 Turn off the **Horizontal Ground** and the select the **Isometric View** and then the **Fit All In** icons. Click two times on the **Zoom In** icon then place your cursor directly over the hole of the front wheel and then click on your **Middle mouse button** to center the wheel on your screen.

5.26 Place your cursor over the Red dot on the **Compass** and drag it over the outside of the front wheel as shown in **Figure 9.9.5**. Next, select the **horizontal axis** with your cursor and pull the wheel away from the body as shown in **Figure 9.9.6**. If you make a mistake, simply press **Ctrl Z** undo the previous movement. Then try again. Return the compass to its default position, select **View – Reset Compass**.

Figure 9.9.5

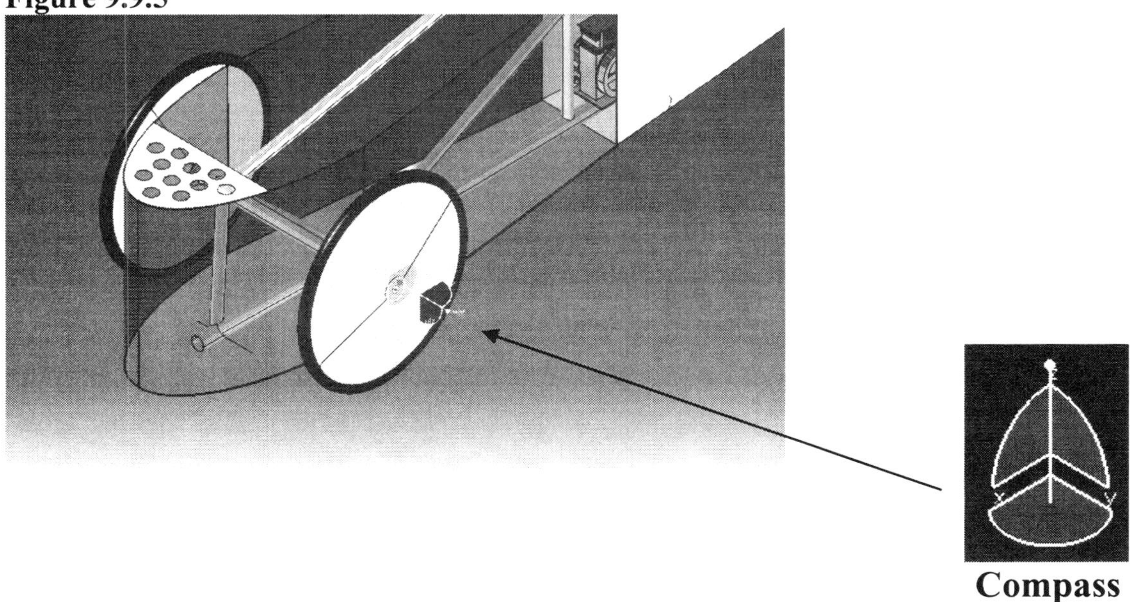

Compass

Figure 9.9.6

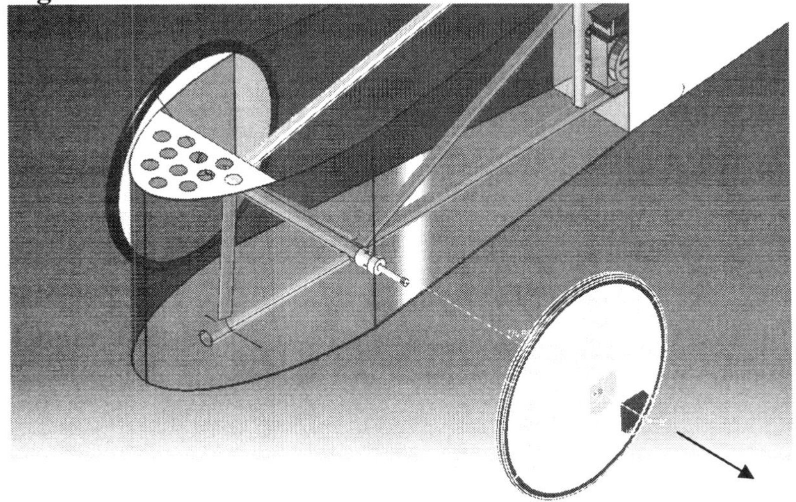

5.27 Select the **Translation or Rotation** icon from the **Move** toolbar. Notice that the **Compass** moves from the upper right corner of the screen to the front wheel area of the vehicle. The positive X-axis direction runs from the back to the front of the vehicle. Enter **-750** in the **Offset X area** of the **Move** window as shown in **Figure 9.9.7**. Now use your cursor and select the **Back Cover** of the vehicle and the Apply button will activate in the **Move** window. Click on the **Apply** button and watch the back cover move away as shown in **Figure 9.9.8**.

Figure 9.9.7

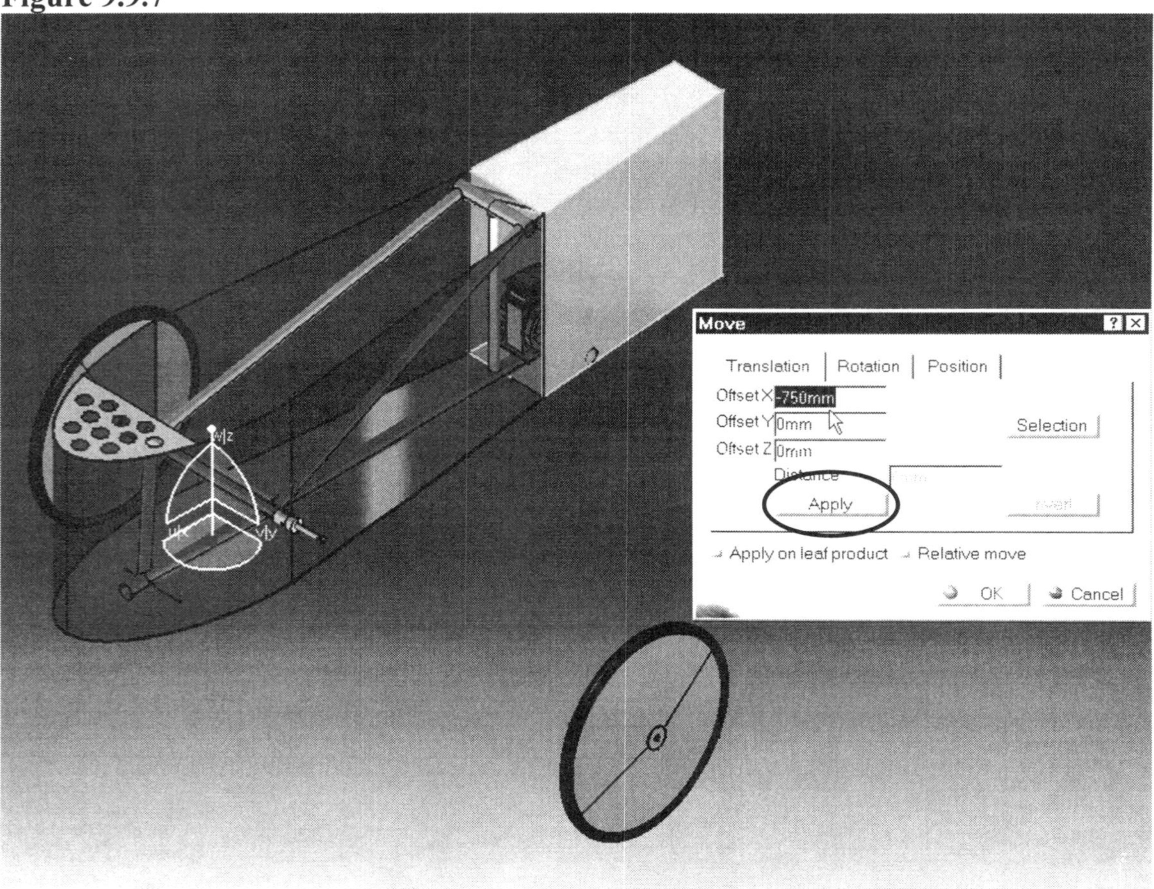

Figure 9.9.8

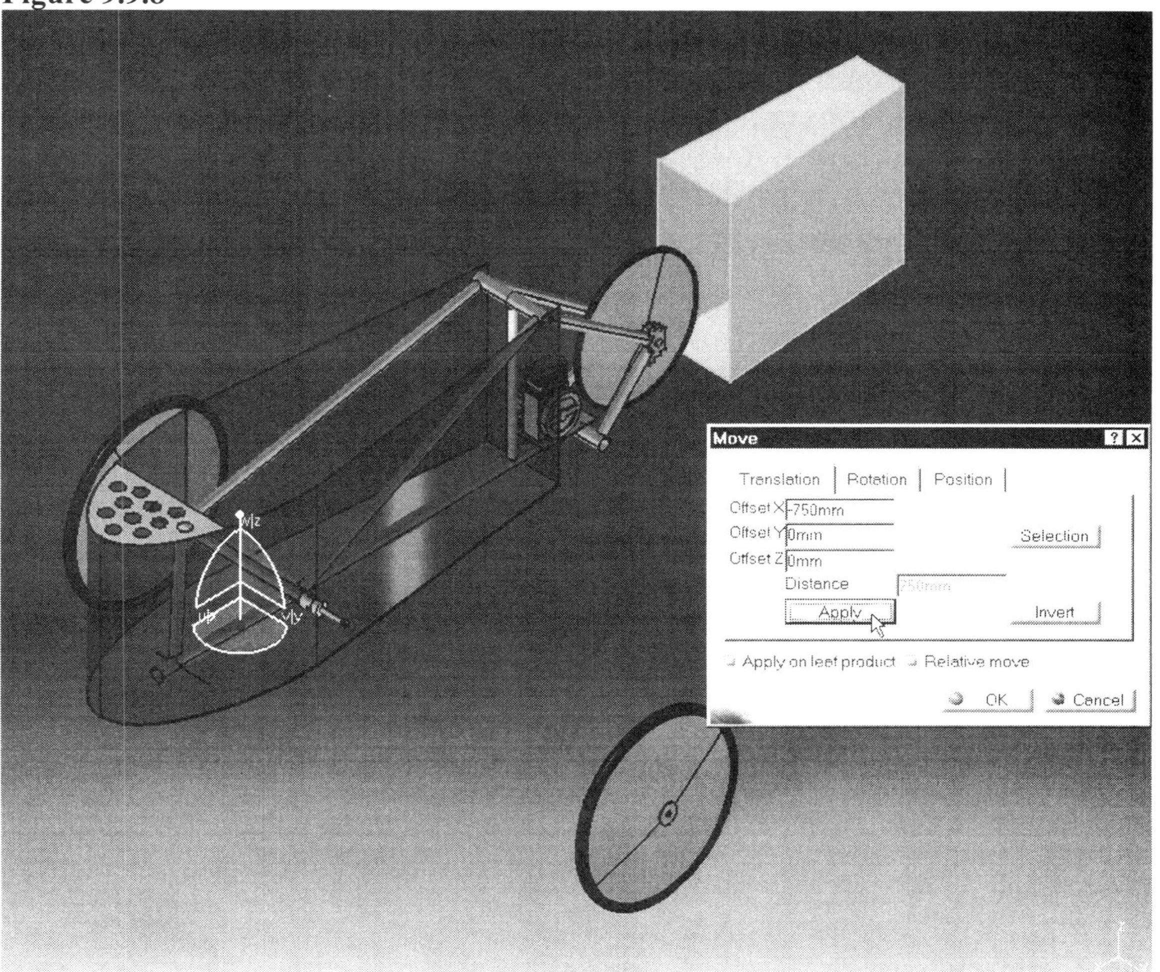

5.28 To move the assembly back to the original positions select the **Reset The Position Of Products** tool found in the **Move** tool bar. You could also use the **Ctrl Z** twice, although you must be careful not to undo too many times. Congratulations on using the DMU Navigator.

6 Creating An Annotated View

Annotated views can be a useful tool for design reviews. It allows you to create notes to yourself or others that are directly connected to the exact part (area) being reviewed yet with out interfering with the actual model files. The following steps will briefly take you through creating a simple annotated view.

6.1 Start by selecting the 2D **Create Annotated View** tool. This brings up the **DMU 2D Marker** tool bar. These tools are very similar to paint programs the tools create bitmap images. Notice when you select the **Create Annotated View** tool a **Annotated Views** branch appears on your **Specification Tree** under the **Applications** branch. The view by default is called **View.1**; you can customize the name by selecting on the **View.1 Properties** and editing the name. Reference Figure 9.10.

6.2 Select the **Arrow** tool and create an arrow pointing to the center of the wheel similar to the one shown in Figure 9.10.

6.3 Select the **Add Annotation Text** tool and add the note as shown in Figure 9.10.

6.4 If you rotate the view all the annotations disappear and you are back to the normal **CATProduct** view. You can return to the view (with all the annotations) by selecting the view from the **Specification Tree** or selecting the **Manage Annotated Views** tool. This will bring up the **Annotated View** window listing all the views created. Double click on the annotated view that you want to review and it will reappear.

6.5 To clear the view of all annotations select the **Delete All Annotations** tool. Before you use this tool make sure you are in the correct view.

Figure 9.10

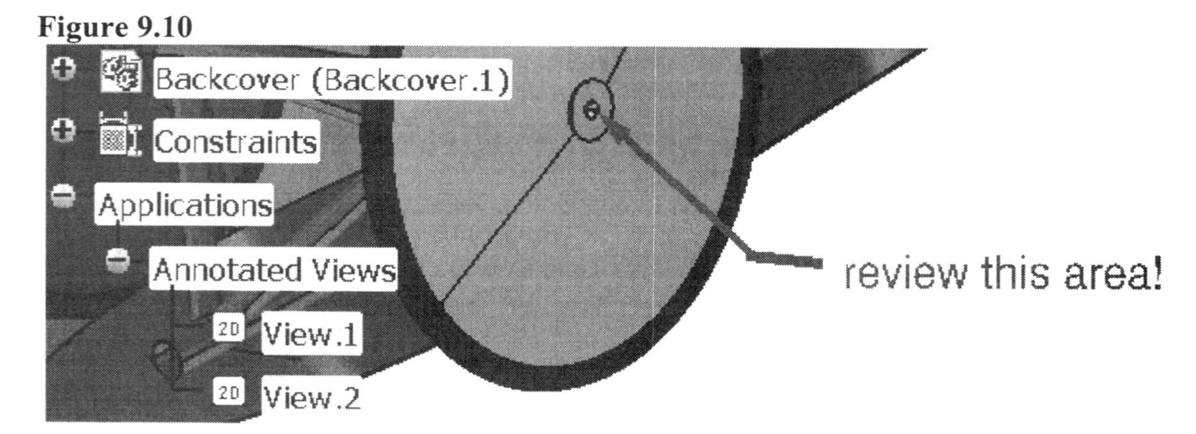

7 Creating 3D Text

The **DMU Navigator Work Bench** also allows you to create 3D text. The 3D text is different in that the text stays with the model, it does not disappear. The text will stay attached to the selected entity even if you switched back to another work bench. The text will stay normal to your point of view as you rotate the model. To temporarily hide the text you can put in **Hide/Show** mode. To create the 3D text, complete the following steps.

7.1 Select the **3D Annotation** tool.

7.2 Select the entity that you want the text attached to. This will bring up the **Annotation Text** window, which is a simple text editor.

7.3 Click the mouse in the text editor window to activate the editor. Type in the desired 3D text.

7.4 Select the **Apply** button to preview the text. If it is satisfactory select the **OK** button to complete the 3D text. Notice that the 3D text appears on the **Specification Tree** under the **Applications** branch.

8 Creating A Scene

Creating a scene saves the view and orientation of the model. This step will get you started in creating several simple scenes. It will also show you what can be done in a created scene. To create a scene, complete the following steps.

8.1 Get the model oriented the way you want it. For this step orient your view similar to what is shown in as **Scene1** in Figure 9.11.

8.2 Select the **Scene** tool . This will bring up the **Edit Scene** window. You can accept the default name or modify the name.

8.3 Select the **OK** button to create. The created scene will show up on the **Specification Tree** as shown in Figure 9.11 (Figure 9.11 has four scenes represented).

8.4 The scenes show up on the bottom left of the screen as shown in Figure 9.11. To enter a scene, double click on the selected scene. For this step select **Scene1** (the scene you created in the previous steps). As CATIA V5 loads the scene the background color changes along with **Specification Tree** and the tool bar. In a scene you can **Search**, **Explode**, **Publish**, **Reset** and **Save Viewpoint**, reference the **Scene** tool bar.

Figure 9.11

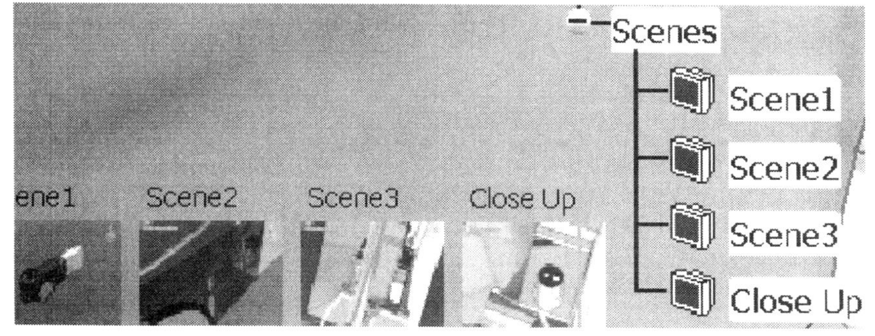

8.5 Now that you are in **Scene1** select the **Explode** tool. This will bring up the **Explode** window as shown in Figure 9.12. The Explode window allows you to modify a few settings. For this step accept the default settings. Select the Apply button. The entire assembly will automatically disassemble to an exploded view. You can use the scroll bar to animate the reassembly and back to the disassembled position.

8.6 Select the **OK** button. The Explode window will disappear and the exploded assembly will remain.

8.7 To reassemble the **SuperMileage Product** (assembly) select the **Reset Select Products** button. The **SuperMileage Product** will be restored to full assemble position.

8.8 This is all of the tools this lessons uses from the scene environment. Two other tools that would be worth your investigation are the **Star Publish** tool and the **Save Viewpoint** tool. The **Start Publish** tool allows you to create html pages. The **Save Viewpoint** tool allows you to save the view point created in the scene.

Figure 9.12

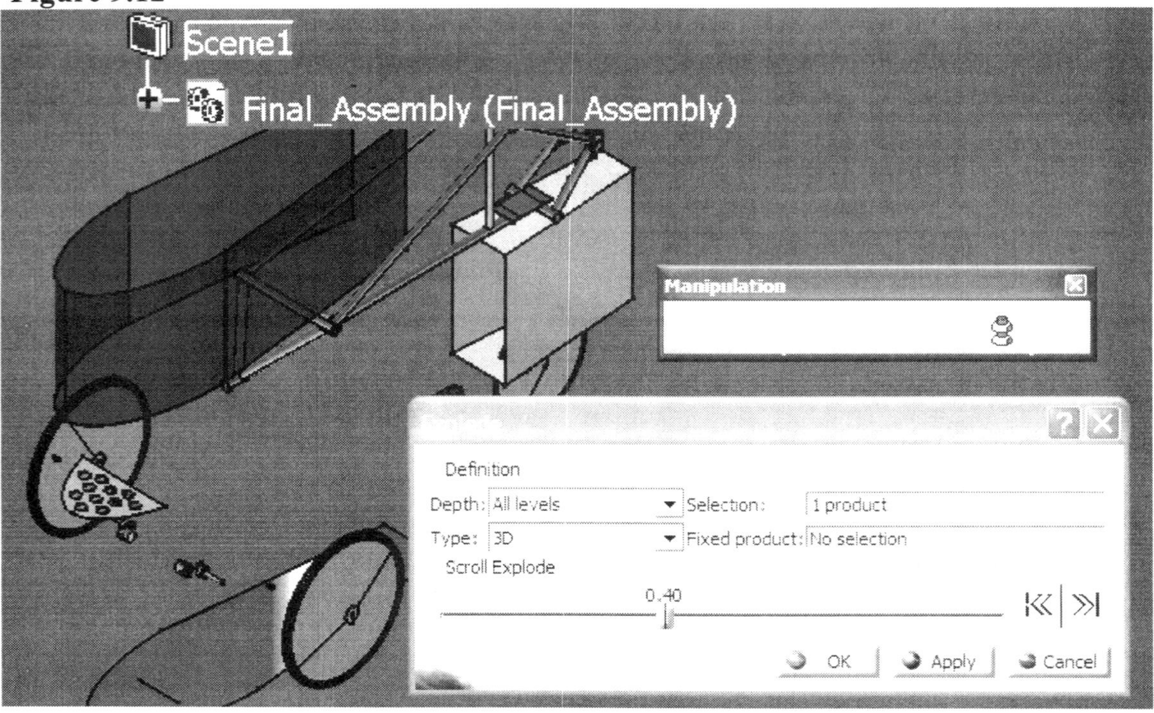

8.9 One last and very important tool you must know about the scene environment is the **Exit From Scene** tool. This tool returns you back to the normal **DMU Navigator Work Bench**. The tool looks just like the sketcher **Exit** tool. For this step select the **Exit From Scene** tool.

9 Recording and Replaying A Fly Thru

The DMU Navigator Work Bench has a lot of tools that lets your creativity define your limits. Step 3 taught you how to use the Fly Mode. This step will show you how to make a recording of the fly thru using the DMU Generic Animation tool bar. To make a recording of your fly thru complete the following steps.

9.1 Complete the steps described in 3.1 through 3.8. At this point you should be ready to begin the **Fly Mode**.

9.2 Select the **Viewpoint Animation** tool. This will bring up an additional **Viewpoint Animation** tools as shown in Figure 9.13.

Figure 9.13

9.3 Select the Recording tool. This will bring up the **Resulting Replay** window. This window by default will name the replay **Replay.1**. For this step rename the **Replay.1** to "**Test Flight**."

9.4 Once you select the **OK** button the recording will start. When you are ready to fly thru the **SuperMileage Assembly** as you did in **Step 3**, select the **OK** button and go!

9.5 To stop the recording, select the **Stop** tool. Notice that the **Test Flight** is added to the **Specification Tree** under the **Application Replay** branch.

9.6 To replay the **Test Flight** select the **Replay Simulation** tool. CATA V5 will prompt you to select a **Simulation Object**.

9.7 Select **Test Flight** from the **Specification Tree**. This will bring up the **Player** tool bar as shown in Figure 9.14.

Figure 9.14

9.8 To replay the recording, select the **Play** ▸ tool. You now recorded proof of you flying skills.

The **DMU Generic Animation** tool bar has additional tools that offer more power and control when creating animations. This brief introduction to creating a simple animation is enough to get you started.

Lesson 9 Summary

The **DMU Navigator Work Bench** is so much fun it is easy to forget what a powerful analysis tool it is. This work bench brings the assembly to life. This lesson has given you the basic skills to get around in the **DMU Navigator**. You are encouraged to continue exploring the additional **DMU Navigator** tools.

Lesson 9 Review

After completing this lesson you should be able to answer the questions and explain the concepts listed below:

1. How could the DMU Navigator be used in the product review process?

2. How many tool bars are available in the **DMU Navigator Work Bench**?

3. What type of file did you insert as an "Existing Component" during this Lesson?

4. What file extension does CATIA V5 give to **DMU Navigator** files?

5. What option under the **Start** pull down menu gives you access to the **DMU Navigator Work Bench**?

6. What tool inserts a part or assembly file into the **DMU Navigator Work Bench**?

7. While navigating through an assembly in the **DMU Navigator**, what button do you press to reverse the direction?

8. What two keys do you use to increase and decrease the speed of your fly through?
9. How do you move the horizontal ground grid if it is activated?

10. What is the primary difference between using the **Fly Mode** versus the **Walk Mode**?

11. How do you switch between the **Fly Mode** and **Walk Mode**?

12. **T** or **F** You must be in the **Perspective** view when working in the **Examine Mode**.

13. What icon can you activate that will allow you to enlarge a component on the screen in a separate window?

14. What is the name of the window that appears on the screen when you select this icon?

15. What tool can you use to pull parts away from the assembly along an axis?

16. **T** or **F** The **DMU Navigator Work Bench** creates automatic exploded 3D views of assemblies.

17. **T** or **F** The annotations created using the **2D Annotations** tool disappear when the assembly is rotated from the plane the annotations were created in.

18. **T** or **F** The text created using the **3D Annotations** tool disappear when the assembly is rotated from the plane the annotations were created in.

19. How do you exit a **Scene** environment?

20. **T** or **F** Animations created and recorded in the **Generic Animation** tool bar do not show up on the **Specification Tree**.

Lesson 9 Practice Exercises:

Now that you have navigated one assembly in CATIA V5, you can strengthen your newfound knowledge by completing the following practice exercises using **CATPart** and **CATProduct** files.

1 Insert the "**Catiaframe.CATPart**" as an **Existing Component** into a new DMU Navigator session. First, while in the **Fly Mode,** activate the **Horizontal Ground** and zigzag your way through the exterior surface of the frame. Secondly, attempt to fly through the interior of the frame tubing as though you were entering a tunnel.

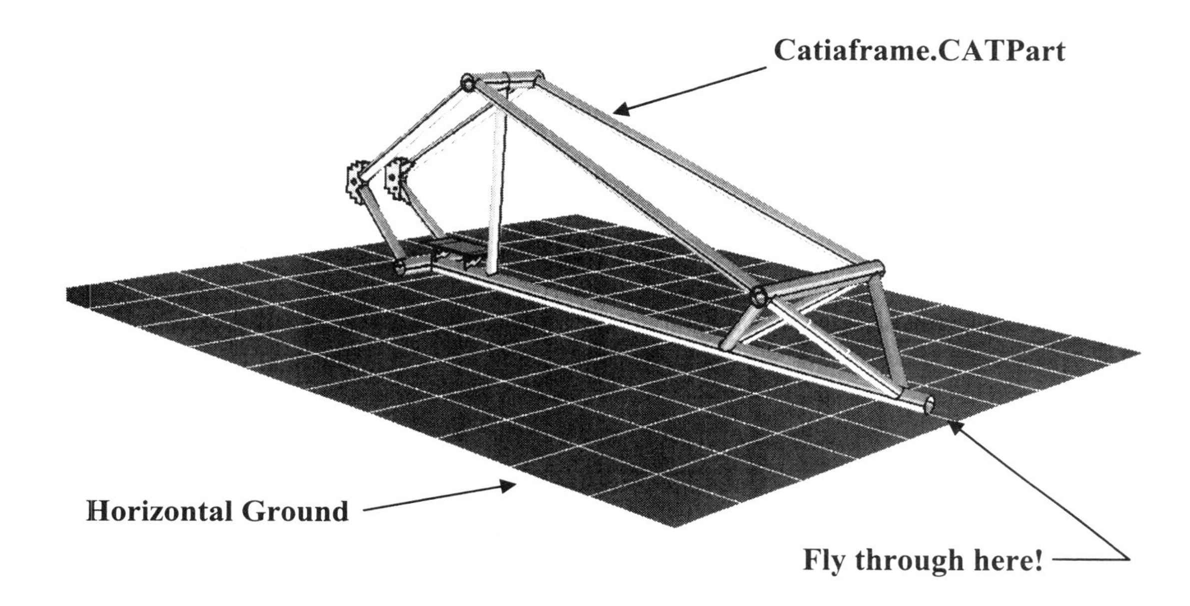

2 Bring in a part or an assembly file in IGES format (.igs) from another CAD software package into the **DMU Navigator** and use the **Fly Mode**.

Suggested steps:

2.1 Open a file in your other CAD software package and look for either **File Export** or **File Save As** to see if you can save or export your file in ***.igs** format. Remember where you save this file so you can retrieve it in CATIA V5.

2.2 Open the *.igs file in CATIA V5 and wait for the file to be translated into CATIA V5 then click on **The Fit All In** icon to make sure the object appears on the screen. Save the file as a **CATPart** file.

2.3 Apply a material to the surface of your part and activate the **Applies Customized View Parameters** to display the material you selected.

2.4 Now enter the **DMU Navigator** and use the **Fly Mode** with your translated part from another CAD software package.

3 Go to the www.schroff1.com/catia web site as explained in this lesson and down load the **Basic House.Zip** Unzip the Basic House.CATProduct file and save to your directory.

4 Practice **Flying** thru the **Basic House.CATProduct** (assembly). Using the **Generic Animation** tool record a fly thru that meets the following requirements.

4.1 Start with an isometric view of the entire house.
4.2 Fly through the front door or window.
4.3 Once you are in the middle of the living room stop and use the Look Around tool to view at least 180 degrees of the living room.
4.4 Take a close up look at the picture on the wall.
4.5 Take a look at the ceiling.
4.6 Fly under the coffee table.
4.7 Proceed out the back door out to the deck and down the stairs.
4.8 Save the recording for playback.

5 Practice **Walking** thru the **Basic House.CATProduct** (assembly). Using the **Generic Animation** tool record a walk thru that meets the following requirements.

5.1 Start with a front view of the entire house.
5.2 Walk through the front door.
5.3 Once you are in the middle of the living room stop and use the Look Around tool to view at least 180 degrees of the living room.
5.4 Take a close up look at the picture on the wall.
5.5 Take a look at the ceiling.
5.6 Proceed out the back door out to the deck.
5.7 Reverse the direction and back to inside the house.
5.8 Save the recording for playback.

6 Use the fly mode to enter the living room of the **Basic House.CATProduct** and create a **2D Annotation** similar to the one shown in the following figure.

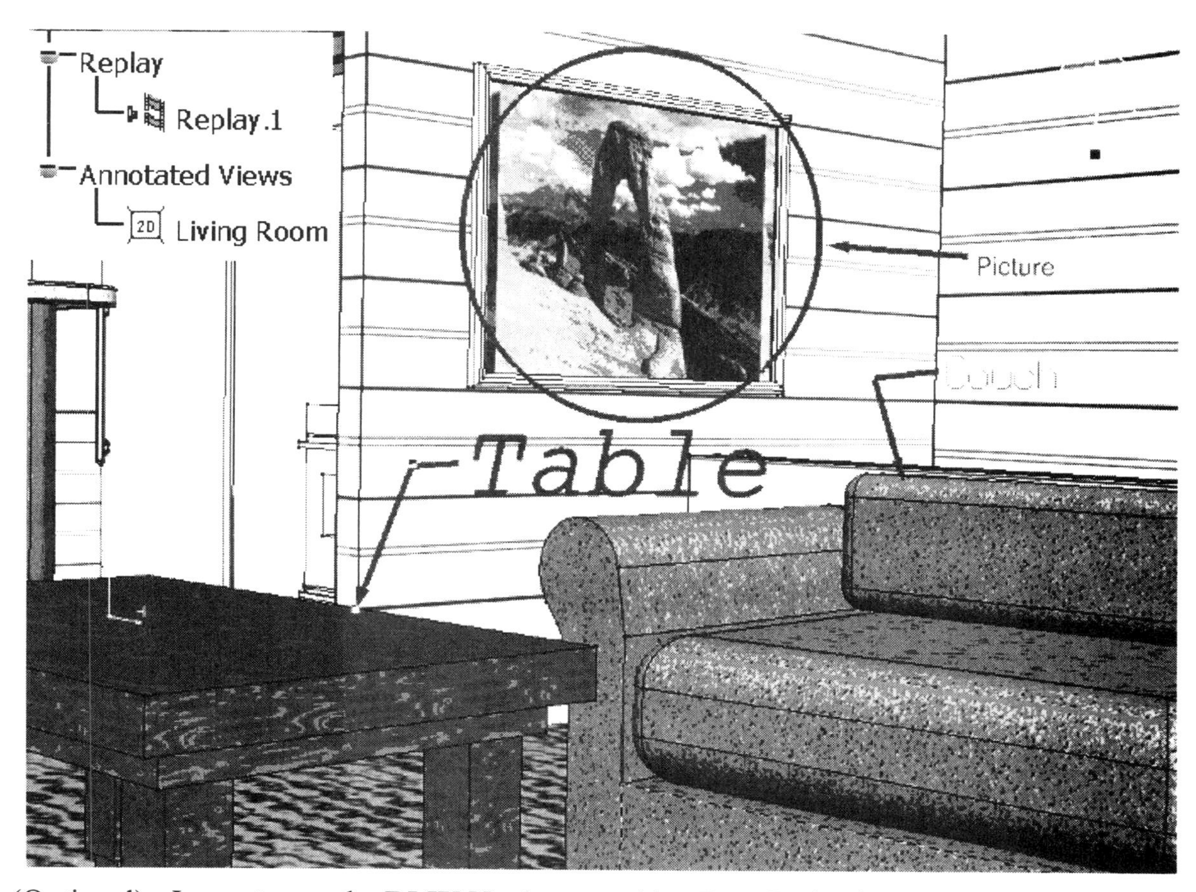

7 (Optional) - Learn to use the **DMU Navigator** with a Data Projection system connected to a computer that has **CATIA V5** loaded on it.

Suggested steps:

7.1 Bring the **SuperMileage.CATProduct** file into the DMU Navigator.

7.2 Turn on the Data Projection system and project the **SuperMileage** file for a simulated design review session. Use the **Measure Between** and **Magnifier** windows to double check measurements of individual parts.

Use the **Hide** function to hide various parts during the design review. Example: hide the back cover of the vehicle to expose the engine!

7.3 Use the **Turning Head** icon to look around the inside of the vehicle.

7.4 Place the **Compass** on the side (use the side with the shaft) of the engine and try to slide it off the mounting platform away from the frame.

NOTES:

Lesson 10 Real Time Rendering Work Bench

Introduction To The Real Time Rendering Work Bench

The **Real Time Rendering Work Bench** is a great presentation/marketing tool. This work bench gives you the tools to create realistic renderings to present your design. The work bench also has tools for animations and avi file creation. In order for you to complete this lesson, you will again use the **SuperMileage Assembly** files as you did in Lesson 9. Using this assembly within the **Real Time Rendering Work Bench** will allow you to better understand how create environments (see Figure 10.1), add a camera, and define lighting in order to produce a realistic rendering. You will also learn how to add image files and/or materials to the walls of the environment

Figure 10.1 SuperMileage Rendered in a Box Environment

Lesson 10 Objectives

This Lesson will show you how to do the following:

- Introduce you to the **Rendering Work Bench** tools
- How to load a model into the **Rendering Work Bench**
- How to create an **Environment** by defining lights, cameras, images and shootings
- How to create animations such turn table and simulations
- How to create AVI files
- How to export a final rendering images
- How to apply your own image to environments

Real Time Rendering Work Bench Tool Bars 

There are five standard tool bars found in the **Rendering Work Bench**. The Apply Material tool bar from the Standard Bottom tool bar is also included since it adds some special tools for the **Rendering Work Bench**. The individual tools found in each of the tool bars are labeled to the right of the tool icon. The far right column is a definition of the tool.

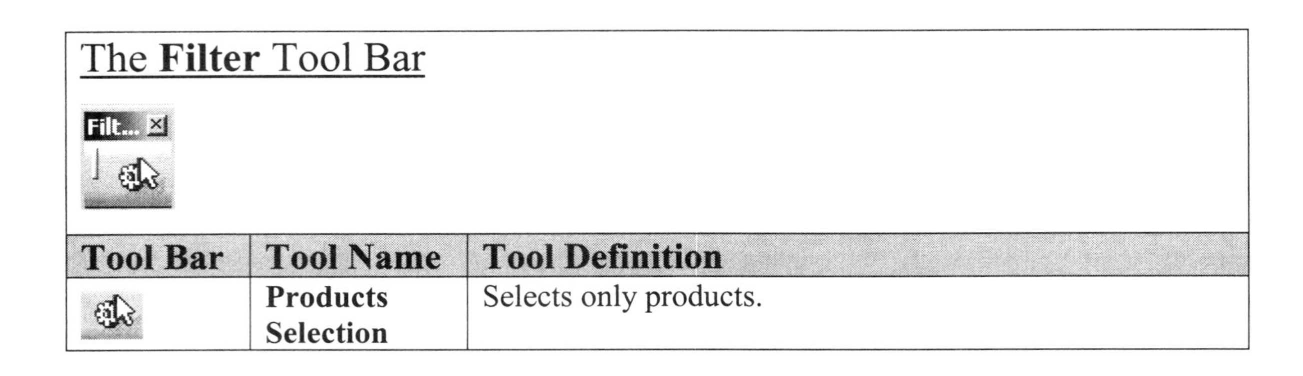

The **Filter** Tool Bar

Tool Bar	Tool Name	Tool Definition
	Products Selection	Selects only products.

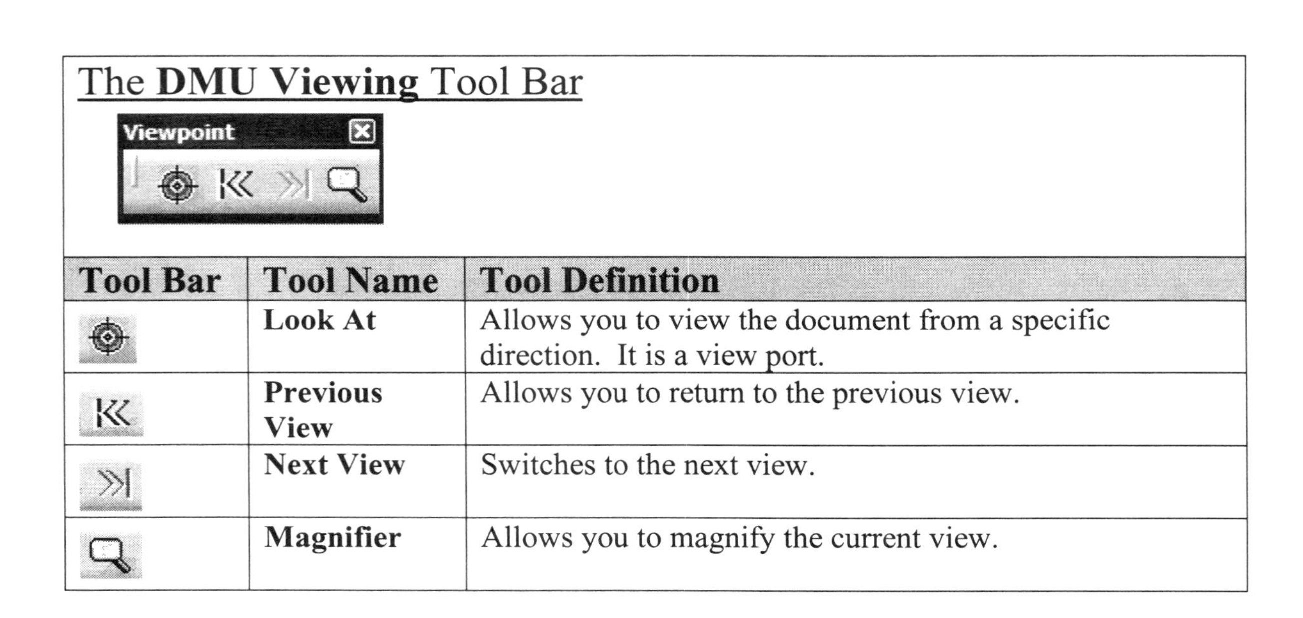

The **DMU Viewing** Tool Bar

Tool Bar	Tool Name	Tool Definition
	Look At	Allows you to view the document from a specific direction. It is a view port.
	Previous View	Allows you to return to the previous view.
	Next View	Switches to the next view.
	Magnifier	Allows you to magnify the current view.

The **Scene Editor** Tool Bar

Tool Bar	Tool Name	Tool Definition
	Create Camera	Creates a Camera.
	Create Environment Tool Bar	
	Create Box Environment	Creates a box environment.
	Create Spherical Environment	Creates a spherical environment.
	Create Cylindrical Environment	Creates a cylindrical environment.
	Create Light Tool Bar	
	Create Spot Light	Create a spot light source.
	Create Point Light	Create a point light source.
	Create Directional Light	Creates a directional light source.
	Create Rectangular Area Light	Creates a rectangular area light source.
	Create Disk Area Light	Creates disk area light source.
	Create Sphere Area Light	Create sphere area light source.
	Create Cylindrical Area Light	Create a cylindrical area light source

The **Animation** Tool Bar

Tool Bar	Tool Name	Tool Definition
	Create Turn Table	Creates a turn table simulation.
	Simulation	Edits shuttles or kinematics simulations according to clash specifications.
	Generate Video	Generate Video.
	Play Simulation	Plays simulation.

The **Render** Tool Bar

This tool bar is found in the **Photo Studio Work Bench**.

Tool Bar	Tool Name	Tool Definition
	Create Shooting	Creates a new shooting.
	Quick Render	Renders the scene.
	Render Scene Tool Bar	
	Render Shooting	Renders a shooting.
	Redo Render	Renders last used shooting.

The **Apply Material** Tool Bar

Two of the tools in this tool bar have been covered in previous lessons; the new tool is detailed below.

Tool Bar	Tool Name	Tool Definition
	Apply Sticker	Applies the sticker to the contextually edited product.

Bringing The Assembly Into the Real Time Rendering Work Bench

This lesson will follow the same process used in Lesson 9 for inserting a component. You will be inserting an entire assembly as an existing component into the **Real Time Rendering Work Bench**.

1 Entering the Real Time Rendering Work Bench

Follow the steps below to activate the **Real Time Rendering Work Bench**.

1.1 Prior to inserting an existing component (**SuperMileage.CATProduct**) into the **Real Time Rendering Work Bench** the first thing to do is enter the **Real Time Rendering Work Bench**. To do this, select the **Start** pull down menu from the top of the screen as shown in Figure 10.2.

1.2 From the **Start** pull down menu select the **Infrastructure** option (Figure 10.2).

1.3 From the **Infrastructure** menu select the **Rendering** option. This will bring you into the **Real Time Rendering Work Bench**.

NOTE: The name at the top of the **Specification Tree** is "**Product 1**." You might have noticed that this applied to the **Assembly Design Work Bench** in Lesson 6 and the **DMU Navigator Work Bench** in Lesson 9.

Figure 10.2

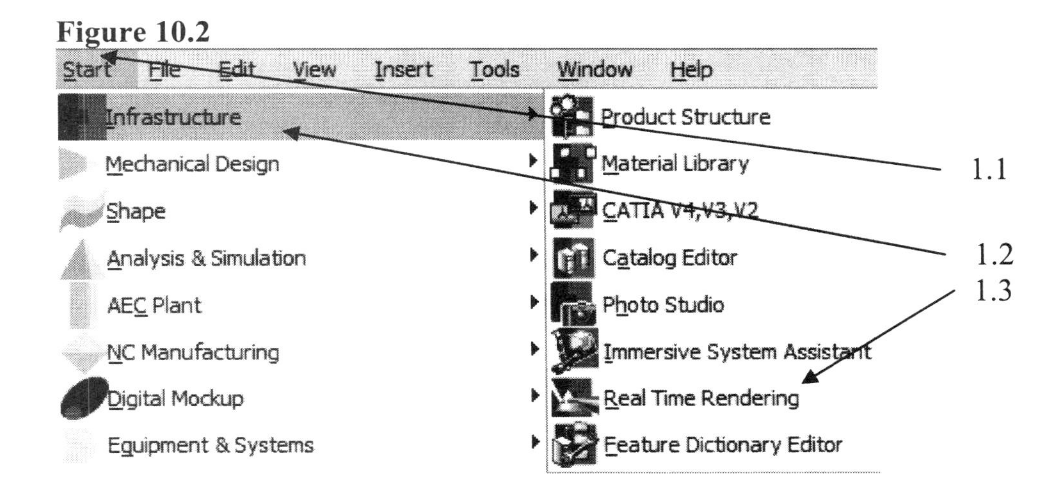

2 Loading The Documents Into The Real Time Rendering Work Bench

In Lesson 6 you learned how to insert detail parts to make an assembly and in Lesson 9 you brought in the same assembly into the **DMU Navigator Work Bench**. You can insert individual part files, sub-assemblies and final assemblies into the **Real Time Rendering Work Bench** session. For this lesson your learning activity will focus on an entire assembly (**CATProduct**) brought in as an existing component into the **Real Time Rendering Work Bench**. Follow the steps below to bring the **"SuperMileage.CATProduct"** assembly file into the **Real Time Rendering Work Bench**.

2.1 The top of the **Specification Tree** must be selected before objects can be inserted into the **Real Time Rendering Work Bench**. The top of the **Specification Tree** is labeled as "**Product1.**"

2.2 Use your right mouse button to activate the menu then scroll down and select the **Existing Component** tool as shown in Figure 10.3.

Figure 10.3

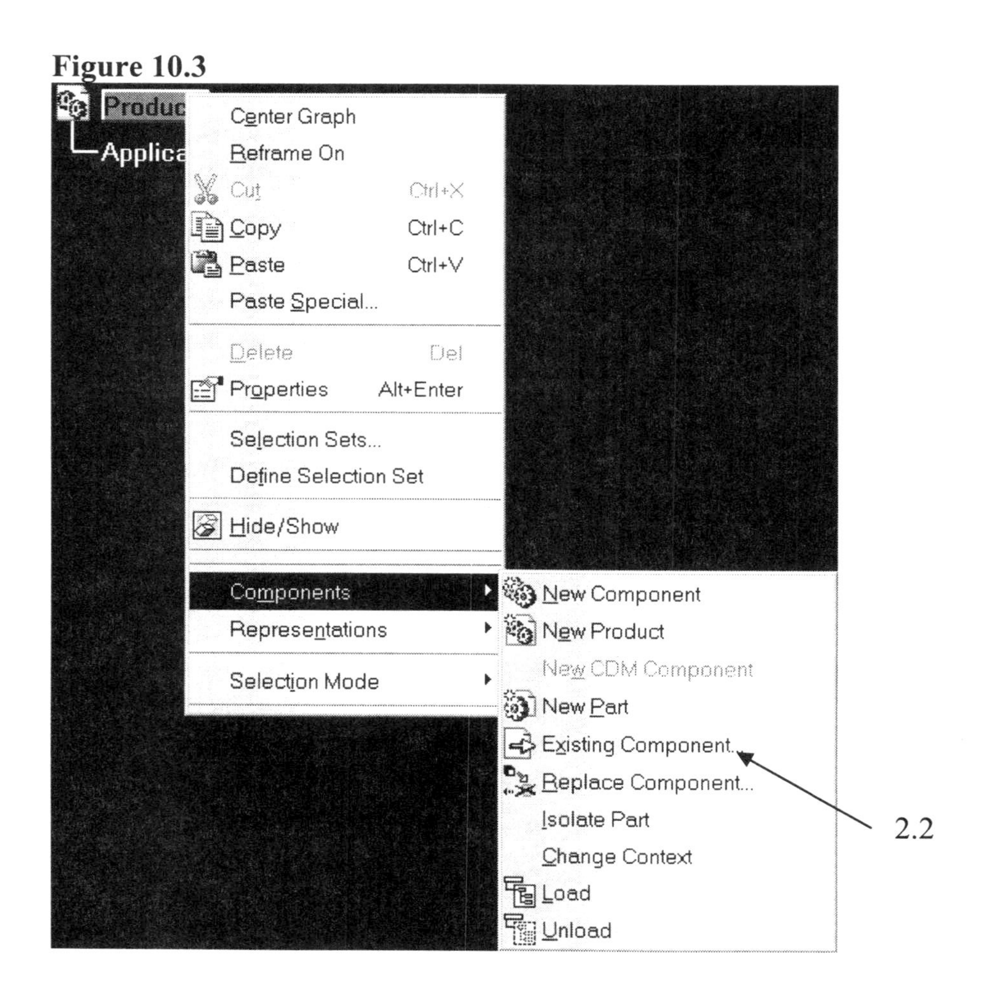

2.3 The **Insert an Existing Component** window will appear as shown in Figure 10.4.

Figure 10.4

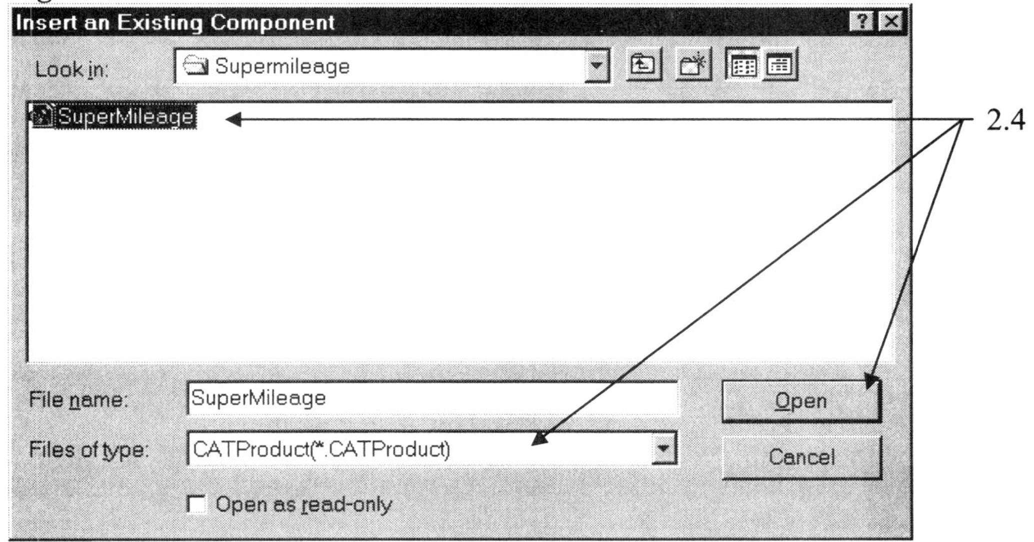

2.4 In the **Insert An Existing Component** (Figure 10.4) window find the file named "**SuperMileage.CATProduct.**" Select the **Open** button. The assembly will appear on the screen. If it does not, then select the **Fit All In** icon and then select the **ISO** tool from the **View** tool bar. Your screen should show the **SuperMileage Assembly** similar to what you worked with in Lesson 9.

3 Creating An Environment

To better understand how to insert a camera and to apply light sources you should first create an Environment around the object you are rendering. There are three (3) types of Environments. The three types are: **Box**, **Spherical** and **Cylindrical**. Figure 10.5 shows an example of each type of environment. Follow the steps below to place the assembly in each of the three environments.

Figure 10.5

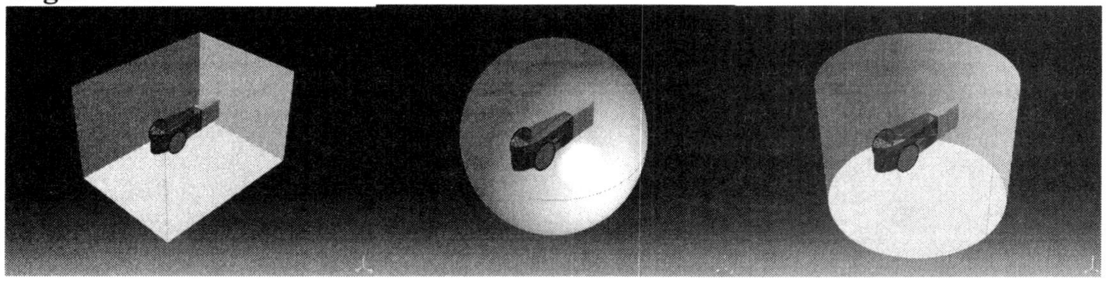

Box **Spherical** **Cylindrical**

3.1 Select the **Create Box Environment** icon in the **Scene Editor** toolbar.

3.2 Select the **Create Spherical Environment** tool in the same **Scene Editor** toolbar. If you have trouble locating it, remember to pick the black down arrow on the **Create Box Environment** tool!

3.3 Select the **Create Cylindrical Environment** tool . To see the three environments within your **Specifications Tree**, you must expand the tree by clicking on **Applications** (+) and then on **Environment** (+).

3.4 With the tree expanded, you need to return to a **Box Environment**. To do this simply place your cursor over **Environment.1** in the **Specifications Tree**, click your Right mouse button and scroll down and select **Environment Active**.

NOTE: The active environment appears in yellow while the other two environments are grey.

Figure 10.6

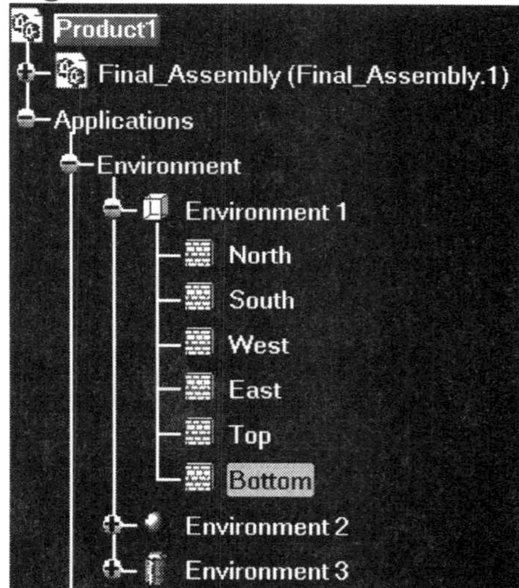

3.5 You will now use the **Box Environment** to experiment with managing walls, define lighting and materials within the environment, setting up a camera, and rendering a view.

3.6 Make sure you are in a **Perspective View**. If you are not, select the **View**, **Render Style**, and pick **Perspective View**.

3.7 With the **SuperMileage Assembly** in the **Box Environment**, select the **Right View** tool and select the **Fit All In** tool . Pick the **Zoom In** tool three times to enlarge the view.

3.8 Expand the **Specifications Tree** to reveal the six walls of the **Environment 1** (Box) as shown in Figure 10.6.

3.9 Click on the **Bottom** wall to make it active. The active wall will appear orange in color.

3.10 Notice that the perimeter of the **Bottom Wall** is hi-lighted as shown in Figure 10.7. You can simply click on the **Bottom Wall** or any wall to activate it.

3.11 Now move your cursor over the **Bottom** area and look for a vertical arrow to appear as shown in Figure 10.7.

Figure 10.7

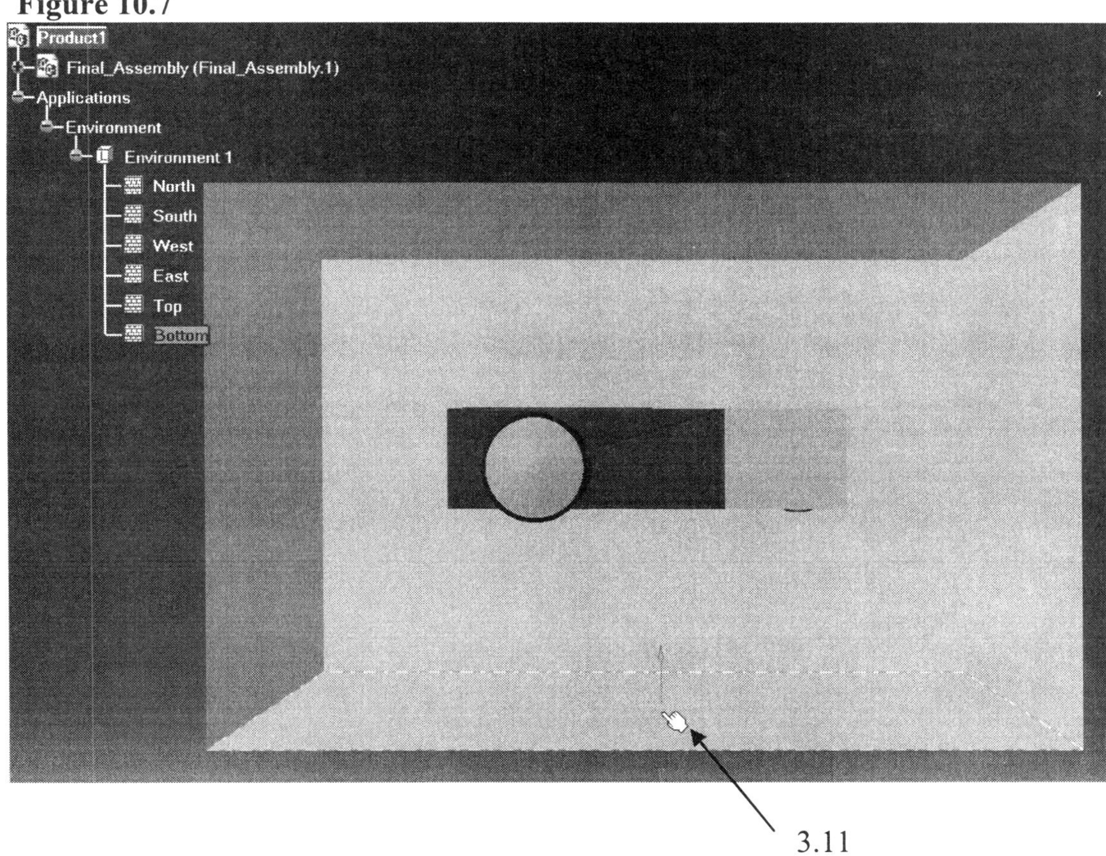

3.12 With your cursor over the **Bottom Wall**, hold down on the left mouse button and drag the **Bottom** upward until it almost touches the wheels of the vehicle.

3.13 Select the **Isometric View** tool. Place your cursor over the **Bottom Wall** to continue moving the bottom closer to the wheels. If you get too close, the bottoms of the wheels begin to disappear as shown in Figure 10.8. Position the **Bottom Wall** so it is touching the wheels.

Figure 10.8

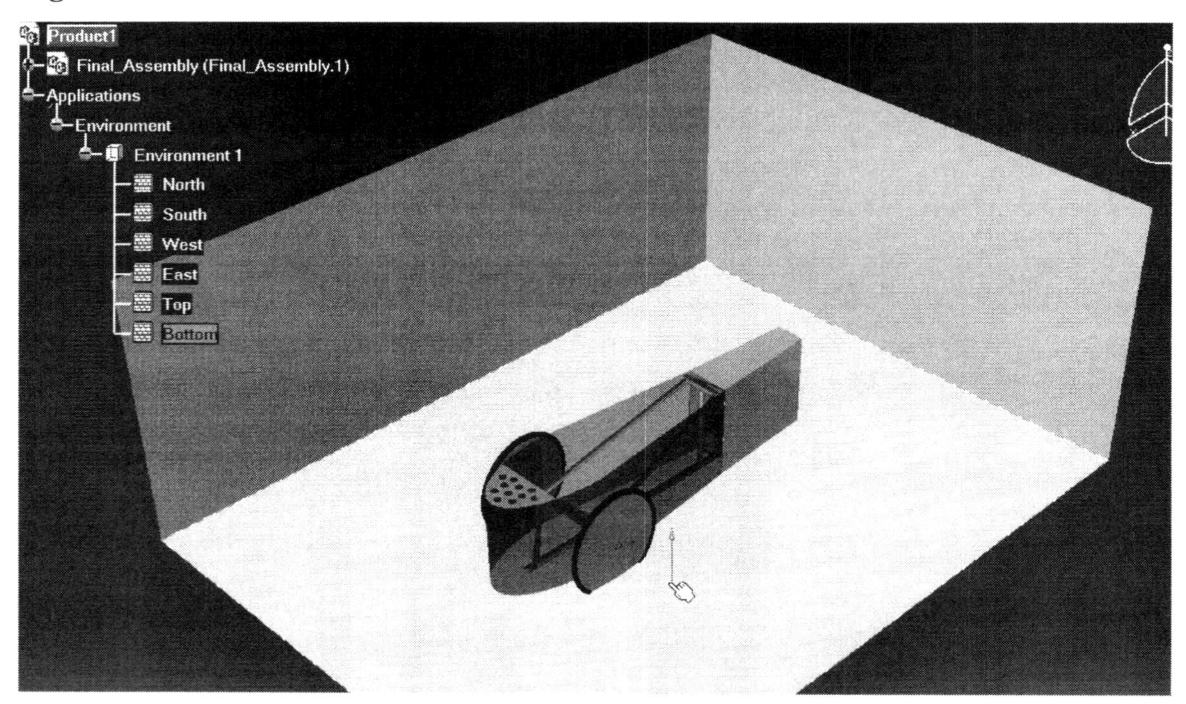

4 Defining A Light

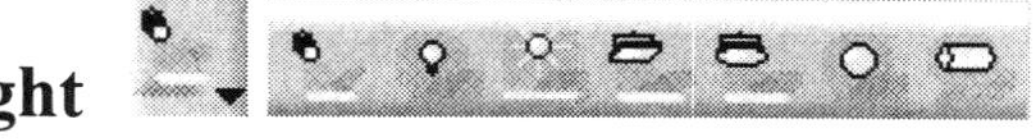

Before we add a textured material and/or image to the walls of the **Box Environment**, you will need to create a spot light and position the light using the Compass. The following steps will take you through defining a light source.

4.1 Select the **Right View** tool.

4.2 Select the **Fit All In** tool.

4.3 Select the **Spot Light** tool from the **Scene Editor** toolbar. Your screen should look similar to Figure 10.9.

Figure 10.9

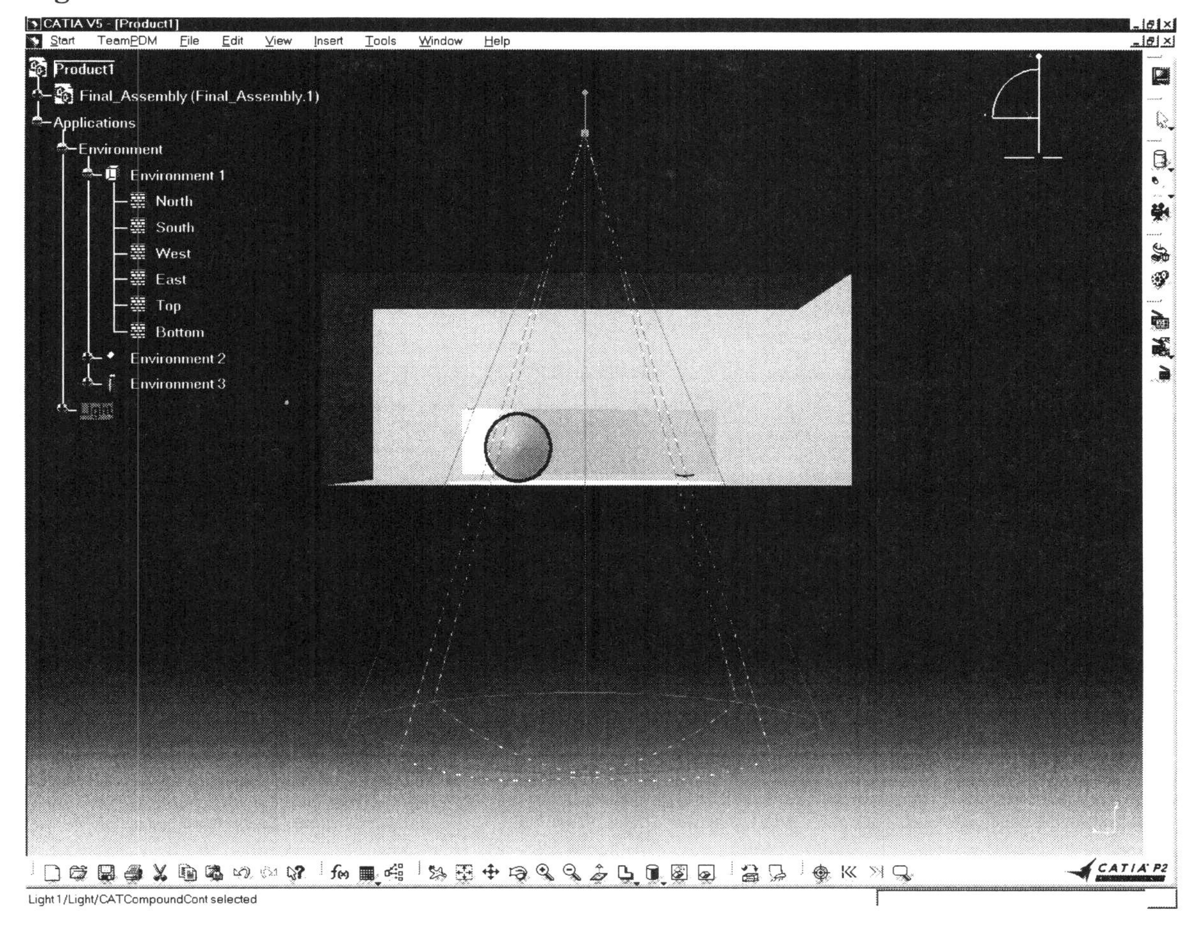

4.4 Use your cursor and grab the **Red Dot** at the bottom of the **Compass** and place it directly over the **Origin** of the light as shown in Figure 10.10.

Figure 10.10

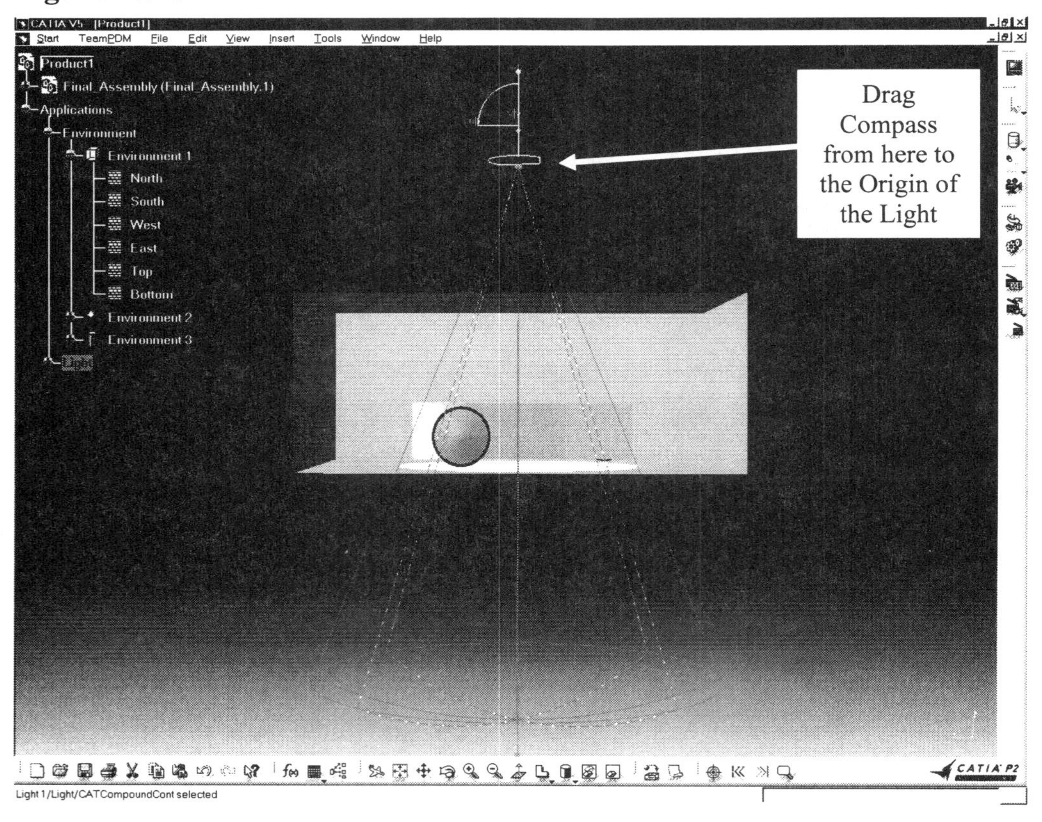

4.5 After placing the **Compass** on top of the **Spot Light**, click in free space and this lighting symbol will appear above the environment under the **Compass**.

4.6 Change to the **ISO View** and notice the light projected the on the bottom of the environment. It may not be directly over the vehicle.

4.7 Use your cursor to click on the light symbol to allow you to use the **Compass** and move one axis at-a-time (see **U** | **X** and **V** | **Y** on bottom of Compass) to position the light directly over the top of the **SuperMileage Assembly** as shown in Figure 10.11.

NOTE: When you are done positioning the light, click on freespace to get your light symbol to appear and then use **View**, **Reset Compass** from the pop down menu to return the **Compass** to its original location in the upper right corner.

Figure 10.11 Activate **Light** and move using **Compass**

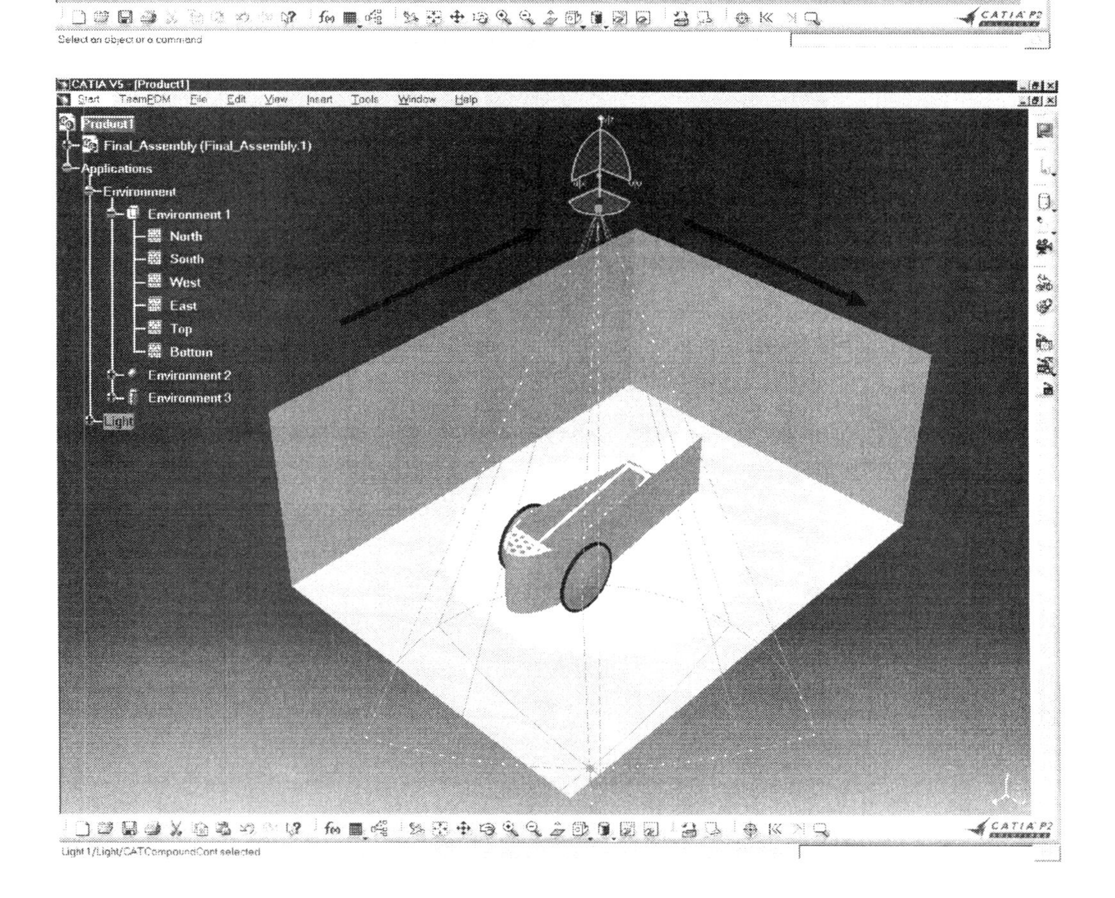

4.8 With the **Compass** removed from the light symbol, reselect the light symbol. Move the cursor half way down the cone of the light until you see a curved arc with arrows at each end similar to what is shown in Figure 10.12. Drag and enlarge the cone to saturate the entire box environment with light.

Figure 10.12

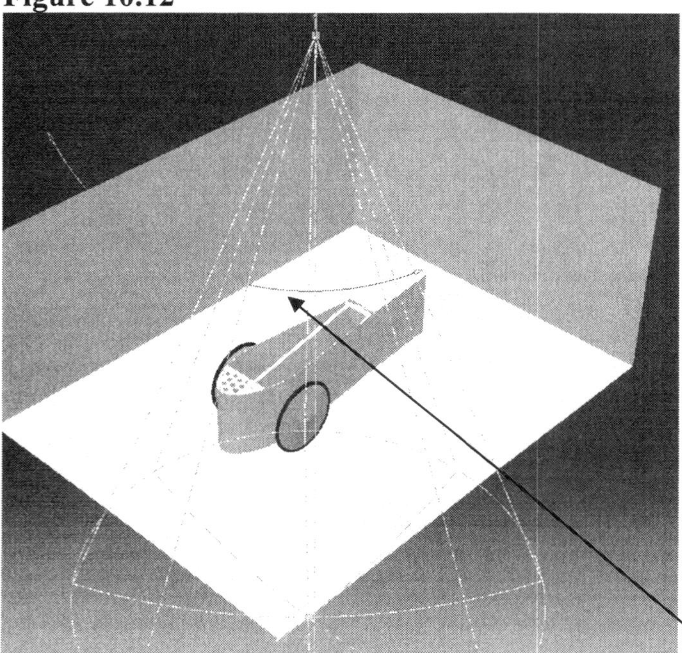

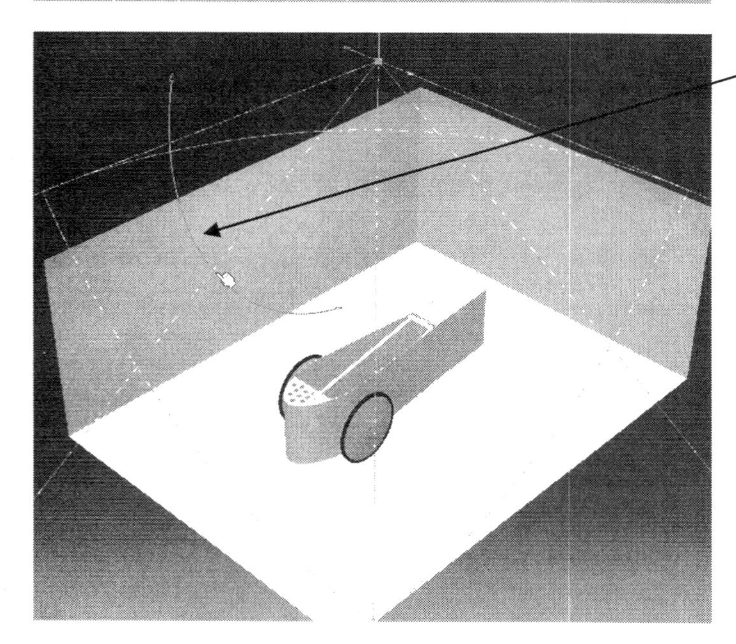

Use the **Curved Arrow** to drag the cone of light in order to expand the coverage in the entire environment!

5 Adding Images To The Walls 

This step shows you how to add materials or images to the walls of the rendered environment. In order to see materials and images within a rendered environment, you will need to activate the **Applies Customized View Parameters** tool.

5.1 Select the **Apply Materials** tool.

5.2 Select the **Other** tab reference Figure 10.13.

5.3 Select the **DS Star** icon and drag this image onto the three walls (**North, South**, and **West**).

5.4 Select and drag the **Chessboard** image onto the **Bottom** wall.

5.5 Close the **Library** window.

Figure 10.13

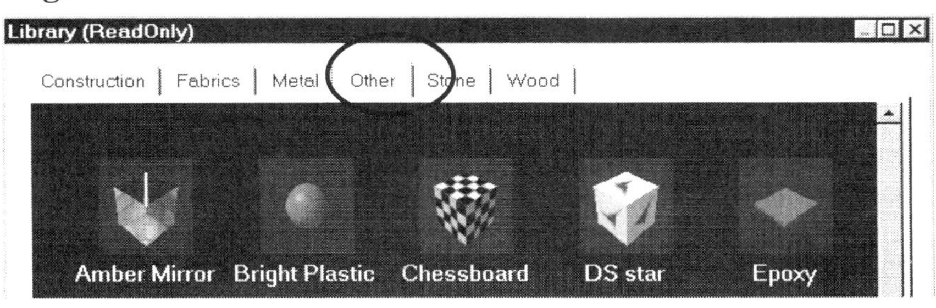

5.6 Place the cursor over the **Bottom** wall and use your **Right** mouse button to bring up the "**Contextual Menu**." Scroll down to either **Properties** OR **Bottom object/Definition**. This will open the **Bottom Properties** window.

5.7 Select the **Texture** tab and then click on the **Fit All In** Wall box. The **Chessboard** image should now become more visible on the bottom as shown in Figure 10.14.1.

5.8 Repeat this process for the **North**, **South** and **West** walls to reveal the **DS Star** image. The box environment should look similar to you see in Figure 10.14.2.

Figure 10.14.1

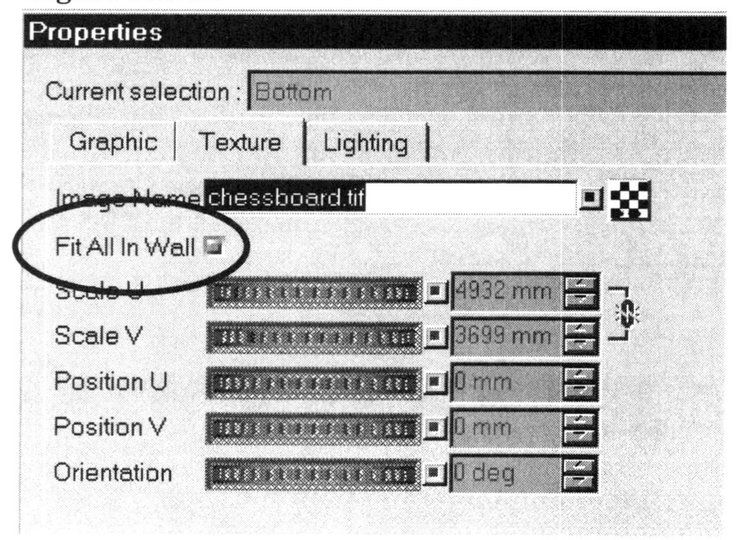

Figure 10.14.2 Environment After Images Fit Walls

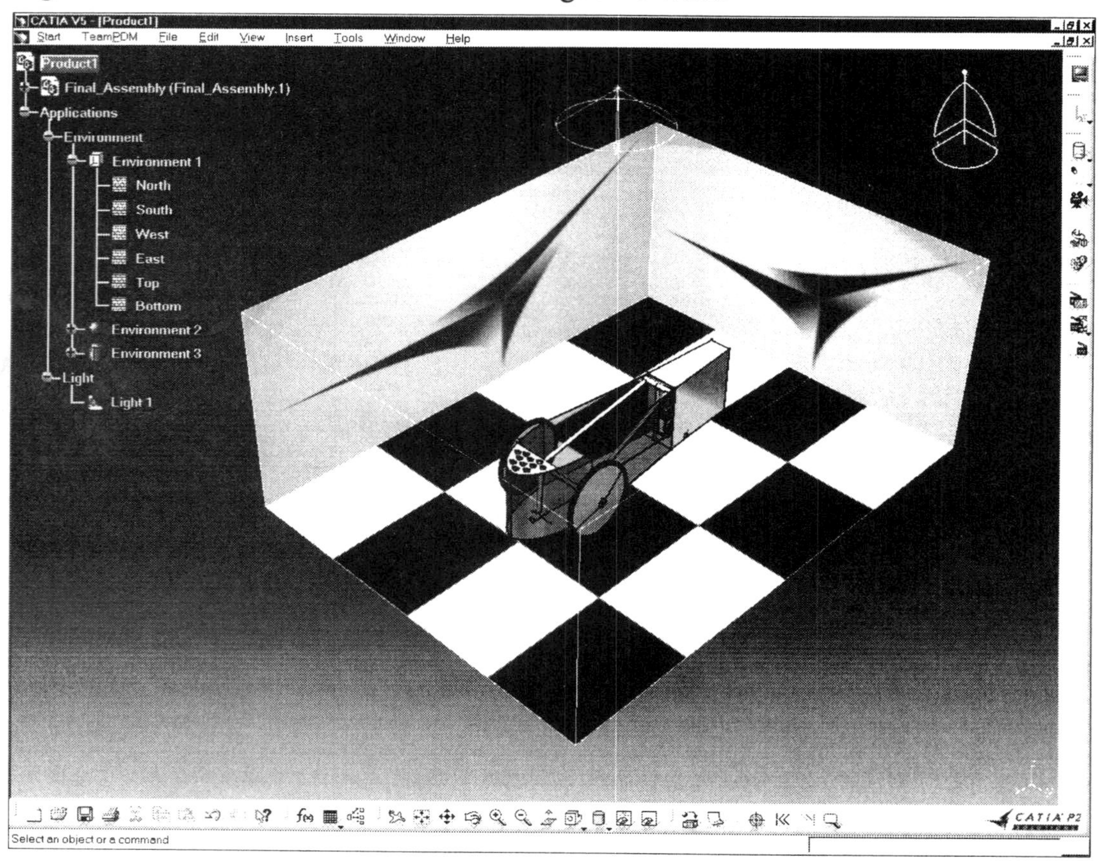

NOTE: If you want to apply your own image file (TIF) to the walls complete the following:

a.) Select the **Wall** to apply the image to (so it is highlighted).
b.) **Right** click on the wall to bring up the **Properties** window.
c.) Select the ***wall Object** and then the **Definition** option.
d.) Select the **Image Name** icon. Browse the directories to your image.
e.) Select your image file from the **File Selection** window.
f.) Use the **Properties** window to set the direction, size and lighting of the image.

6 Create A Camera And A Shooting

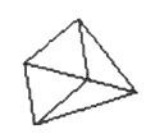

You now have the **SuperMileage Assembly** in a **Box Environment** with a light source. Before rendering, we will add a **Camera** and create a **Shooting**. The Camera will be added using the Real Time Rendering Workbench. The **Shooting** and **Rendering** will be completed in the **Photo Studio Workbench**. This will allow you to select rendering parameters such as light sources, the environment of your choice, camera or current view. Let's begin by adding a Camera.

6.1 With the **SuperMileage Assembly** positioned similar to what is shown in Figure 10.14, select the **Camera** tool from the **Scene Editor** tool bar. A large orange-colored rectangle will appear around the perimeter of your screen and the **Specifications Tree** will now show the Camera.

6.2 Expand the **Camera** (+) in the tree and the use your Right mouse button to activate the **Contextual Menu** as shown in Figure 10.15.

6.3 Select **Properties**.

6.4 From the **Properties** window select the **Lens** tab. This will show you the **Camera** view of what you are about to render.

Figure 10.15 Camera Properties and **Lens View**

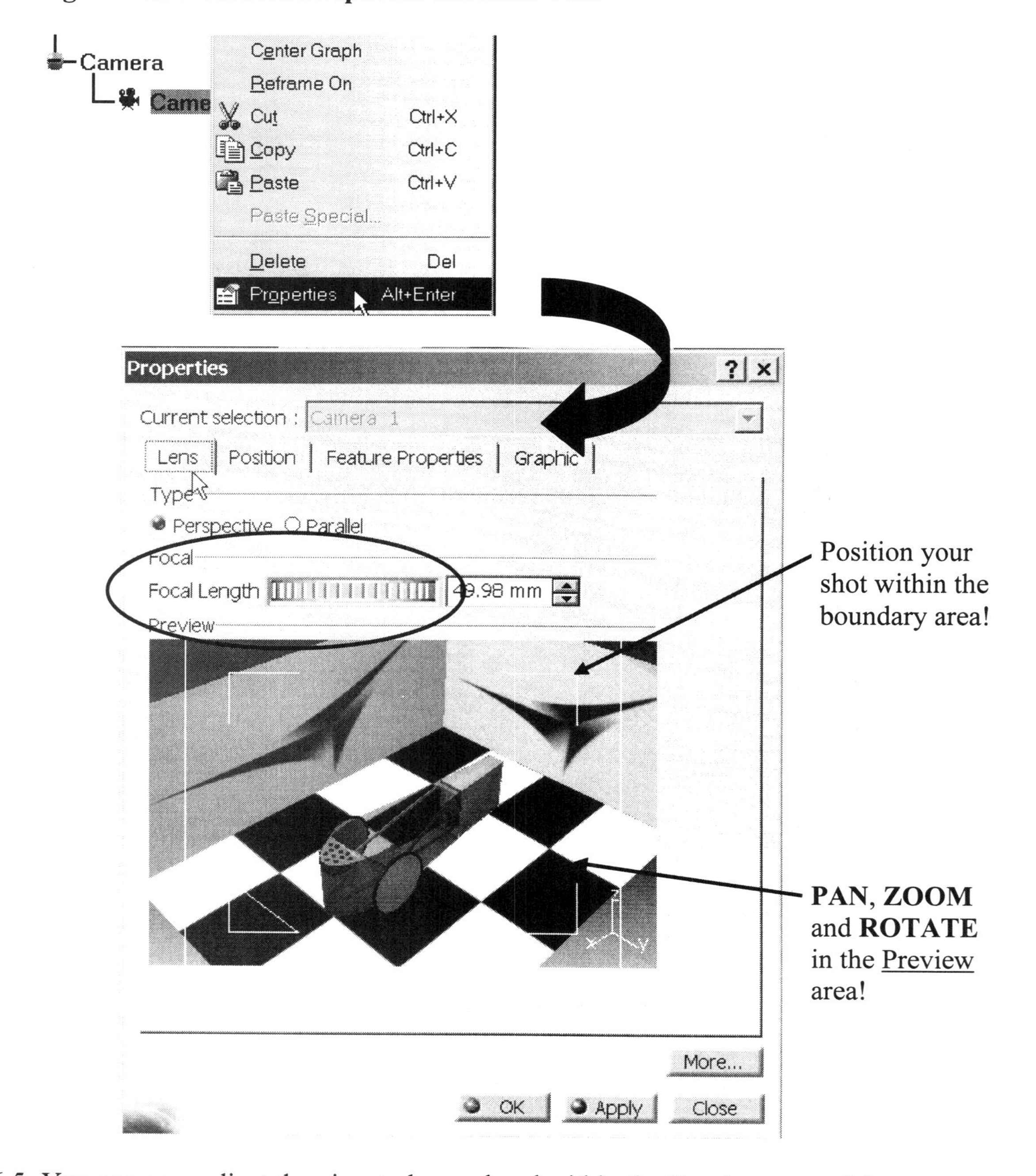

6.5 You can now adjust the view to be rendered within the **Preview** area of the **Properties** window using several methods.

6.5.1 First method, **spin the dial** using your mouse to adjust the **Focal Length** as shown in Figure 10.15.

6.5.2 Second method, you can simply treat the **Preview** area like a regular CATIA V5 screen and use the mouse to **PAN**, **ZOOM** and **ROTATE** the view to achieve the desired result.

6.6 Fine adjustments can be made if you select the **Position** tab. You can tweak one axis (X,Y,Z) at-a-time for either the **Origin** or the **Target** of the Camera. If you use this function, be sure to reference the axes on the **Compass** as your guide. You can make a change on an axis and then click on the **Lens** tab to see the result. You can go back-and-forth between the **Position** and **Lens** tabs until you are satisfied. Once you have the shot you want click **OK** in the **Properties** window.

6.7 Save the file.

Figure 10.16 Fine tune **Position** using the **X,Y,Z Dials**

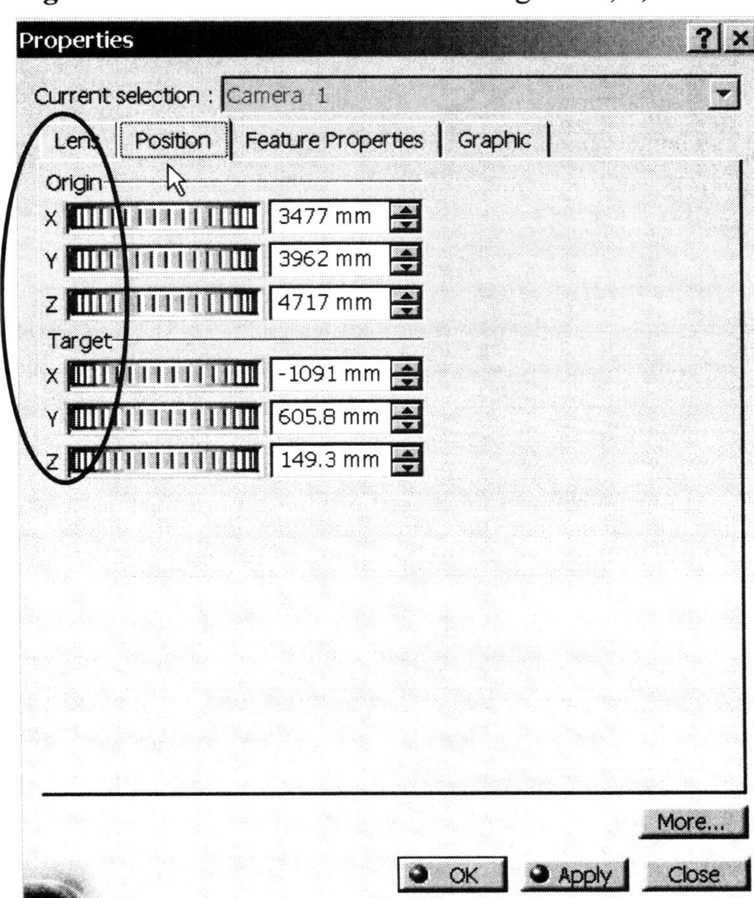

6.8 You are ready to create a **Shooting** prior to rendering so select the **Shooting** tool from the **Render** tool bar. **Note:** The **Shooting** and **Rendering** tools are located in the **Photo Studio Work Bench**. This will bring up the Shooting Definition window.

6.9 Select **Camera 1;** do not select the **Current View** option.

6.10 Select the Output to be **On disk** as shown in Figure 10.17. You could change the Directory and the Frame (file) if you so chose.

6.11 Look at the **Quality** tab as seen in Figure 10.18. You can turn on/off textures and shadows, control the image output size and increase/decrease the geometry size used to calculate the model.

6.12 Select the **OK** button.

6.13 **Save** the file. Notice **Shooting 1** appears in the **Specification Tree**.

Figure 10.17 Shooting Definition - Frame Tab Parameters

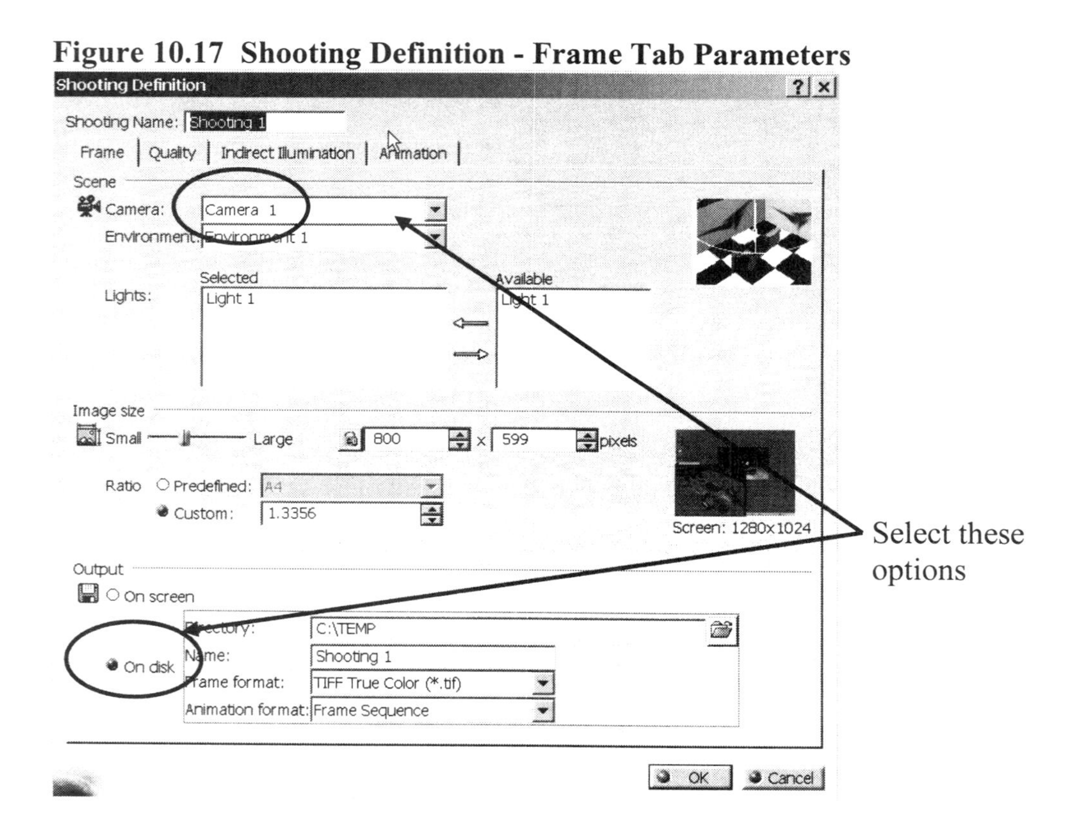

Figure 10.18 Shooting Definition - Quality Tab Parameters

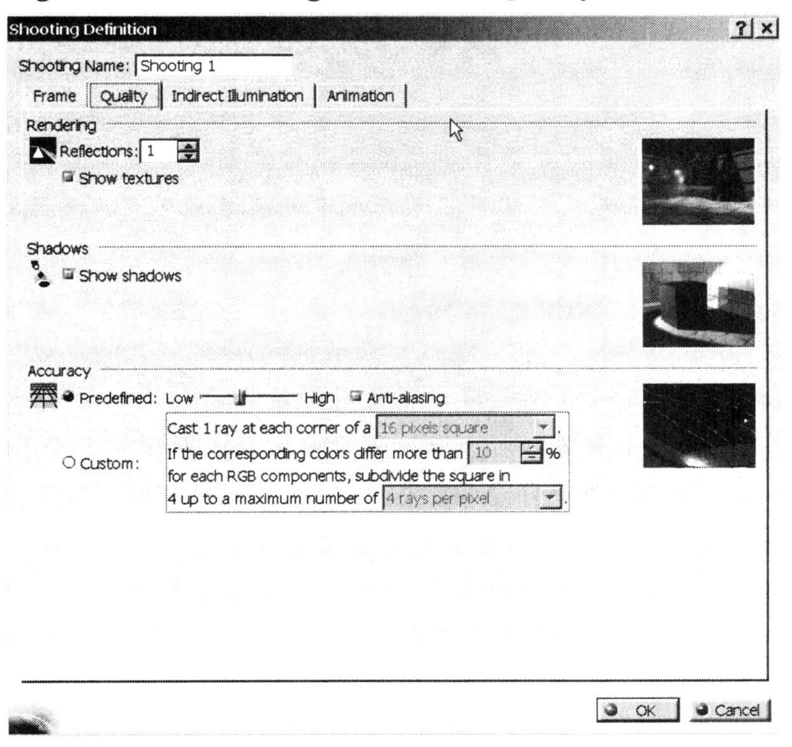

6.14 Now, you are ready to **Render!** **Se**lect the **Render Shooting** tool from the **Render** tool bar.

6.15 When the **Rendering** process is complete and the image appears, look for your saved image file. You can refer to Figure 10.17 to remind yourself what directory and file format you used.

7 Animating The Rendering 

The **Real Time Rendering Work Bench** is one of several work benches in CATIA V5 that give you animation tools. In the **Real Time Rendering Work Bench** you have several options. To create a simulation, complete the following steps.

7.1 Select the **Simulation** tool. This will bring up the **Select** window. The window will have a list of objects you can select to animate.

7.2 Select the **Environment** you created in the previous steps.

7.3 Select **OK**. This will bring up the **Edit Simulation** window as shown in Figure 10.19.

7.4 Select the "**Automatic Insert**." Selecting this button has CATIA V5 automatically save (insert) the animated movements. The **Compass** will appear on the object you selected in Step 7.2.

Figure 10.19

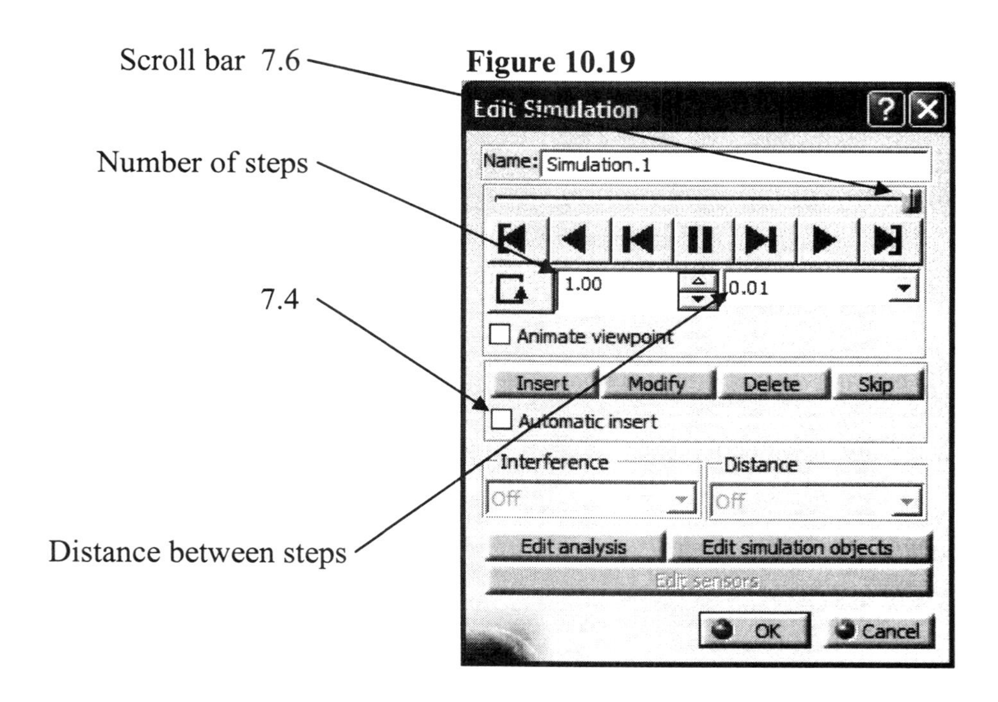

7.5 You use the **Compass** to move the selected object (**Environment**) the same way the **Compass** was used to move parts in previous lessons. Move the **Environment** up half way through the **SuperMileage Assembly** and then rotate the environment about 90 degrees. As you move the **Environment** click the mouse about 10 times during the complete move. Clicking the mouse determines the number of steps that define the animation. Notice the each time you click the step number increases. Reference Figure 10.19.

7.6 To preview the simulation select the **Play** button. Notice the location of the **Scroll** button. If the animation is at the last step you will have to select the **Play Backward** button. If the animation is too fast you can change the distance between the animated steps, reference Figure 10.19.

7.7 If the simulation looks like you wanted select **Ok** to create the simulation. Notice that a simulation branch is added to the **Specification Tree**.

7.8 To view the newly created simulation, select the **Play Simulation** tool.

7.9 This brings up the **Play** tool bar, the same tool bar covered in Lesson 9. The tool bar will stay dimmed until you select the object that you want to play. For this step select the **Simulation.1** you just created.

NOTE: Even though you are in the **Real Time Rendering Work Bench** you can at this point select the **Replay.1** animation created in the **DMU Navigator Work Bench** (previous lesson).

7.10 Select the **Play Forward** button. This will play the simulation. If the simulation runs too fast select the **Parameters** tool on the **Play** tool bar. This will bring up the **Players Parameters** window. Change the **Time Step** parameter to "**.05**." Play the simulation again. Try several of the parameters till you get a simulation to look smooth.

7.11 Once you get the simulation the way that you want select the **Generate Video** tool. This tool creates an independent **AVI file**. The AVI file can be played outside of CATIA. This is a great marketing/presentation tool.

7.12 You will be prompted to select the object you want to use to generate a video. For this step select the simulation you just created (by default **Simulation.1**, unless the previous steps took numerous attempts or you renamed the simulation).

NOTE: You can also create an AVI file from the animations created in the **DMU Navigator Work Bench** (Replay.1).

7.13 You will be prompted to select the directory to save the avi file in. You will also be prompted to name the file. Once you select the **OK** button the file creation will begin. Simulations usually take some time to compile so be patient.

7.14 One last animation tool is the **Turn Table** tool. This tool allows you to create a quick animation of the selected **Environment**. The path of the animation is predefined, it rotates about the axis. Once the **Turn Table** is created you can play it back the same as **Simulations** and **Replays**. You can also create AVI file from the Turn Table.

Lesson 10 Summary

The **Real Time Rendering Work Bench** is a fun work bench, it allows you to complete your design. It allows you be creative. You will need to practice again-and-again the process of setting up and modifying your rendering. You won't get it right the first time, and that's o.k. **So practice, practice, practice!**

Lesson 10 Review

After completing this lesson you should be able to answer the questions and explain the concepts listed below:

1. How many different types of environments are available in the **Scene Editor** tool bar?

2. How many tool bars are available in the **Real Time Rendering Work Bench**?

3. What other work bench prior to this lesson uses the same process for inserting an **Existing Component**?

4. What are the three types of lighting available in the **Scene Editor** tool bar?

5. What option under the **Start** pull down menu gives you access to the **Real Time Rendering Work Bench**?

6. After rendering an object you can save the image in a variety of formats. One example of a format is known as a TIFF file. What are the three (3) formats available to you?

7. **T** or **F** The animations (**Replay.1**) created in the **DMU Navigator Work Bench** can be played in the **Real Time Rendering Work Bench**.

8. **T or F** All types of animations (Replay, Simulation and Turn Table) can be used to create an independent AVI file.

9. When creating a simulation what tool do you use to move the object?

10. When creating a simulation what do you do to make the simulation record a step?

11. **T or F** In the **Photo Studio Work Bench** a **Camera** and a **Shooting** is the same thing.

12. What branch of the **Specification Tree** do the animations (replay, simulation and turn table) show up on?

13. What tool do you use to manipulate the location and direction of a light source?

14. **T** or **F** You use the **Apply Material** tool to change the representation of the part/product being rendered.

15. **T** or **F** The **Shooting Definition** window allows you to create a **Shadows**.

16. How many **Cameras** can a rendering have?

17. When using the **Play** tool to view a simulation and the simulation animates too fast what tool do you use to slow the simulation speed down?

18. **T** or **F** CATIA V5 allows you to use your own image on the walls of the Environments you create.

19. **T** or **F** **The** size and orientation of the images applied to the walls of the environment can not be modified.

20. **T** or **F** **The Real Time Rendering Work Bench** is found in the **Digital Mockup** section of the **Start Menu**.

Lesson 10 Practice Exercise:

Now that you have rendered one assembly in CATIA V5, you can strengthen your newfound knowledge by completing the following practice exercise using the **Engine.CATPart** file, a component of the **SuperMileage Assembly**.

1 Insert the "**Engine.CATPart**" as an **Existing Component** into a new **Rendering** session. Create a rendering session that meets the following requirements.

1.1 Apply **Aluminum Material**.
1.2 Create **Box Environment**.
1.3 Apply **Checkerboard Texture** to the bottom of the Box.
1.4 Import and apply a texture of your choice to three of the side walls.
1.5 Apply a single light source with a direction of your choice.
1.6 Create a **Camera** that has a **ISO View** and is **Zoomed** into the part.
1.7 Create a **Turn Table**.
1.8 Create an AVI file from the **Turn Table** animation.
1.9 **Render** and **Save**.

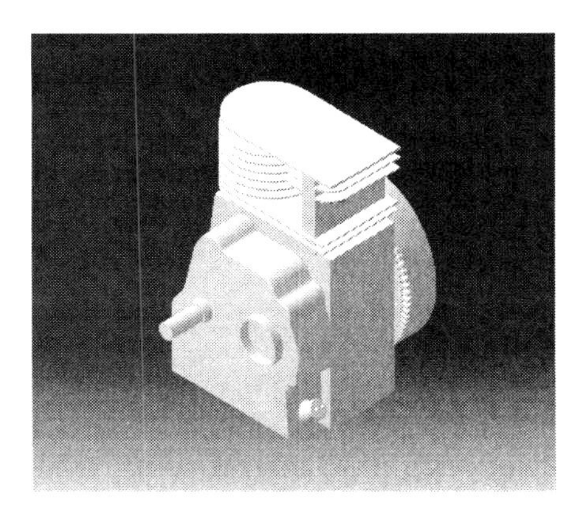

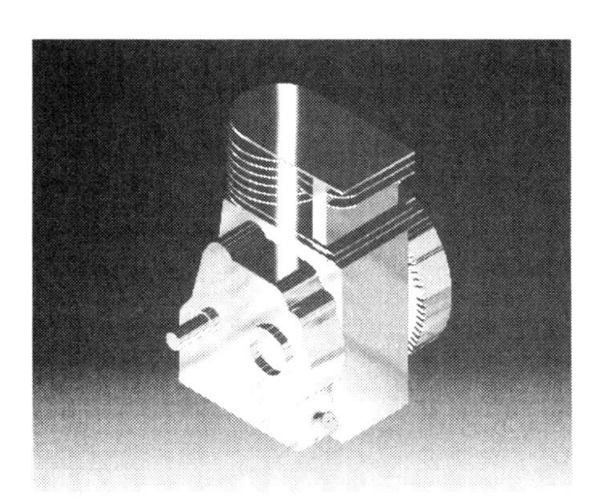

2 Insert an assembly or part model from another CAD software package into a new **Rendering** session as an **Existing Component**. Be sure to look at all of the file types available (Pro/Engineer, Solidworks etc.). Create a rendering session that meets the following requirements.

2.1 Apply a material of your choice (whatever is appropriate).
2.2 Create **Cylindrical Environment**.
2.3 Apply **Checkerboard Texture** to the bottom of the Box.
2.4 Import and apply a texture of your choice for the side walls.
2.5 Apply a single light source with a direction of your choice.
2.6 Create a **Camera** that has an **ISO View** and is zoomed into the part.
2.7 Create a **Turn Table**.
2.8 **Render** and **Save**.

NOTES:

Lesson 11 Parametric Design

Introduction To Creating and Maintaining Basic Part Intelligence

In Lesson 1 you created a sketch and in Lesson 2 you extruded the sketch into a three-dimensional part. Lesson 5 and 6 taught you how to create multiple sketch and multiple bodied parts. In varying degrees all these lessons used constraints to better define the part/parts. This lesson brings everything you learned in the previous lessons together and

Figure 11.1

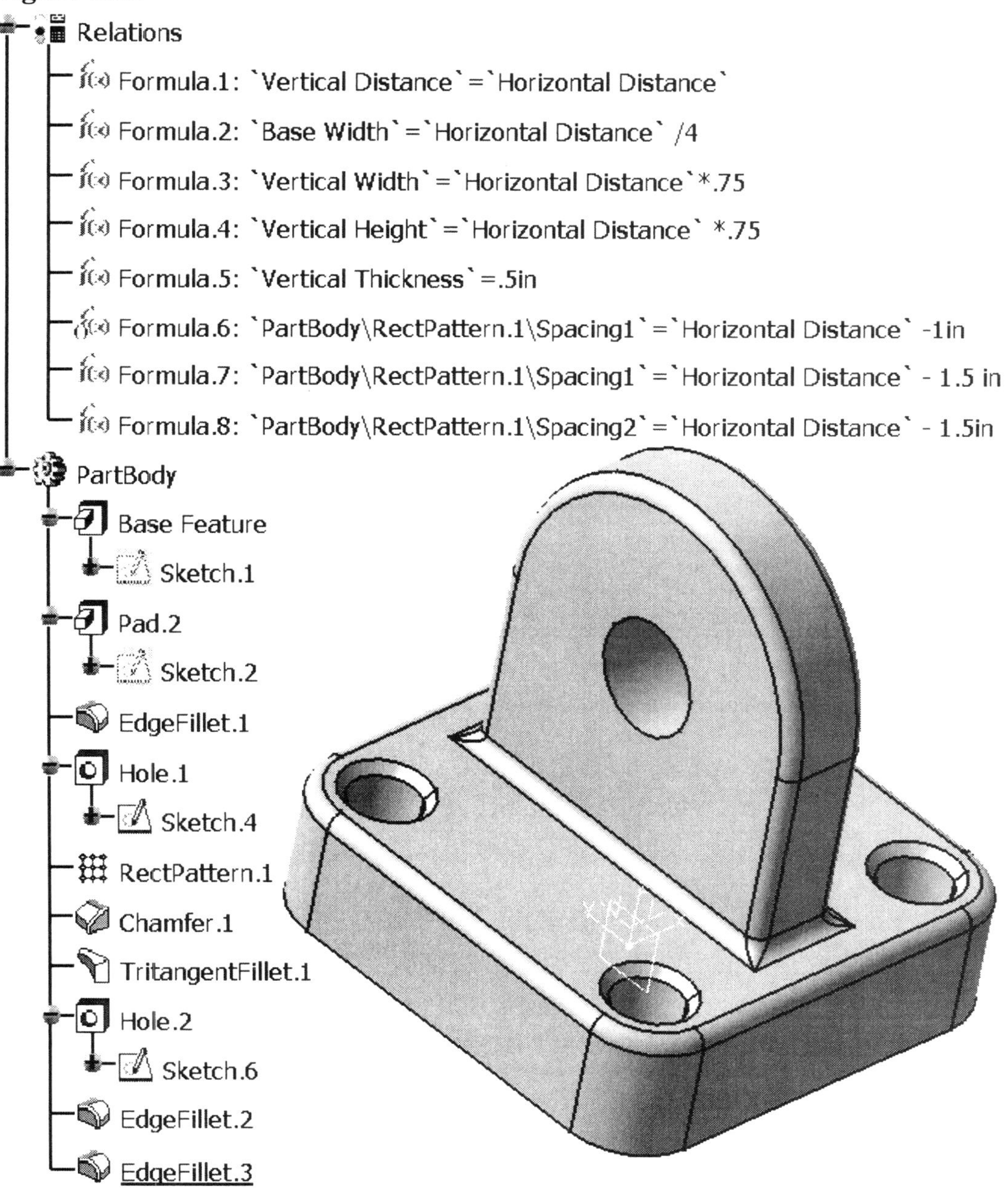

takes it one step further. Constraints will be applied to a part so the "**Design Intent**" of the part will be maintained even when revisions to the constraints are made.
For example, study the part shown in Figure 11.1, how would you create this part? It is not a difficult part. At this point you should be able to quickly sketch and extrude the different features to completion. You could constrain it to control the relationship of the different features. But, could you constrain it in such a way that the vertical attachment feature would stay centered no matter what value was used to define the width and depth of the base feature? This Lesson will teach you not just how to add information to a parametric design, but how to organize that information into basic part intelligence. With basic part intelligence the part could be quickly and easily updated with all the design relationships remaining in tact.

Lesson 11 Objectives

The objective of this lesson is to teach the user to take the time to determine the logical process of constraining the part to meet the design intent. This lesson will show you how to do the following:

- Re-enforce what was learned in the previous lessons.
- Analyze the part. Break the part down into simpler sub components. (This lesson refers to the sub components as features.)
- Determine the primary feature. This is the feature that best defines the part, the feature that the remainder of parts is built around.
- Create the part starting with the primary feature, secondary and so on.
- Determine the design intent of the part being created.
- Organize the parametric information so that the part contains actual intelligence, not just information.
- Not only create the part using the supplied dimensions but also constrain the different features and the relationship between the different features in such a way that the design intent of the part is not compromised.
- Update the part by revising only one constraint.

The step-by-step instruction will not be as detailed as in previous lessons. It will be assumed the tools used in the previous lessons are now common knowledge. If they are not, you are encouraged to go back and review the subjects you have questions on.

Tool Bars Used In This Lesson

There are two main tool bars used in this lesson. The first one is the **Constraints** tool bar and is found in the **Sketcher Work Bench**. The second is the **Knowledge** tool bar found on the bottom horizontal tool bar. The basic tools found in the **Constraint** tool bar should be review.

The **Constraints** Tool Bar

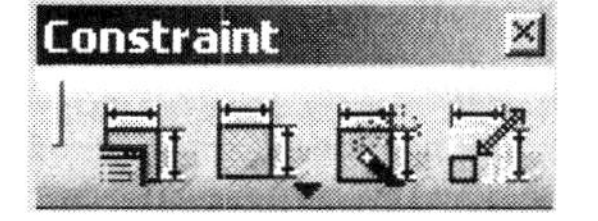

This tool bar has been covered in previous lessons, but since it is the primary focus of this lesson it has been included with some additional information.

Tool Bar	Tool Name	Tool Definition
	Constraints Defined In Dialog Box	Allows the user to select the type of geometrical constraint to apply to the selected entity/entities. The entity/entities must be selected for the tool to be selected. Only the constraints that apply to the selected entity will be selectable. This tool brings up the Constraint Definition window as shown in Figure 11.2.
	Auto Constraint	Creates geometrical and dimensional constraints automatically.
	Animate Constraint	Animates dimensional constraints. The user specifies a lower and upper limit and the number of steps between the limits.
Constraint Creation Tool Bar		
	Constraint	Creates dimensional constraints from entities selected by the user.
	Contact Constraint	Creates a contact constraint between two different entities.

The **Knowledge** Tool Bar

This tool bar allows you to use formulas and spread sheets to parameterize your sketches, parts and assemblies.

Tool Bar	Tool Name	Tool Definition
	Formula	This tool allows you to use a formula to drive parameters.
	Report Header	Allows the user to define design standards and check parts against the standards.
	Knowledge Inspector	This tool allows you to preview a design change prior to committing to the change. This is an advanced tool.
	Relations Sub Tool Bar	
	Design Table	This tool allows you to use data from an existing spread sheet to drive assigned parameters within a design.
	Law	Accesses the law editor.

Geometrical Constraint Symbols & Definitions

The following shows an example and gives a brief description of the Geometrical Constraints listed in the Constraint Definition window as shown in Figure 11.2.

Constraint Symbol	Constraint Name	Constraint Definition
	Symmetry	Makes two different entities symmetrical about a center line. There are many applications for this constraint. One is to make two different lines equal distance from a center point as shown in the symbols column.
	Midpoint	Intersects an end point to another entity
	Equidistance	Creates an equal distance between selected points

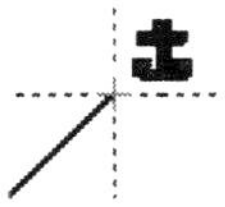	**Fix**	This "anchors" the entity to that exact location. The entity being fixed cannot be moved.
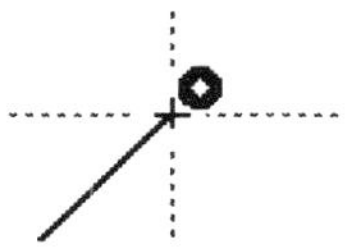	**Coincidence**	Coincidence means that two different entities share the same location. The entities are and will remain connected no matter what other constraints are applied to the entities.
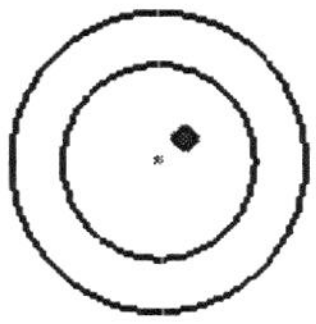	**Concentricity**	Allows circular geometry to share a common center point.
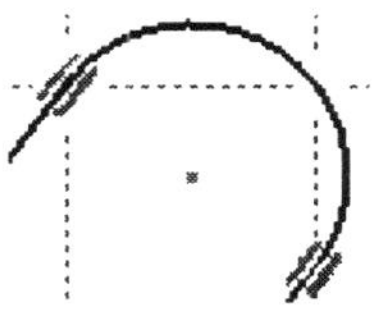	**Tangency**	Keeps two separate entities tangent to each other no matter what other constraints are applied to the entities.
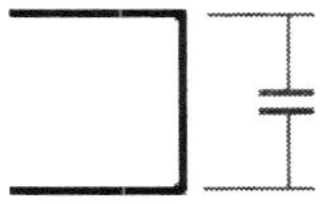	**Parallelism**	Keeps two separate lines parallel to each other no matter what other constraints are applied to the lines. If another constraint is applied that counters the parallel constraint, the sketch will be over constrained.
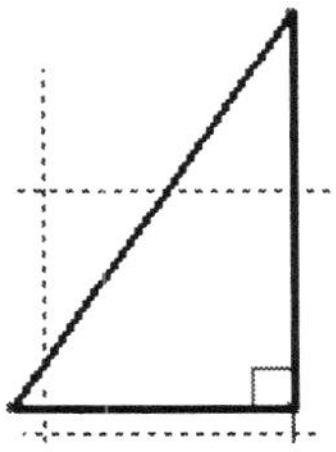	**Perpendicular**	Keeps two separate lines exactly 90 degrees to each other no matter what other constraints are applied to the lines.
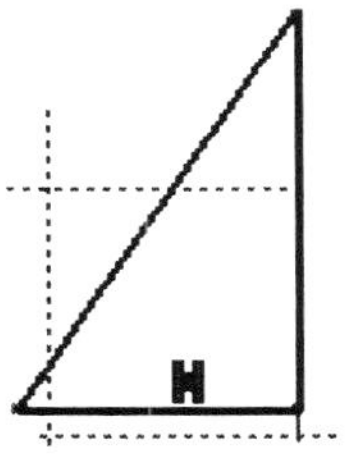	**Horizontal**	Locks a line down to be parallel with the horizontal axis.
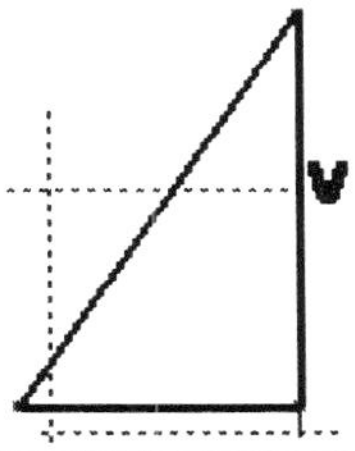	**Vertical**	Locks a line down to be parallel with the vertical axis.

Dimensional Constraint & Definitions

The following shows an example and gives a brief description of the Geometrical Constraints listed in the Constraint Definition window as shown in Figure 11.2.

Constraint Symbol	Constraint Name	Constraint Definition
0.76	**Distance**	Distance is the measurement between two different entities. The key is that two separate entities must be selected. This constraint locks down the distance between two selected entities.
0.76	**Length**	Length is the measurement of one entity from one end to the other end. The key is that only one entity is selected. This constraint locks down the length of the selected entity.
75.156°	**Angle**	Locks down the angle between two different entities.
D 0.45	**Radius / Diameter**	Used on circles and/or arcs. Locks down the size of a circle and/or arcs.

Figure 11.2 Constraints Defined In Dialog Box tool brings this window up.

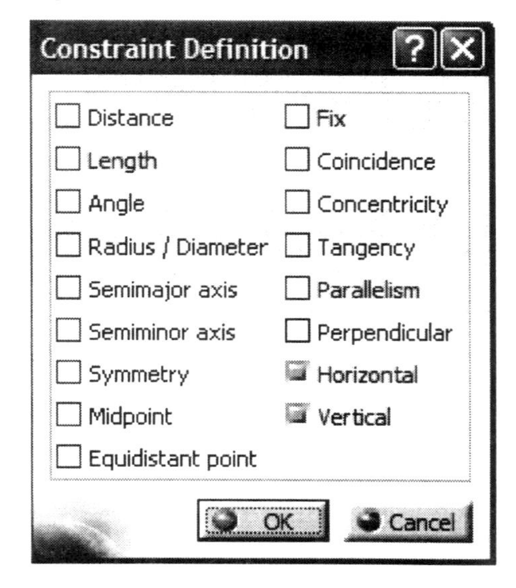

1 Stop! Take Time To Analyze The Part

This step is the easiest step to ignore. It's not required to complete the part, yet it is the most critical step in the entire process. Stop and take time to analyze the part! It will take more time up front to plan the process, but will save you time in the long run, especially if the part has to go through multiple modifications. As you analyze the part you need to be able to answer the following question:

I. What is the design intent of the part?
II. What relationship will the part have with other parts, if any?
III. Is the part the first of a series of similar parts?
IV. How many Sketch features are required to create this part?
V. Which sketch feature is the primary or dominate feature?
VI. What is the order of importance of the remaining features?
VII. What features are best created as **Sketch Features**?
VIII. What features are best created as **Dress Up Features**?
IX. Are there any other restrictions and/or requirements?

Use Figure 11.1 to answer as many questions as possible. Figure 11.1 may not communicate all the information needed to answer all the questions, answer as many as possible.

NOTE: So, what is a feature? A feature is a sub-component of the total part. For example the "**L Shaped Extrusion**" created in Lesson 1 contained one feature, it was created in sketch 1. It was the sketched shaped similar to the shape of an "**L**" and extruded using the **Pad** tool (Lesson 2). What is a Part? A part is referred in this lesson as the result of multiple features, a solid. Figure 11.1 shows the **Ring** Attachment Part.

2 Develop A Plan Of Attack

Develop a plan of attack based on the answers to the questions in Step 1. Armed with the answers to these questions you can create and apply the geometrical and dimensional relationships. Driving constraints and functional constraints control the size and shape of the each feature and ultimately the entire part. The following is a written analysis of the Ring Attachment shown in Figure 11.1.

Question I: What is the design intent of the part?
Analysis/Answer: Without a production drawing and other part documentation it is difficult to determine the design intent. So, answer the questions according to the information listed below.

a.) The **Base** is to remain square.

b.) The 0,0,0 point is to remain in the center of the **Base**, regardless of the size of the **Base**.

c.) The **Base** thickness and the **Vertical Ring** Thickness are to be ¼ the thickness of the **Base** width.

d.) The **Vertical Ring** is to remain square (total width equals height).

e.) The **Vertical Ring** is to be centered on the base.

f.) The diameter of the hole in the **Vertical Ring** is determined by the outside radius on the **Vertical Ring**.

g.) This is the standard for an entire family of parts. The family of parts has the same relationship requirements but is different in size.

Question II: What relationship will the part have with other parts, if any?
Analysis/Answer: This part is a stand alone part. It will not be placed in an assembly. No other part will be driving the part constraints. The part will not be driving any other part constraints and/or parameters. So, the answer is none. This obviously makes the planning process easier. There are no external variables to deal with.

Question III: Is the part the first of a series of similar parts?
Analysis/Answer: If the part is a stand alone one time design, then the extra time up front will not have as much value. Defining the constraints and relationships are not as critical. In this case the part is a stand alone part but will have a series of parts generated from the design. This makes it even more critical that the constraints and relationships be well thought out. The idea is to be able generate multiple parts by modifying only a few parameters/constraints. A Design Table could even be used to drive the size of the family of parts.

Question IV: How many Sketch features are required to create this part?
Analysis/Answer: Using the methodology (CATIA V5) used in the earlier lessons how many sketches would it take to create this part? This is a simple part, it should be clear that two sketches will be required to create this part. Remember the fewer the sketches the easier the part is to manage and modify. Use the Dress-Up features when possible (i.e. Fillets, Chamfer, Threads/Tap ect). Creating these features using the Dress-Up tools will save several steps in the design process and they can be quickly created, copied and modified.

Question V: Which sketch feature is the primary or dominate feature?
Analysis/Answer: The requirements given stated that the 0,0,0 point was to be the center of the **Base**. Another requirement states that the **Vertical Ring** is to be located from the **Base** (centered). This means the **Vertical Ring** is dependant on the **Base**. This would be enough information to declare the **Base** the primary feature. This means the other features would be built around the base.

Question VI: What is the order of importance of the remaining features?
Analysis/Answer: Since the **Ring Attachment** is a simple part there is only one other feature, the **Vertical Ring**. If there were additional features you would need to prioritize the features. Determine which features require driven dimensions and which features drive other dimensions.

Question VII: What features are best created as **Sketch Features**?
Analysis/Answer: Again, this is a simple part so the answers should be somewhat obvious. For the **Base** feature the square profile needs to be a sketch feature (profile created in **Sketcher** to be extruded in the **Part Design)**. The four holes will be created using the **Hole** tool (which creates its own sketch). The **Vertical Ring** feature has the same requirements.

Question VIII: Which features are best created as **Dress Up Features**?
Analysis/Answer: Again, this is a simple part so the answer should be somewhat obvious. The radius and fillets should be created after the sketch feature is complete. If the radius and fillets are the same, they can all be created using the same constraint, that way only one constraint would have to be modified to update all the radii and fillets.

Question IX: Are there any other restrictions and/or requirements?
Analysis/Answer: None known at this time.

All the questions have been answered but there still is no formulated plan of attack. The plan listed below is a suggested plan of attack (based on the information discussed so far in the lesson).

Step a: The **Base** will be the primary feature, the feature that all other features are built around.

Step b: The sketch of the primary feature (**Base**) will be created on the **XY Plane**.

Step c: Adequately rename the **Specification Tree** as you develop the part. This makes it easy to follow now and later. This is a continuous process and a good habit to get into, regardless of the simplicity of the part.

Step d: The primary feature will be constrained to remain square at all times.

Step e: The center of the primary feature will be at the 0,0,0 point.

Step f: The primary feature will be extruded to the required thickness using the **Pad** tool.

Step g: The sketch of the secondary feature will be created on the **YZ Plane**.

Step h: The bottom of the secondary feature sketch will be in contact (coincidence) with the top profile of the primary feature.

Step i: The secondary feature sketch will be constrained to the required dimensions. The feature sketch will be created as square. **Dress Up** tools will be used to create the round.

Step j: The holes will be created to meet the required locations, using the **Hole** tool. The four **Base** holes must be constrained so the hole to edge distance remains the same no matter what size the base is changed to.

Step k: Use the **Tritangent Fillet** tool to create the outside radius of the **Vertical Attachment**.

Step l: Create the **Vertical Attachment** hole and constrain it to the rounded edge created in step 11.

There really is not a right or wrong in developing a plan to building a part. Better and best become less obvious with the complexity of the part. The plan has been created. Now is the time to execute the plan.

3 Execute the Plan

Put the plan to the test! Create the part. The remaining steps will help guide you through the creation and constraining of the **Ring Attachment** so that the stated design intent is maintained. The creation steps are general instructions. Specific instructions are given when dealing with constraints and formulas that are required to create and maintain the part intelligence (design intent).

4 Create the Base Feature

You need to create the base feature per the "plan of attack." This will complete **Step a** and **Step b**. Create the **Base** as shown in Figure 11.3.

4.1 The sketch plane is the **XY Plane**, per **Step b** of the plan.

4.2 Select the **Sketch Tools** tool bar and verify that the **Geometric Constraints** toggle is on (highlighted). If the **Sketch Tools** tool bar is not visible go to **View**, **Toolbars** and check the box in front of the **Sketch Tools**. This will automatically create **Horizontal** and **Vertical Constraints** as you create the rectangle.

4.3 Create a rectangle approximately 3″ × 3″ with the axis near the center of the rectangle. Notice a parallel symbol is added to the **Specification Tree** under the **Constraints** branch as shown in Figure 11.3. This is the Horizontal and Vertical constraints automatically create because of the toggle turned on in step 4.2.

Figure 11.3

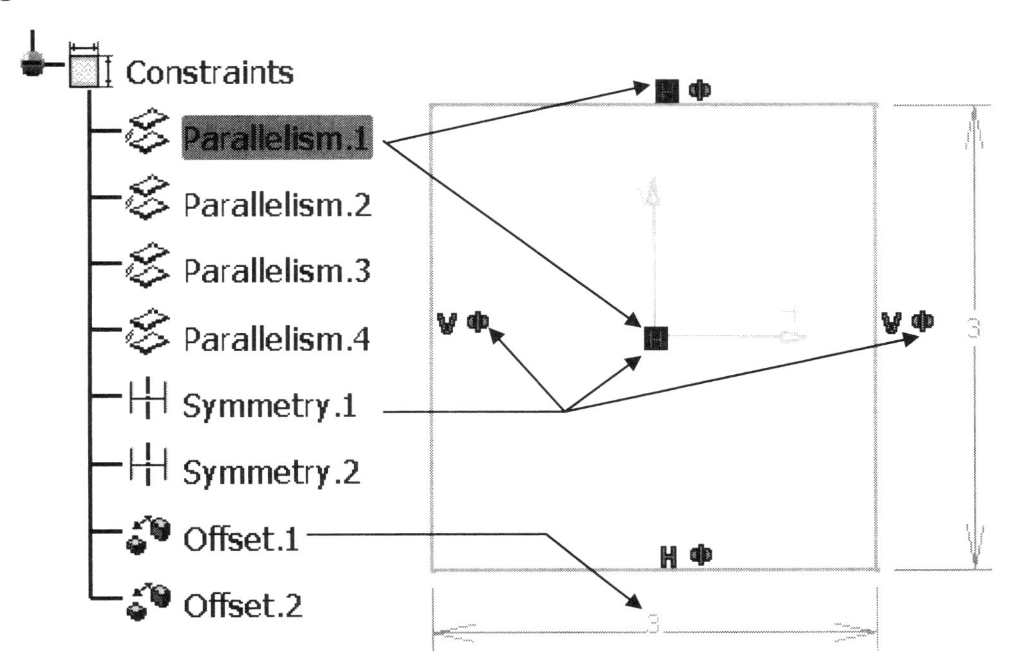

4.4 The next step is to create a **Symmetry Constraint** on the vertical and horizontal lines. This will constrain the rectangle to the axis. In other words, the axis will be in the center of the rectangle (0,0) which is one of the requirements of the part. This constraint will keep the base feature centered to the (0,0) even after modifications to the rectangle. Complete the following steps to create symmetry about the axis.

4.4.1 Multi select (hold the **Crtl** key while selecting) the following in exact order: Select the left vertical line of the rectangle, then the right vertical line, and finally, the **Vertical Axis**.

4.4.2 Select the **Constraints Confined In A Dialog Box** tool.

4.4.3 From the **Constraint Window** select **Symmetry**.

4.4.4 Select the **OK** button to create the symmetry between the two lines. This will place the two vertical lines symmetrical about the vertical axis.

4.4.5 Complete the same process for the horizontal lines. Except make the lines symmetrical about the horizontal axis.

Notice that two **Symmetry** symbols now appear in the **Specification Tree** in the **Constraints** branch. Sometimes it is hard to tell which constraint is which; you can move your mouse over one of the **Symmetry** symbols in the tree and see the constraint highlight on the geometry. Again, reference Figure 11.3.

4.5 Create the **Distance Between Constraint** both in the horizontal and vertical direction. Use the **Distance Between Constraint** to tie both horizontal lines together and both vertical lines together. The **Length Constraint** only constrains one horizontal line and one vertical line.

4.6 Modify both constraints to **3″**. Notice, no matter what size you make the rectangle the center is always 0,0. This accomplishment satisfies one of the requirements for maintaining design intent. You have just defined the relationship between the horizontal axis and the horizontal lines and the vertical axis and vertical lines.

4.7 Another requirement for maintaining design intent was to keep the **Base** square. This means that right now you would have to modify two constraints to maintain a square (the horizontal and vertical constraint). This step will show you how to create and apply a formula so the horizontal constraint drives the vertical constraint. All you will have to do is modify one constraint and the other will update automatically.

Figure 11.4

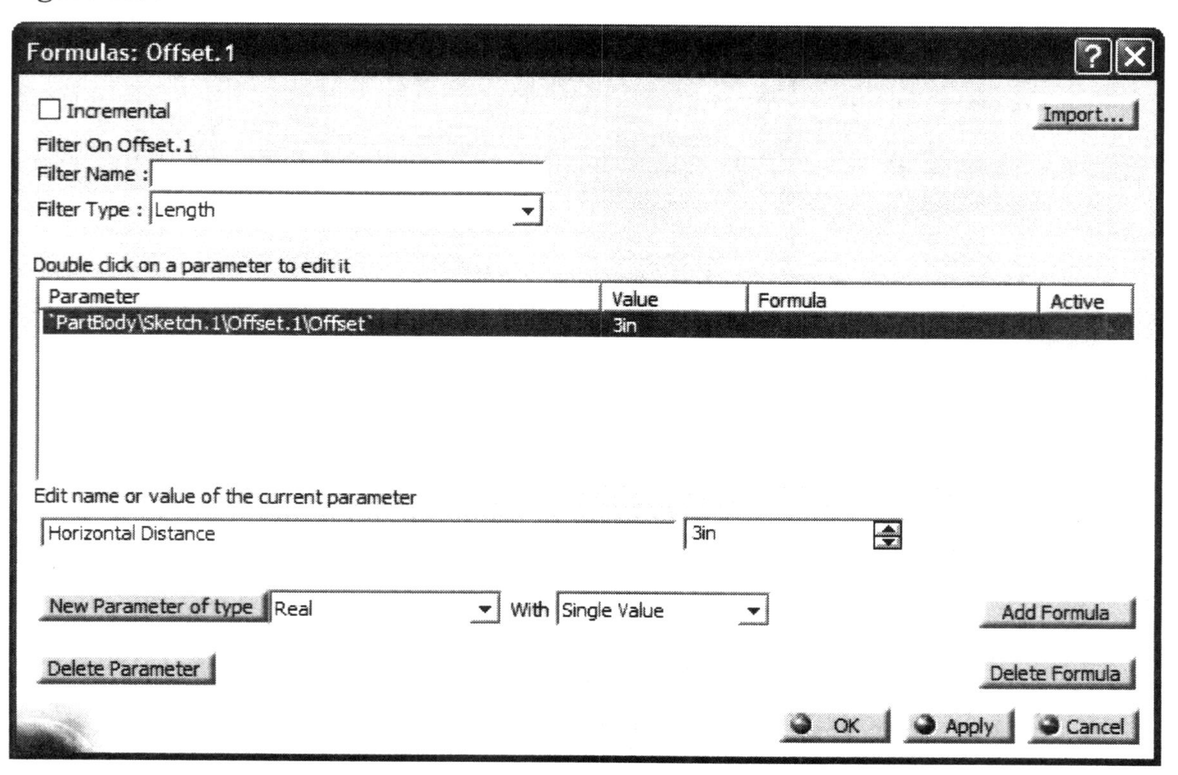

4.7.1 Select the **Horizontal Constraint**.

4.7.2 Select the **Formula** tool f(x). This will bring up the **Formula: Offset1** window as shown in Figure 11.4.

4.7.3 Since you selected the **Horizontal Constraint** prior to selecting the **Formula** tool, CATIA V5 automatically loaded the information for that constraint. CATIA V5 labeled that constraint **Offset.1**. If you did not pre select the formula you would have to search through a list of constraints (of type **Length**). CATIA V5 label "**Offset.1"** doesn't mean much to you, so let's give it some personal meaning by renaming it. In the box labeled "**Edit name or value of current parameter**" replace the "**….Offset.1**" with "**Horizontal Distance**" as shown in Figure 11.4. Select **OK** to implement the change.

4.7.4 Complete the same process for the **Vertical Constraint**, name it "**Vertical Distance**."

4.9 Now that both constraints are renamed to a label that you will recognize select the **Vertical Constraint** again.

4.10 Select the **Formula** tool again.

4.11 Verify that the **Vertical Constraint (Vertical Distance)** is in the **Parameters** window. Now select the **Add Formula** button.

4.12 This will bring up the **Formula Editor: Vertical Distance** window as shown in Figure 11.5.

4.13 Select "**Length**" option in the **Members of Parameters** window. This will reduce the number of options given in the **Members of Length** window, reference Figure 11.5.

Figure 11.5

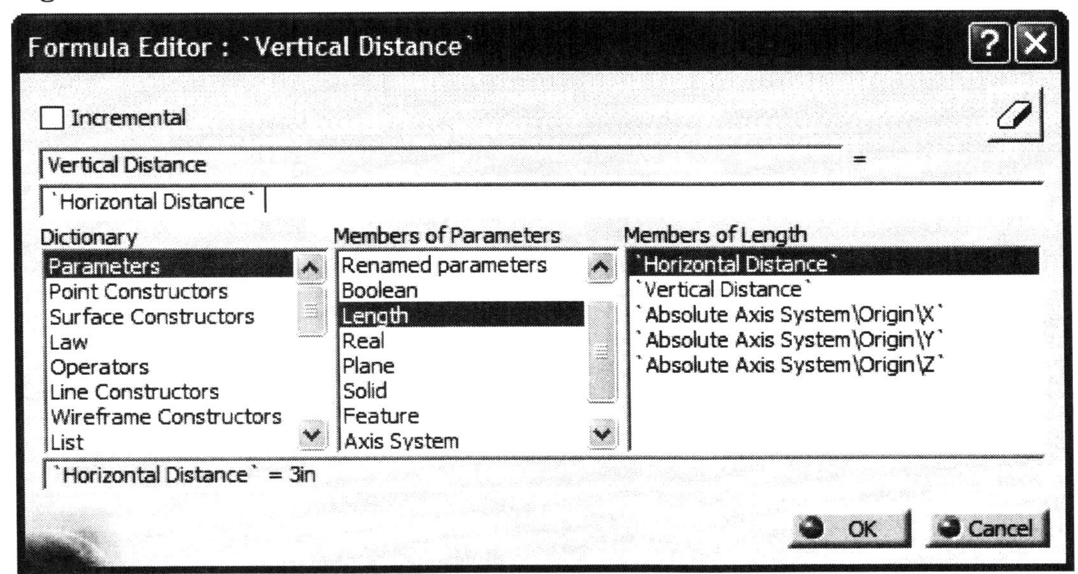

4.14 Double click on **Horizontal Distance** located in the **Members of Parameters** window. This will place the **Horizontal Distance** variable in the **Equals** window. If you didn't rename the constraints it would be very difficult to select the correct constraint. You would have to know (verify) that the vertical constraint was labeled as **"...Offset.1**."

4.15 Referencing Figure 11.5, the top two windows should read the formula as **Vertical Distance = "Horizontal Distance**." The bottom window shows the current value of the **Horizontal Distance** to be **3″.**

4.16 If your formula looks like the one shown in step 4.14 select **OK** to create the **Formula**. Figure 11.6 shows the **Vertical Constraint** with a formula symbol signifying that the value is driven from a formula. The **Specification Tree** shows the **Formula** in the **Relations** branch. The representation of the formula can be modified several different ways. One way to select the constraint is to right click to get the properties window, select **Offset2.Object**. This will give you the display options as shown in Figure 11.7. To get the **Relations** branch to be displayed in the **Specification Tree** you will have to select the following; **Tools**, **Options**, **Infrastructure** branch, **Part Infrastructure**, **Display** tab and select the **Relations** option.

Figure 11.6

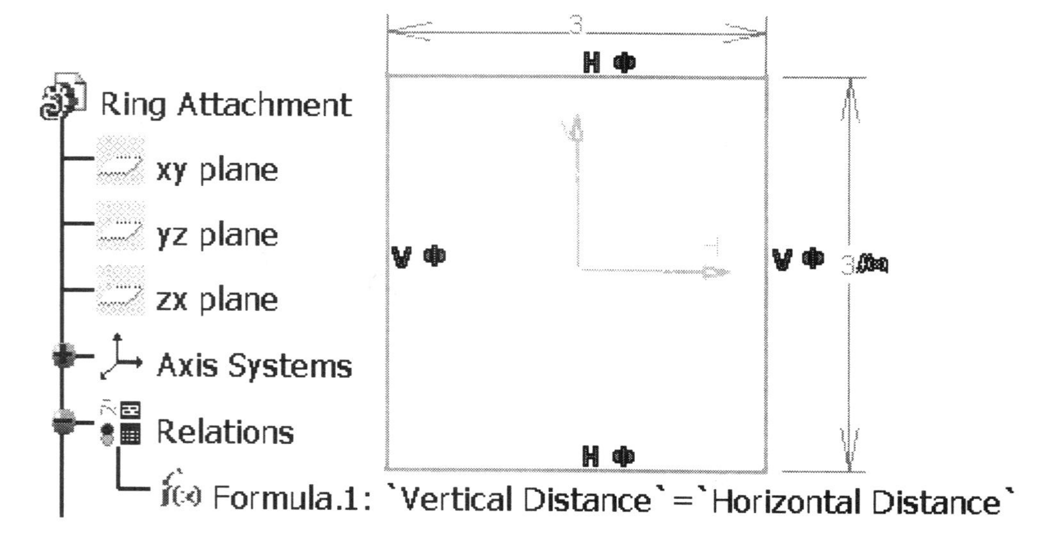

Figure 11.7

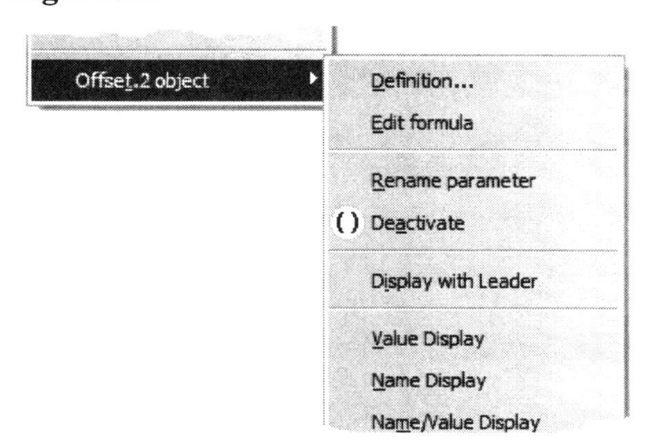

4.17 Now would be a good time to test your design, modify the value of the **Horizontal Constraint** to **5″**. The **Vertical Constraint** should update to 5″ and the sketch should remain symmetrical about the X and Y Axis. Change the **Horizontal Constraint** back to **4″**. You have now met more of the design requirements listed in the previous steps.

4.18 Exit the **Sketcher Work Bench** and extrude the **Base** profile **.75″** in the positive **Z** direction.

4.19 You have one more formula to apply to the **Base**. Per part requirement the **Base** thickness is suppose to be ¼ of the base width. The following steps will show you how to create a formula that will satisfy this requirement.

4.19.1 Select **Pad.1** from the **Specification Tree**.

4.19.2 Select the **Formula** tool. This will bring up the **Formulas: Pad1** window.

4.19.3 Rename the **"...Pad.1/Length"** **Parameter** to "**Base Thickness**." Refer to Step 4.7 on how this is done.

4.19.4 In the same window select the **Add Formula** button. This will bring up the **Edit Formula: Base Thickness** window.Set the **Base Thickness** equal to the **Horizontal Distance**. You do this by selecting the **Length** option in the **Members of Parameters** window. This limits the options to **Length** type members (in the **Length type Member** window). Double click on the **Horizontal Distance** option. This places the parameter in the **Equal** (=) window. Place the cursor in the space immediately after "**Horizontal Distance**" and type the rest of the equation which is "**/4**." Your formula should look like the one shown in Figure 11.8.

Figure 11.8

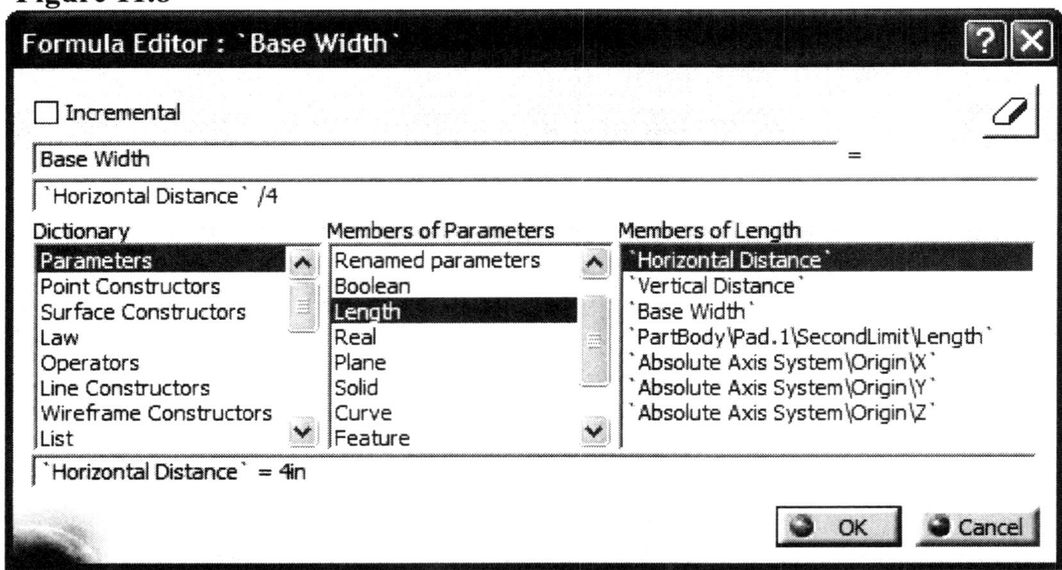

4.19.5 Select **OK** to create the formula. Select **OK** again to exit the Formula window.

4.19.6 Test your formula. Go back to the base sketch and modify the **Horizontal Distance** constraint. **Exit** Sketcher. The part should update to a square and the **Pad** thickness should update to ¼ the of the **Base** width (**Horizontal Distance** value).

The **Base Feature** is done except for the dress up features which will be completed after a secondary feature sketch is completed.

NOTE: This is a simple part, so it does not have a lot of components in the **Specification Tree**. Regardless of the complexity of part it is good practice to name the components in the **Specification Tree**. The level of customization is up to you and/or the company standard. This step proved how much easier it is to find a renamed constraint. The constraint will come in handy when implementing a **Design Table**. Rename the **Pad1** to **Base Feature** as shown in Figure 11.1.

5 Create the Secondary Feature

This step will take you through creating and constraining the **Secondary Feature**. This will also take you through the process of creating and applying formulas to meet design requirements.

5.1 Use the **YZ Plane** to create the **Secondary Feature** sketch.

5.2 Create a rectangle just above the **Base** part as shown in Figure 11.9.

5.3 Create a **Coincidence Constraint** between the bottom horizontal line and the top edge of the Base. You can do this using the **Contact** tool or Multi selecting the two lines and selecting the **Coincidence** button in the C**onstraint Defined In A Dialog Box**. Either way it will create a **Coincidence Constraint** in the **Specification Tree** as shown in Figure 11.9.

5.4 Create a **Symmetry** constraint about the **Vertical Axis** using the two vertical lines. Refer to the previous steps on how to create the symmetry. This constraint will keep the sketch centered to the vertical axis. Figure 11.9 shows the **Symmetry Constraint** highlighted, all the components of the constraint are highlighted.

Figure 11.9

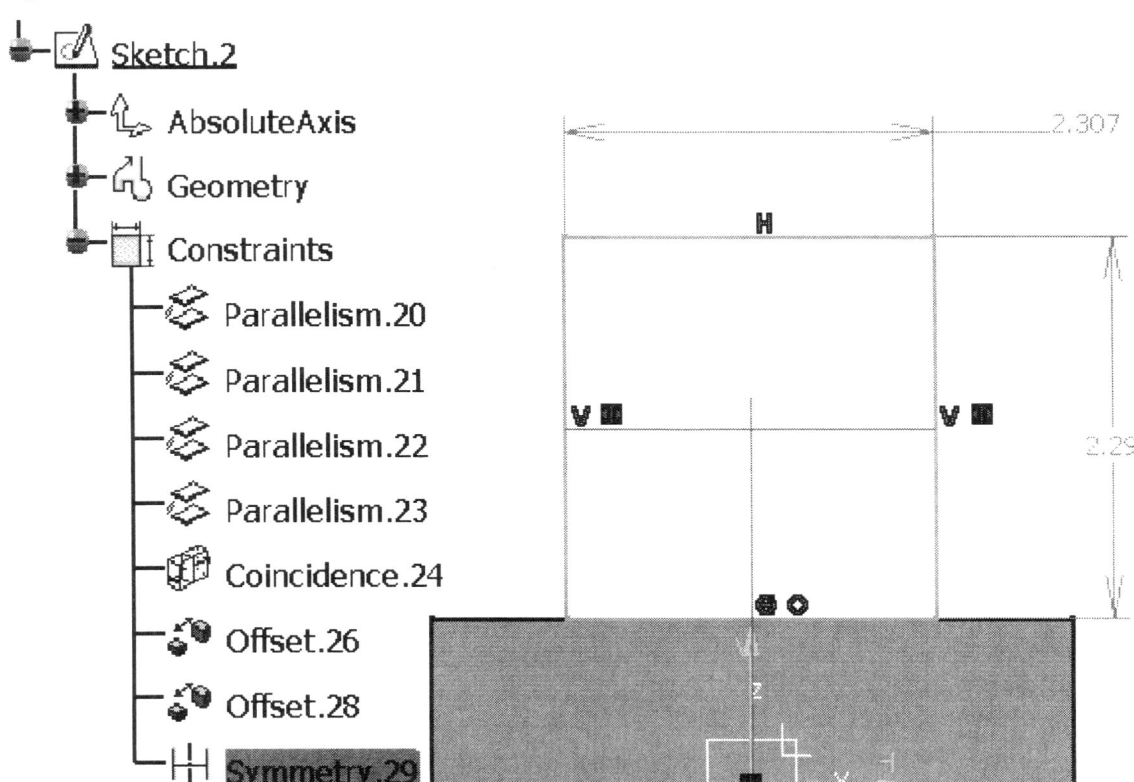

5.5 Create a **Distance Constraint** on both the **Horizontal** lines and the **Vertical** lines similar to the **Base Sketch**. Figure 11.9 shows the constraints created at random distances. The following steps will show you how use the formula tool to drive the constraint values.

5.5.1 Select the **Vertical Constraint** and rename it “**Vertical Height**.” Refer to step 4 on how to rename a constraint. Remember these values are of type **Length**.

5.5.2 Select the **Horizontal Constraint** and rename it “**Vertical Width.”**

5.5.3 With the newly named **Vertical Width** constraint active, select the **Add Formula** button. Create and apply the following formula the Vertical Width constraint: **“Vertical Height” = “ ‘Horizontal Distance’ * .75”** Reference Figure 11.10.

5.5.4 Create the following formula for the Vertical Width:
Vertical Width = **Horizontal Distance * .75**

5.5.5 If you make a mistake on a formula or just need to modify a formula double click on the constraint containing a formula, the **Constraint Definition** window will come up. Select the **Formula** tool . This will take you into the **Formula Editor** window.

5.6 Your sketch is fully constrained; you are ready to create the pad. Exit the Sketcher Work Bench.

5.7 Create a **Pad** about **.25″** thick and **Mirror** the **Pad** about the **Sketch**.

Figure 11.10

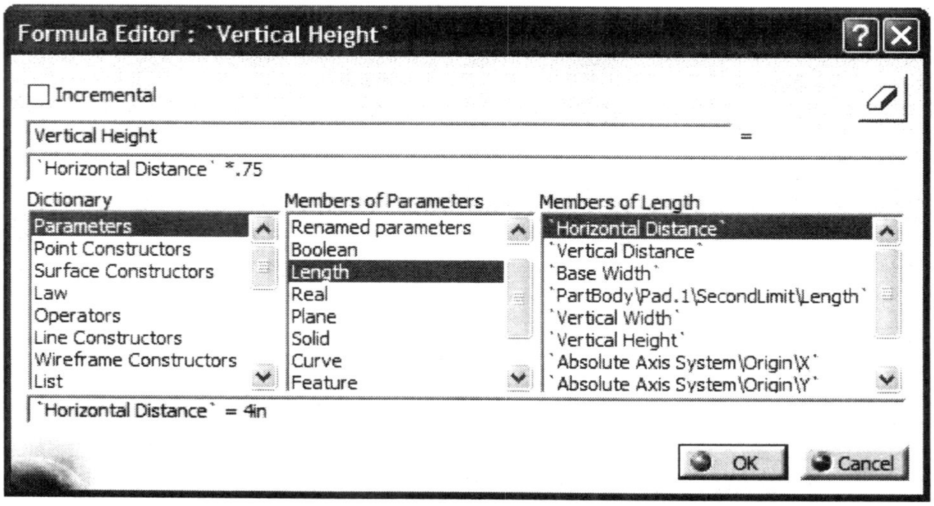

5.8 Select **Pad.2** (the Pad just created).

5.9 Select the **Formula** tool.

5.10 Rename the **Constraint** to "**Vertical Thickness**."

5.11 Create a **Formula** that makes the **Vertical Thickness** always equal to **.5″**. In the **Equal** (=) window be sure to specify the units as **.5in**, otherwise you will get an error. Creating and applying this constant satisfies another design requirement. The **Base** size can change but the thickness of the second feature will always be 1in.

Both the Primary and Secondary features have been created and constrained to meet the design criteria. It is now time to dress up the solid.

Figure 11.11

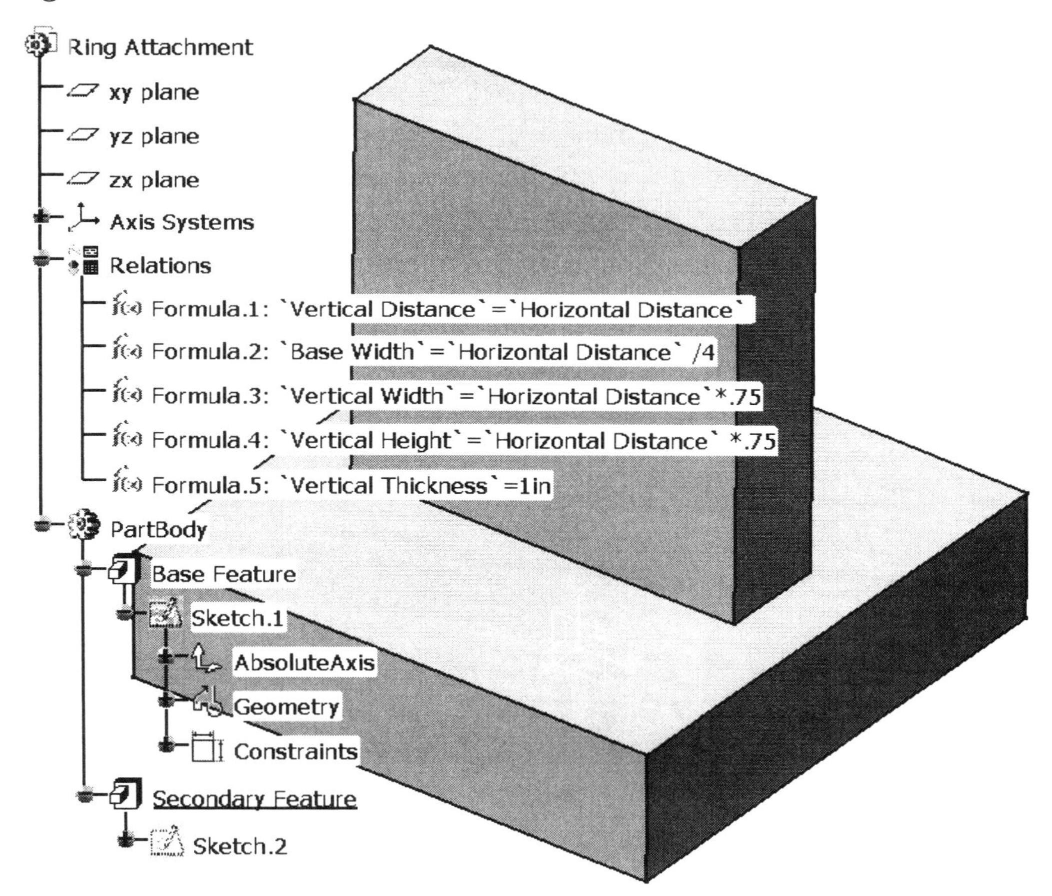

6 Creating The Dress Up Features

Now it is time to create **Dress Up** Features. What features are most critical? What features are parent features? Review your part, reference Figure 11.11.

6.1 Again there are several "right" choices but for this step you will start by creating an **Edge Fillet** on the four vertical edges of the **Base Feature**. Create the **Fillet** using the **Multi Select**, this way only one value is required. One variable applied to all four edges, thus you only have one variable to modify. Use **.75″** radius for the **Edge Fillet**.

6.2 The next dress up feature you need to create is **one** of the four vertical holes located in the **Base Feature**. It does not matter which hole you create, the other holes will be created using a **Rectangular Pattern**. Create the hole concentric to the **Edge Fillet** you created in the previous step. Make the hole a through hole with a radius of **.375″**.

6.3 Next create **Rectangular Pattern** from the hole created in the previous step. Use the values shown in Figure 11.12 to create the **Pattern**. For the **Second Direction** information use the same values. To make sure the direction is correct you can select the edge of the **Base** for **Reference Direction**.

6.4 Now that you have a **Rectangular Pattern** of four holes you need to add a formula to pattern. This will allow the hole locations to be automatically updated with each modification to the **Base Feature**. To accomplish this complete the following steps:

6.4.1 Select the **Pattern** branch of the **Specification Tree, so it is highlighted**.

Figure 11.12

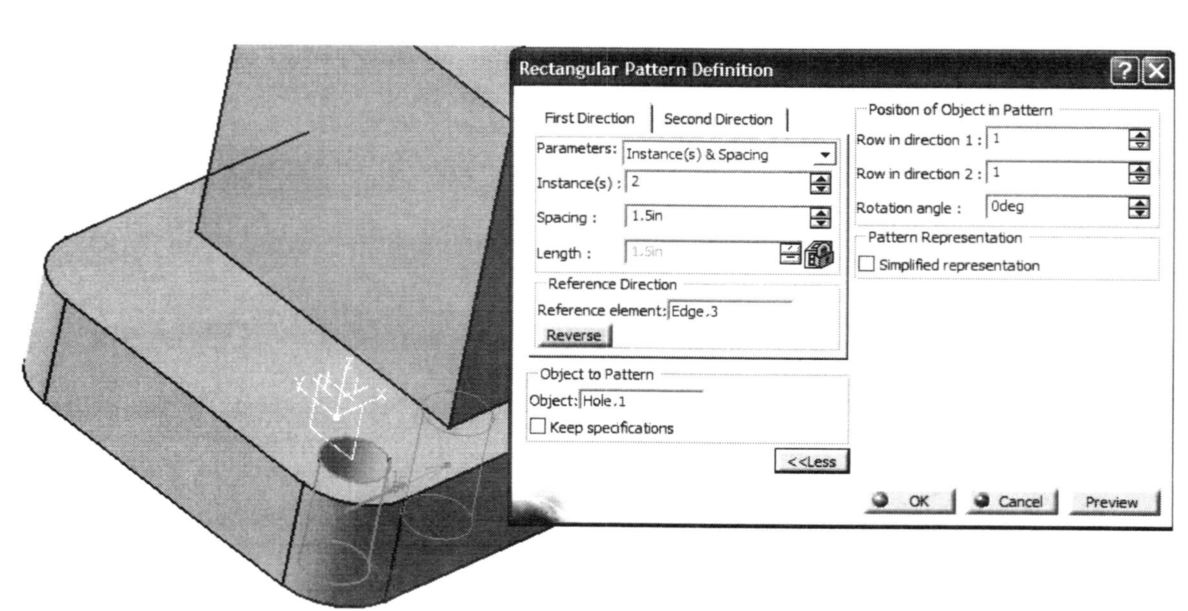

6.4.2 Select the **Formula** tool. This will bring up the **Formula: RecPattern1** window. Since the **Pattern** branch was selected the **Formula** window lists only the constraints related to the **Pattern Feature**. There are 2 variables, **"...\Spacing1"** and **"...\Spacing2,"** both require a formula.

6.4.3 Create a formula for both variables listed in the previous step. Both variables will have the same formula The formula is:

"PartBody\RecPattern1**spacing1**" = "Horizontal Distance" – 1.5in
"PartBody\RecPattern1**spacing2**" = "Horizontal Distance" – 1.5in

You can type in the formula manually or you can use the **Formula Editor** window as in the previous steps. If you type in the formula make sure you get the variable names exact otherwise you will get an error. When you type in the "**1.5in**" be sure to type in the unit of measurement otherwise you will get another error. The formula constrains the distance between the pattern to the **Horizontal Distance** and then subtracts 1.5 inches (half the diameter of the outside radius of the **Base Edge Fillet**).

6.4.4 Double click on the Hole branch of the Specification Tree. This will bring up the Hole Definition Window. Select the Type tab. Select the down arrow, select the Countersunk option. Type in **.1″** for the depth value. Select the OK button. Notice this places a countersunk hole on all four holes because to the Rectangular Pattern tool.

6.4.5 If your part looks similar to the one shown in Figure 11.13 you are ready to go to the next step.

6.5 Create a **Tritangent Fillet** on the **Secondary Feature**. Select the two narrow sides of the feature for the **Faces to Fillet**. Select the top surface for the **Surface to Remove**. This will put a radius on the **Secondary Feature**.

Figure 11.13

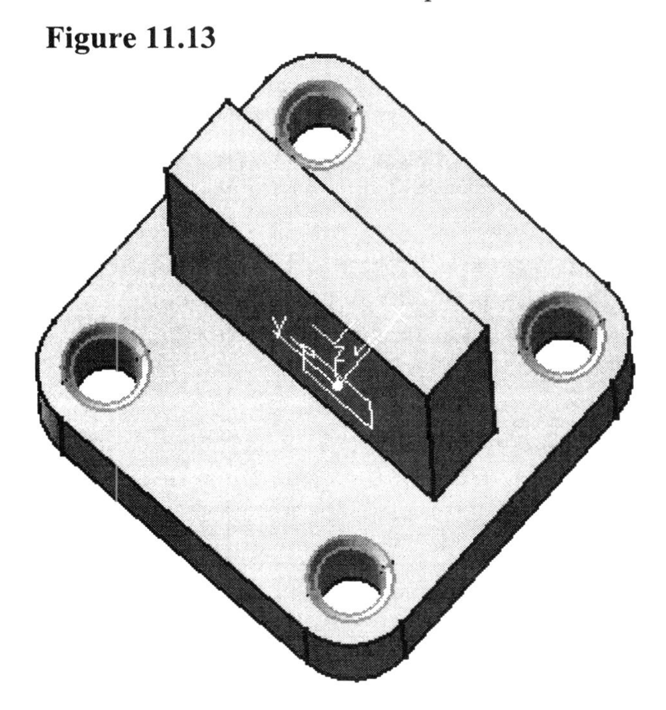

6.6 Create a hole through the **Secondary Feature** that is **Concentric** with the radius created in the previous step. To do this select the edge of the curve, select the **Hole** tool and then the surface the curve is on. Use **.563″** for the radius.

6.7 The last **Dress Up Feature** is the **.15″ Fillet** applied to almost all of the remaining edges. Create the fillet .15″ to all edges as shown in Figure 11.14.

Figure 11.14

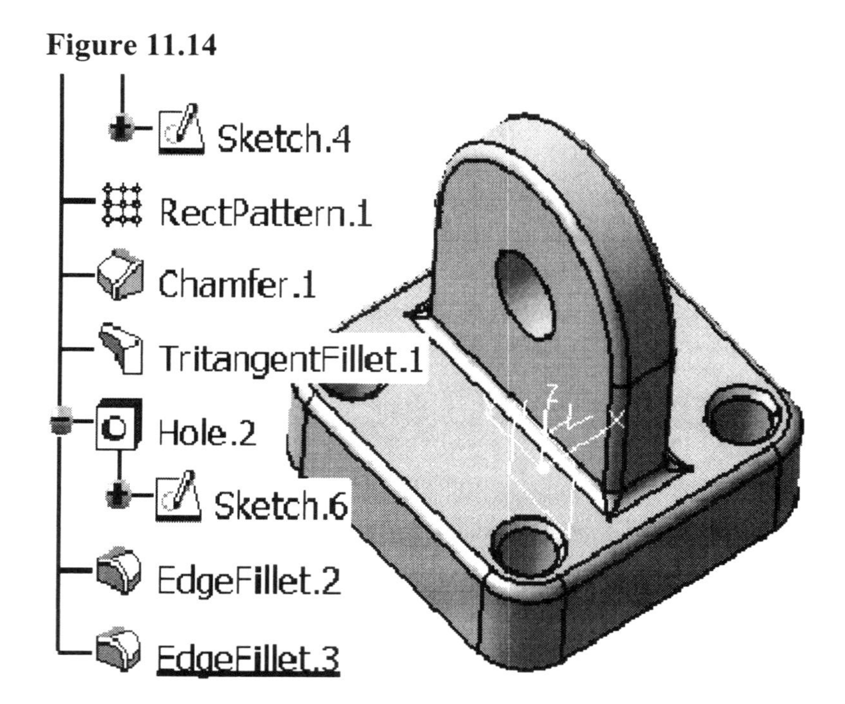

7 Putting The Plan To The Test

It is a good idea to test your formulas during every step of the creation process. That way the problems are discovered as they are created and can be solved before adding any more variables to the problem. Test the part by modifying the driving constraint. If the part updates you are half way there. The next question is, did the update compromise the design intent? If the design intent was compromised you will need to go in and modify the constraint and/or formula that caused the compromise.

For this step go to **sketch1** and double click on the **Horizontal Distance**. Modify the constraint to 8″. Exit the **Sketcher Work Bench**. Update the part. If all the features are updated and the design intent is still met you have successfully completed this lesson.

8 Additional Notes

You have created a solid with basic part intelligence. It meets the requirements given at the beginning of the lesson. The part does use formulas to drive other constraints but the part is far from full proof. If you modified the **Horizontal Distance** to a value of 1.5″ the update would cause some of the geometry to be lost. You can use additional Knowledge tools to set limits of the acceptable range.
You could use the Design Table to import values for the constraints. This would allow you to select the configuration that you want rather than manually modifying the driving constraint every time.

Lesson 11 Summary

This part is basic. The constraints and formulas are basic. Hopefully through the simplicity of the part you can see the power of properly applying constraints and formulas. Hopefully you can see where it would make sense to add additional constraints and more involved formulas. Think of the possibilities, if you can accomplish so much on such a basic part what could you do with an entire assembly?

Taking time up front to fully understand the requirements and application of a part will always save time in the long run. It is very possible that a part could be over analyzed and over formulated, in most cases the reason would be that the designer did not fully understand the application of the part. If you practice the process you will get faster and faster. Properly constraining a part could become second nature.

Figure 11.15 Formulas used in this lesson:

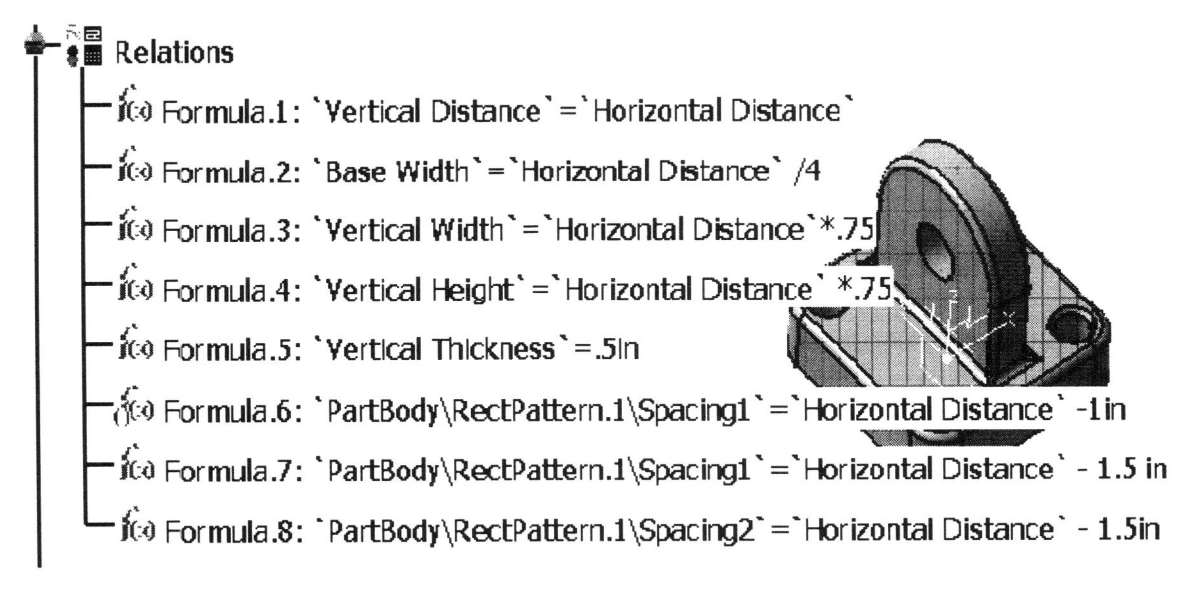

Lesson 11 Review

After completing this lesson, you should be able to answer the questions and explain the concepts listed below.

1. Define part intelligence.
2. Define parametric design.
3. What is a **Geometric Constraint**?
4. What is a **Dimensional Constraint**?
5. **T** or **F** When constraining a part it is a good practice to constrain the sketch with **Geometrical Constraint** before using **Dimensional Constraints**.
6. **T** or **F** CATIA V5 allows you to rename **Dimensional Constraints**.
7. How is a formula used to drive other constraint values?
8. **T** or **F** CATIA V5 allows you to rename **Geometrical Constraints**.
9. **T** or **F** CATIA V5 supplies tools to check a design to a companies predefined standards.
10. Can a **Formula** be used to define **Dress Up** features?
11. Can a **Formula** be used to define the spacing in a **Pattern**?
12. **T** or **F** The Anchor tool is and example of a **Dimensional Constraint**.
13. List five different types of **Geometrical Constraints**.
14. **T** or **F** When entering a constant value in a formula you must include the units of measurements.
15. What branch of the **Specification Tree** does a **Formula** appear on?
16. **T** or **F** The **Knowledge** tool bar supplies tools that make it possible to create a family of parts from data located in spread sheet (**Design Table**).
17. Why is it important to analyze a part prior to starting the design process?
18. **T** or **F** The process described in this lesson for planning and designing a part only applies to simple parts, it will not work for complex parts and/or assemblies.

19. If you have an option of creating a feature in **Sketcher** or creating the same feature using a **Dress Up** tool in the **Part Design Bench** the better practice would indicate which method?

20. The constraints used in this lesson to create formulas were of what type of Parameter?

Lesson 11 Practice Exercises

Now that your CATIA V5 tool box has some tools in it, put them to use in the following practice exercises. The shapes are simple and can be completed in one **Sketch**.

1 Create a sketch similar to the one shown below. Exact dimensions are not critical. Using the least amount of constraints possible make the sketch meet the following requirements:

- All angles 90 degrees.
- All lines either horizontal or vertical.
- The standing leg and base leg are of equal length.
- The thickness of the standing leg and base leg ¼ of the length.
- The profile is constrained so no elements can be moved on their own.
- The outside corner is always touching the 0,0 point.

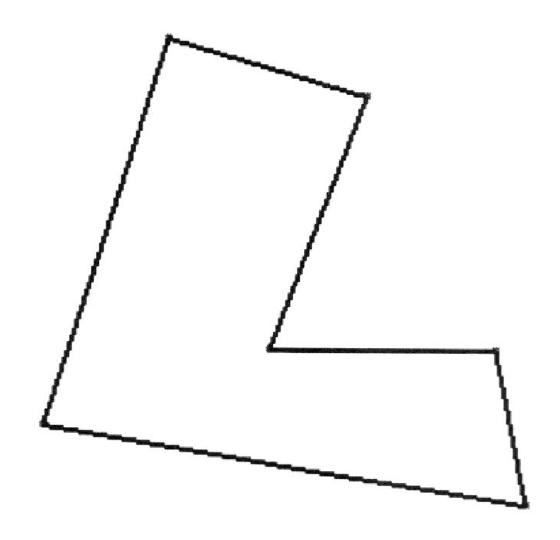

How many constraints did it take? Save the sketch as **Lesson 11 Exercise1.CATPart.**

2 Use the sketch from Exercise 1 to meet the following requirements:

- Extrude the part to the length four times the base leg width.
- Create an inside radius (Edge Fillet) that is 1/6 of base leg with.

How many constraints did it take? Save the sketch as **Lesson 11 Exercise2.CATPart.**

3 Create a sketch similar to the one shown below. Exact dimensions are not critical (relationships are). Using the least amount of constraints possible make the sketch meet the following requirements:

- All angles 90 degrees.
- All lines either horizontal or vertical.
- Make the part symmetrical (equal distant and same length).
- The thickness of the legs 1/6 of the over all height.
- The profile is constrained so no elements can be moved on its own.
- The center of the figure is always touching the 0,0 point.

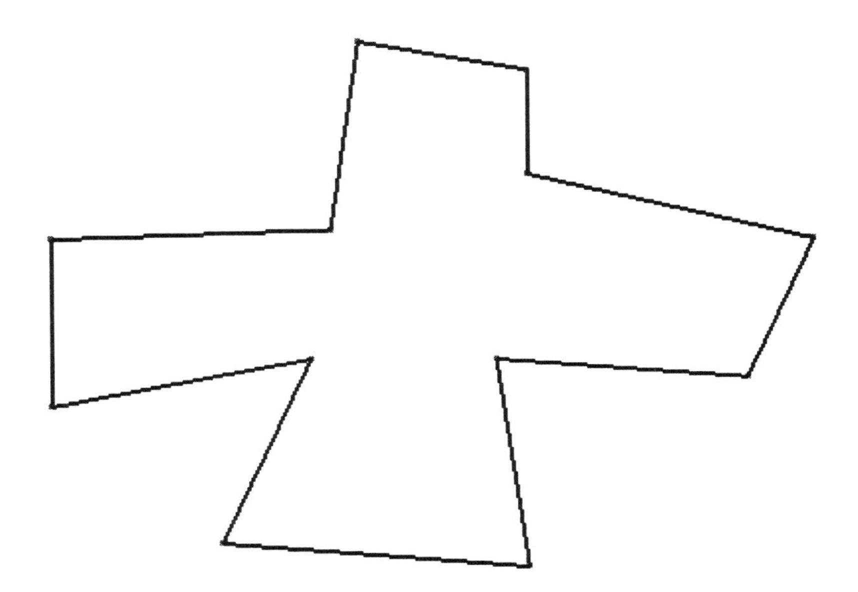

How many constraints did it take? Save the sketch as **Lesson 11 Exercise3.CATPart.**

4 Use the sketch from Exercise 3 to meet the following requirements:

- Extrude the part to the length four times the base leg width.
- Create an inside radii (Edge Fillet) that is 1/6 of base leg with.

How many constraints did it take? Save the sketch as **Lesson 11 Exercise4.CATPart.**

5 Yes, this part again! With your new found knowledge use the planning process to design this part. Can you create it more robust and efficiently? This part can be a constraining nightmare if not done correctly. You might consider mirroring and or possibly using pattern.

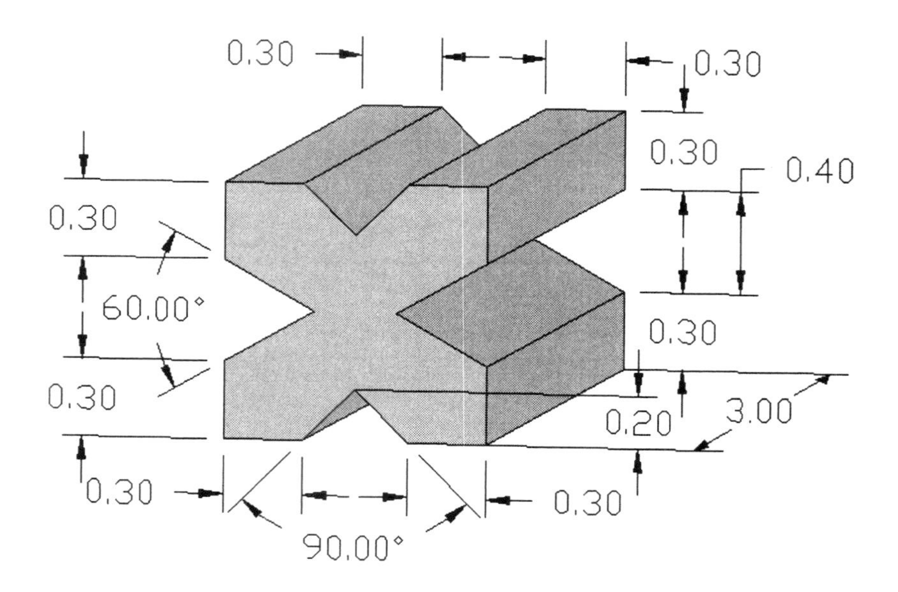

Save the file as "**Lesson 11 Exercise 5.CATPart**."

Workbook/CATIA V5 Terms And Definitions

The terms and definitions below are not your regular dictionary definitions. The definitions are based on how they are used in this workbook. The workbook definitions are included to help lessons to clearly communicate the meaning to the reader.

Assembly: A CATIA V5 session or file (document) that has multiple CATParts assembled together. The file/document would be saved using the CATProduct extension.

Blue Compass: A CATIA V5 tool used in the **Drafting Work Bench** to orient the orthographic layout of a part.

Body: A profile that has been extruded, revolved or swept to create a single solid. Body can be used in the place of solid. In the **Part Design Work Bench** a part (**CATPart**) can have multiple bodies.

Boolean: Solids that share space. If you have two separate solids and they overlap (share some space in 3D) the shared space can be subtracted out (Boolean math).

CATDrawing: This is CATIA V5's file extension for files created in the **Drafting Work Bench**. The different file types (extensions) that CATIA V5 creates can be and are linked together. A **CATDrawing** is usually linked to a **CATPart** file or **CATProduct** file. **CATDrawing**s are referred to as documents.

CATIA V5: Computer Aided Three-Dimensional Interactive Application, Version 5. CATIA V5 is a product of Dassault Systemes and marketed by IBM.

CATPart: This is a CATIA V5 file extension. **CATPart** files are created in the **Sketcher** and **Part Design Work Benches**. This workbook at times refers to the part and/or file as a **CATPart**. **CATPart** files are referred to as documents

CATProduct: This is CATIA V5's file extension for files created in the **Assembly Design Work Bench**. The different file types (extensions) that CATIA V5 creates can be and are linked together. **CATPart** files assembled together are saved as a **CATProduct** file. **CATProduct** files are referred to as documents.

Compass: A tool CATIA V5 uses to help move and/or manipulate the orientation of CATIA V5 parts. The location of the compass is by default the top right of the working screen.

Component: In generic terms it is used the same as entity, a sub set of the whole. A component is sometimes referred to as a simple part. The **Product Structure** and **Assembly Design Work Benches** use component in specific terms, "**Insert New Component**" and "**Insert Existing Component**".

Constraints: A method of applying parameters to a profile sketch, part entity or assembly. The parameter determines the relationship of one entity to another entity, or a particular property of just one entity. A constraint looks and acts much like two and/or three-dimensional dimension. There are **Geometrical Constraints** and **Dimensional Constraints. Dimensional Constraints** can be determined by a formula as well as a numerical value.

Document: Any type of file containing data. The type of data contained is usually designated by the file extension. For example, a CATPart, CATProduct and CATDrawing are all CATIA Document. Files, models and/or parts are often referred to as documents.

DMU: Digital Mock Up.

Dress Up Features: Features used to dress up (modify) an extruded, revolved or swept solid. The features are created using dress up tools such chamfer, corner and break.

Entity: Any kind of single, two or three-dimensional CATIA V5 creation. Entity is sometimes used interchangeably with element, geometry and/or component. Some examples of entities are; a line, an edges, a surface and even a constraint.

Extrude: A 2D profile/periphery created in the **Sketcher Work Bench** that is made into a 3D solid. This is accomplished in the **Part Design Work Bench** using the **Pad** tool. Lesson 1 and 2 talk about extruding the part or profile sketch, but they are referring to the **Pad** tool.

File: This is what the CATIA V5 creation is saved in, also referred to as a document. The type of file CATIA V5 creates depends on the workbench it was created in. There are **CATPart** files, **CATDrawing** files and **CATProduct** files. CATIA V5 has more types of files, but this workbook only covers the three listed. CATIA V5 refers to a file as a document.

Formula: Math formulas can be used to drive the constraint values.

Geometry: Any kind of single or multiple, two or three-dimensional CATIA V5 creation. Geometry is sometimes used interchangeably with the word "element" and/or "entity."

Hide/Show: A toggle tool that places selected entities in a workspace that is not visible from the typical workspace. The selected geometry is not deleted but just place out of view for later reference. Entities placed in the **Hide/Show** will be dimmed on the **Specification Tree**. You can use the **Swap** tool to view **Hide** workspace.

Highlight: Highlight an element by moving the cursor onto the element and clicking the left mouse button so the element is turned red which is the default color. The element must be selected to highlight, so the word "select" is sometimes used interchangeably with highlight.

Icon: Represents the tool. To start a process the tool must be selected, you select the icon that represents the tool.

Insert Existing Component: A tool found in the **Product Structure** and **Assembly Design Work Bench** that allows you to insert an exiting document, this tool is basically the same as inserting a **CATPart.** Reference the Menu 2 (**Specification Tree**) in the Introduction.

Insert New Component: A tool found in the **Product Structure** and **Assembly Design Work Bench** that allows you to create a part that is exclusive to that product, it does not stand alone as a its own document. What this really means is you can not use the newly created component in another **CATProduct** document. Reference the Menu 2 (**Specification Tree**) in the Introduction.

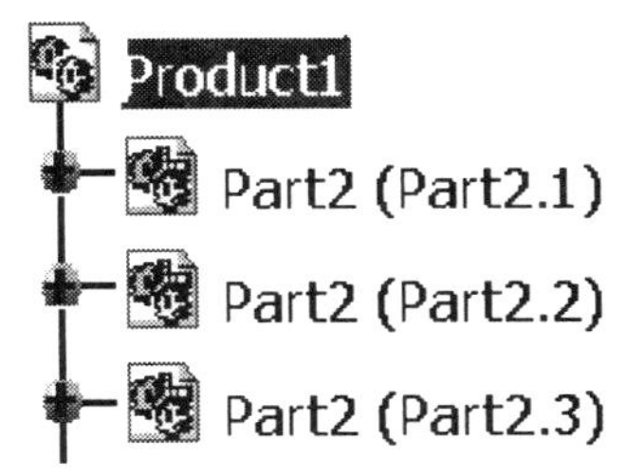

Instance: In the **Assembly Design Work Bench** the same existing entity (usually **CATPart** or **CATProduct**) can be inserted into the assembly an unlimited amount of times. The instance keeps track of how many times the part was inserted into the assembly. The instance shows up in the **Specification Tree**. In this example **Part2** was inserted 3 times. If **Part2** is modified, the modification will be propagated in all instances.

Knowledgeware: A design and application process implementing industry knowledge and standards. Physically it is a tool bar containing tools that allow you to create and manipulate knowledge based information, creates relational design. There are numerous knowledgeware tools throughout CATIA V5. The tools used in this workbook are **Formula** and **Design Table**.

L Shaped Extrusion: The name of a specific part created in Lesson 1. It is also referred to and includes its file extension, "**L Shaped Extrusion.CATPart**."

Normal: Perpendicular, 90 degrees to the reference element.

Open Body: A branch that is created under the **PartBody** branch of the **Specification Tree**. Wireframe and surfaces entities are found in this branch (**Generative Shape Design Work Bench**).

Parametric: Geometric entities that contain intelligence. For example a line has a start and end point and is defined by a slope. A parametric line contains the length, location and its relationship to other entities such as parallel and tangent. So, as the entity is modified all the related entities automatically adjust depending on the predefined relationship.

Parent/Child Relationships: This is directly related to the **Specification Tree**. A **Parent** is at the base branch of the tree. The **Child** components are everything on the lower (smaller) branches. Multiple **Child** components define the **Parent** component. Deleting the child component can create problems with the **Parent** component.

Part: The term "part" is synonymous with CATPart. Part is used to refer to a specific CATPart such as "**L Shaped Extrusion.CATPart**." For example in the workbook the "**L Shaped Extrusion**" could be referred to as "the solid" or "the part."

Periphery: The outside edge and/or surface of a part. Periphery is at times used interchangeably with the word profile.

Plane: A graphical entity CATIA V5 uses to represent a defined two-dimensional slice of the CATIA V5 three-dimensional working space.

Profile: The profile is the cross sectional outline of a part. Profile is at times used interchangeably with the word periphery. Profile and periphery are used to define a sketch that is ready to be extruded, revolved or swept.

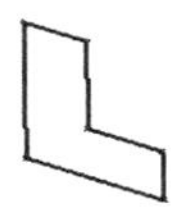

Prompt Zone: The bottom left of the CATIA V5 screen that has changing text, depending on the tool being used. You select a particular tool and CATIA V5 will prompt you on what selection and/or input is required.

Select an object or a command

Reference: The original **CATPart** that instances are linked to. Reference the Menu 2 (**Specification Tree**) in the Introduction.

Relational Design: Relational design could refer to one individual part but is usually used in reference to an assembly. An assembly consisting of parametrically designed parts. The individual parts share predefined relationships usually in the form of constraints, formulas and design tables. Correctly predefining the relationships could result in quick and easy modifications to the assembly. This is opposed to being required to individually modify each part and part entity to the new requirement.

Select: Move the cursor over the entity and click the left mouse button. The result of selecting an entity is the entity being highlighted or a process being started.

Sketch: A two dimensional drawing consisting of lines, points, arcs, etc. A sketch is created in the **Sketcher Work Bench** and/or the **Drafting Work Bench**.

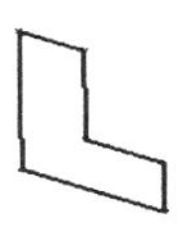

Sketch Features: 3D geometrical entities based on a 2D sketch (**Sketcher Work Bench)**. Solids are created from sketch features by extrusion (Pad tool), rotation or sweeping. If a body is created using a tool shown in the Sketched-Based Features tool bar.

Sketcher: Short reference for **Sketcher Work Bench** .

Solid: A type of part. A 3D entity that a material type can be applied to such as a CATPart. A solid can also be an imported solid using STEP or other solids translator. In the workbook a specific part such as the "**L Shaped Extrusion.CATPart**" is referred to as a solid.

Specification Tree: An organized history of everything required to create the part /product. Every step to creating a part is documented in a tree structure. There are icons that represent different branches to the tree such as Partbody, Product, Openbody and Sketch. Lesson 2 shows how to use replay to review the design process.

Surface: An entity that describes a boundary in space but has no thickness. A surface can be stretched over a wireframe or extracted from a solid. A surface can look very similar to a solid. A surface can be shaded and enclosed to create a solid. A document containing a surface/surfaces has a **CATPart** or **CATProduct** extension.

Tool: The operation that accomplishes a particular task and/or process, such as the **Pad** tool. Tools are represented by icons. One or more tools that share a similar task/process are grouped together to make a **Tool Bar**.

Tool Bar: A designated bar containing one or more individual tools. The tool bar will be labeled according to the functions of the tools it contains. For example the Sketch Based Feature tool bar.

T Shaped Extrusion: The name of a specific part you create in Lesson 2. It is also called out with its file extension, "**T Shaped Extrusion.CATPart**." The "**T Shaped Extrusion**" is sometimes referred to as the "part" during the particular lesson. The "**T Shaped Extrusion.CATPart**" is also referred to as a document.

Wireframe: A combination of three-dimensional **Open Body** elements. Usually consists of points, lines and arcs. A wireframe cannot be shaded. The workbook will capitalize and bold the word wireframe when it is referencing a tool and/or a specific wireframe.

Work Bench: A specific set of CATIA V5 tool bars packaged together to accomplish a particular task for example the Part Design Work Bench has tools to create individual parts where the Drafting Work Bench supplies tools specific to drafting and creating production drawings. Work benches have their own screen layout, default tools bars/tools and files extension (document). Usually the work bench name signifies what its particular task is.

CATIA V5 Index

Note: Pages referenced in this index are ones that have significant use of the word. Not all occurrences of the word are listed.